叶家东 编著

耕云集

云降水物理学和人工影响天气
云对流动力学中尺度对流系统

南京大学出版社

图书在版编目(CIP)数据

耕云集/叶家东编著.—南京:南京大学出版社,2020.12

ISBN 978-7-305-23776-8

Ⅰ.①耕…　Ⅱ.①叶…　Ⅲ.①气象学-文集　Ⅳ.①P4-53

中国版本图书馆 CIP 数据核字(2020)第 167947 号

出版发行　南京大学出版社
社　　址　南京市汉口路 22 号　　　　邮编　210093
出 版 人　金鑫荣

书　　名　耕云集
编　　著　叶家东
责任编辑　王南雁　　　　编辑热线 025-83595840

照　　排　南京开卷文化传媒有限公司
印　　刷　徐州绪权印刷有限公司
开　　本　787×1092　1/16　印张 25　　字数 621 千
版　　次　2020 年 12 月第 1 版　　2020 年 12 月第 1 次印刷
ISBN　978-7-305-23776-8
定　　价　198.00 元

网　　址:http://www.njupco.com
官方微博:http://weibo.com/njupco
官方微信号:njupress
销售咨询热线:(025)83594756

荣誉证书

中 国 气 象 学 会

CHINESE METEOROLOGICAL SOCIETY

荣誉证书

叶家东 同志：

在中国人工影响天气事业五十年中您做出了突出贡献，特予表彰。

二零零八年十月

1959 年苏联专家基留辛来华讲学时送的小纪念品，时隔 60 年翻出来看看，恰似新的一般。

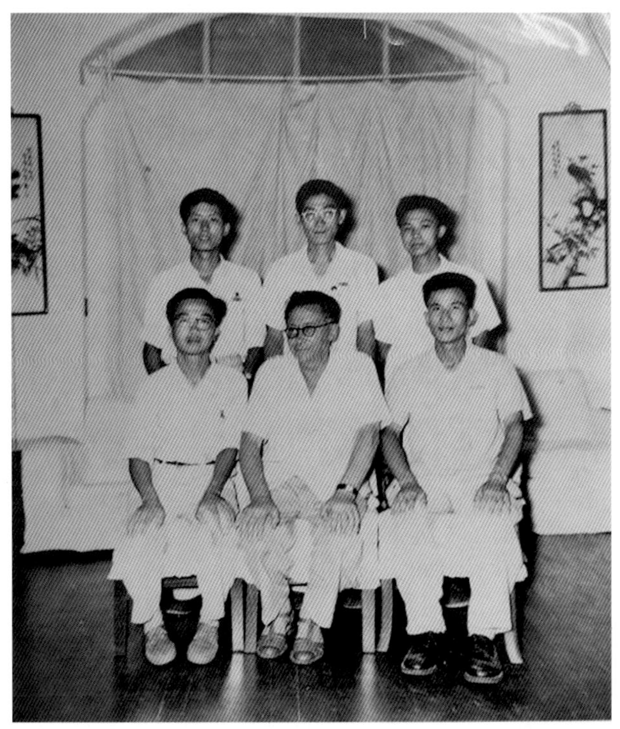

1960 年基留辛回国前夕在其寓所（南京萨家湾专家楼）合影留念，前排左起：许绍祖、基留辛、邹进上；后排左起：叶家东、苏寿祁、罗秀卿（1962 年罗秀卿不幸去世）

Have a good time on your visit to Utah and a pleasant return to China

叶老师 范老师：祝您们一切顺利！

Goodbye ...

We'll miss you.

Best Wishes Bob

It was very nice to know you. Good luck in your career.

Have a pleasant trip home and I look forward to seeing you again.

Roger

Good luck Mr & Mrs Yeh

Good Luck, Dallas

It was good working with both of you. Good luck wherever your research takes you next.

Good Luck

Best wishes for good science in Nanjing and best wishes to your daughter!

1988 年离开美国科罗拉多州立大学 CSU 前，研究组的伙伴们临别赠言和签名留念。Cotton 和 Pielke 两位教授还郑重其事地盖上他们珍视的中文图章，显然，他们以拥有一枚中文私章而感到荣幸，人情味挺浓的。上图为 CSU 大气科学厅五楼前合影，二排后排分别为 Cotton 和 Pielke。

1978 年黄山始信峰
天都莲花五彩桥，罗汉十八竞踊跃，
仙人欲指攀天路，始信峰云自逍遥。

1998 年纽约中国城孔子大厦前的孔子雕塑像前留影，不远处还有一尊林则徐雕像。都说海外华人华侨热爱祖国，他们在海外奋斗拼搏，谋生存，求发展，事业有成，不忘在异国他乡传播中华情，弘扬中国魂，此情此景，令我肃然起敬。

Cotton 的住处合影,左二范蓓芬,左四 Cotton 夫人 Volice,右二 Cotton,右一叶家东。Cotton 的家有些特殊,建在落基山海拔 2 800 米的山包上。据说,是在他们几个学生的帮助下自己建的一座三层楼房;Cotton 的起居也有些特别,Volice 说他晚上 9:30 睡觉,早晨 4:00 起床,4:30～8:30 是他的工作时间,备课、看书、写作都在这一时段。中国有句古训:实功夫何处下,三更灯火五更鸡,却在这位西方人身上得到了体现。他 8:30 吃早饭,9:00 骑着摩托车沿着盘山路去上班,历时 40 多分钟。

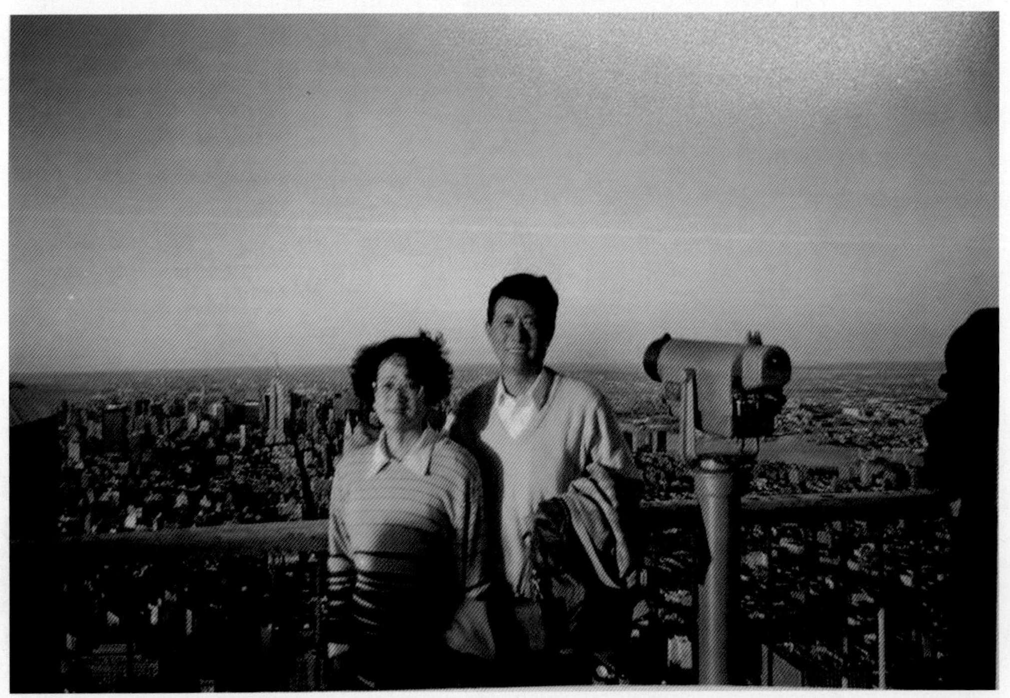

1998 年纽约世贸大厦顶上合影留影。时隔 3 年,在“9·11 恐怖袭击事件”中此楼毁于一旦。事后打开此相片看看,感触很多,如梦似幻,不可名状。

作者简介

叶家东，男，1936 年出生。南京大学大气科学学院（系）教授，浙江省东阳市东江镇茜畴村象山干人。1952 年东阳巍山北麓中学初中毕业，1955 年杭州一中（杭高）高中毕业，1959 年南京大学大气科学系（气象学系）本科毕业，留校任教。1985～1989 年访问美国科罗拉多州立大学和犹他大学，1997 年退休。在职期间兼中国气象学会大气物理委员会委员，全国人工影响天气科技咨询评议委员会委员，英国皇家气象学会会员。主要从事大气物理学、云降水物理学、人工影响天气及其效果统计评估技术、云动力学以及中尺度气象学等领域的教学科研工作。

主要业绩：国内首次设计并与福建省气象局合作推行我国第一个人工降雨随机化试验——古田试验，在数理统计上首次提出方差不相等的双样本回归分析方法。古田试验历时 12 年，平均增雨 23%，统计显著度优于 0.01。古田水库发电增值 2 100 万元。该项研究成果先后获得 1978 年全国科学大会奖、江苏省重要科技成果三等奖、中国气象局科技进步二等奖，并获国内外同行专家广泛好评。中国气象科学研究院名誉院长周秀骥院士评价该试验：这是我国迄今科学水平最高、最严格的人工降水试验，在国内是领先的，在国际上也是先进的人工降水科学试验。古田试验前期阶段成果在国际会议上宣读后，联合国世界气象组织主管人工影响天气的副秘书长 List 教授在来函中指出：你们的研究工作引起了相当一部分人的关注。我们都希望你们能将这一项研究坚持下去。2008 年在长春举办的庆祝中国人工影响天气科技事业五十周年的纪念大会上，本书作者获中国气象学会颁发的荣誉证书：在中国人工影响天气科技事业五十年中，您做出了突出贡献（由于本人在国外探亲，此证书由专人送来南京大学）。

在访问美国科罗拉多州立大学期间，从事云动力学、云物理学和中尺度对流系统的结构分析和数值模拟研究工作，提出在中尺度对流复合体（MCC）的层状区存在冰晶聚合带概念，其中冰晶聚合效率很高，聚并系数是变数，最高可达 1.4。在 MCC 的对流区与层状区之间存在过渡区，并解释其物理成因。发现层状区冰质点谱分布函数的两个参数均非常数，诊断分析表明它们是高度（温度）的函数。回国后主持和参加多项国家自然科学基金项目和国家科技攻关项目，在云物理学、人工影响天气、云动力学、中尺度对流系统等领域发表论文五十余篇，其中多篇论文在国际会议文集和国外刊物上发表。著作有《人工影响天气的统计数学方法》《积云动力学》和译著《风暴和云动力学》。

参看《中国专家人才库》(7)(634 页)

《中国专家人名词典》(10 页)

《浙江古今人物大辞典》(续编)(167 页)

《东阳当代人物》(84 页)

代序一

叶家东同志是我国云降水物理与人工影响天气的主要学术带头人之一。他取得的优秀科研成果主要有两个方面：

（1）他在与福建省气象科学研究所主持合作进行的福建古田人工降水随机试验中，首次提出区域回归随机试验设计方案，该试验历时12年，取得了显著成绩。这是我国迄今科学水平最高、最严格的人工降水试验，在国内是领先的，在国际上也是先进的人工降水科学试验。

（2）他根据观测资料对中尺度对流复合体（MCC）层状区微物理与动力结构进行的分析研究，是现代云降水物理学前沿的科研工作。他在冰晶聚合体以及冰质点分布函数等方面取得的成果具有创新特色的国际先进水平。

他把云降水物理与中尺度气象学相结合，是当前发展的方向。

中国科学院院士

中国气象科学研究院名誉院长

周秀骥

1994年11月22日

代序二

叶家东同志长期以来从事云物理学和人工影响天气的研究和教学工作，早在1959年就在徐尔灏先生的指导下参与了人工降水试验研究。自20世纪70年代起他指导并设计、实施了福建古田人工降水随机试验计划。该试验持续进行了12年，获得人工增雨23%的效果，其显著性水平高于0.01信度，多次得到国际上的肯定和重视。此外，他还参与或指导我国不同地区的多项人工增雨、防雹效果统计检验的方案设计，对我国人工影响天气效果检验的科学化、客观化起到了重要的推动作用；在国内，这一领域的科研与实践工作具有权威性影响。

叶家东同志勤于著述，20世纪80年代编著的《人工影响天气的统计数学方法》，系统地介绍了有关人工影响天气的统计学概念和数学处理方法；他作为第一作者编著的《积云动力学》是我国系统介绍这一学科的唯一教材；译著的《风暴和动力学》是国外20世纪90年代出版的具有权威性的论著。这些著作有的已在国内广为引用，有的已成为这一领域中的必读文献。

叶家东同志曾多次在全国性大气物理或人工影响天气研讨会和培训班上做专题报告或讲授有关专题，在培养我国这一科技领域人才方面做出了重要贡献，在国内赢得了很高声誉，是我国这一领域的学科带头人之一。

叶家东同志在云降水物理和人工影响天气领域的进展和面临的科学问题等方面具有很高的综合分析和洞察能力。他近期在国内外所从事的研究工作中，如粒子聚并效率、粒子谱参数化、云与环境的相互作用和暴雨的抑制等，大都是云物理学进一步发展的重要疑难问题或是具有重大潜在效益的科学探索的前沿性研究。

中国工程院院士

国家气候中心研究员

章基嘉

1994年12月27日

注：原国家气象局副局长、原中国气象学会理事长、中国工程院院士章基嘉教授不幸于1995年10月6日在北京因车祸去世，终年65岁！这是我国气象界的一重大损失，我们表示深切哀悼与怀念。

前　言

那天，在南京大学（简称南大）食堂用餐，同桌两位博士生出于好奇，轻声问道："您俩是南大老师吧？""是的，不过已经退休了。"当我得知他们一个来自广州，一个来自哈尔滨，便笑着说："你们来自天南地北，我们可是西洋东归，刚从海外回来。南大是我们的家，所以总是在南大食堂用餐，与年轻人混在一起，仿佛自己也年轻化了似的，感觉挺好。"我意犹未尽：在我心目中，我这一生有三个家。一个是我生长的老家，在浙江省东阳市一个叫象山干的小山村，隶属东江镇茜畴村。我们习惯自称为东阳人（"东阳侬"）。东阳市前些年还定期寄来故乡的杂志《东阳人》，图文并茂，印刷装帧十分精美，使我不时感受到故乡的勃勃生机和温馨。前些年腿脚还较为灵便，清明节，只要身在国内，我们都会回乡扫墓，"亲人埋骨是故乡"，难以忘怀的故乡。另一个就是南大。我和老伴蓓芬在这里已经学习、工作、生活了六十多年，我们是南大人。在国外校友聚会时，我们会戏称彼此为"南大老乡"。在我的心目中，我还有一个"家"，就是我从事的云降水物理和人工影响天气科研教学领域。那位来自暨南大学的博士生插了一嘴："精神家园？""对，是精神家园。"难怪人家是现当代文学博士生，才思敏捷。

我国的云降水物理和人工影响天气科技事业始于1958年，而我恰巧在1958年12月作为预备教师参加了南京大学气象学系（现大气科学学院）大气物理专业的成立会议，自此开始投身于我国的云降水物理学和人工影响天气的教学科研事业。作为一名年轻学子，难免会染上一些浪漫情怀和对未来彩虹般的憧憬，但我还是勉励自己力求脚踏实地、勤奋耕耘、量力而行、开拓创新。我历来自认为是我国"人影村"的一名村民，这种感觉萦绕心头，挥之不去。2008年中国气象学会在长春举办了庆祝中国云降水物理和人工影响天气科技事业五十周年纪念大会。我因身在国外，未能与会。但会议组织者仍将我早些年的一篇纪念徐尔灏教授的文章《高山仰止，景行行止》列入会议文集，并为我颁发了一份荣誉证书：在中国人工影响天气事业五十年中，您做出了突出贡献。还派专人将证书送来南大，这使我感动，也受到鼓舞。我看重这份情谊，将它放在本文集的扉页。如果这文集能为我国"人影村"的年轻村民们有些助益，就自感欣慰，这也是一个老村民的一份心意。

大气物理专业成立之初，专业发展方针定为以云雾物理为中心，边界层气象和雷达气象为两翼。为了开展云雾物理学专业方向的建设，在徐尔灏教授的领导下，采取了两项措施：一是开设一个全国性的云雾物理讲习班，聘请苏联专家基留辛来南大讲授云雾物理学。学员来自北京大学、中科院大气物理研究所(地物所)和兰州高原大气研究所等全国十多个高校及院所的二十多名进修教师以及南大气象学系61届毕业班的学员。基留辛是一个务实谦和具有长者风度的老师,大家都尊敬他,相处都比较融洽。1960年回国前他说了一句我始料不及的话:十年以后我们将要向你们学习。二是筹建云雾物理实验室并开展野外云雾物理高山云雾站考察研究。我担任了云雾物理实验室主任12年。20世纪60年代我主要承担云物理实验课和带领学生赴衡山、庐山等高山云雾站实习考察研究工作。这对培养学生实地了解云雾降水的宏微观物理过程,增强感性知识无疑是十分有益的。但是,由于当年的技术条件所限,从科学研究的角度看,收效并不十分显著。"不识云雨真面目,只缘身在云雾中。"但我们毕竟开创了这一领域,而且,高山云雾考察实践带给我们的不仅仅是知识的增长,而且对于培养学生包括我们自己面对艰难勇往直前的探索精神,大有裨益。"咬定青山不放松,立根原在破岩中,千磨万击还坚劲,任尔东西南北风。"每当我念诵这首郑板桥的《山竹》诗时,脑海里不时会浮现出当年在衡山、庐山等高山云雾站学生实习奔忙的情景,而且心目中认定,那山竹就应该生长在庐山太乙峰北极的乱崖上,为了进行梯度观测,当年中央气象局庐山云雾站在太乙峰顶端的巨岩上,搭建了一座小木屋和云雾观测平台,我们有两位62届的同学周笃诚和黄云坚守在这里进行云雾观测。雷电来袭时,小木屋顶上的避雷针上火花直冒,吓得两位同学趴在岩石上不敢动弹。事后想起这情景我心里发毛,要真出点什么事,那可怎么得了。每当想起这两位同学,不免有些歉疚,可他们却并无怨言,一直坚持到实习结束。"山竹"的品格在他们身上得到了生动的体现。不过,为了安全起见,在以后几届学生的实习中,太乙峰顶端这个云雾观测站就撤销了,只在含鄱口和望鄱亭观测,后来又搬回到天控所驻地日照峰观测。庐山云雾站当年有一个挺牛的名称:庐山天气控制研究所,简称天控所,带有浓烈的时代气息,那时候不仅我们自己年轻,我们的时代也还年轻,青春的脚步就这样沟沟坎坎、兢兢业业走过来了,谈不上豪情满怀,却也无悔。

1964—1965年,南京大学和上海民航局以及中科院大气物理研究所(地物所)等单位协作开展上海地区冬季层状云结构考察和人工降雪试验研究,应用老式的安-二型双翼飞机作为观测平台和催化剂(干冰)播放平台,取得了较为明显的成果。我在《科学画报》1964年第三期写了一篇《耕云播雪》,文章简单介绍了这项试验研究。该期的封面图是我在飞机上拍摄的层状云双机飞行试验研究的状态。1965年夏开展的夏季对流云结构考察研究采用伊尔-14飞机在浓积云多个高度观测,途中飞机颠簸异常强烈,一次在一块5 200米厚的浓积云里面穿云观测,在云的核心区,颠簸强烈,我一只手握住固定在机顶上的长枪式滴谱仪的把手,另一只手抓住座椅靠背,整个身子腾空而起,就像玩高低杠似的,更有甚者,坐在椅子上的许绍祖猛然腾空而起,头碰到机舱顶上又猛地落在座椅上,就在对流核心区域内我们观测到13.6克/m^3的高值含水量区,还表明浓积云里有多个对流核心区。只可惜,"文革"一开始，这类科研活动一并中止，不禁令人扼腕。

尽管某些具体项目可能会时过境迁，但是，走出书斋、走出课堂、直面大自然、探索大自然本身的奥秘，这种治学之道无可厚非。我们怀念徐尔灏教授，是他指引并带领我们沿着“气象物理化”思路开拓创新谋发展的。

20 世纪 70 年代我的注意力转向人工降水外场试验设计和效果统计检验研究工作。1971 年，我在长沙会议上做了一个开展人工降水随机试验的报告，得到了南方诸省湖南、福建、安徽、四川、江西等的积极响应，福建古田试验就是在这样的背景下开展起来的。我根据古田水库对增水发电的社会需求和流域的地理状况设计了一个区域回归随机试验方案。福建省气象研究所的程克明和他的同事们在所长孙岳云的全力支持下攻坚克难，奋力开拓，开展了我国首个有科学设计的人工降水随机试验。前期成果在法国举办的国际会议上宣读以后，得到了国际同行的广泛好评和关注。联合国世界气象组织（WMO）主管人工影响天气事业的副秘书长 List 教授在给我的来函中说，你们的研究引起相当一部分人的高度关注，我们全都希望你们把这一研究坚持下去。在程克明离开福建调回常州工作后，曾克平和这个团队的其他同事继续坚持将这一项研究开展了十二年，取得了较为丰硕的成果。增水 23%，统计显著度达到 0.01 水平，水库增水发电经济效益 2 100 万元，这项研究获得 1978 年全国科学大会奖和中国气象局科技进步二等奖等多项奖项。

1985—1988 年，我访问美国科罗拉多州立大学（CSU），从事云对流动力学和中尺度对流系统结构分析和数值模拟研究。我原先打算学习 CSU 的区域大气模拟系统（RAMS）进行数值模拟研究，这也是当时的一种思潮，年轻人都热衷于此。由于 CSU 前不久与国家海洋大气管理局（NOAA）合作开展一个叫 PRE－STORM 的研究计划，NOAA 装备了两架 C－130飞机作为空中实验室进行中尺度对流系统的考察研究平台，积累了相当多的云内层状区的微结构资料。CSU 的 Cotton 教授希望我去分析这部分的微结构资料，由于 20 世纪 60 年代有过高山云雾考察研究和飞机观测云结构的历练，我深知这批资料十分珍贵，便欣然应诺。每周六由 Mike. Fortune（Cotton 的博士生）开车送我到博尔德 NOAA 研究总部的资料库，我一头扎进资料堆，把相关的资料都打印下来，带回学校做分析。在 MCC（中尺度对流复合体）的层状区存在冰晶聚合带，从 0 ℃层向上延伸至－5 ℃层，其中冰晶聚合效率很高，可以超过 1，最大可达 1.4，这个聚合层是雷达回波亮带的主要贡献者，所以说层状区（云）回波亮带。与其说是冰晶融化带不如说是冰晶聚合带造成的更为切贴，是可以从 0 ℃层向上延伸至－5 ℃层，在有中尺度辐合上升区域，过冷云滴附着在冰晶聚合体表面时更增大了冰晶聚合体的回波反射率。

这项 MCS 层状区微结构分析研究在 1986 年落基山 SnowMass 举行的学术会议上得到了明显的好评。我在会上做了报告，会后有好几个国际同行前来表示祝贺，其中有夏威夷大学的 Takahashi(他告诉我他已回到东京大学任教),犹他大学的 Fukuta(这为我日后去该校访问奠定了基础),以及多伦多大学的 List,在他任 WMO 副秘书长期间曾为古田试验给我来过信函。List 到中国工作访问时在北京我俩也曾见过面,算是熟人了。我问他还记得我吗?他说当然,随手从口袋里翻出一个小本子,翻开当年在北京时我在上面签的名,挺友善的,还有几位记不清名字的同行。我随后将微结构诊断所得的冰晶分布函数以及冰晶聚合效率改进 RAMS 模式的微物理过程的模块进行数值模拟试验,得到了一些具有特色的模拟结果,并指明微物理相变潜热的效应在 MCC 中尺度系统组织化过程中起到关键作用(中空暖心结构和中低压的形成,使得 MCC 在对流活动中止后,系统尚能维持 6 小时或更长时间。)。不过这里有一个插曲:我回国后将相关的 MCC 数值模拟结果著文投稿到英国皇家气象学会的季刊(Q.J.),我是以南京大学的身份和 Cotton 以 CSU 的合作方式投稿,Q.J.编辑部说他们询问过 Cotton,Cotton 回话说不知道这事,好像并不支持我以南大人的身份投稿,所以编辑部不便发表。我想可能是为了安慰我的失落,以英国皇家气象学会的名义邀请我参加该学会并寄来入会申报表,我将申报表寄去以后不久就收到该学会寄来的外籍会员的会员证,并按月寄来学会的会刊《Weather》,我则每年需交 35 英镑的会员费,也许这满足了我当时小小的虚荣心。不过,总的来讲我与 Cotton 相处得挺好的。我回国时承诺将他和 Anthes 合著的《风暴和云动力学》一书译成中文在中国出版,后来中译本寄给他后,他十分兴奋,来函说他每次上课都将中译本也带去并展示给学生看。我和蓓芬两口子在他那里访问长达三年之久,这段时间是我业务生涯中比较充实也比较自在的时期。一家三口在这落基山麓小山城生活,有点其乐融融之感。当然,这可能只是一种主观的感觉,因为其实我们并没有深层次地介入美国社会,我们只是外来的客人,而且只限于学校科研氛围里面,女儿叶黎在普陀高中念了一年高三,1987 年高中毕业时,在 CSU 的体育馆里举行一次盛大的毕业典礼,连同家属有数千人参加,校长在大会上做报告时,点名表扬了叶黎,并示意坐在第一排的叶黎站起来,说:"从中国大陆来才一年,刚来时语言尚有些困难,一年下来却达到每门课都全优的成绩,实在令人惊奇。"而且女儿还挣得了 12 个大学的学分。但是她却高兴不起来,因为前不久在申请大学奖学金过程中,女儿原本是一项数额较大的奖学金(6 千美金)的两名候选人之一。该集团一退休高管来家访时还表示叶黎很有竞争力,但最后结果却落选了。原因很简单:这项奖学金只奖给本州人,女儿不仅不是本州人,连美国人也不是。女儿平生第一次品尝到了人生的辛辣,她不无感慨地说了一句与她十八岁年龄颇不相称的话:我们是捆着双手与他们竞争的。自此开始走上了一条艰辛的打工求学勤奋创业之路,一条在异国他乡无依无靠不懈拼搏的开拓创新之路。

前面两篇代序是当年提升职称时国内同行专家的评议书。事后系主任将它们发放给当事人,我看后颇为感动。这两位院士平素都是我钦佩的学者。周秀骥院士早年也是研究云物理学出身。20 世纪 60 年代初期，他领衔的“暖云降水起伏理论”给我留下的印象尤深。记得 1961 年我毕业不久，曾带领比我低一届的洪钟祥、姚克亚、龚乃虎等 10 名毕业班同学赴衡山云雾站实习，住在祝融峰西侧的上封寺。其间，顾震潮先生率领周秀骥、巢纪平等一批青年专家赴衡山进行积云宏观考察，并上山给同学们做学术报告。“暖云降水起伏理论”“积云动力学”等让我耳目一新，印象甚深。此后两年我差不多将暖云降水起伏理论的有关文章，包括别列也夫的文章都通读了。后来温景嵩在随机函数等方面做了修正和发展。所以我在 Cotton 和 Anthes 著的《风暴和云动力学》一书的中译本中注明：暖云降水的起伏理论是由周秀骥、顾震潮等人开创发展的，赵九章向国外做了介绍(《风暴和云动力学》中译本 91 页)。我与顾震潮先生在南岳是第二次见面，前一年他陪舒拉克维里茨访问南大时见过，在斗鸡闸开座谈会，当时我颇为惊讶：大名鼎鼎的顾震潮先生怎么看上去竟是这么年轻的一个白面书生！这次在南岳上封寺联欢会上，他还为同学们演唱了一首《毕业歌》。歌唱得挺流畅，但他是面斜对板壁，背斜对着大伙唱的，有点像学生上堂背书模样，挺动人的。一天晚上，他还抽空来我房间看我，这使我喜出望外，得以有机会当面请教做学问之道和搞研究之法。他快人快语，主传勇于探索、不要怕出错、错了就改、精益求精。他自己出文章，有时废文稿叠起来厚可盈尺。末了，他还说：徐(尔灏)先生是很好的老师，他也是我的老师。我不明就里，但却难忘。为了研究气溶胶，顾先生在从北京至湖南的火车上(那时火车慢，途中约需花费一周时间)将一部胶体化学啃下来了，一时传为佳话。他那种冲锋陷阵、一往无前的开拓创新精神和身体力行、务实求真的治学风格激励着多少青年学子拼搏奋进，只可惜在大动荡年代里，常年透支过度，英年早逝，令人扼腕。后来人可否汲取点教益：善待事业、善待他人，但也不要太亏待自己，留点空间给自己，力图身心健康。现代社会强调可持续发展，人也一样，需讲究可持续奉献。极度劳顿的事，偶尔为之、适可而止，切不可持续过劳，需知人家总统也是有假期并且有任期的。

章基嘉院士的评议书写得十分具体详细。我颇感诧异。他在南京气象学院任职时由于我老伴范蓓芬与他同事过几年而有过一面之交，他调至国家气象局任副局长后就很少见面，加上各自的专业方向业务交叉也比较少，对我的科研业务他不可能了解太多。直至数年后，在一次会议期间，国家气象科学研究院人工影响天气研究中心的游来光研究员在闲谈中提到前些年章基嘉副局长曾找过他，请他写一份关于我的评议材料。我这才恍然大悟，原来这份评议书出自游来光之手。老游是我的好友，也是我钦佩的一位同行学长。记得 1975 年在北京大学举办的云雾物理和人工降水训练班讲课期间，我与他同住一室，相处甚笃，几乎无话不谈，老游诚恳朴实，勤奋敬业，严谨治学，无怨无悔地奉献着自己的才智学识，加上他那淡定幽默、机智风趣、广结人缘的处世风度，成为我国人工影响天气领域口碑最佳的少数专家之一。特别在我国北方层状云人工降水试验研究(1981—1992)项目中，殚精竭虑，做出难以替代的主导性贡献，令人难以忘怀。在他晚年患病期间，我和蓓芬出差北京，曾去医院看望过他，原本瘦削的身体显得格外苍白虚弱，但神情依然淡定平和，想不到这竟是最后一别！

走进南京大学鼓楼校区北园的大门，在东侧图书馆的外墙上赫然刻着南大校训“诚朴雄伟　励学敦行”八个大字。细细品味，若有所悟，乃作一小诗《校训吟》解读之。诚信为本树人生，朴实无华修身性。雄健擎托凌云志，伟宏创业天地新。励志攀登科学峰，学海无涯汲华精。敦睦和衷齐奋搏，行路万里济苍生。（此诗曾刊登在《南京大学报》2004 年 12 月校报上）谨以此诗自勉并望与新老南大人互勉。

本文集的整理和出版，得到了南京大学大气科学学院潘益农书记和杨修群院长等院领导的热情鼓励和大力支持。谨此表示衷心的感谢！

本文集所列论文不少是多人合作或多个单位协作的成果，凡列入本文集的均系本人执笔撰写负责。

衷心感谢南京大学出版社吴汀、王南雁等诸位编辑辛勤操劳，细心审校，使这一文集得以问世。谨以此文集纪念我从事云降水物理学、人工影响天气及云对流动力学和中尺度对流系统科研教学事业六十周年。人生征程无坦途，沟沟坎坎，在所难免。现在我已年逾八旬，携手老伴范蓓芬副教授，栖居仙林南大和园，颐养天年。我不时勉励自己沿着“平静、平淡、平和、平安”这“四平路”走好每一天，嘴里轻轻地哼着：马儿哟！你慢些走喂慢些走，我要把这迷人的景色看个够……蓦回首，离天三尺三，复有何憾！

目 录

1 概述

1.1 大气物理学概要

（《当代新知识大系》大气物理部分书稿）

条 目

大气圈	大气物理学	高层大气科学
大气成分	大气分层	大气热力学
大气层结稳定度	大气辐射学	大气辐射传输
反照率	云的辐射特性	辐射收支
地球系统热量平衡	大气光学	大气光象
大气能见度	大气电学	雷电物理学
大气声学	云雾降水物理学	云
云降水微物理学	云动力学	人工影响天气
云水资源	人工增雨	人工防雹
人工消雾	人工遏制雷电	人工削弱台风
播云设计	播云技术	播云效果评估
人类影响区域天气和气候	人类影响全球气候	

大气圈 覆盖地球表面的大气层，是地球系统的组成部分。地球系统是由大气圈、水圈、岩石圈、冰雪圈、生物圈及人类圈六大子系统组成的巨系统。通常将人类圈并入生物圈，鉴于人类的现代生产活动及创造的科学技术，正日益显著地影响全球系统的发展与变化，将其从生物圈中分离出来很有必要。全球系统的六大圈之间存在着力学、物理、化学及生物过程强耦合的非线性相互作用，而且是一个开放系统，太阳、月亮等天体和地球内层的地核、地幔等构成全球系统的外部环境。这个复杂的巨系统具有一定的脆弱性。向大气排放不到空气亿分之一的氟氯烃（CFC）就可能破坏大气臭氧层形成危害生物和人类圈的臭氧洞；向大气排放 CO_2 等温室气体引起全球增暖对人类生存环境的影响更是众所关注的热点问题。研究大气圈必须着眼于全球系统这个大背景。

大气是由氮、氧、氩、氖、氦、氪、氙等常定成分和水气、二氧化碳、臭氧等可变气体成分及固、液态气溶胶组成，总质量约 5.3×10^{18} 千克，大约是地球总质量的百万分之一。大气圈的底界为地球表面，海平面的大气压力平均为 1 013 百帕，当温度为 273.16 K 时（标准状况）相应的空气密度为1.293千克/米^3。大气密度和气压均随高度按指数律递减，大气总质量的99.9%集中在48千米高度以下气层中。大气圈上界有的以磁层顶为界，向阳一侧因受太阳风挤压离地约6.4万千米，背阳面则可延展至数十个地球半径以上；也有将极光出现的最高高度作为大气上界，约1 200千米；据卫星探测，大气上界在2 500～3 000千米高度。但是，人类活动和天气过程多发生在20～30千米以下的大气低层。大气圈按照物理、化学属性的垂直分布特征可以分为若干层，如按温度分布特征可以将大气圈分为对流层、平流层、中间层、热层和外逸层，按电离状况和光化学反应特征还可从大气圈中分出电离层、磁层以及臭氧层等(参看“大气分层”)。

大气物理学 研究地球大气中各种物理过程和物理现象的学科。它综合应用流体力学、热力学、统计物理学、电磁学及量子力学等基本物理学原理，结合地球大气的特点，研究大气结构、运动特征、云雨相变、湍流交换、辐射传输和大气热力学过程(大气静力学、云雾降水物理学、大气湍流和边界层大气物理学、大气辐射学和大气热力学)，以及大气中发生的光、电、声等物理现象(大气光学、大气电学、大气声学)，也包括人工影响天气、大气遥感物理学和中层大气物理学等现代大气物理分支。大气物理学是大气科学中涉及面最广的基础学科，20世纪80年代大气物理学论文发表率占大气科学论文总数的26%，居三大核心学科之首(大气物理学、大气动力学和大气探测学)。自20世纪70年代以来我国大气物理学研究进展包括：(1) 大气边界层物理学：对边界层效应参数化，大气湍流形成机制、特殊下垫面上的湍流结构、边界层与自由大气相互作用、夜间边界层结构特征等方面进行分析研究、外场考察和数值模拟试验，并取得明显的进展；(2) 云降水物理学和人工影响天气试验：外场探测和实验设备有所加强，对北方层状云结构和云水资源进行较系统的外场考察，在冰雹云的结构考察，云降水数值模拟研究，人工增雨外场随机试验研究，高效复合型碘化银催化剂研制和相应的飞机、地面火箭、增程高炮等播云工具研制和应用，以及人工影响天气试验基地建设和以计算机为控制中心的试验决策指挥技术系统的筹建等方面都取得重大进展；(3) 大气辐射学：开展了青藏高原综合辐射特性观测研究，东亚春季尘暴特性研究，气溶胶光化学和化学特性研究，云-辐射反馈作用及其在天气、气候动力模式中的应用研究，辐射与天气、气候及农业生态环境的关系研究，对大气吸收和衰减特性，红外透过率和大气冷却率，云中含多次散射的大气辐射传输特性及反射率，卷云中冰晶散射相函数，卫星遥感的温度反演等进行了开拓性研究，并用辐射耦合模式模拟辐射雾、层积云的辐射收支等。随着气候变化、大气环境和灾害性天气的预报和防治等热门专题领域趋向深入，对大气物理学基本规律的认知需求愈益提高，推动着大气物理学向更广泛的领域和更深的层次推进。

高层大气科学 研究对流层顶以上的大气中各种物理-化学-动力学过程的学科。对流层顶以上的大气温度结构主要由辐射过程确定。平流层和中间层以及热层的一部分合称中层(10～100千米高度)，其中的高温主要由臭氧吸收太阳紫外辐射引起，中层的上界是中性大气的顶。热层中空气强烈吸收紫外辐射而处于高度电离状态，气温可达500～2 000 K。中层大气研究是高层大气科学的核心领域。中层大气本身存在一系列复杂的物理-化学-流体力学耦合过程，并与高、低层大气有着复杂的相互制约作用，促使中低层大气成分和进入低层大气的太阳辐射特别是紫外辐射发生重要变化，从而有可能使全球生态平衡和人类生存环境受到影响。20世纪70年代后期开展中层大气研究计划(MAP)，研究中层大气潮汐、中层大气扰动结构、流星探测系统、平流层微量气体收支、平流层气溶胶探测、中层大气重力波与湍流、中层大气电动力学、臭氧时空分布、太阳入射辐射量、上下层之间能量输送和相互作用、中层大气物理过程数值模拟等。当前中层大气研究主要着眼于三个方面:(1) 以臭氧变化为中心的辐射-动力-化学问题,中心是臭氧和微量气体长期变化和控制因素研究;(2) 中层大气动力学和环流特征及其与上下层的耦合,涉及各种尺度波动、湍流及其相互作用和控制因子分析;(3) 太阳活动与中低层环流及气候关系的观测分析和物理过程研究,如中层大气对太阳活动及相关事件的响应方式和可能的响应机理的探索性研究。

大气成分 地球大气是由多种气体和悬浮颗粒混合组成。气体成分有氮、氧、氩、氖、氦、氪、氙等常定成分和二氧化碳、水汽、臭氧、甲烷、二氧化硫、一氧化碳、氧化氮、氢等可变成分;悬浮的固态和液态气溶胶颗粒有尘埃、烟粒、水滴、冰晶、盐粒、花粉和孢子等。不含水汽和气溶胶颗粒的大气称为干洁大气,也称干空气。90 千米以下由于湍流混合作用,干洁大气的各成分比例随高度不变,可看成"单一成分"的气体,平均分子量为 28.966。在温度 273.16 K 和气压 1 013.25 hPa 的标准状态下,干洁空气的密度为 1.293 千克/米^3。其中氮、氧、氩的体积混合比分别占 78.084%、20.946% 和 0.934%,可变成分中的二氧化碳(CO_2)近地面平均体积混合比约占 0.033%,其他所有成分的含量合计不到干洁空气的万分之一。90 千米以上高度分子扩散占主导作用,质量大的分子数密度随高度递减比轻分子快,使得较轻气体在高空占较大比例并造成干洁空气平均分子量随高度减小,故称 90 千米以上气层为非均匀层,其下大气层称均匀层。从 60～80 千米开始,氧和氮因太阳紫外辐射而有不同程度离解,100 千米以上氧分子几乎完全离解为氧原子,250 千米以上氮分子也基本离解。从 1 000 到 2 500 千米高度,大气稀薄接近真空,有一氦层,再上有一层更稀薄的氢"质子层"。

大气中水汽含量变化最大,体积混合比变化范围在 0～4%,最大可达 7%。水汽是大气中唯一有三相变化的可变成分,是云、雾、雨、露、霜、雪、雹的原料。二氧化碳、水汽、臭氧、甲烷等能吸收地球长波辐射而具有"温室效应",其含量虽然极微,却具有影响地球气候和生态环境的潜力。极地冰芯采样分析表明近 200 年来大气中 CO_2 含量增加 20%以上,甲烷增加 200%以上。数值模拟估算 CO_2 浓度倍增时,因温室增暖效应全球平均气温将升高 2～4 ℃,结果会引起极冰融化、海平面升高,并导致一系列生态环境变化。大气可变成分的监测是全球气候和大气环境基准监测的重要内容。

大气分层 根据大气温度或其他物理性质随高度的变化特征将大气分层。最常见的是按温度随高度分布特征将大气层分为对流层、平流层、中间层、热层和外逸层。另外,按大气成分可分为均匀层和非均匀层;按空气电离状况分为电离层和中性层;按光化学反应特征从大气中分出臭氧层;按电离气体受地球磁场制约状况可分出磁层。(1) 对流层:大气最低层,由于密度和气压因重力场而随高度递减,空气抬升膨胀冷却,故温度随高度递减,铅直递减率平均约为 $\gamma=-\dfrac{\partial T}{\partial Z}=0.65$ ℃/(100 米),随时空变化很大。对流层厚度随纬度和季节而变化,赤道 17～18 千米,中纬度 11～12 千米,高纬 8～9 千米,夏季比冬季厚。对流层集中了整个大气层质量的 3/4 和几乎全部水汽。对流层受地表影响强烈,对流活动旺盛,风、云、雨、雪、雷电等绝大多数天气现象都发生在此层。(2) 平流层:从对流层顶至 55 千米左右高度,温度最初随高度略增,约 30 千米以上随高度迅速增温,到平流层顶达 270～290 K,主要由臭氧(O_3)吸收太阳紫外辐射所致。平流层气流平稳,对流微弱,与对流层之间很少混合,其中水汽稀少,能见度大。有时对流层中强对流云顶可伸展到平流层底部,并可形成少量卷云。从对流层顶向上水汽含量陡减,O_3 含量却增加一个数量级,火山喷发进入平流层的火山尘可滞留一年以上。中、高纬度 20～27 千米高度可见到色彩绚丽的珠母云。(3) 中间层:从平流层顶向上,温度再次随高度降低,至 80～85 千米高度低达 160～190 K。层内有强光化学反应,在太阳紫外辐射作用下部分大气电离。高纬度黄昏在中间层顶部可见到银白色夜光云。(4) 热层:从中间层顶向上伸展至 500～600 千米高度,空气十分稀薄,处于高度电离状态,它们吸收波长小于 0.175 微米的紫外辐射使温度随高度迅速增高,可达 1 000 ～2 000 K,故称热层。(5) 外逸层:热层顶向上一直延伸到 1 600 千米高空,地球引力场的约束很小,空气粒子自由程又很大,一些高速粒子会射向星际空间而外逸。1 600 千米高空空气密度只有海平面的千万亿分之一。(6) 臭氧层:O_3 主要分布在 10～50 千米,20～25 千米高度含量极大,是由波长介于 0.1～0.24 微米的太阳紫外辐射使氧分子分解为氧原子($O_2+h\nu \rightarrow 2O$),O_2 与 O 及另一中性分子(N_2 或 O_2)三体碰撞生成 O_3($O_2+O+M \rightarrow O_3+M$,M 是第三体)。臭氧层吸收绝大部分太阳紫外辐射,保护人类和生物免受过多紫外辐射侵害。20 世纪 70 年代后期发现南极上空出现臭氧含量陡减(冬季减少 3.4%～5.1% 以上)的"臭氧空洞",据推论是人类排放氟氯烃(CFCs)和氮氧化物(NO_2,NO)等所造成的。1987 年各国签署限制生产和消费 CFCs 保护臭氧层免遭破坏的《蒙特利尔议定书》,规定发达国家 1998 年起 CFCs 的生产和消费水平逐渐减少到 1986 年水平的 50%。(7) 电离层:60～500 千米高度的空气分子吸收太阳辐射而电离为电子和离子。电离层能反射无线电波,对电波通讯很重要。按电子数浓度将电离层分为 D、E、F(F_1,F_2)层。D 层(60～90 千米)中午时最大电子浓度为 1.5×10^4 厘米$^{-3}$,E 层(90～140 千米)白天最大电子浓度为 1.5×10^5 厘米$^{-3}$,F_1(140～200 千米)最大电子浓度为 2.5×10^5 厘米$^{-3}$,F_2(200～500 千米)最大电子浓度达 1×10^6 厘米$^{-3}$,出现在 250 千米高度附近。各层的高度、厚度和电子浓度随时间、季节、纬度及太阳活动强弱而变。一般白天电子浓度增大,夜间 D 层消失,F_1 和 F_2 合并成 F 层。太阳活动强时电子浓度随之增大。(8) 磁层:500 千米以上高空粒子非常稀薄,相互间很少碰撞,荷电粒子运动强烈受地球磁场约束。磁层受太阳发射的超声速等离子体流(质子、电子、氦原子核)-太阳风作用,向阳面受缩磁层顶离地约 6.4×10^4 千米,背阳面则扩展到 2×10^6 千米以上形成看似彗星状的长尾。高能等离子体被地球磁场捕获形成 2～3 个辐射带:内辐射带中心离地约 3 200 千米,多数是质子,其能量 20～40 兆电子伏,带内强度稳定少变。外辐射带中心离地9 600～16 000 千米,主要是 1 兆电子伏以上的电子。太阳活动强烈时 5 万公里高空可出现第三辐射带,由低能粒子组成。

大气热力学 根据能量守恒和转换的热力学第一定律和第二定律研究地球大气的热力学特性和过程的学科,包括:大气中能量的传输和各种能量相互转换的条件和规律及其与大气动力过程的关系;干空气和湿空气的绝热过程和温度直减率以及大气层结稳定度;热力学图解、不稳定能量和热力学温度在大气能量学分析和天气预报中的应用;辐射和云雾的水相变热力学以及湍流热交换等过程对大气运动的非绝热加热效应等。大气热力学是大气科学中的基础性分支学科,与大气动力学、云雾降水物理学、物理气候学及天气预报等分支学科之间关系密切。

大气层结稳定度 大气层结指大气中温度、密度和湿度的垂直分布,大气层结的静力稳定度是表示大气产生对流运动的潜在能力的一种量度。静止大气中任一气块处于静力平衡和热力学平衡状态,当一气块受外力作用而产生垂直虚位移时,若环境空气的状态有使它返回原始平衡位置的趋势,则称该大气层处于稳定平衡状态,或称层结稳定;若气块有继续离开初始位置的趋势,则称大气处于不稳定平衡状态,或称层结不稳定;若气块既不返回又不远离初始位置而是停留在新位置上与环境处于平衡状态，则称为中性平衡大气或层结中性稳定。假定气块受外力扰动上升时既不影响环境大气运动，彼此也不发生混合，气块绝热上升且满足准静力平衡条件，可得气块法层结稳定度判据：环境气温直减率 $\gamma=-\dfrac{\partial T}{\partial Z}>$ 干绝热递减率 $\gamma_a\left(\gamma_a=\dfrac{g}{c_p}\right)$，对未饱和及饱和湿空气均不稳定，称绝对不稳定；$\gamma<$ 湿绝热递减率 γ_m，对未饱和及饱和湿空气均稳定，称绝对稳定；$\gamma_m<\gamma<\gamma_a$ 为条件性不稳定，对未饱和空气为稳定，对饱和湿空气为不稳定，这是自由大气中最常遇到的层结条件。在这种场合，对流云能否发展，主要取决于气块在初始扰动触发下能否上升到其抬升凝结高度。实际上气块上升时与环境空气有夹卷混合效应并有环境空气补偿下沉效应，这时层结稳定度判据需做修正。实际大气常处于运动中，如位于气旋辐合抬升的大气层，若低层水汽丰沛，抬升后首先达到饱和，随后上升发生凝结，气温按湿绝热递减率 γ_m 降温，而上层较干的未饱和空气仍按干绝热递减率 γ_a 降温，整层抬升的结果会使大气层变成不稳定，称这类气层为位势不稳定。低空有西南暖湿气流，中高空有干冷西北气流时，这种高低空的差动平流常会造成位势不稳定层结，有利于激发对流性降水。

大气辐射学 研究大气中辐射传输及转换的基本规律和物理过程及其在遥感技术中的应用，以及地球大气系统辐射能量收支的大气物理分支学科。辐射是大气、地球与太阳能量交换的主要形式，是地球气候形成和演变的基本因素。在气候模式、大气环流和数值天气预报模式乃至中尺度模式中，要考虑非绝热加热项中的辐射效应，其中包括云的辐射效应，涉及一系列复杂的辐射效应参数化问题，是各类模式研究中的关键因素之一。近代卫星和地基遥感技术的开发以辐射传输理论为基础，反过来也促进了现代大气辐射传输理论的研究。驱动地球大气和海洋系统运动的主要能源来自太阳，太阳辐射可看成是温度为6 000 K至5 700 K的黑体辐射，最大强度位于波长0.47微米处，辐射的波谱很宽，但99%的辐射能量集中在0.276～4.96微米波段，99.9%的能量集中在0.217～10.9微米范围内，其中可见光波段(0.4～0.76微米)、红外波段和紫外波段的辐射能量分别占46%、46%和8%。在日地平均距离处大气上界与太阳光线垂直面上的太阳辐照度称为太阳常数。1981年其值定为 $S_0=(1\ 367\pm7)$ 瓦/米^2，因太阳活动强弱而有0.5%的变率。另外，地球和大气放射的辐射近似于温度为250 K的热辐射，辐射能谱峰值位于10～15微米波长之间，98%以上的辐射能集中在4～120微米的红外波段，与太阳辐射波谱很少重合，故常称太阳辐射为短波辐射，地气系统的辐射称为长波辐射，它们在大气中传输衰减特性有很大不同。

大气辐射传输 辐射在大气中传输经大气吸收和散射而衰减，并导致辐射强度、波长、传播方向及偏振状态等发生变化：(1) 大气吸收：晴空大气中短波辐射的主要吸收介质是臭氧、氧和氮。高层大气中氧和氮吸收了大部分波长小于0.3微米的入射太阳紫外辐射而电离，平流层中的臭氧吸收掉几乎全部余下的紫外辐射。长波辐射的主要吸收气体是水汽、二氧化碳和臭氧。水汽的主要吸收带在波长6.3微米和大于12微米波段，另外在1.38、1.87和2.7微米处也有吸收带。太阳辐射因水汽吸收可削弱4%～15%辐射能。二氧化碳的红外吸收带主要在波长14.7、4.3和2.7微米处，臭氧在9.6微米有强吸收，在4.7和14.1微米也有吸收。在红外辐射区波长为3.5～4.2微米以及8～14微米波段，水汽和二氧化碳吸收小，大气透过率高，称为“大气窗区”。另外在0.4～0.7微米的可见光波段，大气吸收也弱，称可见光窗区。卫星遥感主要利用这两个窗区摄取可见光和红外云图。(2) 大气散射：大气散射主要发生在可见光波段。大气分子的散射称为瑞利(Rayleigh)散射，其特性是前向和后向散射最强且对称，为非偏振光，垂直于入射方向的散射辐射最弱，且为线偏振光。散射方向性图呈部分重合的双球状。散射通量密度与波长的四次方成反比，故短波的散射衰减特别强。粗粒散射：尺度 r 与波长 λ 相当的粒子散射称为米氏(Mie)散射，散射辐射通量密度是尺度参数 $\rho=\frac{2\pi r}{\lambda}$、散射角 θ 及粒子折射率 n 的函数，随着 ρ 增大，向前散射迅速增强，前后向散射不对称性愈明显，散射辐射愈集中在前向一小角度范围内。辐射在云和气溶胶中传输时会被云滴、冰晶、雨滴、雪花和气溶胶粒子散射，并伴有部分吸收。散射和吸收特性与粒子大小、粒子数浓度、折射指数、谱分布有关，且要考虑多次散射过程，远比晴空大气散射复杂。太阳辐射从大气上界经过大气层衰减后垂直地传播到地面时的透射比称为大气透射率，其对数值即大气光学厚度是辐射波长的函数。

反照率 物体表面反射的辐射能与入射辐射能之比。地表反照率随入射辐射波长、太阳天顶角、下垫面属性如植被覆盖物状况而有很大差异。平均讲，耕地反照率仅为 0.04～0.08，沙漠 0.21～0.60，冰雪面 0.35～0.82，农作物 0.10～0.28，各类土壤 0.07～0.25。绿色植被对 0.7 微米以下的短波辐射反照率很小，而对 0.7 微米以上的红外辐射反照率很大。根据这一特性采用卫星双通道遥感地面的反照率比可以估测地表植被覆盖状况，对草场监测和农作物估产。水面的反照率强烈地依赖于太阳高度角和风速。风速小于等于 2 米/秒时水面较平缓，太阳高度角大于 50°时，平均反照率约 0.03，随着风速增大海面起伏不平，反照率增达 0.10 左右，当太阳高度角趋于 0°时，反照率可高达 0.50。云作为整体，其反照率随云厚、云微物理结构、云状和太阳高度角而异。一般云愈厚，反照率愈大；对含水量相同的云，其云滴浓度愈大，散射截面积就愈大，故反照率也大。各类云的反照率：层云(0.05～0.85)，层积云(0.35～0.80)，雨层云(0.64～0.70)，高层云、高积云(0.40～0.56)，卷云(0.15～0.20)，卷层云(0.44～0.59)。地球-大气系统的反照率表示被大气、云和地球表面反照(反射)向空间的辐射量与入射的太阳辐射总量的比值，称地球行星反照率。据估算，全球年平均行星反照率为 0.30 左右，赤道至热带地区不超过 0.25，北极和南极地区分别高达 0.57 和 0.70。

云的辐射特性 云滴、雨滴、冰晶和雾粒等水汽凝结体与辐射的相互作用方式相当复杂。云对辐射的吸收和散射特性取决于入射辐射波长、水汽凝结体的大小及其相态和形状。数量众多的小云滴是太阳辐射的强散射体，而雨滴则是强吸收体，其对散射贡献却很小；众多的小冰晶对太阳辐射散射也很强，吸收则很弱，冰晶散射的能量除依赖于冰晶自身形状外，还与冰晶下落时的基本取向与入射辐射的相对取向有关。一般云滴和冰晶对太阳辐射的吸收很小，而对长波辐射则吸收很强，对于含水量适中的云透过不到 50 米路径就能吸收掉 90%的入射长波辐射。它们对长波辐射的反射却很小。深厚积雨云或雨层云对长波辐射可视为黑体。云的“液态水路径”是云的一种重要辐射特性，定义为沿入射辐射方向云厚的累积凝结水量(或冻结水量)，是研究辐射在云中透过性能的重要参量。因为云中散射需考虑多重散射过程，所以云中辐射传输过程变得十分复杂。云能将晴空大气的辐射“大气窗”关闭，因而限制了长波辐射放射到空间而增强大气逆辐射，对地面和近地层大气起保温作用。气溶胶的辐射特性除了与云有类似的复杂性外，还依赖于气溶胶粒子的化学成分，对吸湿性粒子则还依赖于大气湿度，湿度大时其尺度增大，形状和折射率都会改变，所以与云类似，也是大气辐射传输中较难处理的课题。

辐射收支 地面和大气都在不断地以辐射的方式交换能量。物体辐射收支的差值称为辐射差额，是温度变化的基本因素。(1) 地面辐射收支：地面接受的太阳辐射包括太阳直接辐射和大气的散射辐射，两者之和称为总辐射，其通量密度用 Q 表示。令 a 为地面对总辐射的反照率，则地面吸收的太阳总辐射为$(1-a)Q$。同时，地面也放射长波辐射 F_E 并吸收大气层向下放射的长波辐射，后者称大气逆辐射 R，两者之差即地面有效辐射 $F_0=F_E-\delta R$，δ 为地面对长波辐射的吸收率。地面辐射差额 B_0 定义为地面吸收的总辐射与地面有效辐射之差：$B_0=(1-a)Q-F_0$，地面辐射收支是地面热量平衡的基本部分，是气候形成的主要因素，对研究气团变性、近地层大气温度变化、雾和霜冻等有重要意义。地面辐射差额随昼夜、季节、纬度以及大气成分和云量云状等而变化。一般白天为正，夜间为负。夏季我国不同纬度间地面辐射差额差异不大，而冬季则随纬度增加而迅速减小。我国 39°N 以南地区全年各月平均辐射差额为正值，以北地区 12 月为负值。年平均辐射差额介于$(2.8\sim5.2)\times10^8$ 焦耳/米^2 之间，年较差随纬度增高而增大，与气温年较差密切相关。(2) 大气辐射收支：整层大气的辐射差额 B_a 是大气吸收的太阳辐射 Q_a 与地面有效辐射 F_0 之和扣除大气和地面放射透过大气上界射向空间的长波辐射 F_s 即 $B_a=Q_a+F_0-F_s$。由于 Q_a 较小，B_a 总为负值，表明大气是一辐射能汇。对流层中的辐射亏损通过地面向上输送的潜热和感热通量予以补偿。(3) 地-气系统辐射收支：地球-大气系统的辐射差额 B_s 为地面和大气各自的辐射差额之和：$B_s=B_0+B_a=(1-a)Q+Q_a-F_s$。地-气系统辐射差额随时间、纬度、下垫面属性而异，是各类大气运动的根源，但就全球多年平均而论，这种辐射差额为零，全球地-气系统大体上维持辐射平衡。卫星观测 B_s 随纬度变化：从 33°N～37°S 的低纬地区，$B_s>0$，有辐射热量盈余，在此范围以外的中、高纬度时 $B_s<0$，有亏损。这种纬向不平衡通过纬向大气环流(如哈特来环流)和洋流得以补偿。

地球系统热量平衡 全球热量平衡除辐射收支外还包括地面向大气输送的感热和通过蒸发、水汽输送及相变引起的潜热通量。取太阳常数 $S_0=1\ 367$ 瓦/米^2，按全球平均的太阳入射通量为 341.8 瓦/米^2，以此作为 100 单位得表 1 所列的全球热量平衡表：

表 1 地球系统热量平衡表

太阳短波辐射		地球系统长波辐射	
入射太阳辐射能	100	地面放射长波辐射能至大气	115
水汽、O_3、气溶胶吸收	19	水汽、CO_2、O_3 等温室气体及云和气溶胶吸收	106
云吸收	4	温室气体及云和气溶胶放射长波辐射回地面	100
地面吸收	46		
云反射回空间	17	地面长波辐射透过大气层散逸至空间	9
地面反射回空间	6	温室气体向空间放射长波辐射能	40
空气分子散射回空间	8	云向空间放射长波辐射能	20
		地面感热通量输送至大气	7
		地面潜热通量输送至大气	24

由表1可见，地面接受的短波辐射46单位，长波辐射为100单位，合计146单位；地面发射的长波辐射115，输送给大气的潜热和感热通量为24＋7＝31，合计支出也是146单位。大气层接受的短波辐射23，长波辐射106，合计129；而大气层放射的长波辐射为100＋40＋20160，辐射亏损31单位，由地面输送的潜热和感热通量弥补。所以地面和大气各自的热量收支平衡。从中也可看出，地面的热源主要是太阳辐射，而大气的热源主要是地面。大气的热量收支中，云的作用很大，云对行星反照率的贡献占55％，在向空间的长波辐射中云又占了41％；云在潜热通量转化中也起关键作用。只要表1中这种全球能量收支的分配比例保持不变，地球气候就将保持平稳，一旦其中一种或多种分量发生系统性改变，如太阳发光度和地球轨道参数、云量和云微物理结构、气溶胶浓度、温室气体浓度及下垫面属性等发生系统性变化时，就会导致全球热量收支失衡，从而导致地球气候发生变化。

大气光学 日月星辰的光给我们带来宇宙的信息，照亮我们的环境，提供能量，使我们这个世界适于生存，这些功能都是通过光在大气中传播实现的。大气光学就是研究光在大气中传播时，光的特性如颜色、强度、偏振和空间分布等受到大气中各成分的散射、吸收、折射、衍射、色散、衰减和消光等物理效应而发生的变化，从而为解决有关的光学工程问题提供物理依据和计算方法。大气对光的效应还被用来研究大气的某些属性，如利用大气对光的散射和吸收探测大气污染物分布等，某些大气光学现象还能反映大气密度分布、云雨微物理结构及天气过程，常用于目测判断云降水发展和天气演变趋势。可见，光是电磁辐射谱中的一小部分，有关光在大气中传播特性归入大气辐射条目，光受大气作用而产生的大气光学现象参看大气光象条目。

大气光象 日月光通过大气时，会发生散射、衍射、透射、色散、衰减等物理效应，由于光路上空气密度不均匀，还会发生光的折射、反射现象，从而会产生种类繁多、色彩斑斓的大气光学现象，有的出现在晴空，如海市蜃楼、天空的蔚蓝色等，有的出现在云雾降水中，如虹、晕、宝光、华等，有的出现在晨昏时分，如霞、曙暮光、绿闪等。在高纬度高空，有时会出现绚丽的极光，虽然成因与上述光象有较大不同，也是令人注目的大气光象。这里择要阐释几种：(1) 霞：日出前或日落后天边常会出现色彩鲜红明亮或橘黄色的朝霞和晚霞，是大气散射的产物。日光倾斜穿过较厚大气层时，短波长的蓝、紫色光被大气散射较多，抵达观测点上空已削弱殆尽，余下的主要是波长较长的红、黄色光，经空气分子或云、雾微粒，尘粒散射进入人眼，也可经云雾反射或透射入眼，显示出绚丽多姿的霞光。由大气散射造成的光象还有晴空的蔚蓝色和曙暮光。(2) 蜃景：物体发出的光通过密度分布异常的大气层时，会发生折射和全反射，使远处看到的物象与原物相比产生位置、形象改变，上下颠倒，左右反向，时隐时现，飘忽不定的蜃景(海市蜃楼)。物象位于物体上方的称为上现蜃景，位于下方的称为下现蜃景。前者多出现在早晨近地层有极强的逆温层存在，空气密度随高度急剧减小时，来自物体的光穿越其间会发生折射，当入射角达临界值时即发生全反射而抵达远处人目，看上去景物似在空中，所谓“蓬莱仙境”即此类上现蜃景。下现蜃景则出现在近地面灼热时温度直减率超过3.14°/100米，空气密度随高度增加，光线传播时折射向上弯曲所致，在沙漠或夏季灼热的柏油路面上可见此景。(3) 虹：太阳高度角低于42°时，日光斜照在雨幕(或云雾)背景上，光线在半径大于25微米的雨滴(雾滴)中经折射、内反射、再折射而在雨幕(雾幕)上显现出以对日点为中心的鲜艳彩色(或淡白色)圆弧，其视半径角约42％色彩呈内紫外红，称为主虹，是不同色光有不同折射率造成的色散效应所致，水滴愈大，虹的色彩愈鲜艳，色带愈窄。在主虹外侧，可出现霓(副虹)，内红外紫，是日光射入雨滴经二次折射、二次内反射形成的，视半径角约 52°，民间流传的“东虹日头西虹雨”的天气谚语，对天气系统常自西向东移的我国大部分地区有一定预报价值。(4) 晕：出现在冰晶云上，日(月)周围的彩色或白色光圈、光弧、光柱、光斑，统称晕，是由冰晶对日(月)光折射、反射形成的。常见的是视角半径为 22°的折射晕，其色序为内红外紫，是由日(月)光从冰晶的一侧面射入，从与射入面成 60°夹角的另一侧面射出而形成的。晕常与卷层云、砧状云相伴，故能预兆天气的变化，如“日晕三更雨，月晕午时风”的天气谚语便是。(5) 华：日(月)光照在云雾上显现出紧绕日(月)的彩色(内紫外红)或白色光环，视半径角约 1°～5°，较大的有 10°左右，是由日(月)光通过尺度与可见光波长相近的小云滴或小冰晶发生衍射所造成的光象。华环与云滴大小成反比，故有“大圈日头细圈雨”的谚语。(6) 宝光：人站在山头，日光从背后斜射过来将其人影投射到面前的浓云密雾中，在云(雾)幕上头影四周呈现出五彩缤纷的一圈圈光环，俗称峨嵋宝光(或佛光)。其成因：日光斜射到云(雾)幕上经云(雾)滴衍射发生色散，然后又经另一些水滴反射抵达人目而看到彩色光圈，各色光圈的视半径角分别与其衍射角一致，故宝光的色彩排列与华相似。(7) 极光：出现在高纬度高层大气中的一种气体发光现象，俗称“开天门”。极光形状各异，有火焰状极光，明亮的光线相继朝地磁极天顶快速移动，似火焰般一股股自下向上冲，蔚为壮观；有射线结构极光，形如垂幕状、射线状或弧状、带状结构，亮度大，移动快；无射线结构极光，光度暗淡，变化少。极光的色彩有黄绿色、鲜红色、淡红色、灰紫色或蓝色。多数极光持续时间 1 小时左右。极光底部一般在 100 千米高度左右(90～130 千米)，顶部可达 200～650 千米，甚至更高。极光越强出现高度越低。极光是由太阳发射出的高能粒子流(质子、电子)进入大气上层使大气分子、原子激发后发光形成的。这些高能粒子流在地球磁场作用下趋向南北高纬度，故极光发生频率最高的极光带位于南、北磁纬 70°～65°之间。极光出现频率与太阳活动密切相关，太阳黑子数最多年与最少年的极光出现频率比大约为 5∶1。

大气能见度　标准视力的人眼观察水平方向以天空为背景的黑色目标物(视角在 0.5°～5°之间)，能从背景上分辨出目标物轮廓的最大距离称为气象能见距离，将其分成若干等级即气象能见度，如 0～9 级能见度分别对应的能见距离(千米)和典型天气状况为：＜0.05(极浓雾)、0.05～0.20(浓雾)、0.20～0.50(中雾)、0.50～1.00(轻雾)、1.00～2.00(薄雾)、2.00～4.00(霾)、4.00～10.00(轻霾)、10.00～20.00(清朗)、20.00～50.00(很清朗)、＞50.00(十分清朗)。由于对目标物、背景、观察者视力都做了规定，气象能见度就主要取决于大气透明状况，故可反映雾况和大气污染程度。由于大气分子、气溶胶微粒(尘埃、雾滴、雨滴、雪晶等)对光的散射，在目标物与观察者之间的大气散射光就叠加在目标物和背景光中，这部分“气光”是随距离增加而增强的，而目标物和背景光在途中受气层散射而削弱，其结果当目标物距离足够远时，目标物与背景的亮度对比渐趋不明显，此时目标物趋于“视而不见”。当目标物不是黑体而是建筑物或树木等反射能力较强的物体，而背景也不是天空而是其他如森林、草地、雪面等时，相应的能见距离要比气象能见距离小，比值(0.50～0.97)称为能见度系数。能见度对港口、航道的航行，机场跑道起落飞行以及高速公路安全运行关系十分密切。由于目的不同，能见度的名称和观测规定也有区别，如飞机空中斜视能见度、塔台能见度、跑道视程以及夜间灯光能见度等。

大气电学　研究大气电特性和雷暴电学的大气物理分支学科。大气按电离状况分成 60 千米以下的中性层和其上的电离层及磁层，电离层导电性强，与地表面构成两个球形等位面，相当于一个球形电容器；其间的中性大气层实际上也含少量离子，有微弱导电性，强对流天气下能出现雷暴电现象。引起大气电离的因素有：地壳和大气中放射性物质辐射的放射线、宇宙线和太阳紫外线；另外还有种种局地因素，如云雨过程、火山喷发、林火、瀑布、海浪击岸、尘暴、雪暴等。近地面大气中小离子（10^{-8}～10^{-7}厘米）浓度的 10^{2}～10^{3} 厘米$^{-3}$，大离子（10^{-6}～10^{-5}厘米）浓度在洁净地区约 200 厘米$^{-3}$。晴天大气中正负离子浓度之比介于 1.10～1.20，故大气对地面而言总带正电，晴天大气电场指向地面。近地层大气电场强度陆上平均为 120 V/m，海洋上为 130 V/m，场强数值随高度减小。在大气电场作用下，正、负离子反向移动构成大气传导电流，晴天大气传导电流密度平均为 2.7×10^{-12} A/m^{2}。对应于地面场强 130 V/m 的地面导电率为 2×10^{-14} $\Omega^{-1}\cdot m^{-1}$。晴空湍流和对流会产生对流电流，与传导电流合称为晴空大气电流。在晴天大区传导电流作用下，只要数分钟就可使地球上的负电荷消失，但实际上晴天电场变化并不大，必有别的起电机制对地球-大气系统这个球形电容器充电，这就是雷暴起电（参看“雷电物理学”）。维持大气电场的大气电平衡过程有：(1) 晴天传导电流（＋90），降水电流（＋30）；(2) 闪电电流（－20），尖端放电电流（－100）。大气电学对无线电通信、电力、建筑和航空航天事业都有重要意义。

雷电物理学 研究雷暴云电场、电荷分布、起电机制以及闪电和雷击物理特性的分支学科，与云降水物理学关系密切。全球任一时刻有 2 000～6 000 个雷暴存在，每秒钟有 100～300 次闪电发生，约 20% 是云-地闪电，每次向地面输送约 20 C 的负电荷，起到对地-气系统这个球形电容器充电的作用。雷暴云中典型的电分布是上部正电荷区（＋24 C），中下部 0 ℃层以上区域是负电荷区（－20 C），融化层以下云底部有一小正电荷区（＋4 C）。雷暴云下近地面大气电场方向向上，与晴天电场反向。云中电荷产生率约 1 C 千米$^{-3}$·分$^{-1}$。有几种雷暴云起电理论：(1) 温差起电-冰晶热电效应。冰晶两端温度不同时，冷端积累较多 H^{+} 而带正电，热端 OH^{-} 过剩而带负电，一旦冰晶破裂就会造成电荷分离，有三种方式：雾粒与冰晶碰撞时接触温差起电，雾或雹碰冻过冷水滴时碰冻温差起电，雾粒与冰晶碰撞摩擦温差起电。三种机制的结果都是小冰晶带正电，雾粒带负电。(2) 感应起电。云和降水粒子在大气电场作用下极化，下半部带正电，上半部带负电，降水粒子碰撞云粒子又分开时，云滴上半部的负电荷将传至降水粒子上而产生电荷分离。(3) 对流起电。雷雨云发展时，上升气流将低层正离子带至云上部并附着在云粒上，云顶以上大气中负离子则随云侧下沉，气流输向云下部，也附在云粒上。通过地面尖端放电使这种对流起电以正反馈过程发展形成雷暴云中正负电荷区分布。通常认为非感应起电机制是云中碰撞粒子间电荷分离的主要机制，一旦云中电场建立，感应起电就愈重要，最后成为雷暴云中相碰粒子交换电荷的重要机制。至于云底部小正电荷区的成因，提出大水滴破碎起电和融化起电两种机制，结果都是大水滴带正电，小碎滴带负电，云底部大水滴多，故形成正电荷区。云中通过起电机制形成强电场，一旦局地电位梯度达击穿值，就放电造成闪电，近地层干空气击穿电位梯度约 3×10^{6} V/m，云雾中约 10^{5} V/m。云与云、云内、云与云外空间的闪电称为云闪，云与地之间称为地闪。一次闪电持续时间仅约 0.2 秒，却是一种分阶段逐次发生的复杂放电过程，如地闪就由先导闪击、返回闪击和后继闪击几种放电过程组成，每次闪击间隔时间约 50 毫秒，闪电时闪道中气温可达 15 000～20 000 ℃，甚至 30 000 ℃，闪道内气压骤升至 10～100 个大气压，引起空气急剧膨胀产生冲击波，膨胀速度达 10^{3} 米/秒。随着传播距离增大，冲击波转化为声波，即雷声。雷击对人身和建筑物安全、森林安全、飞行安全至关重要，雷暴电学一直是与国民经济休戚相关且引起人们广泛关注的科技领域。近年发展的闪电定位仪可用来监测雷暴发展和冰雹云定位，是防雹作业中的重要监测设备。

大气声学 研究大气中声波的发生、传播、接收及各种声效应的应用的大气物理分支学科。大气中人耳可闻的声音频率在 20～20 000 Hz 之间，低于 20 Hz 的为次声波，高于 20 000 Hz的为超声波。大气中声传播速度与大气状态参数及风有关，干空气温度 15 ℃时声速为 341.0 米/秒，增温 1 ℃声速增加约 0.61 米/秒；湿空气中水汽压每增加 1 hPa，声速约增加 0.04 米/秒；有风时声速是静止空气中声速与风速的矢量和。由于大气结构不均匀性和复杂变化，会强烈影响声的传播，发生反射、折射、散射、吸收及衍射等，如界面上声反射，因温度梯度和风梯度引起的声折射，物体边缘和界面阻抗变化导致的声衍射，温度起伏和湍流起伏以及云、雾、雨、雪等微粒引起的声散射和吸收，都会影响声在大气中的传播。为方便计，仿照光线代表光传播路径，应用“声线”表示声波传播路径。声波在平流层顶及 100 千米高度附近的逆温层中折射会产生声的“反常传播”，如在 100 千米远处听不见的爆炸声在 200 千米以外却能听到声响——“异常可闻区”。利用声传播与气温、风及湍流的关系，研制声学大气探测仪，如声学测风-测温仪，测量近地层温度和风；声雷达测量低层大气温度分布和晴空湍流结构等。当前大气声学研究的活跃领域：(1) 声波在湍流大气中的传播特性研究，是声雷达遥感探测的理论基础；(2) 城市噪声在大气中传播及声污染防治问题；(3) 大气次声波传播及其遥感探测原理和技术研究。由于大气对次声的吸收衰减很少，故次声波可传播远距离并长时间存在于大气中，如果说经典大气声学主要是以低层大气中可闻声为研究对象，那么现代大气声学就涉及高层大气中次声超远程传播及其应用研究。大气次声源分人工源和自然源。前者如高空爆炸、大气核试验等，利用次声被动遥感可监测远距离核爆炸并推断高层大气风、温结构；后者如台风引起的海浪激发的次声波(周期 3～8 秒，振幅约0.5 Pa)，强对流风暴产生的次声波(周期 0.2～2 分)；另外火山喷发、地震、极光、陨石坠落、日食等都可引起次声波，可根据次声波传播与上述各种强天气现象和自然灾害现象之间的联系，通过对它们的接收、定位和分析，提取对灾害性天气现象和灾害事件有预报价值的信息，这类探测技术的原理基础现都还处在探索研究阶段。

云雾降水物理学 研究云、雾和降水形成过程和演变规律的学科，分宏观云动力学和微观云物理学。前者研究云体或云系形成、发展和消亡的动力学和热力学过程，尺度范围为 10^0～10^6 米。表征云的宏观参量有云生命期、垂直厚度、垂直速度、含水量、温度、湍流强度和云底高度等。云降水微物理学研究云滴、冰晶和降水粒子形成、发展的物理过程，这些粒子尺度仅 10^{-5}～10^0 厘米。主要微物理参量有相态、含水量、数浓度、粒子大小及其谱分布(滴谱)等。推动云雾降水物理学发展的主要因素：(1) 人工影响天气的需求。20 世纪 40 年代后期，随着人工影响天气科学技术的发展和航空事业发展对云微物理学、飞机结冰及雾中能见度等飞行安全需求大增，推动着云雾降水物理学迅速发展成为一门独立的大气物理学分支学科，中心是云雾降水的微物理学，而对云的宏观动力过程及其与微物理过程及环境大气的相互作用认识不足。自 20 世纪 70 年代以来，外场考察表明，云或云系中降水形成受制于大尺度天气动力过程、中尺度和云尺度动力学和热力学过程、云微物理过程以及辐射和边界层湍流等诸多尺度十分不同的动力和物理过程之间的相互作用，需要将它们结合起来研究才能提高人工影响天气的科学水平及播云实效。(2) 定量降水预报的需求。自 20 世纪 60 年代以来，在大尺度气压场形势预报已取得显著进展时，定量降水预报准确率提高不快，大雨(日雨量＞25毫米)的24小时预报准确率TS评分一直停留在0.2左右。症结在于强降水过程往往具有中小尺度特征(10^0～10^2千米),常规天气观测网难以分辨。数值天气预报的环流形势与强降水型之间时空尺度差异也很大,特别是夏季对流降水型尺度比天气系统小约两个数量级。要提高定量降水预报水平,需将天气动力学与中尺度动力学、云动力学及云降水微物理学结合起来。(3) 云雾降水物理过程在大气化学、污染物远距离输送及其清除、云的辐射气候学中扮演重要角色。云雾降水化学是对流层大气化学的核心内容,酸雨、气溶胶污染等大气环境课题中有大量云物理和云动力学问题。云量、云高、云厚、云的辐射特性及其与云微结构的关系是大气环流和气候模式辐射效应参数化研究中最为敏感和棘手的课题之一。正是气候变化、大气化学和大气污染这些当代热门专题赋予云雾降水物理研究新的活力,推动其向更深层次和更广的范畴拓展。

各种现代探测技术,如气象卫星、多普勒雷达、飞机观测平台以及各种地基遥感技术,包括双通道微波辐射计、双偏振雷达、多参数雷达、风廓线雷达及机载多普勒雷达等新技术的发展和应用,大大增强了认识和研究云降水过程的能力,为组织有设计的中小尺度降水系统综合考察研究计划奠定技术基础。技术进展还表现在各类探测资料的实时计算机处理和分析、传输技术,以及不同类型资料的计算机四维同化技术的发展并应用到日臻完善的中尺度和云模式中。

云 云及其降水是最重要的天气现象之一,它提供人类和万物生灵赖以生存的淡水资源,也带来一系列诸如暴雨、洪涝、冰雹、龙卷、台风、雷电和雪暴等气象灾害。云和云系在全球大气环流和气候变化中起重要作用。云中相变潜热释放充当着全球大气环流的“引擎”,又是地球水分循环的主要组成部分,能垂直和水平输送水分,改变水的相态,为地球提供降水,并覆盖地球将近60%的地面使其免受太阳直接辐射。全球平均有17%的入射太阳辐射被云反射掉,从而改变地面温度和土壤及植被水分蒸发率,云同时吸收地球出射的长波辐射,调节大气和地面温度。云量、云高、云厚和云含水量乃至云滴大小和数浓度都对地球辐射收支有影响,因而对地球气候有很大影响。云在大气化学中充当一种湿化学反应器,是酸雨制造者。云中短波辐射强度对光化学反应也有重要影响。云,特别是对流风暴云系能将污染物输送到中高对流层及平流层低层,在污染物远距离输送中起关键作用。

《国际云图》按地面观测的云高云形将云分成十属:卷云、卷积云、卷层云、高积云、高层云、雨层云、层积云、层云、积云、积雨云。另外还有地形云未列入上述分类系列。有人将雾归属层云、层积云一类,除了后者离地面有一定高度,其成因也有诸多类似之处。

随着雷达、卫星探测技术的广泛应用,对云的认识加深,有条件从云的发生学和动力学角度对云和云系分类。形成云的基本动力过程有:(1) 对流运动:层结不稳定大气中,在有热力或动力强迫因素触发下会发生对流运动,升速达1～20米/秒,可产生厚度大(10^0～10^1千米)但水平范围不广(10^0～10^1千米)的积状云(淡积云、浓积云、积雨云),生命期一般几十分钟至2小时,分三个阶段:a. 发展阶段,云中盛行上升气流,云顶抬升;b. 成熟阶段,云中除上升气流外,局部出现下沉气流并开始降水;c. 消散阶段,云内盛行下沉气流,云渐消散或转化为层状云,降水渐弱转而停止,通常称之为气团雷暴。冷锋区或锋前暖区辐合带的位势不稳定大气中,可形成100～200千米长的强烈对流带,即飑线云系(团),寿命达6～9小时或更长。中纬度夏季扰动天气下发展的雷暴群有时组织成广达数百千米、生命史长达6小时以上的中尺度对流复合体(MCC),主体是强雷暴群,由层状云砧区覆盖。其中的强对流风暴云团包括超单体雷暴和多单体雷暴，是冰雹、龙卷、突发性暴雨和雷暴大风的主要制造者。热带辐合带内频繁发生的热带云团、热带飑线和热带风暴系统中，对流云是最有活力的主要成分，尽管形成机制各有不同。(2) 气旋辐合和锋面抬升运动：其范围广(10^2～10^3 千米)，时间持久(一至数日)，虽然升速仅为 10^0～10^1 厘米/秒量级，却总能形成深厚而广阔的气旋云系，包括卷云、卷层云、高层云和雨层云，常分为几层，水汽丰沛时往往连成一片，地面降水强度约为 1 毫米/小时，有时冷锋云系中会出现对流云。自 20 世纪 60 年代以来，多普勒雷达和飞机观测等广泛应用，发现中纬度气旋云系中存在各种中尺度雨带结构，雨带宽 35～70 千米、间距 55～100 千米不等，气旋暖区中也有中尺度雨带，升速 10^1 厘米/秒，降水强度约 10 毫米/小时，从而对经典的锋面气旋云系结构提出重大修正，较合理地解释了锋面气旋降水不均匀性。在低纬度由锋面、气旋辐合气流也可形成这类降水性云系。梅雨锋云系和热带气旋(台风)云系通常是对流云嵌套在层状云系中，形成机制和结构特征与中纬度气旋云系有较大区别。(3) 地形抬升和波动运动：气流在山脉迎风坡被迫上升可形成地形性层状云或山帽云，若大气层结不稳定，越山气流会形成积状云甚或强对流风暴，稳定层结下背风坡可形成波状云。大气中存在逆温层时，上下气层风速不同也会激发波动，形成波状云。(4) 边界层垂直混合：可形成层积云和层云。卫星观测表明，全世界海洋面有 34%以上面积由浅薄的层积云覆盖着，它们降水潜力虽不大，却能显著影响辐射收支从而影响区域气候。

云降水微物理学 研究云质点(云滴、冰晶)和降水粒子(雨、雪、冰雹等)形成和增长过程中所经历的凝结、凝华、冻结、繁生、碰并、凇附、聚并及蒸发等物理过程的分支学科，其物理基础主要是水相变热力学和气溶胶力学。当湿空气上升冷却达到饱和及过饱和时，水汽就在某些吸湿性或可溶性气溶胶粒子(称云凝结核 CCN，如$(NH_4)_2SO_4$ 和 NaCl 等)上凝结形成云滴。云滴很小，直径约 10^{-3} 厘米，且较均匀，很难通过碰并等过程增长成降水粒子。降水微物理过程的核心问题是降水胚滴的产生或初始滴谱的拓宽机理，包括暖雨过程和冷雨过程：暖云中少数巨型凝结核(10^{-4}～10^{-3} 厘米)、湍流碰并和电力碰并等所产生的少数大水滴都可充当降水胚滴。一旦云中产生半径大于 19 微米的降水胚滴，因落速差异引起的重力碰并机制就能有效地形成降水粒子。当雨滴直径接近 6 毫米时会因流体动力不稳定而破碎，直径 2～3 毫米的雨滴与小水滴碰撞也会破碎，碎滴可充当次级降水胚滴而继续繁生增长，此即对流性暖雨过程中的连锁反应机理。冷雨过程是指冰相起重要作用的降水过程，包括：(1) 冰晶初始核化过程。温度低于 0 ℃的冷云中云滴常常呈过冷液态水状态。温度更低时，少数大气冰核(IN)活化，可充当冰晶核心而产生初始冰晶。核化机理有凝华核化、吸附核化、接触冻结核化、浸没冻结核化等。(2) 冰晶次级繁生机制。在暖于−10 ℃的冷云中冰晶浓度有时可比活化的冰核浓度大 3～4 数量级，是由冰晶次级繁生机制造成，包括过冷雨滴冻结时产生碎片、枝状冰晶或冰针碰撞碎裂、冰晶碰冻过冷云滴时形成碎屑等过程。(3) 水-冰转化过程，也称贝吉隆过程或冰晶效应。冷水云中一旦产生冰晶，由于冰面饱和水汽压比水面的低，冰晶能快速凝华增长，云中水汽压因而会降低至水面未饱和，于是水滴蒸发，冰晶通过消耗过冷水滴而增长形成降水胚粒。这种“水-冰转化”是重要的冷云降水机制。(4) 增长的冰晶碰冻过冷云滴而凇附增长形成霰粒，冰晶撞冻过冷雨滴则形成冻滴。(5) 冰晶之间聚并形成雪花。多数情况下，霰或冻滴以及雪花降至 0 ℃层以下融化形成雨，这是中纬度乃至低纬度大多数降雨的主要方式。至于冰雹，由于冰雹云中垂直气流和含水量分布十分复杂，霰粒和冻滴充当雹胚，在上升气流垂直切变环境下可经历不同的含水量和温度条件而循环增长形成明暗相间的多层结构冰雹和海绵状雹，也可在含水量累积区直接淞附增长为冰雹。近五十年来云降水微物理学虽然已经取得很大进展，但它的症结是微物理过程在相当大程度上受制于云的宏观结构和云动力学过程，自然环境下云中群滴发展演变及其相互作用十分复杂，微物理过程本身不确定因素也很多，如核化机理、碰并效率、破碎和繁生概率等目前的认识尚较肤浅，实验室模拟仿真性较差，数值模拟则面临合理简化和参数化处理问题，尚有许多改进和未知领域待研究，主要动向是加强有设计的外场考察研究，利用飞机观测平台和采样仪器、多普勒雷达、双通道微波辐射计、双偏振雷达等现代遥感设备探测云内外流场和云微物理结构及其演变特征，为理论分析提供坚实的观测基础。

云动力学　应用流体力学和热力学原理研究云发生、发展的宏观动力过程的分支学科，是云降水物理学的重要组成部分，与云降水微物理过程关系十分密切。云的宏观动力过程为微物理过程提供发展背景，决定了后者的进行速率、持续时间和空间范围；反之，微物理过程中水汽相变潜热释放和降水粒子的重力拖戈作用对云的动力学过程有重要的反馈效应。按时空尺度分为积云和对流风暴动力学及层状云动力学两个分支。

积云和对流风暴动力学：在条件不稳定层结大气中发动对流的启动能源来自锋面抬升和地形扰动、局地加热、边界层辐合、阵风锋、夜间边界层大振幅重力波等局地扰动因素，以及低空急流、干线等次天气尺度扰动。大多数积云对流特别是强对流风暴都是与大、中尺度天气扰动或特殊地形条件相联系。积云对流发展的主要动力是对流泡内空气密度与环境大气密度不同引起的阿基米德浮力效应，所以决定空气密度的大气温、湿度层结条件性不稳定结构是对流发展的主要条件（温度直减率 γ_{-} ＞湿绝热直减率 γ_m）。制约积云发展的因素有：(1) 夹卷效应。环境空气通过湍流夹卷、动力夹卷和云顶穿透性下沉夹卷等方式稀释云含水量、削弱云浮力。建立在夹卷概念基础上的积云理论模式有气泡模式、气柱模式和穿透性下沉夹卷模式，一般只考虑相变潜热效应和液水含量变化，不涉及降水形成微物理过程。在侧向夹卷概念基础上发展的一维定常夹卷模式引入参数化微物理过程，可模拟降水过程并应用于积云动力催化模拟试验。(2) 云外补偿下沉气流效应。积云对流发展时环境空气因流体连续性而产生补偿下沉运动，下沉绝热增暖削弱云的有效浮力，下沉空气与云的混合夹卷效应则稀释云含水量和动量，对积云发展起阻尼作用。(3) 降水微物理效应包括降水重力拖曳和云下蒸发冷却效应，是激发下沉气流促使积云崩溃的主要因素。固态降水的冻结和凝华潜热效应在云上部冰晶化过程中能使云获得附加浮力而迅速发展。(4) 辐射和辐射冷却效应。夜间雷暴常因云顶辐射冷却和云底接收地面长波辐射加热，增大云内热力不稳定度而使雷暴发展加剧。(5) 扰动气压动力效应。深对流中扰动气压梯度力和扰动气压浮力与热浮力同量级，静力平衡假设不适合积云对流。扰动气压效应一般对云中上部起抑制作用，而对云下部则能增加有效浮力，起到一种抽吸作用，可解释强风暴云底观测到的负热浮力空气能加速上升的现象。(6) 环境风垂直切变效应。强风暴云系总是在风有强的垂直切变的扰动天气环境下发展，风垂直切变在导致风暴中形成倾斜上升气流和下沉气流同时并存的准稳态流场结构中起重要作用，并影响风暴系统移动特征和持续性。所以，积云和对流风暴云系（包括浓积云、积雨云、气团雷暴、超单体雷暴、多单体雷暴、飑线云系、中尺度对流复合体和热带云团）的发生、发展涉及一系列宏微观过程之间、云与环境之间复杂的非线性相互作用，一般采用三维非静力平衡积云模式加以描绘和模拟。

层状云动力学：层状云可分两类，一是边界层层积云和层云等结构较均匀且成因较简单的气团性层状云，主要通过垂直混合过程形成，促成因素有近地层自由对流、中高空冷平流、边界层强的风切变、红外辐射效应、云顶夹卷不稳定效应、大尺度下沉运动效应和毛毛雨的影响等；另一类是气旋云系，包括温带气旋云系、热带气旋云系（台风云系）和梅雨云系等，它们的成因、结构和有关的天气现象十分复杂，其中有层状云，也有积状云，且包含各种中尺度组织如中尺度对流雨带结构。水平伸展数百至上千千米，持续时间 6 小时至数日之久。在这类云系中不同尺度动力过程、辐射和相变热力学过程以及湍流交换和微物理过程都能在不同程度上以十分不同的方式影响云系的发展和结构特征，又存在复杂的相互作用和转化关系。因这类云系时空尺度大，可采用静力平衡假设，但需考虑地转效应，通常用三维水平非均匀初始场静力平衡模式加以描绘和分析。

人工影响天气 根据天气演变的物理学原理和动力学规律，对某些天气过程施加人为干预使之朝有利于人类的方向演化的科学技术措施，包括人工增雨、人工防雹、人工消雾、人工消云、人工遏制雷电、人工削弱台风、人工抑制局地暴雨（雪）和人工防霜冻等。鉴于自然天气过程涉及的能量十分巨大，人类很少能与之直接相抗衡，人工影响天气的策略构想是利用自然云雨过程中有时存在的相态不稳定或胶性不稳定平衡状态，向云中播撒合适的催化剂改变云微结构，进而影响云的微物理过程和热力、动力结构及其所产生的天气现象。20 世纪 40 年代末由于发现通过播撒诸如干冰和碘化银这样的冷却剂或人工冰核可使过冷云滴变为冰晶的人工引晶技术，现代人工影响天气才得以开创。在有利天气条件下，这是一种耗资少、效益高的科学技术，其理论基础已通过实验模拟和数值模拟得到证实，并在雾、地形云、层云和积云等结构较简单的自然系统中进行了验证。但这类方法对云的自然不稳定平衡条件依赖性较大，播云技术是否适宜，也极大地制约人工影响的效果，加上天气过程自然变异性大，人工影响的效果常常被巨大的自然变差所掩盖而难以鉴别，这也为这一新兴科学技术的进展增添了客观难度。

当今全世界有数十个国家和地区实施着一百多个人工影响天气试验计划。我国人工影响天气始于 1958 年，自 20 世纪 80 年代以来进展明显：(1) 引进和研制了一批先进探测技术，除卫星云图及数字化雷达等常规装备外，还包括机载多尺度粒子测量系统（PMS）、识别冰雹的双线偏振 C 波段雷达、遥感云含水量的双频道微波辐射计等；(2) 开展人工增雨云水资源、增雨效果随机试验统计检验、冰雹云特征和冰雹微结构、防雹效果评估、雾结构考察及消雾消云、闪电监测和遏制闪电等外场考察和试验研究，为播云设计提供物理判据；(3) 云降水数值模拟研究开始应用于优化播云试验和人工影响天气效果的理论预测；(4) 播云技术研究为外场作业提供多种高效播云手段；(5) 建立了一批汇集多种信息的实时监测、传输、分析判断及通信联络的综合性作业指挥技术系统，在抗灾斗争中发挥了积极作用。人工影响天气是一项多学科、多部门、多层次的系统工程，为加强对这项高科技事业的组织、协调和指导，1994 年国务院批准建立跨部、委、局的全国人工影响天气协调会议制度，这将促进我国人工影响天气事业稳健发展，提高其科学技术水平和总体效益。1991 年世界气象组织（WMO）制订的大气研究和环境计划（AREP）提出加强人工影响天气的科学基础：发展遥感技术（可变波长雷达、激光雷达和微波辐射仪），对云中水凝物的演变进行四维测量；提倡开展确定云和云系降水效率及判别支配降水量的因子的方法研究；拟定最佳播云方案；发展物理模拟，模拟云动力学和微物理学相互作用、污染物的化学变换，以及预报自然的和人工影响的云和云系降水量；发展评估试验效果的统计技术和物理技术。

云水资源 温度低于0 ℃的冷云中常含有液态过冷水滴，它们主要通过凝结和碰并机制增长，比较缓慢，其中大量过冷云水不能及时变成降水而最终会蒸发掉，致使云的降水效率很低。这种不能及时变为降水的过冷水可看成是自然云中一种无法泄出的空中水库，这就是“云水资源”。人工引入“最佳”浓度冰晶可以打开此过冷水库的汇出口，以增加降水效率和降水量。不同的云自然降水效率差异很大，如：孤立冰雹云3%、小雷暴云20%、地形性对流雨带25%、低空雨带30%、冷锋云系60%、高空雨带(70～80)%。可见，地形性云、低空雨带、小雷暴云和孤立冰雹云降水效率都较低。关键是如何判断含有大量过冷水而并未参与降水过程的云或云中部位，尚缺乏快捷有效的方法。通常采用飞机携带液态含水量仪或粒子测量系统(PMS)的相应探头对云做气候性观测，探明特定云系过冷层中液态含水量的分布特征。我国北方层状云考察表明，液态含水量随温度下降大致呈指数递减，0～−15 ℃层一般低于0.2～0.3克/米^3，空间分布很不均匀。在自然降水开始阶段常有较丰富的过冷水含量。近年发展的双频道微波辐射计可遥感云中垂直累积液态水含量和大气总水汽含量尚处于研制试用阶段，有待进一步完善。

人工增雨 对某些微物理结构处于不稳定平衡状态的云播撒合适的催化剂使本来不能降水的云产生降水，或使能够自然降水的云提高降水效率或发展动力而增加降水量的科技措施，也称人工降水。有两类播云原理：静力催化和动力催化。冷云的静力催化是向过冷云中播撒干冰、液氮等冷却剂或碘化银等人工冰核，使缺少自然冰晶的过冷水云产生“最佳”浓度的冰晶，破坏云的相态不稳定平衡状态，激活水-冰转化机制使更多的云水及时转化为降水以提高降水效率。暖云的催化方法是播撒吸湿性微粒或大水滴提供初始降水胚滴，破坏云微结构的胶性稳定状态，激活暖云降水的碰并机制，使较多云水转化为雨水而提高降水效率。动力催化是在冷云部位过量播撒冷云催化剂促使云迅速冰晶化，释放大量相变潜热而增加云的浮力，有利条件下能使对流云发展更旺盛、更持久，从而增大云的水汽输送量而增加总雨量。国内外一些成功的冷云静力催化试验表明可增加雨量10%～30%，取决于云的自然条件是否合适和播云技术是否适宜。这一领域尚有一些科学技术问题待进一步研究解决。

人工防雹 采用人工引晶或爆炸等方法抑制冰雹云中大冰雹的形成以减轻雹灾的措施，其科学依据是对云中过冷水“有利竞争”的概念。在冰雹源区大量引晶，制造人工雹胚与云中原有的自然雹胚“争食”过冷水，抑制雹胚凇附增长速率，从而减小冰雹最终尺度以减轻雹灾。也有同时或单独在雹云的暖区播撒吸湿性微粒促进暖雨过程，形成早期降水，削弱进入雹胚源区的过冷云水供应。近年发展的一种防雹构想：在主体雹考外围供给水分的新生馈云单体中播云，使过冷水尽早冻结提前降雨，削弱向主体雹云的水和能量供应，并防止馈云单体本身发展成为新的雹暴。由于馈云与主体雹暴之间存在复杂的相互作用，不适当的引晶也有可能导致雹灾加重。实践中还有用火炮等轰击冰雹云的爆炸防雹法，其作用机制尚不清楚。防雹作业的科学技术难点如雹暴结构和动力学、雹云预报、防雹原理、雹胚源区识别、有效的引晶技术及效果评估都待深入研究。但国内外一些防雹实践提供的证据表明有利条件下可使雹灾减轻30%～70%。

人工消雾 采用播撒催化剂、动力扰动混合或直接加热等方法清除雾的措施。机场、港口、高速公路或江河航道等地有雾时能见度降低，需进行人工消雾以确保交通安全运行。比较成熟的是消过冷水雾，播撒干冰或液氮等冷却剂在过冷雾中产生适量冰晶，借助水-冰转化过程使雾滴蒸发冰晶增长并降落而使雾消散。在国外一些冷雾较频繁的重要机场已作为业务措施应用。20 世纪 90 年代北京地区采用地基液氮装置消冷雾已获得初步成功。对暖雾曾采用播撒吸湿性微粒（如氯化钙或尿素）使雾滴蒸发消雾；也曾用动力扰动混合法消暖雾，用直升机在雾顶上搅拌干空气产生下沉气流使干暖空气混入雾中降低湿度使雾蒸发；还有用报废的喷气发动机燃烧汽油加热法消机场暖雾。这些方法或由于影响范围小，受环境风、湿场干扰大而效益不佳，或代价昂贵而难以推广。对于冰雾，目前尚无合适的消除方法。

人工遏制雷电 人工改变雷雨云中电场结构或起电过程以遏制或削弱雷电活动，减少雷击造成的森林火灾，保护重要设施（如核电站、导弹发射基地等）免受雷击破坏的措施，目前尚处于探索研究阶段。遏制雷电的方法：(1) 影响云中起电机制。大量播撒成冰核，产生大量冰晶，使过冷水蒸发，改变云中电荷分布，增加传导电流，削弱云中起电过程。(2) 向成熟的雷雨云中播撒大量细小的金属丝或导电纤维（如镀铝尼龙丝），通过电晕放电增加云体的导电率，削弱云中电场强度以遏制雷电的产生。外场试验表明这两种方法都有减少闪电频数的效果。也曾用小火箭携带细金属丝射入云中触发早熟的闪击，获得一定成效。我国 20 世纪 60 年代林业部门就提出遏制闪电的需求，80 年代随着航天事业发展的需求，开展用小火箭携带金属丝播云的方法遏制闪电试验，取得一些令人鼓舞的结果。

人工削弱台风 通过向台风（飓风）特定部位播云改变云的热力结构和动力结构，减小台风中心区气压梯度从而削弱最大风力，以减轻台风灾害，因风险度较大，只做过一些探索性试验研究，基本途径：(1) 在台风眼壁对流区播撒大量碘化银激活水—冰转化过程，释放相变潜热，使眼壁区增暖而降低气压，减小台风中心与眼壁区之间的气压梯度，从而削弱最大风速；(2) 在外围螺旋对流云带播云，使低层入流的暖湿空气被人工激活的对流云“截流”并转向上空，减小台风中心区的水汽和角动量供应，从而削弱眼壁区对流和中心区最大风速。这些途径尚带有一定的推论性。在 20 世纪 60 年代后期进行的外场削弱飓风试验中，曾取得削弱最大风速 10%～30%的效果。由于台风风力自然起伏大，上述效果的可信度尚有争议。20 世纪 80 年代飞机观测发现成熟台风对流区 0 ℃层以上的过冷水含量相当少（少于 0.5 克/米^3），而冰晶浓度却较大，可播度不大，因而人工削弱台风的可行性有待讨论。不过台风中强对流核心区的飞机观测资料尚欠缺。

播云设计 播云作业是一项复杂的、牵动面广的系统工程。为了在外场作业中针对合适的播云对象，采用恰当的和适量的催化剂，在合适的时机播撒到云中合适部位以求有实效，播云措施需仔细筹划设计，包括物理设计、统计设计和技术设计。物理设计主要保证播云的科学合理性，包括在云水资源调研基础上制定可播度判据、预报和检测可播云出现的时机和部位，云降水数值模拟预测最佳播云方案和模拟效果出现的时间、地点等所需的方案和实施程序制定；统计设计主要从效果评估要求出发为播云试验（作业）提供统计分析资料系列和合适的控制参量，包括试验区和试验单元设置、雨量站网配置、雷达等遥感仪器布局、统计试验方案制定等；技术设计包括播云实施方案、播云器设置和播云剂量控制方案、探测仪器配置和观测、采样方案，以及数据采集网络、信息传输系统、通信网络、指挥技术系统的制定和筹建。我国一些省级人工影响天气业务机构初步建成以计算机网络为中心的播云决策指挥技术系统，为外场作业科学化、现代化奠定了技术基础。

播云技术 将合适的催化剂及时适量地播撒到云中特定部位，是播云措施能否奏效的关键技术环节，包括催化剂和播云器两个方面。研究较多的是冷云催化技术。有两类冷云催化剂：一是干冰、液氮等冷却剂，能使局部过冷云滴骤冷至-40 ℃以下而均质核化成冰晶；二是碘化银（AgI）一类的人工冰核，使过冷云滴通过异质核化产生冰晶。干冰在-10 ℃时成核率可达10^{11}克$^{-1}$，-6 ℃以上其成核率高于碘化银。碘化银成冰阈温为-4 ℃，具有效率高、用量少、毒性低和使用方便等优点，应用最广泛。曾试验过包括有机冰核和生物冰核在内的上千种物质的成冰活性，大都不如碘化银。采用复合型碘化银剂可大大提高成冰效率并减少银的消耗量。播云器以催化剂而异。冷云人工增雨除采用飞机投放干冰或喷射液氮等冷却剂外，多数采用机载碘化银丙酮溶液燃烧器、碘化银焰剂燃烧器及焰弹；地形性冷云常采用地面燃烧碘化银丙酮溶液，靠迎风坡气流将碘化银气溶胶扩散播入云中，效率不高但成本低廉；对防雹和对流云增雨，我国多采用高炮或火箭发射碘化银弹入云爆炸分散播云，其成冰效率很低。近二十年来国内外着力研制和改进高效复合型碘化银焰剂。我国研制的复合型碘化银焰剂在-8 ℃云室中检测成冰率达10^{15}个/克碘化银，以此为基础研制出多种型号的碘化银焰弹火箭，机载碘化银焰剂发生器和焰弹、焰弹式增程碘化银高炮弹等，为科学作业提供较有效的催化手段。近年还探索研究雹暴和超声射流膨胀冷却产生冰晶的机制，可望提供一种无污染的廉价催化技术。国外还探索对气溶胶进行活化处理的试验，可使任何气溶胶成为冰核而无须碘化银参与，是值得关注的新动向。

播云效果评估 降水或降雹自然变率很大，目前还无法做出准确的定量预报，况且播云只影响微物理过程的某一环节，这种"以巧破千斤"的策略能奏效的机会自然是有限的，所以效果评估本质上是一种高噪声低信号的鉴别问题。有三种途径。(1) 统计评估。以增雨为例，寻找控制参量估测自然雨量，再与播云后实测雨量比较确定效果。其中有两类方案。(i) 随机化试验方案。将适于播云的机会随机地分成两组，一组播云，一组不播云作为控制，比较两组雨量以检验播云的统计效果。这比较客观，缺点是为获得有统计意义的样本试验周期较长，且要放弃一部分机会不播云，在抗旱作业中难以接受。(ii) 非随机化作业方案，包括历史回归、区域回归和气象协变量统计检验等方法。防止人为主观偏倚干扰效果评估的客观性是这类方案的要害。(2) 物理评估。对播云后从冰晶产生、宏微观结构演变到降水粒子降落的各个微物理链节上进行物理监测，追寻播云物理效应。近年迅速发展的飞机观测和数字化雷达等遥感探测设备已用于优化播云措施并为统计效果评估提供物理证据。(3) 数值模拟效果估算。云降水数值模拟可用来估算播云潜力、选择最佳播云方案以指导外场作业，也可预测播云效果，及其出现的时间、地区，据以验证统计效果的科学合理性。

人类影响区域天气和气候 众多的证据表明，人类活动可以影响局地和区域尺度（$\leqslant 10^3$千米）天气和气候。主要有两种途径。(1) 排放气溶胶和气体。矿物燃料和植物燃烧排放的大量气溶胶和气体（如硫化物和水汽）可影响云微结构、降水过程及辐射平衡。例证：甘蔗产地燃烧渣叶排放大量小凝结核可使下风方雨量减少25%；纸浆厂排放众多巨型凝结核会使局地雨量增加30%；高空喷气飞机排放大量水汽形成凝结尾迹云，据估计在欧洲和美国它们可分别覆盖天空的1%和2%；船舶航行时排放大量气溶胶和水汽形成导致海洋层积云反照率增强的"船舶航迹云"，可影响航道上的辐射收支；在都市严重污染地区气溶胶可直接削弱太阳辐射15%；高纬度工业区排放可形成持续的北极霾，影响能见度和交通，也影响辐射收支，从而影响气温。证据很多，但要确认这些效应的物理机制，尚需组织有设计的考察研究。(2) 改变下垫面属性。日益扩展的都市区建设、过分开垦和过渡放牧使土地沙漠化，干旱地区过分截流使河道断流、湖泊干涸，过量开采地下水引发海水倒灌使土地盐碱化，滥伐森林、毁坏热带雨林等不仅严重破坏生态环境，而且也改变局地反照率影响辐射收支和气温，特别是改变地面粗糙度及蒸发、蒸腾等水分循环过程，会引发中尺度环流和对流性天气。美国圣路易斯城的试验表明下风方雨量增加 25%，雷暴发生率增加 45%，雹暴增加 31%。据分析，主要由下垫面粗糙度和水分循环改变引起的城市“热岛效应”所致。其他如中亚的咸海和我国的塔里木河面临干涸的危险，南美亚马逊河流域热带雨林正以每年 2 万平方千米速率递减等，其所产生的区域天气气候效应和机理及其全球气候内涵以及防范对策都是需要深入研究的领地。

人类影响全球气候 地球气候依赖于地球系统（大气圈、岩石圈、水圈、冰雪圈、生物圈、人类圈）的辐射和热量平衡。地面辐射收支主要取决于下垫面的反照率，而大气的辐射收支则依赖于大气中温室气体、气溶胶和云量。任何改变反照率、温室气体浓度、气溶胶含量、云量和云微结构的人类活动都可能影响气候。南极冰心采样分析表明，20 世纪 90 年代大气中 CO_2 浓度比工业革命前增加约 26%，并以每年 0.5% 的速率递增；甲烷（CH_4）和氟氯烃（CFC）等也以每年 0.9% 和 1% 的速率增加。1990 年第二次世界气候大会提出告诫：如果不设法减少上述温室气体排放量，预计 21 世纪末全球将升温 2 ℃～5 ℃，海平面得上升 65（±35）厘米。这将严重威胁沿海低地、港湾和岛屿的生存环境，并将严重影响森林、农业、渔业和水资源。不过，上述预测是根据气候模式估算的，目前的气候模式尚较粗糙，未能恰当地处理由温室气体浓度增加引起的辐射响应，包括水汽、云、地表反照率、海洋环流、植被呼吸和光合作用以及有关的生化过程等方面的反馈效应，其中有正反馈，也有负反馈。即使是最复杂的三维大气环流模式（GCM），要定量地恰当处理这些反馈效应，也还有许多方面需加改善，况且人类影响气候不只限于温室气体增暖效应。例如，人类排放气溶胶不仅能影响局地和区域天气和气候，而且通过影响云的微结构而具有影响全球气候的潜力。有证据表明，近百年来大气中 SO_2 和硫酸盐气溶胶含量已有很大增加，它们可充当云凝结核（CCN）增大云滴浓度，从而使云反照率增加并冷却地面，这种效应对光学厚度薄的卷云和层积云最明显，其作用与温室气体增暖效应相反。粗略估算，全球海洋性层积云平均云滴浓度增加 15% 所造成的辐射冷却效应可与当前估计的温室气体增加的增暖效应相抵。同样，要使这种 CCN -云反照率假设建立在合理的科学基础之上，也需做深入的基础性研究。

1.2 重要天气现象

1.2.1 风

空气流动便形成风，风通常指空气相对于地面的水平运动，而空气的垂直运动称为对流。自然界中空气的冷热分布不均匀，一般情况下，热的地方气压低，冷的地方气压高，产生一种自高压指向低压的力，即气压梯度力，使空气从气压高处向气压低处流动。英国的蒲福经多年的海上航行观察，于 1805 年根据风对地物的影响力制定了分为 13 个等级（0～12 级）的风力等级表。该表至今仍在使用。风在水汽和能量的水平输送过程中起着重要作用，它是影响天气、改变气温和湿度的重要因素。早期的航海事业就因发现信风而得到较大的发展。风又是一种可再生的能源，风力发动机可用来发电、排灌等。但风也会对建筑物构成威胁，是设计桥梁、大坝及高层建筑的抗风强度时所必须考虑的因素。

风的分类

空气的运动有三个等级，规模最大的是大气的全球运动，如信风带、西风带等，总称行星风带；其次是季风、气旋、反气旋等；最小的是如海陆风、山谷风等范围较小的风。受科里奥利力影响，大尺度气流如温带西风急流、信风、季风等大体上沿等压线流动。北半球背风而立时左侧是低压，南半球则正好相反。边界层内因摩擦力影响，风偏向低压一侧斜穿等压线。因局地昼夜温差引起的热力环流，如海陆风、山谷风等则带有地方性特色。由雷暴引发的雪暴大风、龙卷乃至台风所产生的狂风则是综合大气动力、热力及云物理过程引起的灾害性大风。

海陆风

海陆风是因海陆热力质差异明显，造成温度差异而形成的热力环流。陆地的比热容比海水小，白天陆面温度比海面温度高，陆面上气压较低而海面上气压较高，在气压梯度力作用下，低层产生由海面指向陆面的海风；陆面低层暖空气辐合上升，海面上空较冷，空气下沉以补充低层流走的空气，至 1～2 千米高度，气压梯度力变成由陆面上空指向海面上空，气流也从陆面指向海面，从而形成海风环流。下午陆面增温最强，海风最强，可深入内陆数十千米，在大尺度流场较弱时尤为明显。夜间相反，海面空气较暖，陆面辐射冷却降温较多，空气较凉，形成低层由陆地指向海洋的陆风。白天海风环流常在岛屿或半岛上引发低层辐合，触发热力对流发展，如中国海南岛和美国佛罗里达半岛白天积云降水较多，夜间则大多晴朗无云。在较大湖泊的边缘地带也会产生类似海陆风的湖陆风。

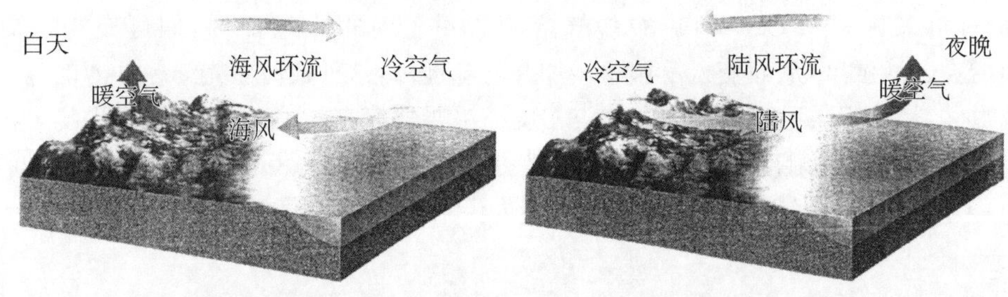

注：陆风和海风出现于沿海地区。日间风由海面吹向陆地，夜间风由陆地吹向海面。海陆风是由于陆地和海洋温度升降的速率不同而造成的局地热力环流。

陆风和海风

山谷风

山谷风是由于昼夜间山谷与山坡受热不均而引发的热力环流。白天山坡光照充足，增暖比山谷快，气压较低，风由山谷吹向山坡，形成谷风。山坡暖空气上升，气压则随高度缓慢降低，在 1～2 千米高度处可高于谷地上空同高度气压，在该处风从山坡吹向山谷上空，然后下沉构成谷风环流。夜间山坡上辐射冷却快，降温比山谷迅速，山坡上的空气比山谷上空同高度空气凉，风由山坡吹向山谷，形成山风。白天谷风引起山坡空气辐合，空气湿润时会引发积云对流发展。

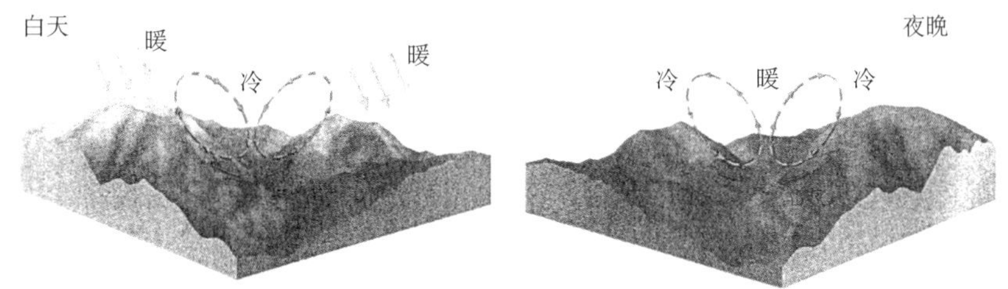

注：山风和谷风常见于山区。山谷风的形成是由于山坡与山谷的受热差异引起的。日间风从山谷吹向山坡，夜间风沿山坡吹向山谷。

山风和谷风

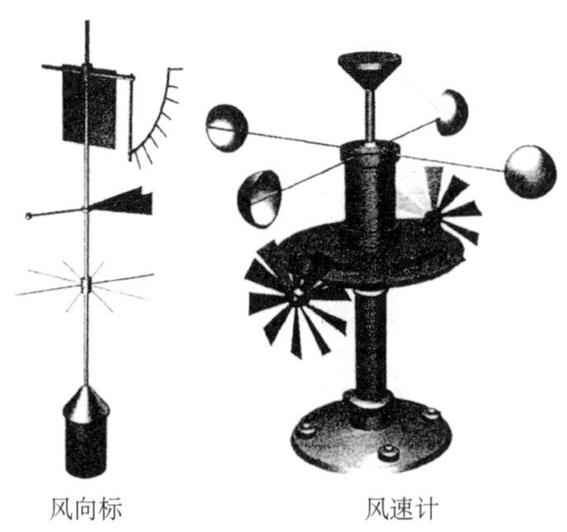

注：风向标用来测定风向。风向标的箭头指向哪个方向就表示当时刮什么方向的风。风速计是测量风速的仪器。风速计由三个或三个以上的风杯构成，风杯固定在一支绕垂直竖杆旋转的杆臂末端。风杯受风时杆臂旋转，风速就被记录下来。地面的风是在开阔无障碍地面上10～12米高处测得的。

风向标和风速计

焚风

焚风是过山气流在背风坡下沉而变得干热的风。当气流翻越较高大的山岭时，在迎风坡因地形抬升，空气中的水汽常常凝结产生地形云和降水，使空气变得较干燥。当干燥的空气越过山顶向下流动时，因气压增加空气压缩而增温，到了山脚下形成干热的焚风，可引起农作物或草场干枯并易引发山火。欧洲阿尔卑斯山区、美国落基山脉东麓常有焚风发生，美国称之为钦诺克风。中国太行山东麓石家庄一带也常有焚风引发的干热天气。

注：暖湿空气沿着山坡爬升，空气中水汽因冷却而凝结，在迎风坡形成云和降水，然后比较干燥的空气过山后下沉增温，便形成焚风。

焚风的形成

雷暴大风

从积雨云中发展起来的雷暴云到成熟阶段，因雨水的重力拖曳和云外干冷空气夹卷引起蒸发冷却效应，使云中产生下沉气流，当雷暴云中的冷空气迅速下沉到达地面后，近地面气温急剧下降、气压急剧上升、风向突变、风力迅速增大，完全替代了原来的暖湿空气，其前锋与低层暖湿空气辐合形成很大的水平气压梯度，可产生风速超过40米/秒的灾害性强风，还可能伴随雷雨或冰雹。这种低层向外流的强风顶面可以造成强风切变，对飞机起降安全构成严重威胁。当低层空气很干燥时，下沉气流中的水滴可以全部蒸发掉，形成强下击暴流。由于雷达也很难监测到它，所以对飞行安全威胁特别大，许多飞机起降时失事均与这类强风切变和下击暴流有关。

龙卷

龙卷是强烈发展的雷暴云中出现的强烈旋转的小涡旋。当雷暴云中上升气流与下降气流之间垂直切变非常强烈，可产生具有水平轴的涡旋，涡旋的两端弯曲并伸出云底后，形成一股具有垂直轴的快速旋转的气流，旋转风速度可达100米/秒以上。由于离心力作用，旋转气流中心气压降得很低，卷入其中的水汽快速凝结形成可见的漏斗云。随着龙卷涡旋加深，漏斗云向下伸展，抵达地面时由于高速的旋转风（旋转周期为1秒量级）和极低的气压两者的联合作用，造成拔树倒屋及将人畜物件乃至汽车席卷吸入空中抛至远处的灾害。龙卷本身水平尺度为数十米至数百米，很少超过1千米，它随着强雷暴母体云移动可造成数十千

注：龙卷形成前，天空会出现较大的雷雨云。其中一个区域的云特别黑且稠密，此区的空气会快速旋转，然后形成漏斗状的云向下延伸。龙卷常伴随闪电、雨及冰雹。漏斗状云如果到达地面，就会掀起大片的尘土及对其他物体造成巨大的破坏。

龙卷的形成

米乃至数百千米的受灾地带。

1.2.2 云和雾

云是湿空气上升冷却达到饱和后的水汽在大气凝结核上凝结而成的产物。地球有近 60% 的面积为各类云所覆盖。云产生的降水不仅为人类和生态系统提供了淡水资源，也带来了一系列诸如暴雨、冰雹、龙卷、雷暴大风、雷电和雪暴等气象灾害。云是地球水分循环和大气能量输送及转换的主要角色之一，能垂直和水平输送水分，改变水的相态，释放潜热，产生降水并提供驱动大气环流的能源。云能将投射到地球上的太阳辐射反射掉约 17%，同时吸收地球放射出的长波辐射，调节大气和地面温度，对气候有重要影响。

雾是由贴近地面的空气冷却或混合达到饱和后由水汽凝结而成的。它影响大气能见度，对航空、航天、河海航运及高速公路等各类交通安全构成威胁。大气污染引起的光化学烟雾会损害人类健康。

注：积云呈棉絮状或馒头状，进一步发展可变成宝塔状的浓积云。浓积云再发展就是积雨云了。积雨云体态高大，顶部可伸展到十几千米的高度。发展旺盛的积雨云顶部常常会出现砧子形状，称为云砧，由冰晶组成。层云结构均匀，云层较薄，低的层云如果与地面相连就是雾。层积云介于层状云和积状云之间，由较大条状、块状和片状云块组成，结构不太均匀。雨层云云层较暗，水平分布较广，结构均匀，高积云由白色或灰色块状或片状云块组成，常呈波浪状。高层云由较均匀的层状云幕组成，卷云有的像一缕一缕的头发丝，有的呈鱼钩状。卷云一般遮不住太阳和月亮的光芒。卷积云是由冰晶组成的云，像鱼鳞一般。卷层云常呈薄纱状云幕。

云系

云的分类

地球上同一时间不同地区会出现不同形状的云，有的像棉絮，有的像宝塔，有的像城墙，有的像幕布。19 世纪初，英国的科学家首先尝试给云分类，不过当初的分类相当简单。云是气象观测的重要项目之一，为了使全世界能统一观测和使用云的资料，1891 年世界气象组织根据云的形态、组成和性质等，把云归纳成 4 族 10 属。

第一族为直展云族，包括积云和积雨云两属，一般先从底部开始发展，然后逐渐向上积累增厚，故名。除了积雨云顶部有冰晶外，都是由水滴或过冷却水滴组成。由于对流作用，云体垂直发展旺盛。云底比较平坦，离地面较低，一般为几百米，最多也只有一二千米，顶部发展较高。

第二族为低云族，包括层云、层积云和雨层云三属。云层较低，云体大多由水滴组成，雨层云中上部则常由水滴与冰晶混合组成。

第三族为中云族，包括高积云和高层云两属。高度在低云和高云之间，在温带地区一般为 2～7 千米，在热带略高，在极地略低。

第四族为高云族，包括卷云、卷积云和卷层云三属。高度较高，大多由冰晶组成。丝缕结构明显，洁白纤细，卷云体尤其如此。

从云的形状和性质区分，层云、雨层云、高层云和卷层云都是层状云，积云、积雨云、高积云和卷积云都是积状云（孤立、分散、垂直发展的云块），层积云属于层状和积状并存的云，卷云则为纤维状的云。云的变化很复杂，常常有好几种云同时出现。即使是一种云，也不是固定不变的。

云的形成

一定温度下，空气能够容纳的水汽量是有限的。在地面气压条件下，当气温为 30 ℃、0 ℃和−30 ℃时，每立方米空气最多能容纳的水汽量分别为 30.38 克、4.86 克和 0.334 克，称饱和水汽含量。随着温度降低，饱和水汽含量迅速减少。在对流层，气温随高度降低，每升高 1 千米，约下降 6 ℃。湿空气上升冷却迟早会饱和。空气过饱和后，多余的水汽易在大气中悬浮的小尘粒上凝结形成云滴，温度很低时则可凝华成冰晶。这些小尘粒称为凝结核或冰核，通常在零点几微米以下，云滴的直径从几微米到几十微米不等，在空气中下落速度约为 1 厘米/秒，可长时间悬浮在空中形成云。由水滴组成的云叫水云，由冰晶组成的云叫冰云，高空的卷云便是冰云。也有水滴和冰晶共存的，叫混合云。

注：太阳辐射使地面升温，地面又使周围的空气变暖，变暖的空气上升。暖空气上升过程中逐渐冷却，所含的水汽凝结成小水滴，小水滴积聚形成云。在白天，随着越来越多的水汽凝结，云块也就越来越大。

云的形成

上升运动是湿空气团形成云的关键因素。不同的上升运动形成的云的类型也不一样。大范围的气旋辐合抬升及冷暖气团交汇时，暖空气沿着锋面爬升可形成锋面气旋云系，包括卷云、卷层云、高层云和雨层云等层状云。接近地面的锋区的云层加厚变低，是温带地区主要的降水性云系。它的降水强度不大（每小时 1 至数毫米），但范围广（水平尺度数百至数千千米），时间持久（1 至数日），常能产生较大的总雨量。在冷锋云系中，当大气层结不稳定（即气温随高度升高而降低的速度大于 6 ℃/千米）时，湿空气上升的温度递减率仍为 6 ℃/千米，上升的湿空气比周围空气暖，有利于对流，也会出现对流云和对流性降水。锋面气旋云系内部有时会形成长达 100 千米左右的中尺度对流云带，从中产生较强的对流性降水。中国的梅雨锋云系主要是由于冷暖气团势均力敌、梅雨锋长期停滞在江淮流域而形成的，常造成大范围持续性暴雨和洪涝灾害。

对流性云系包括淡积云、浓积云、积雨云和强对流云团等，后两种云又通称雷暴云，是热带地区和温带地区夏季的主要降水性云系，大气层结不稳定是形成对流云的基本条件。这时，如果局部地区空气受热造成热力抬升，或者地面气流辐合造成动力抬升，就会触发对流运动，形成对流云或对流云系。这种过程非常快，也常常很强烈，上升速度比暖气团沿锋面爬升速度快 100 倍以上，垂直高度可达十几千米，有时会产生突发性局地暴雨及冰雹、龙卷、雷暴大风和雷电等灾害性天气。

湿气流遇到山脉爬升会产生地形性对流云或波状云、荚状云。另外，边界层湍流混合则会产生边界层层积云或层云，它们的降水潜力虽不大，但能阻隔太阳光照射地面，从而对局地或区域气候产生一定影响。

云层水平方向的大小差别很大。全球云系的水平距离最大可达 1 000～10 000 千米；温带气旋云系的水平距离为 500～3 000 千米，台风云系的水平距离大约为 500 千米，比较小的积云的水平距离在百米以下。这些大大小小的云系和单体，组成了遍布全球的复杂多变的云雨系统。

雾的形成和分类

近地面空气冷却增湿，或冷暖空气混合导致水汽饱和后凝结成的大量水滴或冰晶悬浮在近地面大气中，使大气能见度降低至 1 千米以下的天气现象称为雾，能见度在 1～10 千米的则称为轻雾。雾分四大类，有辐射雾（包括地面雾、平流-辐射雾、上坡雾、山谷雾）、锋面雾（包括暖锋前雾、冷锋后雾、锋面过境雾）、平流雾（包括海雾、热带气团雾、海陆风雾、蒸发雾）以及冰雾（由悬浮空气中的大量微小冰晶组成）。

雾的垂直厚度从 1 米到 1 500 米不等。单纯的辐射雾厚度较薄，日出后不久消散，而平流-辐射雾可厚达 1 千米以上，存在时间也长得多。雾滴半径一般为 2～15 微米，空气含水量为 0.03～0.5 克/米^3，视气温和雾的浓度而异。

雾对交通安全构成严重威胁，特别是在机场、港口和高速公路，许多交通事故都与雾影响能见度有关。现代高速运输系统经常由于浓雾的笼罩而完全停顿。为减轻雾的危害，一些繁忙机场采用人工消除冷雾的方法，播撒干冰或液氮制冷，产生冰晶清除过冷却雾。对暖雾，价廉而有效的消除方法尚不多。大气污染造成的光化学烟雾是人为原因形成的雾，会严重影响人类健康。1952 年英国伦敦发生的烟雾事件，死亡人数达 3 500～4 000 人。控制工业污染物排放是克服这类危害的根本途径。

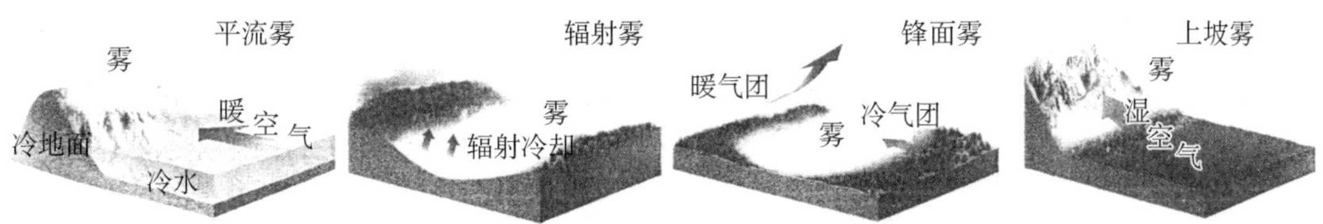

注：暖湿空气流经较冷的下垫面时冷却而凝结成平流雾。夜间近地面空气随地面辐射冷却达到露点温度时形成辐射雾。冷暖气团相遇时也常有雾产生，称为锋面雾。上坡雾是由于湿空气沿山坡向上移动温度降低而形成的。大多数雾在形成过程中是由几种机制共同起作用的，只是其中某一机制可能起着主导作用。

雾的形成

从云雾中降落到地面的液态水和固态水统称为大气降水，包括雨、毛毛雨、雪、米雪、冰丸、霰、冰雹等。直接凝聚在地面物体上的露和霜则称为地面降水，也属降水范畴。降水是人类生存发展和生态系统得以维系的淡水资源供应源。中国淡水资源人均占有量仅占世界平均水平的 1/4，且时空分布很不均匀，华北地区淡水资源人均占有量仅为全国平均占有量的 1/6。节约用水、合理利用有限的淡水资源是一项日益突出的社会需求。

液态降水和固态降水

雨主要由直径大于 0.5 毫米的水滴组成，下降速度为 2～9 米/秒，大多从雨层云、积雨云和降水性高层云中降落。过冷却水滴降到温度在冰点附近的地面或地物上时冻结成冰，成为雨凇，这种降水称为冻雨，对交通和输电系统有较大危害。

毛毛雨由直径小于 0.5 毫米且大小较均匀的小水滴组成，主要从层积云和层云中产生。

雪是由冰晶组成的固态降水，而冰晶有六角板状、枝状、柱状及针状等多种形状，其形状取决于云中温度和过饱和度等特征。较大的冰晶称为雪晶。当云下部温度高于－5 ℃时，雪晶常聚并形成雪团，称为雪花，它是冬季中纬度降水的主要形态。即使是低纬度或夏季积状云降水，多数也是在云中上部的冷云部位形成雪花，降至 0 ℃层以下融化成雨。

米雪由层状云或浓雾中形成的直径小于 1 毫米的白色不透明冰粒组成。米雪下降时与过冷却水滴碰撞合并，或部分融化后再冻结，可形成透明或半透明冰粒，称冰丸。雨滴冻结也可形成冰丸，或称冻滴。

霰由白色不透明冰粒组成，也称软雹或雪丸。冰粒呈球状或圆锥形，直径为 2～5 毫米，落地反跳易碎，主要从对流性云中降落。积雨云上部常形成霰，降至 0 ℃层以下融化成雨。

冰雹由直径大于 5 毫米的冰块组成，最大直径可达数厘米，呈球形、锥形或圆盘形，由透明冰或透明与不透明冰层交替组成，中心常有一雹胚，由霰或冻滴组成。还有一种内含液态水的海绵状冰雹，反映出成雹过程中云内不同温度、湿度、含水量及流场特征。冰雹从强雷暴云中降落，是一种局地灾害性天气，对农作物和人畜生命及财产等都有很大破坏力。

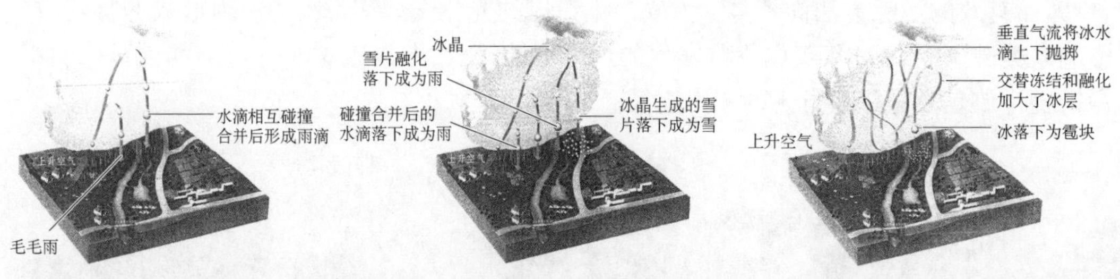

注：上升空气到高空冷却，当达到饱和或过饱和后，水汽在凝结核上凝结成云滴，经碰撞合并作用形成雨滴降下。在高层温度很低时，水汽还能在冰核上直接凝华成冰晶，冰晶相互聚并形成雪花，雪花降落到大气 0 ℃层高度以下时融化成雨。雹块是由于冰冻的雨滴在积雨云中反复抬升回落，使得其冰层不停加厚，最后落到地面成为冰雹。

降水的形成过程

降水量

降水量以地面累积水层厚度(以毫米计)表示。日雨量大于10、25、50、100和200毫米的分别称为中雨、大雨、暴雨、大暴雨和特大暴雨。按雨强分,每小时雨量大于2.5、8和16毫米的分别称为中雨、大雨和暴雨。中国有记录的最大雨强为1971年7月1日山西梅桐沟的640毫米/时;日雨量最大值为1967年10月台湾新寮的1 672毫米,其三天总雨量达2 749毫米。非洲法属留尼汪岛的赛路斯日雨量最高达1 870毫米,为世界之最。

降水分布

地球上降水的空间分布很不均匀,受纬度、季节、海陆分布、地形,特别是大气环流和天气系统的影响十分强烈,加上产生大量降水的云系往往具有中尺度特征(10~100千米),致使降水时空分布起伏特别大。南北纬30°附近是干燥带,主要受副热带高压控制,空气大范围下沉,降水稀少,世界上主要沙漠区都集中分布在那个地带。中纬度和接近极地的西风带地区是锋面气旋活跃地区,降水比较丰沛,其中大陆西海岸的山地迎风坡,降水量尤丰,如挪威山区年降水量可达6 000毫米以上,而同一纬度深入内陆的俄罗斯乌拉尔地区只有490毫米。

中国南方多雨,北方少雨,如广东、广西地区年降水量超过1 500毫米,而华北只有300~500毫米,1 000毫米雨量等值线位于长江北岸,几乎与长江平行。中国的降水也表现为东南多雨,西北少雨,如东南各省年降水量均在1 500毫米以上,台湾高达3 000~4 000毫米,而西北地区则不到250毫米。中国地处季风气候区,夏季多雨,冬季少雨,如广州夏季雨量占全年46.8%,冬季只占7.3%,北京夏季占76%,冬季只占1.8%。中国降水量的年际变化也比较大,是旱涝频发的国家。

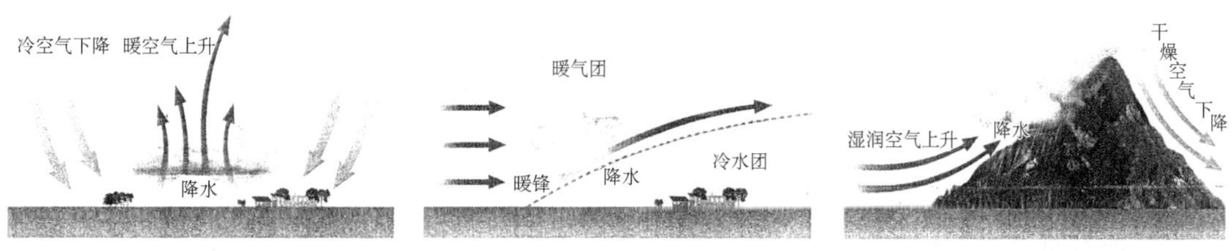

注:按降水性云的动力特性,降水可分为对流性降水、锋面气旋系统性降水和地形降水等。对流性降水从对流云系中降落,一般为阵性降水,强度大,持续时间短,常伴有雷暴,称雷阵雨,强对流风暴中有时会产生冰雹、龙卷、突发性暴雨等局地灾害性天气。锋面气旋系统性降水从锋面气旋云系中降落,一般为连续性降水,伴随着系统中的中尺度对流雨带,降水强度时有起伏,可造成大范围连续性暴雨灾害,地形降水指湿空气沿地形抬升引起的或加强的降水,带有一定的地方性特色,迎风坡多雨,背风坡少雨。

降水类型

降水形成过程

当湿空气上升冷却达到饱和及过饱和时,多余的水汽就在某些吸湿性或可溶性的气溶胶微粒上凝结形成云滴,但云滴很小,直径约10微米,它在空气中下落速度经常不到1厘米/秒,出云后很快蒸发掉,不可能落到地面形成雨。通常雨滴直径约为1毫米,与云滴相比直径约大100倍，体积更相差几十万倍。从云滴长成雨滴十分不易。云层产生降水需具备一定的条件。云体大、云层厚、云的维持时间长，以及云中含水量大等都有利于降水的形成。

在由水滴组成的云中，雨滴主要依靠大小不同的云滴因落速不同而相互碰撞合并长大而成，称为暖雨形成过程，云滴大小不均匀的云较易产生雨。另一种重要的降水形成过程与云中出现冰晶有关。当云层很厚、上部气温很低，如低于－10 ℃时，水汽就可能在一些称之为冰核的微粒上凝华（物质直接由气相转变为固相）形成冰晶，一些含有冰核的云滴也可能冻结成冰粒。当冰晶与过冷却水滴并存时，在同一温度下，冰面的饱和水汽压小于水滴的饱和水汽压，于是水滴蒸发并向冰晶上凝华。长大的冰晶下落途中与水滴碰撞合并进一步增长形成雪晶或霰粒，雪晶相互聚并就形成雪花。霰粒或雪花降落至0 ℃层高度以下融化成雨，这是中纬度乃至低纬度大多数降水形成的主要方式，称作冷雨形成过程。冬季地面气温较低时，就直接成为降雪。冰雹是由霰或冻滴充当雹胚在强对流云中进一步增长形成的，过程更为复杂。

注：雪花是冰晶的聚合体，而冰晶则由水汽在冰核上凝华增长而成。冰晶的形状很多，但均由六角形晶体衍生而成，其形状取决于云中的温度、高度和含水量等因素。

冰晶形状

1.3　特殊大气现象

1.3.1　发狂的龙卷风

六合读者王：“7・13”特大龙卷风肆虐苏北宝应、高邮大地造成数以千计的人员伤亡和数千间房屋倒塌的巨大经济损失。我也是农民，我能理解这对农民来说是多么惨痛的损失。请问Q博士，龙卷风究竟是什么东西呢？为何那么凶猛无比？它能不能预防呢？

Q博士回答：龙卷风是破坏性极强的一种小尺度的气象灾害。它是转性强雷暴猛烈的产物。龙卷环流从强雷暴母体云中孕育后以漏斗状云向下伸展，这是一股快速旋转的气流，旋转速度可达100米/秒以上。由于离心力作用旋转气流中心的气压降至很低，这低压导致卷入其中的水汽快速凝结形成目视可见的漏斗云。龙卷造成毁灭性破坏力的主要原因是高速度旋转风和极低气压两者的联合作用，它像一只发狂的猛兽吞噬一切所遇到的物体，造成拔树倒屋及将人畜物体乃至汽车卷吸入空中的惨烈灾祸，但龙卷本身水平尺度并不大，仅十几米至数百米，很少超过1千米。它随着强雷暴母体云移动可造成长达数十千米乃至数百千米的受灾地带。

不过，龙卷毕竟是种稀有气象灾害事件，它发生发展的条件比较苛刻。“7·13”苏北强龙卷爆发时该地区处于副热带高压边缘，苏北又频降暴雨，暖湿气流湿静力能供应充沛。在这种有利形势下容易爆发旋转性强风暴，也称超单体风暴。世界上美国中西部大平原和北非撒哈拉大沙漠地区都是龙卷频发的地区。由于旋转性母体强雷暴中气旋与次级龙卷涡旋（水平尺度仅数十至数百米）在尺度上存在量级差异，加上龙卷出现稀少，生存时间短暂（仅数十分钟至数小时），对它的结构特征和形成机制尚有一些环节和问题待解决。

龙卷的预报是一个比强雷暴预报还要困难的课题。龙卷途经水面时“龙嘴”低压，且能吸起大量白花花的水珠，所以说“龙嘴吸水”倒是真有其事而非妄言。1999年6月22日傍晚发生在武汉东湖风景区的“水龙卷”便是例证。

龙卷破坏力极强，但其水平尺度很小，如能及时稳定并判断其运行路径，可以漏斗云的来路向侧方向飞速逃离，趴在低凹处以免被旋风和低压吸起，还是有可能免受其害的。切忌跑进屋内或屋旁树下，以免被倒塌的房屋树木压倒或砸伤（本次回答由南京大学大气科学系叶家东教授撰文，特表感谢）。

1.3.2 飞机尾迹云

近日，又有“不明飞行物”频频亮相。“不明飞行物”究竟为何物？众说纷纭，其中一种说法是飞机尾迹。那么飞机尾迹又是怎样形成的？对人类有什么影响？

飞机尾迹又叫飞机凝结尾迹，或尾迹云，主要是高空喷气飞机排放的水汽在空中凝结成水滴或直接凝华成冰晶体形成的稀薄云带。通常飞机尾气中含有大量水汽，一架波音727飞机巡航速度飞行时每小时可排放3.7吨水汽，在高空很低的气温下，空气的饱和水汽压很低，在自然水汽接近饱和的潮湿气层中加入少量水汽就能使空气过饱和而形成尾迹云，通常由小冰晶组成，有时可维持数十分钟乃至更久。在繁忙的高空喷气飞机航线上有时可出现多条飞机尾迹云（如图）。

可见，形成飞机尾迹云需要一定条件。一是低温(一般低于－40 ℃)；二是气层中自然水汽含量较丰富，空气潮湿接近饱和。所以低空飞机较少形成尾迹，即使出现也瞬息即逝，看上去就像燕尾似的后延数百米而已。由于高空气流各异，太阳光照射角度不同，飞机尾迹也会形状、色彩各异。因高低空大气结构有差异，有时高空有尾迹而低空飞机无尾迹也不足为奇，有时高空飞机有尾迹有时则没有也不足为怪，取决于高空气象条件。

飞机尾迹云这种人类活动造成的高空冰晶云对人类自身有什么影响呢？随着航空事业发展，一些重要航线上空飞机尾迹云有时会合并成连片的卷云覆盖在广阔的天空中，在欧洲和美国它们大约可覆盖 1%和 2%的天空，能改变大气的辐射收支，从而影响地面温度和局部气候。500 米厚的这种高空冰晶云能削弱 15%的入射太阳辐射，而它本身又使向下的红外长波辐射增加 21%，不过仍不足以弥补削弱太阳辐射的损失。结果白天地面温度会有所降低，夜间则会增暖，减少了地面温度日变差，使局部气候稍稍趋向温和。自 20 世纪 60 年代以来美国中西部特别是繁忙的航空通道地区已发现云量有所增加，日照减少而地面极值温度趋于缓和的证据。至于地面平均温度，飞机尾迹云的影响是净增暖还是净冷却，则取决于云厚及其持续性，以及地理纬度和季节。

南京大学　叶家东

1.3.3　路遇冻雨

有人说雪是云中雨滴冻结而成的，其实并不对。相反，大多数雨倒是云中形成的雪花或霰粒降落到 0 ℃层高度以下融化的产物。不过也有例外，冻雨就是过冷雨滴或接近 0 ℃的“暖”雨滴降落至冷地面上冻结而成的，气象上也叫雨凇。一场冻雨，树木、草地、电线乃至各类地物上都是亮晶晶的冰棱，树枝沉甸甸的像冰雕，阳光下闪闪发光，景色倒真是绚丽多姿(如图)。只可惜电停了，路面滑溜溜的难以启步，野外冷清清空旷寂静，夜晚则一片漆黑，一种苍凉阴沉的氛围笼罩着大地。这都是冻雨的杰作，它是挺厉害的一种气象灾害，对交通安全、输电线路以及人们的日常生活构成严重威胁。冻雨凇结在道路上形成坚硬光滑的路面，可造成重大交通事故；输电线上凇结的冻雨可形成粗粗的冰柱，电线或电杆承受不住便会断线倒杆，不仅中断输电且会造成火灾及人畜伤害；树木承受不住冻雨冰棱的重力会折断倒伏。除毁坏林木外还会阻断道路交通，笔者就曾亲历冻雨之害。

那年圣诞节期间与家人、友人一同驱车去美国佛罗里达的奥尔兰多迪斯尼世界旅游，从华盛顿哥伦比亚特区沿 85 号公路至亚特兰大途中遇见了冻雨，在弗吉尼亚山区冻雨将道路铺上一层坚冰，两旁树木不时被冻雨冰挂压倒，横躺在公路上，路警来不及清除，汽车只能以 10 千米的时速蜿蜒曲折地在光滑的路面上绕着倒伏的树丛爬行。那路段人烟稀少，这种天气下行车也不多，不时便能听到林中树木折断的劈啪声，平添几分蛮荒可怖的气氛。我们亲眼目睹一辆吉普车被一棵碗口粗的松树压住动弹不得，几个墨西哥人模样的旅客爬出车一筹莫展，跺着脚干等警察前来救助。于是我们格外小心，隔窗盯住右前方路旁的树木是否有倒伏的危险。好容易见到一个加油站，可是电断了，加油机无法工作，好在我们出发前加足了油，只是水电停了，厕所门已锁上无法启用，男人好办，摸到旁边对着树林就方便起来。女同胞可犯难了，她们相互搀扶着，踩着薄冰覆盖的雪走到屋后，几个人站成一排挡住视线，轮流在那雪天冰地的露天“厕所”里方便，领略了一番“回归自然”的风味。站里热食自然是没有了，好在我们中国人精细，事先准备了牛肉、咸水鸭以及面包、饼干、冷饮之类的干粮，不敢懈怠，赶紧上路，一路走一路用餐，开车人则由身后的人一口一口喂食以便节省时间。如此这般直至深夜 10 时许方才抵达亚特兰大。原本九小时的车程却走了十三个多小时。所幸尚算平安，未被那倒伏的树压住。

1999年1月15日，一夜冻雨冰灾，大面积断电。

1.3.4 南极臭氧空洞

最近，有媒体报道南极上空的臭氧层空洞面积已膨胀至 2 930 万平方千米，相当于美国国土面积的三倍以上，比 1998 年增加了将近 8%，且曾一度延伸至智利南端城市居民集中区上空，严重威胁着人类的健康和其他生物的生存条件。臭氧层是怎样形成又如何遭到破坏的呢？为什么臭氧空洞首先出现在南极上空而不是别处？

臭氧层主要集中在地球上空 20～50 千米的平流层，在 25～30 千米高度上浓度最大。全球臭氧总量约 30 亿吨，仅占地球大气的数百万分之一，将它集中起来铺在地面上形成的臭氧层厚度，在标准状况下平均 3 毫米厚。大气中的臭氧是氧分子(O_2)在太阳紫外线和宇宙线辐射以及闪电作用下发生光化学反应生成氧原子，氧原子在第三种中性分子 M 参与下与氧分子发生三体碰撞结合形成臭氧。臭氧生成后能吸收紫外线而离解，离解率随温度升高而急剧增加。在正常情况下臭氧的生消过程达到动态平衡形成臭氧层。

臭氧层虽然含量很少，却是地球生命的重要保护层。它起着两种保护作用：一是能滤掉 70%～90%的太阳紫外线，防止过量的紫外线穿透大气层到达地面，保护人类和地球生物圈免遭过量紫外线伤害。人们知道，过量紫外线辐射是皮肤癌、白内障、免疫功能衰退等的重要致病因素，也影响植物特别是豆科植物的正常生长结果，而且还影响海洋浮游生物的生存条件，破坏海洋生物食物链从而影响鱼类繁衍。二是臭氧也像二氧化碳一样是一种温室气体，能吸收地球长波辐射，调节地球气候。所以，人们将臭氧比喻为像水和空气一样对地球生命不可或缺。没有臭氧层保护，地球生命就将面临灭顶之灾。

自 20 世纪 70 年代以来，全球臭氧层浓度减少的趋势日益明显，南极上空尤为显著，南纬 39°～90°减少量达 5%～10%，称作“南极臭氧空洞”，主要是由于人类生产活动排放的氯氟烷烃(如氟里昂)和溴氟烷烃(哈龙)等消耗臭氧分子所致。氯氟烷烃在大气中十分稳定，可存留数十年乃至更久。所以，1987 年各国通过《消耗臭氧层物质的蒙特利尔议定书》限制上述人工化合物的生产。

人们或许会问：既然氟里昂等消耗臭氧层物质主要源自工业发达的北半球中纬度，为什么臭氧空洞会首先出现在南极上空呢？初步研究认为，这主要取决于气象条件。氟里昂等消耗臭氧的光化学反应需要一定的界面载体参与，且在－68～78 ℃低温下最活跃。南极春季(9～11 月)上空由于极地低涡导致较多高云产生，符合上述条件，所以能较多地消耗该地区的臭氧，以后云减少，气温升高，臭氧因紫外线辐射增强而又会恢复一定的浓度。北极因无大陆，高云比南极少，所以臭氧浓度减少相对较少。不是南极上空氯氟烃类分子浓度高，而是那里气象条件有利导致南极臭氧“空洞”的形成。

南京大学　叶家东

2　综论

2.1　云和降水研究的进展和趋势

叶家东
（南京大学大气科学系）

云和降水是大气中最重要的天气现象。云，常常组织成中尺度云系，为地球提供降水，并产生暴雨、冰雹、龙卷、下击暴流、雷电、雪暴等灾害性天气。云，特别是对流风暴是大气中最有效的一种能量转换器，它输送热量、水汽、动量以及化学污染物质，影响大尺度天气和环流系统，云系还强烈地影响大气辐射输送，从而影响全球热量收支和海陆表面的热量平衡。所以，云和降水的研究对于改进短期天气预报，特别是定量降水预报，对提高人工影响天气的科学水平，及改善区域和全球天气模式、大气环流模式和大气化学模式都具有重要意义。

近四十年来，限制云物理研究和人工影响天气科学水平进一步提高的症结在于对云的宏观动力过程及其与云微物理过程和环境大气的相互作用认识不足。近二十年来，一系列人工影响天气试验研究计划，原来都寄予较大的希望，结果却大都不尽人意。例如，在佛罗里达单体积云动力催化试验获得初步成效的基础上开展的 FACE，结果并不理想。NHRE 以及 GROSSVURSUCH－Ⅳ等为验证苏联建立在“累积带”概念基础上的防雹原理假设而开展的外场试验，也未达到预期目标。再如 PEP 或 SCPP，前者因实际情况离预期的相差较远而中断，后者虽然持续试验许多年，就播云效果而论，至今尚无显著的结论。鉴于现有的人工影响天气的原理是建立在实验室实验和简单云体的外场试验验证基础上的，诸如单块积云的动力催化原理和建立在“累积带”概念基础上防雹原理等，都是基于某种封闭的定常的云模型基础上发展起来的，一旦用之于复杂云系，考虑云与云、云与环境大气的相互作用以及云发展的时变特征，情况就大不相同。由于复杂云或云系中降水的形成受制于大尺度天气动力过程、中尺度和云尺度的动力学和热力学过程，以及云微物理过程和辐射及边界层等诸多尺度十分不同的物理过程之间的非线性相互作用，这里有许多尚未被认识的知识领域。目前人工影响天气的措施所能牵动的只是微物理过程中的某些环节及相应的一部分潜热效应，它们在什么情况下才能左右云或云系的发展进程，确实是很复杂的并且难以预料的。因此，一些大型的涉及复杂云或云系的人工影响天气计划的效果难以检测出来并不奇怪。要提高人工影响天气的科学水平，十分需要加深对云和云系降水形成过程的全面认识，拓宽云物理研究的尺度范围，才能对云或云系的可播性以及相应的播云措施做出正确判断的决策。

另外，在大尺度气压场形势预报上已取得显著进展的情况下，定量降水预报准确率提高不快的症结则在于降水过程往往具有中尺度特征，而常规的天气观测网一般并不能充分地分辨这类重要的降水性中尺度系统。统计表明，从1961年到1978年0～24小时的定量降水预报准确率没有显著变化，大于等于2.5 cm降水的TS评分停留在0.2左右，至于24～48小时的预报准确率，由于自1971年以来引用了有限区细网格模式（LFM），把预报时效延伸至48小时，逐年稍有改善，TS评分达到0.1左右。但定量降水预报准确率的这种改善比同时期大尺度数值天气预报的环流预报准确率的改善要小得多，其主要原因就在于用数值模式预报的环流形势与观测到的大量降水型之间在尺度上有很大差异。此外，对可测降水和大于等于2.5厘米大量降水的定量预报，在暖季比冷季困难得多。因为暖季大部分降水是由对流性环流造成的，其水平尺度比大尺度系统小约两个量级，用常规资料网及数值预报模式的网格常常完全不能分辨。所以，要使定量降水预报取得进展，需要集中力量加强对降水性中小尺度系统的研究。

云物理学和人工影响天气研究需拓宽研究的尺度范围，把目光指向降水性云或云系；定量降水预报研究的焦点则集中在中尺度领域。这是当今中小尺度气象学得以兴起并取得重大进展的主要动力。这也规定了云和降水研究的基本尺度范畴是降水性中小尺度系统。云物理学研究要为提高人工影响天气的科学水平服务，也要为改善定量降水预报和局地灾害性天气的检测和预报服务。云物理学在大气化学和云辐射气候学研究中同样起着重要作用。

对流风暴是重要的降水性云系。它是热带主要的降水产生系统，也是中纬度夏季主要的降水系统，并且是许多局地灾害性天气，诸如局地突发性暴雨、冰雹、龙卷、下击暴流，以及雷电和强烈雪暴等强天气的制造者。所以，对风暴的研究是近三十年来云和降水研究的重要论题，它涉及不同尺度的大气物理过程和天气动力过程之间的相互作用，包括边界层的影响，环境大气流场和热力学特征对风暴结构的效应，云微物理过程的作用，风暴尺度流场在降水发展、类型和强度中的作用，环境干空气夹卷的性质和效应，风暴的再生和传播机制，中尺度对流系统的形成和发展机制，以及风暴系统通过热量、水汽、动量等物理量的输送以及改变辐射收支等对环境大气的效应。所有这些，都是当前云和降水研究的重要课题。观测证据日益明显地表明，降水性中小尺度系统是大气物理过程与天气动力过程交互作用的热点。中小尺度数值模式研究的发展进程也表明，云微物理学、辐射和边界层湍流等大气物理过程的参数化研究是中小尺度数值模式研究的核心课题。

各种现代探测技术，如多普勒雷达、气象卫星、飞机探测平台和各种地基特殊遥感技术，包括双通道微波辐射仪、偏振雷达、偏振激光雷达、多参数雷达，以及风廓线雷达和机载多普勒雷达等技术和设备的发展及应用，使人们认识和研究云和降水过程的能力大为增强。技术的进展还表现在各种遥感探测和飞机探测资料的实时计算机处理和分析技术，以及不同类型资料的计算机同化技术也取得了重大进展。

鉴于降水性云系常常组织成中小尺度系统，为了研究这类系统，需要针对具体的研究目标组织专门的中小尺度综合考察研究计划，进行外场试验。从20世纪70年代的GATE、NHRE、MONEX、SESAME，到80年代的CCOPE、AIMCS、PRE-STORM、EMEX、TAMEX等一系列中小尺度外场试验，积累了一批中小尺度降水性云系的结构特征、发展演变规律及与大尺度环境场的相互作用的实时观测资料，促进了中小尺度降水系统和强天气的研究。现代中小尺度气象学中许多重要的观测事实和研究成果几乎都是来自这些有设计的、目标明确的、设备精良的、各方面协调的综合外场研究计划。我国“七五”计划期间筹建的几个中尺度试验基地为开展相应的中小尺度外场试验奠定了良好的基础。

近年来云和降水研究中的一些重要发现和进展：

（1）积云对流起源的研究强调边界层热力涡动与自由大气中重力波的相互作用。关于锋面抬升触发强风暴的问题，一个重要发现是常常观测到冷锋的水平尺度只有 1 千米量级，锋面上强的垂直运动可高达 4～20 米/秒，边界层中锋面则近乎垂直，表现出地面冷锋具有密度流性质。锋面上强的垂直运动常常足以触发积雨云对流，产生一宽度仅为 5 千米左右的窄冷锋雨带。边界层中冷锋的这种结构特征的成因是一个新的研究课题。

（2）关于夹卷效应。曾经有一种观点认为夹卷混合能促进云滴凝结增长，从而增强暖雨形成速率。但小尺度观测发现，在积云塔发展旺盛阶段，夹卷混合是分阶段进行的，包括体积夹卷和小尺度混合两种过程，前者产生一种粗混合云体，其中包含小片晴空，各自保持原先属性大体不变，在这种状态下云体上升相当一段距离后小尺度混合才使属性均匀化，这时新活化的云滴能参与消耗过饱和水汽，混合云体的云滴不能分享过多水汽而长得更大。

（3）对流风暴中冰雹的起源。对“馈云单体”的作用重新引起重视；初始霰粒下降时融化和溅散产生的次生水滴的再循环可以成为重要的雹胚源；采用示踪剂伴随播云物质的试验设计来评价雹胚形成和冰雹增长理论，为防雹提供可信的原理假设基础。

（4）冰晶聚合体在降水过程中的作用。冰晶聚合体是冬季降雪的重要形态，也是夏季对流风暴和中尺度对流系统层状区过冷层降水的主要形态。冰晶聚并过程发源于较高较冷的气层，当聚合体下降接近融化层时聚并效率大为增强，诊断的聚并系数可以大于 1，从而在 0～－5 ℃层形成一冰晶聚并带。冰晶聚合体在一些雹暴中能充当雹胚源。湖效应风暴研究发现，聚并过程是一种重要降水质点增长过程。故此，有理由怀疑美国 NOAA 等机构关于湖效应风暴的降雪再分布的播云原理假设是否可行。

（5）降水的动力学效应——微下击暴流。多参数雷达与三重多普勒雷达观测湿环境下孤立强风暴下沉气流的强迫机制。观测证据表明，高含水量的降水重力负荷可以发动下沉气流，固态降水的融化冷却效应会进一步增强下沉气流达到微下击暴流的强度。观测还表明早期降水增长是通过暖雨碰并机制的。

（6）多普勒雷达、微波辐射仪、偏振雷达和偏振激光雷达综合观测地形性风暴，发现风暴中可播区（“液态水机会”）是有的，但并不是在整个风暴生存期内都存在的。播云措施需仔细筹划方能奏效。

（7）中尺度对流系统（MCS）层状区中尺度环流特征及其成因。对中纬度飑线系统的多普勒雷达和其他资料分析获得三维流场、扰动气压场和扰动密度场，表明层状区中强的中层辐合，特别是中层环境空气后向入流的加速，是由于层状区的中层气压有系统性偏低引起的。层状区中这种中层低压是由于高空潜热释放和低空雪的融化和蒸发冷却效应所致。分析还发现，与已往的概念不同，层状区内中尺度上升气流与下沉气流的界限不在 0 ℃层，而位于－6 ℃层附近。我们根据飞机垂直加速变化观测的资料统计分析也发现，在 MCS 层状区－5 ℃层以下盛行下沉气流（－0.2 米/秒），其上出现上升气流（＋0.3 米/秒）。这一现象表明，中层辐合是引起中尺度上升和下沉的主要因素，而水汽凝华潜热释放导致中尺度暖心结构和相应的中低压，则是中层辐合的根源。这里中尺度环流动力学与微物理过程潜热效应之间存在CISK型的正反馈相互作用,加上层状云盖的辐射冷却和云底辐射增热效应,使MCS层状区有时能在系统的对流活动中止后维持数小时或更长时间。美国国家STORM计划是20世纪90年代一项重要的中尺度研究计划。

(8) 云的辐射特性和卷云微结构特征。FIRE 1986年卷云强化外场观测发现:卷云中温度低达−35 ℃时还有液态水存在,分布范围达10千米左右。另一个发现是在−45 ℃低温下,有异常多的雪花或冰晶聚合体。它们会影响卷云的辐射特性及辐射效应的参数化研究。但这种聚并机制尚未了解,卷云内存在液态水的原因也不清楚。

(9) 云模式研究的发展趋势。

① 微物理过程参数化:观测表明,负指数分布律或对数正态分布都能较好地拟合降水滴谱分布,但分布参数不是常数,对固态降水质点尤其如此。因此,从模式预报的含水量不能诊断质点数浓度,在云模式中需要建立一套考虑自碰并或聚并过程及碰撞破碎效应的降水质点数浓度预报方程。另一种参数化方案是采用多重基本滴谱分布函数的参数化方案来描述全滴谱的演变过程。

② 云模式与中尺度模式相结合,采用套网格方案在需要仔细研究的核心区域采用细网格模式,可以显式表达微物理过程。新近又发展了自调适套网格方案,根据研究需要在粗网格模式运行时由模式体系自行选择需要嵌套的细网格区域。

③ 模式系统针对不同的物理过程和数值计算方案可以有一系列不同的选择,大大增强模式应变机动能力。

④ 模式的结构采用模块结构,引进预处理器和Job file等构架,便于模式的调试、修改、更新和多通道运行。

2.2 浅论人工增雨

叶家东

(南京大学大气科学系)

我国幅员广大,大部分地区属水资源贫乏地区。淡水资源人均占有量为每年2 670米^3,仅为世界人均值的四分之一,加之降水的时间和空间变率大,干旱频繁发生。近两千年来共出现过一千多次大旱,平均每两年就有一次大旱。1994年春夏季节,在华南洪涝的同时,江淮、黄淮广大地区持续高温少雨,至8月份受旱面积达1.62×10^{11}平方米!面对如此频繁的干旱灾害,人们除了通过兴修水利等方法进行抗灾防灾以外,还运用现代科学技术广泛开展人工增雨作业,为抗旱防灾开辟了一个新领地,并在缓解局部旱情中发挥了积极作用。

一、无动于衷的云

唐朝诗人来鹄有一首题为"云"的诗:

千形万象竟还空
映水藏山片夏重
无限旱苗枯欲尽
悠悠闲处作奇峰

面对旱魔肆孽而悠悠闲处无动于衷的云，诗人的愤懑之情跃然纸上。但是，诗人可曾知道，云也有难言之隐。并不是什么云都能降雨的，得具备一定的条件。在一定的天气形势下，空气上升会因气压降低、体积膨胀而冷却，大约每上升 1 千米降温 6 ℃左右。当冷却到一定程度，空气中的水汽就会达到饱和，再上升，多余的水汽就会在悬浮于空气中的微小尘粒（气象上称为凝结核或云核）上凝结成水滴而形成云。云滴很小，大约只有千分之一厘米，随风飘移。通常雨滴直径有一毫米左右。从云滴长成雨滴体积需增大一百万倍左右！这是一个漫长而复杂的进程。

二、冷雨

20 世纪 30 年代提出的冷云降水的“水-冰转化”理论是现代人工增雨的理论基础。飞机观测发现，当含云的空气上升至温度低于零度时，云滴一般并不冻结，而仍保持液滴状态，这种情况可维持到−20 ℃甚至更低。这种云叫过冷水云，其中的云滴增长主要通过凝结和水滴之间碰撞合并两种过程，通常都比较缓慢。这种云不容易降雨，它的微物理结构呈相对稳定状态。只有当大气中含有某种微尘，其晶体结构与冰晶可能有些相似，能在不太低的负温度下充当冰晶核心而让水汽凝华上去形成冰晶或使过冷水滴冻结形成冰晶。一旦过冷水云中出现少数冰晶，云的微物理相态结构就立刻呈现不稳定状态。由于冰面上的饱和水汽压比水面上的低，所以水汽就会因冰晶吸收而降低到低于水面饱和状态，于是水滴就会蒸发，这就是“水-冰转化”过程，是一种十分有效的冷云降雨机理。少数冰晶能迅速长大，进而通过碰并等其他过程形成霰粒或雪花，降落到暖区融化变成雨。

三、依然是一个梦

从这里人们发现了一种人工影响冷云降水的契机：缺乏自然冰晶的过冷云中大量过冷云水不能变成降水而最终会蒸发掉，致使云的降水效率很低，这种不能及时变为降水的过冷却水可以看成是自然云系统中一种无法泄出的空中水库。在这种云中，人工引入“最佳”浓度的冰晶来打开此过冷却水库的汇出口以增加降水效率，从而可达到增加降水量的目的。然而，要在短时间里将百万亿个冰晶或冰核及时播撒到云的过冷却部位，谈何容易！有什么物质能产生如此众多的冰晶呢？在这种高效催化剂找到以前，人工影响云以增加降水的构想依然是一个梦！

四、意外的发现

为了寻找高效的播云催化剂，人们花费了大约十年时间，却在 1946 年一次意外的发现中迎刃而解了。真是踏破铁鞋无觅处，得来全不费功夫。但是且慢，这决不意味着这一科学发现靠的是侥幸。著名物理化学家兰格缪尔领导的研究小组在研究飞机结冰问题时，他的助手谢费尔将许多物质研成粉末撒进冰箱里，结果都未发现冰晶。一天，冰箱温度不够低，他想让温度再低一点，于是就丢进一块干冰。忽然间，冰箱里马上充满了冰晶，足有几百万个！接着，他取一根在液态空气中浸过的针在冰箱里划一下，结果也能产生无数冰晶，证实起作用的是超低温，而不是干冰本身。兰格缪尔是一位大科学家，在 1932 年获得过诺贝尔奖金，他以敏锐的科学洞察力抓住谢费尔这一意外的发现，锲而不舍进行深入的研究，并在其晚年全身心地投入人工影响天气的科学事业中，直至 1957 年逝世。他们很快就发现，有一个临界温度，约－39 ℃，即水银冻结的温度，在这个温度下会自然地产生冰核。1946 年 11 月 13 日谢费尔进行了一次历史性的外场飞机试验，在一块云顶温度为－20 ℃的层积云中播撒了 3 磅干冰，在 5 分钟内云就冰晶化，并在云下降落了 700 米长的雪幡。谢费尔兴奋异常，情不自禁地对着同伴高呼："我们成功了！"谢费尔功不可没，他为人工播云开拓了一条切实可行的途径。这次试验后的第二天，另一个合作者冯奈古特在实验中又发现了晶体结构与冰相似的碘化银是一种高效的人工冰核。直到今天，干冰和碘化银仍是人工影响天气中用得最为广泛的冷云催化剂。

五、静力催化和动力催化

上述这种立足于人工引进"最佳"浓度的冰晶以提高云的降水效率，从而增加降水量的途径，通常叫作"静力催化"，它在冬季地形性层状云的增雨效果是明显的，能增加 10%～30%山脊上的降雪，以提供额外的灌溉水源。静力催化最成功的例子要数以色列的冬季过冷积云的增雨试验，长期试验结果表明能增加降水 13%～15%，并有迹象表明扩大了灌溉耕地面积。

20 世纪 60 年代后期在美国佛罗里达的热带暖底积云的人工增雨试验中，提出了"动力催化"的播云原理，着眼点放在增加积云发展的动力。在云的过冷却部位大量播撒人工冰晶(核)，使云迅速冰晶化，进而释放大量冻结潜热(1 克水结冰放出 80 卡热量)，云体增温而使浮力增加，在有利条件下能使云爆发性增长，云体变得更大、更厚且更持久，从而能降下更多的雨。降水的形成可望引起更强烈的下沉气流及与环境大气更强的相互作用，使对流更活跃，促使区域雨量有所增加。不过这后一个效应尚未得到明确的证实。

"静力催化"和"动力催化"的区别在于，前者要求人工引进适量的"最佳"浓度冰晶(核)，每升空气 1～10 个冰晶，多了反而不利，众多的冰晶争食有限的过冷云水，结果都长不大而难以成雨；后者则不同，热带暖底积云一般并不缺降水元，动力催化大约需在每升空气中引进 10^3～10^4 个冰晶，方能使云迅速冰晶化。

六、人工降水还是人工增雨

从人工影响天气的历史和现状看，对于结构单一的云体，譬如在自然条件下没有产生降雨的孤立冷积云、过冷层积云或层云，用播云的方法人工引晶可以促使它降水，播云后云的微物理结构和最终的降水效果都比较容易验证，这就是原先意义上的人工降水，它的成功表明基本的播云原理是合理的，这为人工影响天气科学事业奠定了基础。但是这类云的降水潜力并不大，为了获得有实效的增雨效果，人工播云的尝试主要集中在那些在自然条件下能产生降雨的云系。所以通常就叫作人工增雨。

七、寻找"空中水库"

人工增雨能否奏效的关键是云中是否存在过冷云水区，或是否有足够的过冷水含量可以通过人工播云的方法令其及时泄出变成降雨，气象上叫作可播区。随着观测事实的积累，人们发现，并不是所有的过冷云中都如同播云原理原先认为的那样，没有或只含有很少的自然冰晶。例如，台风或飓风云系中层状云区的过冷却部位在不太低的负温度下就有许多自然冰晶；另外，20 世纪 70 年代后期联合国世界气象组织(WMO)曾计划在西班牙开展一个代号为PEP的示范性人工增雨计划，成功后向发展中国家推广。在三年预试验期间发现，该地区各季层状云系的过冷却部位含有大量自然冰晶，而过冷水含量却较少，因而播云潜力不大，不得不中止PEP计划。

云物理学理论的发展使这个问题更为复杂化，原先认为云中自然冰晶的多寡主要取决于所属气团的大气冰核数浓度，因而可以对不同气团的大气冰核含量进行气候性测量，从而判定相应云系中自然冰晶的含量。但是后来的观测发现，在有些云系中，特别是低纬度海洋性云底温度较暖的云，自然冰晶浓度可以比大气冰核浓度大3个数量级（1 000倍）以上，从而提出了冰晶的次级繁生机制。于是大气冰核在有些情况下就失去其预示能力，需要直接测量云中的过冷水含量和自然冰晶浓度。所以，什么云适合于催化？在云的什么部位、什么发展阶段催化？一次播撒多少催化剂方能达到增雨的目的？以及用什么样的播撒技术能及时准确地实施上述播云任务？这些都是人工增雨面临的困难抉择。近年来发展了一系列诸如多通道微波辐射计、云多普勒雷达等遥感探测新技术和飞机云粒子测量系统，能实时监测云中过冷水含量及其空间分布，对寻找人工影响的"可播区"、找出"空中过冷水库"的泄出口有指导意义。一般的人工增雨作业通常难以做到这一点，作业及效果评估中的不确定性是在所难免的。鉴于人工增雨的播云原理是建立在破坏自然云中有时存在的微物理相态不稳定平衡的构想基础之上的，是一种"以巧破千斤"的策略，因此能奏效的机会自然是有局限的，且需要精心设计才能取得预期的增雨效果。

八、"这雨是他们的还是我们的?"

在人工降水蓬勃开展的时候，出现过一幅漫画：两个从事人工降水的牧师通过教堂的窗子凝视着正在往下掉的雨滴互相问道："这雨是他们的，还是我们的?"这是人工增雨效果检验的核心问题。由于自然降水的变率很大，目前还无法做出准确的定量降水预报，而且人工影响引起的降水变化幅度常低于自然起伏量，属于一种高噪声低信号的鉴别问题，因此人工增雨的效果检验是相当困难和复杂的工作。目前各种随机化统计试验方案仍是最被人们接受的效果检验方法，其基本思想是这样的：将适合于播云的机会随机地分成两组，一组播云，一组不播云作为控制，然后比较两组的降雨结果以检验播云的统计效果。这比较客观，但缺点是需要在同等条件下进行许多次试验方能获得有统计意义的播云效果。福建古田水库地区的人工增雨试验采用区域回归随机试验方案，试验12年，得出在锋前或锋区天气形势下，云顶温度为－5 ℃至－15 ℃之间的积层混合云和层状云系。当对比区自然雨量为0.3～7.0毫米/小时，增雨量可达20%～32%，统计显著性水平达到0.05以上。近年来人们十分关注将播云效果的物理检验、统计检验与数值模拟试验结合起来，力求将效果检验的客观性与物理机理上的合理性相统一，综合评估人工影响天气试验。

九、现状和展望

尽管人工影响天气涉及复杂云系以求有实效时，遇到的科学和技术问题相当多，但由于人工增雨本身具有重大的潜在社会和经济效益，与生产及人民生活关系密切，人们仍孜孜不倦地运用从实验室和简单云体中获取的知识来试图解决增雨抗旱的实际需求。当前全世界有几十个国家，特别是干旱和半干旱地区的国家正实施着100多个人工影响天气试验计划。我国近年来人工影响天气规模日益扩大。根据世界气象组织1992年公布的世界各地人工影响天气登记表统计，我国人工影响天气(主要是增雨和防雹)所影响的区域面积占世界人工影响天气总面积的18%，仅次于美国，居世界第二位。作业经费近五年来大约增加了两倍，在抗旱防灾中发挥了积极的作用。粗略估计，1992年的人工影响天气(包括增雨和防雹)的经济效益在16亿元以上，已成为气象服务的一种重要服务手段。在继续开展人工增雨和防雹试验研究的同时，人们正在探索人工削弱局地暴雨等新研究领域。

当前的关键是要结合人工增雨作业开展深入的科学研究，以提高其科学水平和增雨实效。“WMO关于人工影响天气现状的声明”(1992年)中指出：“应实施多国综合外场计划，以加强通过播云进行人工影响天气的物理基础，采用先进的多参数多普勒雷达、机载仪器、自动遥测地面中尺度网，以及空气示踪等现代技术，使人工播云的物理效应得到切实的验证。这将使人工影响天气有关的研究与作业取得迅速进展，近十年来我国气象系统自身的现代化建设提高了对大气进行监测的科学水平，为人工影响天气技术现代化提供了较好的技术基础。近年来，迅速发展的云降水及其人工影响的数值模拟研究在论证播云机理、优化播云方案以及验证播云效果等方面正起着日益重要的作用。一些省、区的外场试验基地建设和作业指挥系统现代化等方面也取得了长足的进展。因此，实现人工影响天气科学化、业务化的基本条件已初步具备，在本世纪末以前如能适当增加这一领域的科技投入，到2000年我国在这一领域实现科学现代化的目标有可能实现，并将带来更大的社会效益和经济效益。

1995年2月于南京大学

2.3 人工影响天气效果客观检验技术的发展

叶家东

(南京大学大气科学系)

一、背景——人工播云原理中的基本科学和技术问题

Brier在Hess编写的《人工影响天气和气候》一书中是这样开始他的关于人工影响天气效果评价的讨论：一幅应时漫画上画着两个牧师，通过教堂的窗子凝视着正在下落的雨滴互相问道：“这雨是他们的，还是我们的?”这个“是他们的还是我们的”效果检验问题，伴随着人工影响天气科学事业发展的整个历史，争论了40余年。对于结构单一的云体，譬如没有自然冰晶的孤立过冷积云、过冷层积云、过冷层云或过冷雾，用播云的方法进行人工影响，其物理结拘和降水的效果都是比较容易验证的。早期的和近期用先进测量技术进行的这类试验，从物理结构和最终的降水效果上都提供了证据，人工影响云雾促使降水或消雾是可能的，基本原理和概念是合理的，这为人工影响天气科学事业奠定了基础。但是，这类云的降水潜力并不大。现实的问题是：能否对更大、更复杂的云或云系进行人工影响以增加有经济效益的降水？抑或能否从物理上证实可以通过人为改变复杂对流风暴的微物理学和动力学结构以抑制冰雹？在这方面，有过一些统计检验的证据和数值模拟的有利解释，但直接的物理测量，在降水形成的整个微物理过程的各个链节上提供直接的物理证据还是不多。到目前为止，全世界的人工影响天气试验计划，在统计检验、数值试验和物理检验诸方面都获得显著的、合理的、令科学家和使用者都满意的试验，寥若晨星，也许以色列的人工降水试验是

仅有的一个。

鉴于人工影响天气的原理是建立在破坏自然云中有时存在的微物理相态不稳定平衡或胶性稳定结构的构想基础之上的，因此，不确定的、不能控制的或不能预料的因素很多，它们都能影响甚至破坏播云措施的效果。我们举冷云催化的例子来说明播云原理假设的局限性。自然云当温度暖于−10 ℃左右时，自然冰晶一般较少，云常常呈现过冷却液态水云的状态，当向这种云或云的这种部位播入适量的人工冰核或冷冻剂时，会产生额外的冰晶，加速水-冰转化的冷雨过程，从而促进降水形成过程或提高降水效率以增加降水量。这里隐含了两种基本假设：(1) 某些云的降水效率低是因为缺少足够的自然冰晶，人工引晶应能增强降水过程并增加降水量；(2) 过冷云中水-冰转化时会释放冻结潜热，从而能增加对流云的浮力，使云能发展得更大，能凝结出更多的水汽从而产生更多的降水。但是人们都知道，大量降水的产生取决于大尺度动力过程、中尺度和云尺度过程以及云微物理过程几种不同尺度的物理过程之间的非线性相互作用。冰相播云原理的直接效果是改变云的微物理学，制造或增加冰晶并释放数量不大的相变潜热，这样一些微小的效应在什么场合下才能左右云或云系的发展趋势或显著地提高其降水效率呢？这实在是一种“以巧破千斤”的策略，能得逞的机会理所当然是有限的，但需要精心设计才有可能取得预期成效。所以，目前对人工影响天气的一般提法仍然是：在合适的天气条件下，当用恰当的和适量的播云剂，在正确的时间播撒到云中正确的部位时，是能够获得预期的效果的。但要确定什么是“正确的、恰当的”，则是十分困难的事。因此，当人工影响涉及复杂云系以求有实效时，遇到的科学和技术问题就很多，诸如：所播的云是否合乎可播性条件，或者云中可播的部位(即所谓过冷液水机会)出现在何时何处？如何才能及时有效地将合适的播云剂适量地播撒至云中恰当的部位等等，这些都是较难实时判断和实施的。因此，复杂云或云系的人工影响计划效果难以检测出来是不奇怪的，它是与云系结构及人们对可播云的预报能力、探测技术和作业水平等一系列科学和技术问题紧密相关联的。

人工影响天气的科学事业虽几经起伏，但总体上仍是经久不衰的。1986 年在 WMO 注册的人工影响天气计划有 84 个，其中 3/4 是业务性播云作业计划，1/8 是研究性试验计划，另外 1/8 则是研究与作业兼而有之。从人工影响天气的历史和现状看，虽然手段尚未完善，但却具有不能熄灭的旺盛生命力，因为人工影响天气本身具有重大的社会和经济的潜在利益，与生产和人民生活关系密切。所以，尽管人工影响复杂云系的原理基础和技术手段尚未稳固地确立起来，人们仍孜孜不倦地运用从实验室和简单云体中获取的知识来试图解决一些生产和生活上迫切增加降水和减防雹灾的需求。社会生产的需求是迫切的，面临的科学挑战又是严峻的，两者推动着人工影响天气事业沿着业务性应用和研究性试验两种不同的但又相互关联相互促进的途径发展。研究性播云试验和业务性播云作业具体目的不同，采用的技术水平不同，效果检验的要求和水平也不相同。下面我们着重就研究性播云试验的设计和效果检验方法的进展做一概述。

二、随机化设计和效果的统计检验

早期的一些人工影响天气计划急于避开复杂的云物理过程的基础性研究，过早地单纯依赖于随机化统计试验来检验播云效果，人们批评这类试验是“黑箱型试验”，这对那些忽视人工影响天气的物理基础研究的人无疑是一种中肯的告诫。但如果将这种观念推广到针对随机化试验设计和相应的统计检验方法本身，则显然是一种误解。随机化设计的基本思想是这样的：将适于播云的机会随机地分成两组，一组播云，一组不播云作为控制，然后比较两组的结果以检验播云效果。这里有一个很重要的前提：适于播云的机会，这包括选择合适的云或云系中合适的部位，确定恰当的播云时机等一系列与播云原理密切相关的物理决策。再者，效果的分层统计又涉及按照云或云系的物理条件进行分类，这仍然要运用云的物理知识，也不是在“黑箱”里乱摸的。应该说这是从现实出发的科学但是比较软弱的一种研究手段。

严格地讲，没有控制的试验不是科学试验，而人工影响天气恰恰是在自然条件无法控制甚至难以监测和预报的大自然环境中进行的试验，以随机化设计分出一部分自然群体作为控制样本，是不得已而为之的一种科学手段。所以，Braham 在“增加降水——一种科学的挑战”(1986)一文中提出“继续进行随机化的探索性播云试验是绝对重要的。”“在完善的数值模式发展以前，随机化的外场试验仍是确定某一种播云技术是否能可信地并可重复地改变降雨量的唯一途径。”不仅如此，随机化的验证性播云试验同样也是必要的。

在效果的统计检验方法上，方差不相等的双样本回归分析对双样本检验和回归分析而言是一种发展。不过大多数经典的统计检验方法都要求统计组群服从一定的分布规律，例如正态分布等等。但降水量并不总是满足这个条件，雨量样本也常常不符合独立性的要求。近 20 年来随着计算机技术的发展，统计数值模拟试验有所发展。人们开给广泛采用复随机化试验和自然复随机化试验等技术来检验随机试验的统计效果和试验功效。这种数值试验不要求统计组群服从特定的分布型式，因而具有更强的适应性，即所谓“稳健性(Robustness)”。

随机化设计和总效果检验虽然是必要的科学手段，但它的基本弱点是需要大量样本，为此需进行长时期的播云试验，时间和经费的代价可能太大，故而难以引起人们的实际兴趣。人们十分希望寻找一种更有效的途径，缩短在外场试验中为获得可信结论所需的时间。人们更希望寻找更为灵敏的方法追寻播云效果，从初始冰晶的形成，通过各种质点增长的微物理过程，直至降水及地。这须求助于数值模拟试验和直接的物理测量。

三、数值模拟试验在试验设计和效果检验中的作用

云和降水的数值模拟试验可用来估计播云增雨和减雹的能力，选择最佳的播云方案和最好的播云剂以指导外场试验，从而增强人工影响天气的针对性，提高播云效果。美国佛罗里达州早期的单体积云动力播云试验中，用一维积云模式实时预报积云的可播度，在提高单体积云播云效果中起了重要的作用。近 20 年来，云模式研究有了长足的进展。各种水平的云模式都相继程度不同地运用到人工影响天气试验计划中。不过，多数模拟试验都还是探索性的。下面我们概括几点与人工影响天气的试验设计和效果检验有关的进展：

(1) 国内外云模式试验都表明，人工影响冷底积云会增加总降水量，而催化暖底积云则有可能减小雨量，虽然也有例外。这是因为两者的降水形成机制有明显的差异。在冷底大陆性积云中，降水形成速率基本上是云中液态含水量、云中上升气流速度、厚度和寿命以及活化冰核数浓度的函数。在暖于−10 ℃的温度条件下，降水效率是很低的。人工影响这类冷底大陆性积云是以色列试验获得成功的基本条件，大致可增水 15%～20%。相反，在暖底积云中，暖雨过程与冷雨过程是同时起作用的，这就大大增加了降水过程的复杂性，从而使播云效果更难预料。二维积云模式试验表明，对体积较大的暖底积云，播云后冰化的物理效

应是明显的，表现在冷雨过程发动较早，雷达反射率增强，其最大值位置抬高，云体发展加速。但是，降水并未增加，因为较早形成的雪花和霰粒多数输进云砧中而不能形成降水。另外，对云顶不超过 8 千米的小积云，由于云的动力过程较弱，播云可以产生较多的雨，增水大约 12%。这表明，同样是暖底积云，强度不同，播云的降水效果可以是不一样的，而且降水效果也并不总是与微物理效应同步。

（2）二维地形云模拟表明，风的垂直切变对播云效果影响很大，在泰国的地形和天气条件下，当风垂直切变的风向逆转高度较低且中低层水汽充裕时，播云促使云在风转向高度上有动力播云效应而能增加雨量 30%，不过仍比无切变时的自然雨量要小。

（3）二维湖效应风暴数值模式试验表明，雪晶聚并过程是一种重要的降水过程，观测也证实了这一点。因此人们有理由怀疑美国 NOAA 等机构的播云假设是否可行。这一假设是：播云产生大量冰晶以减缓淞附过程，雪晶能更深入内陆才降落，从而达到降雪再分布的目的。如果播云结果虽能减缓淞附过程却加速了聚并过程，就不能满足原来播云的基本假设。这种例子进一步阐明了，在对播云假设付诸实施前，需要了解降水形成的自然控制机制。

（4）对人工影响雹暴的二维时变实时数值模拟试验表明，在云底播云可使降雹时间提前，而降雹量可减少 60%。这是令人鼓舞的信息，因为至今还没有直接的观测资料能与防雹的播云效果联系得上的。

（5）一维时变模式试验表明，暖云催化的最佳方案是大剂量小颗粒云顶播撒，剂量大至每次 1 000 千克量级的 $CaCl_2$ 或 NaCl，颗粒直径小至 20 微米。这与我国 20 世纪 60 年代初提出的大颗粒大剂量暖云催化方案相比，除颗粒较小外大体相似。不过，一味地往大气中播撒数以吨计的化学物质，可能潜藏着一种不可逆转的破坏自然生态环境的隐患。早期用喷水的方法影响暖云，虽然效率可能较低，倒是一种干净的播云方法。

四、物理检验——直接的物理观测和遥感新技术的应用

当前，物理检验的一个重要发展趋势是对整个播云进程进行物理监测，包括调研播云潜力，寻找可播区，在降水增长的各个微物理链节上追寻播云效果，从播云后初始冰晶的产生，各类降水质点的增长过程直至降水及地。近 10 年来，新的探测仪器设备，尤其是遥感探测器方面的重大进展为物理检验开辟了广阔的发展领地。目前这些新的仪器设备还处在发展的初期，可用于：（1）测定播云潜力，确定最佳播云时机和部位；（2）为更有效地进行人工影响天气试验的设计、实施及效果的评价提供观测资料。

双通道微波辐射仪的发明提供了一种连续测量液态水和水汽的手段，它预示着以遥感为中心的物理研究高潮即将来临。美国 NOAA 等机构用多普勒雷达、偏振雷达及偏振激光雷达和微波辐射仪来综合研究犹他州南部山区的地形风暴，正在对风暴的环流和降水效率提供一种前所未有的但仍是不全面的知识。这样的研究表明，锋面地形性风暴中，可播区（“液态水机会”）是有的，但并不是在整个风暴生存期内都存在的，而要及时有效地将播云剂播撒到这种过冷液水区，需要仔细地筹划，采用全新的途径。我国北方云资源的调研在天空水资源和播云潜力方面做了大量工作，如北疆冬季层状云探测表明，锋后层积云是较合适的播云对象。

为了监测背风坡水分蒸发损失，需要一种测量云冰的遥感仪器，目前已有的 K 带多普勒雷达只能粗略地估计大冰质点通量。将一种无线电声探测系统（RASS）与一风廓线雷达相组合，可检测对流云和地形云系统的温度廓线。

在对流云内存在大水滴时，很难定量检测初始冰晶的产生，确定影响降水效率各种因子的总体效应也更为困难。但是，新近提出的多参数雷达测量技术提供了一种希望，在混合云中可以区别不同类型的大冰质点和确定雨滴谱分布。不过到目前为止，尚未获得复杂云或云系内将施放播云剂与最终的降水效果直接联结起来的物理证据。有一种叫作 TRACIR 的圆偏振雷达空气示踪技术可检测施放在云中的金属箔屑所反射回来的偏振后向散射信号，这种金属箔以小雨滴的速度下落，从而可充当一种示踪剂，用来模拟或追踪播云引起的初始回波以及最终降水的轨迹。这种示踪技术也可用来确定雹暴中馈云单体对主体云的雹胚源及其输送。

美国北达科他州的防雹试验利用 SF_6 气体作为示踪剂与播云剂一同在云底播撒，研究从云底到过冷却区的输送和扩散过程，检测人工冰核在云中的活化过程。这需借助雷达监测云中初始回波，其研究有助于确定降水机制及播云效果。对自然云的初始回波温度(FET)的频率分布研究表明，自然冰化过程要比人们预料的复杂得多，大约有 1/3 的初始回波出现在低于−5 ℃层的高度，而且 FET 与最大上升气流速度之间有较好的正相关。所以，在估计播云效果时应考虑自然冰化过程的复杂因素。利用雷达通过一种称为面积-时间积分技术(ATI)可以估计对流雨量。根据最大回波高度与风暴雨体积之间的关系可以由回波高度估计雨量。统计分析表明，北达科他州的防雹试验 11 年，播云作业的结果，因雹灾引起的对农作物的保险金少支付了 44%。不过，原先计划在 1989 年开展的 Hailswath Ⅱ 计划，却因经费原因而大大缩减了研究目标和规模。

2.4 人工降水及其效果检验研究的展望

叶家东

（南京大学气象系）

一、国外人工降水发展趋势

大量的人工降水外场试验表明，有的降水增加了，有的效果不明确，还有的得到了反效果——雨量减少，总之存在着一系列问题。首先由于多数人工降水试验的效果主要是统计检验提供的证据，其物理上的可靠性是需要进一步进行验证性分析的。另外，产生上述差异的物理原因和气象学判据是什么？在一般情况下并不清楚，它涉及许多基本的云和降水物理问题。

此外，人工降水还牵涉一些从目前来看可能还不突出，却具有根本意义的问题。譬如某一地区，某一时期播云结果增加了雨量，它对总的水分平衡影响如何？某地增雨会引起其他地方雨量亏损吗？在远离目标区的地方会有什么效应？在短时期内，人工影响天气能否加以控制？对自然生态平衡是否有什么不可逆转的影响等。

目前，对人工降水的一般提法是：在合适的条件下，把适量的、合适的催化剂在恰当的时间里播撒到云中恰当的部位，会收到预期的效果。其中任何一种限制性条件不满足，都有可能导致另外的甚至相反的结果。人工降水实际上是一种精细的物理实验，然而它却是在大气和云不能控制且变化多端的条件下实验的，因此不确定的因素很多。播云措施对云和降水过程的影响又是有限的，所以即使加上上述许多限制性条件，仍难以完全避免结果的不肯定性。况且,限制性条件多了,可播云的选择就更为困难,势必影响人工降水的经济效益。因为一个地区可播云太少,即使从科学上讲人工降水是成功的,其经济效益也可能不大,因而缺乏推广价值。前苏联乌克兰的夏季浓积云试验就遇到这种问题。实际上,什么是合适的催化条件,所播的云是否具备这种条件,多少数量的催化剂才恰当,什么时间播撒、播撒到什么部位才合适,以及如何才能确保上述措施及时地完成等,都还存在着一系列播云原理和工程技术方面的问题。

为了给人工降水的原理和效果建立一个比较健全的科学基础,目前国外研究的主攻方向是:

(1) 降水潜力和有利天气条件。

① 各类云的降水效率、降水资源和播云潜力,寻找"播云窗";

② 对有可能增加降水、减少降水和挪动降水区的特定天气条件进行鉴定、判别和即时预报。

(2) 静力催化的物理机制——云的微物理研究。

① 云中冰晶浓度与降水形态、降水效率及降水量的关系;

② 云中冰晶的起源,次级冰晶繁生过程及其与随后降水的关系;

③ 云中冰晶与冰晶、冰晶与水滴的相互作用;

④ 暖云中雨胚的形成,雨滴碰撞破碎和雨胚的次级繁生机制。

(3) 动力催化的理论基础——云的动力学研究。

① 云中混合和湍流对云动力过程和微物理过程的效应,夹卷对云生长和云生命史的影响;

② 云中动力过程和微物理过程的相互作用,云中冰化过程对云生长和云生命史及降水量的影响;

③ 动力催化对边界层入流的组织化以及对云系组织化的作用;

④ 降水过程对云系发展和维持的影响,对流风暴中湿下沉气流和云砧外流的作用。

(4) 中尺度降水系统的云动力学-云微物理学研究,人工降水的区外效应和催化剂在云下气层和云中扩散等问题的研究,藉以改善人工影响的物理基础。

(5) 研究方法上将外场试验研究与数值模拟和统计模拟研究结合起来,将理论研究、室内实验研究与外场综合考察研究结合起来,制订综合性的研究计划,集中力量解决云物理和中尺度云系的某些关键性问题。如美国 1981 年开展的 CCOPE 计划,着重研究对流风暴的降水效率及风暴与环境的相互作用。

(6) 新的人工降水原理和催化技术的探索性研究。一味地往大气中播撒化学物质,从长远的观点看并不是太理想,因为不干净。有必要探索影响云的热力、动力或微物理结构的干净的人工影响方法。尽管目前一些加热及改变辐射条件的措施收效甚微,但这方面的探索性研究仍在继续。

二、我国人工影响天气的发展趋势

当前我国人工影响天气的研究主要侧重于自然云结构的考察,对云的自然降水进行客观估计,藉以分析各地、各类云的降水资源和催化潜力,并进行选址和"催化窗"的调研分析。国家气象局气象科学研究院等单位近年来用飞机考察,初步查明了新疆天山地区冬季层状云和盆地烟雾天气，及关中地区可能成为我国人工降水试验的重要对象。另外，水库流域的降水试验在我国可能有发展前途，水库蓄水发电，经济效益大，且受天气条件限制少，可以在丰水季节利用较有利的云系进行催化作业。福建古田试验基本上属于这一类验证性试验，这个试验已进行了7年，在统计效果检验方面做了一些工作。目前当务之急是要在催化技术、物理效应的检验等方面开展研究，使统计结果在物理上得到解释。这个试验计划再进行3年，试图在南方雨季水库流域降水试验的效果分析方面得出一些结论和经验。最近吉林气象科学研究所计划在白山水库流域开展有物理设计和统计设计的人工降水试验。

就全国范围来说，目前急需筹建1～2个以人工降水验证性试验为主的综合性研究计划，判明常规的冷云催化技术在我国什么地区什么时期对何种云系能获得明显的效果和经济效益，为有关地区开展人工降水作业提供切实的指导。这样才能将我国这一领域有限的科技力量组织起来，发挥各自特长，保证协调行动。有了这样的基础，我们就有可能在20世纪90年代中期做出与PEP预期目的相当的成果，向我国及第三世界类似地区提供有效的技术经验。

在20世纪90年代，当现代飞机探测设备和多普勒雷达观测系统在我国装备起来并开展工作后，我国有可能组织一个中尺度对流云系降水的综合研究计划。我国中尺度对流性暴雨的研究如果不结合云物理研究，是难以取得实质性进展的。这类研究计划牵涉面要更广一些，云物理只是其中一个方面。这种计划的直接目的还是在暴雨的监测和短期预报方面。但是，如果这种计划开展了，并取得了进展，则到20世末，我国就有可能组织人工影响中尺度对流性暴雨的研究。这是很有中国特色的，并且是有可能做出超水平研究成果的领域。美国20世纪60年代就对越南雨季进行破坏性的人工降水作业，我们有责任在条件大体具备的时候，开展对我国南方雨季的对流性暴雨进行控制性的人工影响试验研究。

2.5　人工增雨试验中的反效果问题*

叶家东　范蓓芬　杜京朝

（南京大学大气科学系，南京 210093）（江苏省气象台，南京 210008）

提　要　该文对人工增雨试验中出现的反效果（减雨）现象及其发生的条件和原因做了概要的评述和分析。根据实例分析，指出人工增雨作业中不适当的催化对象、不适当的催化剂和催化剂量以及不适当的催化部位和催化时机都有可能导致无效或减雨的反效果。在开展人工增雨作业时，应力求在作业有关的各环节提高科学性、减少盲目性，以提高人工增雨实效。文章还简要探讨了人工削弱局地暴雨的可能性，指出这是一个值得认真探索的具有重大潜在社会和经济效益的研究领域。

关键词　人工增雨　人工减雨　静力催化　动力催化

引言

云物理学中通常将地面降水速率与通过云底输入云系的水汽通量之比值或地面降水量

* 本文得到国家自然科学基金项目49575243号资助。

1996-12-21收到，1997-06-29收到修改稿。

与云的凝结水量之比值定义为云或云系的降水效率。不同的云的降水效率差异很大,例如:孤立冰雹云为 3%,小雷暴云为 20%,飑线系统为 40%~50%,冷锋云系为 60%,地形性对流雨带为 25%,低空雨带为 30%,高空雨带则为 70%~80%。可见,地形性云、低空雨带、小雷暴云和孤立冰雹云的降水效率都比较低,其中大量云水不能及时转变成有效的降水,而在空中流失或蒸发掉。人工增雨的静力催化着眼于提高云的降水效率以增加降水,而动力催化则着眼于增加入云的水汽通量。

静力催化原理假设:① 冷云降水是由冰晶发动并通过"水-冰"转化的贝吉隆过程及随后的淞附或碰并过程完成的;② 有些云降水效率不高或根本无降水是因为自然云中缺乏足够的冰晶;③ 通过播撒干冰或碘化银等冷云催化剂的人工引晶技术可弥补冷云中自然冰晶不足,促使降水过程得以有效地发动而达到增雨目的。

动力催化原理假设:① 在孤立积云的过冷却部位过量播撒冷云催化剂(冰核浓度达 10^2~10^3 个/L),使过冷水云迅速冰晶化,并释放冻结潜热,增大云的浮力,空气加速上升,云体增高变大,生命期更持久,从而预期会降更多的雨;② 对积云群体实施过量播撒,促使云体合并,有可能大大增加区域雨量。观测表明,两块或两块以上积云合并,总降雨量可比孤立积云增大一个数量级以上;另外动力催化后的云在强化降雨的同时,也增强降水引起的下沉冷湿气流,它能加强阵风锋并促进邻近地区激发新的对流单体甚至有可能促使对流风暴系统激发成为自传播系统,从而延长对流风暴系统的生命期、增加总降水量。

不论是静力催化还是动力催化的原理假设,都附有一系列有利于播云的限制性条件,即"可播性"条件。静力催化的"可播性"条件包括:① 冷云部分有较丰富的过冷液态水含量;② 过冷云区自然冰晶较少,云顶温度一般介于 −10 ℃~−24 ℃区间(播云"温度窗");③ 云层较厚,而云底温度又不宜太高,云中大水滴较少,水滴碰并及冰晶繁生效率均较低。动力催化除了要求冷云有较丰富的过冷液态水含量外,还要求条件性不稳定层厚,云底暖,云上气层由弱稳定层覆盖,抑制自然对流发展等"可播性"条件,通常用一维积云数值模拟试验得出云的"可播度"表征。这些条件中的任何一条或数条不满足,就可能导致人工播云作业无效或产生反效果——减少雨量。

1 历史上人工增雨试验中的反效果现象

1.1 静力催化试验

一些建立在静力催化概念基础上的随机化播云试验表明在一定条件下播云会导致无效或减少降雨的反效果,虽然统计显著度不高,但已引起人们对云的"可播性",并引起警觉和重视。例如:

(1) 亚利桑那山区积云试验[2]:1955—1960 年和 1961—1964 年期间在美国亚利桑那州开展的成对试验日随机化地形性积云播云试验。在目标区上风方飞机在云底下 300~600 m 高度沿垂直于风向的 48.3 km 长的播云线播云。统计效果为雨量减少约 30%(播云日降水量/非播云日降水量=0.70),统计显著度不高(α=0.16~0.30)。人工播云未能使降雨量增加,产生这种结果的可能原因是该地区山地积云云底温度较高,云中诱发降水可能主要通过暖雨过程,无须冰晶过程启动。

(2) 白顶计划[3]:1960—1964 年在美国中西部密苏里州南部对夏季非地形性积云进行

的按试验日随机化设计的飞机播云试验。飞机在云底下播撒碘化银(播撒率 2 700 g/h)。分层统计结果表明,对云顶高度在 6 096～12 192 m(相应的云顶温度−10 ℃～−45 ℃)的云有正效果;对云顶高度低于 6 096 m 或高于 12 192 m 的云,播云使降水减少。播撒日总降水量减少约 20%,统计显著度 α=0.13。所提出的物理解释是密苏里夏季大部分积云可在 0 ℃层以下的暖区通过碰并形成雨滴,它们被上升气流夹带至−5 ℃～−10 ℃区就冻结成冰粒并有可能促进冰晶繁生作用。这些冰粒尺度达 10～300 μm,数浓度可达 10^4～10^5 个/m^3,这正是播云作业希望达到的浓度。所以,人工播撒额外的碘化银会造成过量播撒而导致降水减少。

(3) 克里麦克斯试验Ⅰ和Ⅱ[4-6]:1960—1965 年和 1965—1970 年冬季在美国科罗拉多州落矶山区对地形性云进行了两次按试验日随机化设计的地面播撒碘化银烟的播云试验(Climax Ⅰ和Ⅱ)。试验效果按 500 hPa 温度(作为试验日平均云顶温度指标)进行分层统计,当 500 hPa 温度为−20～−11 ℃,−26～−21 ℃和−39～−27 ℃时,相对增雨值分别为 75%(α=0.045),10%(α=0.35),−11%(α=0.15)。对上述播云效果的物理解释是云顶温度介于−10～−20 ℃时,自然冰质点浓度常常低于降水所需要的浓度,人工播云增雨可能性大(称之为可播云的“温度窗”);云顶温度<−24 ℃时,自然冰相效应有效,无人工增雨潜力;云顶温度<−28 ℃时,播云后一般降水会因过量播撒而减少;云顶温度≥−10 ℃的暖底云,需考虑冰晶繁生机制的作用,不过冬季这类云出现机会较少。近 20 年来 Hobbs 和 Rangno 等人对 Climax 试验的执行、效果分析及其解释提出一系列批评[7-11],指出:落矶山地区云中冰质点浓度与云顶温度或大气冰核浓度关系并不大,云顶温度≥−20 ℃时也常观测到高的冰质点浓度;以 500 hPa 温度作为落矶山上空冬季地形性云的云顶温度的指标根据不足,自然降雨量也并不随 500 hPa 温度增高至大于−20 ℃而减少;试验中可能并没有严格地执行原定的随机化设计方案;应用商业部公布的雨量资料对试验结果进行再分析表明,500 hPa 温度≥−20 ℃时并无显著增雨效果等。对这些批评中某些论点,试验者和旁观者尚有不同看法[12,13]。Climax 试验是 20 世纪 60 年代开展的早期试验,设计、执行和分析中可能有些粗糙,例如以 500 hPa 温度代表变化的该日平均云顶温度会出现较大误差,试验者也在考虑采用 3 h 时段作为试验单元进行再分析,试图弥补日雨量单元所引发的缺陷。

(4) 澳大利亚播云试验[14,15]:澳大利亚是严重干旱缺水的地区,早在二十世纪五六十年代就开展过一系列飞机人工播云增雨试验,采用 14 日周期或 1 日周期的随机化设计方案,飞机在积云底高度(层状云是−5～−10 ℃高度)播撒碘化银烟。试验效果显示出相当大的变异性,视目标云和大气条件而异,即使同一试验,效果也有逐年变差的趋势。例如:斯诺伊山区试验头两年增雨 26%,统计显著,随后就不显著;新英格兰地区试验头一年增雨 30%,统计显著,第二年以后不显著;瓦拉刚巴地区试验无效;南澳大利亚地区则有负效果;稍后在塔斯马尼亚进行的隔年随机试验则表明秋、冬季增雨 15%～20%,春、夏季无效。针对上述播云效果分析气象原因表明,大陆性气候下冰晶过程在降水形成中起重要作用,播云效果好,但增雨效果有逐年变差趋势,原因不明;海洋性气候下暖雨过程作用大,播云效果差或为负效果。

(5) 以色列人工增雨试验[16-21]:1961—1967 年和 1969—1975 年冬季(11 月～第二年 4 月)在以色列进行了两次按区域随机交叉设计方案进行的飞机人工增雨试验(以色列Ⅰ和Ⅱ)。两个试验除目标区设置有些改动外,其余均相似。以色列Ⅰ试验统计效果为北目标区和中目标区两区平均增雨 15%,统计显著度 α=0.009;以色列Ⅱ采用随机交叉设计双比分析全区的平均效果为−2.5%,统计不显著,而当利用一北控制区对北目标区单独分析时有

13%(α=0.028)的增雨效果，南区由于缺少相应的控制区，难以单独分析南目标区的播云效果。为此在1976—1991年对南区单独进行了一次区域控制随机试验(以色列Ⅲ)[19]，对682个试验日统计分析的估计效果是−4.5%(α=0.42)，表明无任何显著效果；同期在北区进行了业务播云试验[20]，有6%～11%的增雨效果，统计显著度α=0.02。近年来一系列分析研究表明[18-21]，不同的气象条件包括沙尘、气流轨迹和天气系统等对以色列的播云效果有重要影响，特别是沙尘的影响尤为明显。对以色列Ⅱ的北目标区，按尘霾指数进行分层统计，对202个"非尘日"增雨26%，而其余182个"尘日"则减雨2%，但不显著。对以色列Ⅲ和对应的北区业务作业，按沙尘分层统计的增雨效果列于表1。

表1 以色列Ⅲ随机试验(南区)和北区业务作业增雨效果 单位：%

	以色列Ⅲ(南区S)			业务作业(北区N)		
	S	SN	SS	N	NN	NS
尘　日	−6	−10	1	6	10	3
非尘日	5	13	1	12	16	10

注：SN、SS、NN、NS分别表示S和N区中的北、南分区。

由表可见，非尘日的增雨效果NN为16%，NS为10%，SN为13%，而SS为1%；尘日在北区(NN和NS)播云效果要比非尘日减少5%～7%，而在SN分区则要减少23%(由增雨13%变至减雨10%)，SN分区对沙尘指数最为敏感。SS分区，不论何种类型的日子都无播云效果。物理解释是以色列沙漠尘埃是一种很好的冰核源，高冰核浓度与高气溶胶浓度密切相关，以色列南部接近沙漠，偏南或西南气流影响时，从撒哈拉-阿拉伯沙漠吹来的沙尘提供丰富的冰核，能与播云剂"竞争"，从而对播云效果起到一种抑制或削弱作用。以色列的沙尘从南向北减少，这与播云效果从北向南变差相匹配。

自20世纪70年代以来，以色列试验被认为是既有统计显著的增雨效果又有较为合理的物理解释的人工增雨试验。但近年来这一观念受到严峻的挑战[22,23]。文献[22]对以色列Ⅰ和Ⅱ的统计结果进行再分析，发现相应的目标区统计显著的"增雨"效果实际上很可能只是降水的巨大时空分布不均匀性所造成的一种假象。云的物理结构分析也表明，试验者原先认为的以色列云有相当一部分是半永久性的过冷水云，在云顶温度介于−12～−21 ℃之间时冰质点浓度低于10个/L以及滴小、谱窄、高浓度和缺乏冰晶繁生条件等并不确切。有证据表明，以色列云在云顶温度接近−10 ℃时就会产生大云滴和毛毛雨滴，并有较高浓度的冰质点和降水。Rosenfeld和Farbstein(1992)[18]提出在以色列的云中有一半左右在较高的云顶温度下含有较多的自然冰质点是由于沙尘或霾引起的，此时人工播云可能真的减少了雨量，对这一点Rangno和Hobbs[22]并不持异议，虽然他们尚未认可这种可能的减雨效应是由于沙尘或霾引起的。

(6) 福建古田人工增雨试验[24,25]：1974—1986年4～6月份在福建古田水库流域，进行按区域控制随机设计方案的地面高炮或小火箭发射碘化银弹催化的人工增雨试验。其效果为平均增雨23.8%，统计显著度α=0.01；按回波顶温度分层统计：回波顶温度$T>-5$ ℃时效果不显著，T介于−5～−10 ℃时增雨43.6%，效果最佳，T介于−10～−15 ℃时增雨27.2%，效果次好。催化效果有随对比区自然雨量增大而变差的趋势，大雨时的增雨效果不显著。限于条件，古田试验没有取得空中云微物理结构资料佐证地面播云效果，但地面自然冰核观测表明在−12～−15 ℃时活化冰核浓度为 10^{-1}～10^{0} 个/L，浓度不足，影响降水效率；催化后地面雨水中 Ag^{+} 含量峰值出现时间与地面雨强峰值时间大体同步，这在一定程度上支持了播云的统计效果。

1.2 动力催化试验

建立在动力催化概念基础上的一些随机化播云试验也表明在一定条件下播云可能无效或产生减雨的反效果。例如：

(1) 南达科他州来皮德(Rapid)计划[26,27]：1966—1968 年在美国南达科他州进行了一次动力催化播云试验，按区域随机交叉试验设计方案，飞机在云底上升气流区播撒碘化银烟(播撒率 300 g/h)。试验结果对整个大区域(13 730 km^2)而言，目标区雨量减小。按天气条件分类统计发现：阵雨日人工播云可增雨 1 倍多，统计显著度达 0.05。雷雨日当高空西北风时，总雨量减少；而当高空西南风时，催化目标区雨量减少，但下风方 32 km 处增雨。对大区域总雨量减少的物理解释是目标区对流因催化而加强，四周有附加的补偿下沉运动，抑制新的对流发展，从而引起区域降水减少，这种效应在扰动天气下(雷雨日)尤为明显，Dermis 和 Koscielski[28]称之为“动力污染”。

(2) 南达科他州捕云机计划[28]：1969—1970 年对美国南达科他州(北部大平原)积云实施浮动目标区随机试验，碘化银播撒率与来皮德试验相仿(10^2 g/h)，在动力催化中属于轻量播撒方案，目的是使人工冰晶化引起降水效率降低的效应有所缓解。雷达观测表明单体积云试验增雨达 40%，统计上接近显著。协变量分析表明，降雨量的增加与云顶高度增加关系不大，而与降水效率增加有关，原因可能是碘化银主要催化阵雨云附近的新云塔，它们的空间尺度和总降雨量均有所增加，但该区域云的最大高度不一定增加。

(3) 佛罗里达单体积云动力催化试验[29]：1969—1970 年在美国佛罗里达州南部地区开展对单体积云块随机化试验，在目标云过冷却区投射碘化银焰弹。由雷达观测估算催化效果，发现单体积云降雨量增加 2 倍。按天气条件分层统计表明：晴好天气催化云的增雨量大，统计显著度高；阴雨天或天气扰动日增雨量小或为负值。这种天气催化云水平伸展较小，生命期也较短。对扰动日催化效果差的解释是天气扰动日自然对流旺盛，动力催化后较早形成的雪花和霰粒被人工催化动力效应增强了的上升气流输送进入云砧中流失掉，而不能形成有效的降水，至少不能形成局地强降水。

(4) 佛罗里达 FACE 计划[30]：1970—1971 年在美国佛罗里达州进行一次旨在促进积云群合并以增加区域雨量的佛罗里达区域积云试验(FACE)。按随机化试验日设计，飞机投射碘化银焰弹对目标云进行动力催化，结果表明动力催化在有利条件下能够促使邻近积云合并，按浮动目标区分析平均增雨量达 3 倍，但若按固定目标区的降雨量估算，试验基本无效甚至有负效果。分析提出两种可能的动力污染：① 对强大的对流云动力催化，上升气流增强结果会对云外围地区产生额外的补偿下沉效应，从而导致区域雨量减少；② 动力催化一块对流云使其暴发性增长，其云砧可覆盖数千平方千米，因遮挡太阳辐射而使其下风方的积云活动消失，抑制降水。

迄今为止，虽然动力催化试验对孤立的单体积云的生长有一定的动力效应并在一定条件下能使单体积云增加雨量，但对区域降雨量的效果却并不肯定。相反，有迹象表明区域雨量会有所减少。近年来在美国得克萨斯州西部进行的试验表明对复合型多单体系统的雨量和云生长有正的动力催化效果[31]，但在伊利诺斯 PACE 计划的探索性外场动力催化试验中[32]，却发现用碘化银进行动力催化的暖底浓积云比用作对比试验的云生长得略低些，水平尺度、雷达反射率和雨水通量都稍微小些，显示出一定的负效应迹象，不过，采用不同的方法评估结果稍有差异[33]，但总的效果并不显著，这从一个侧面说明，动力催化引起的各种动力效应有时是互相抵触的，从而使最终的地面降雨效果变得捉摸不定。再加上降水本身的巨大时空变率，使得催化效果难以鉴别。

2 人工增雨试验中减雨的可能机理

（1）不适当的播撒对象：云顶过冷（<－24 ℃），自然冰晶充裕；播云有利云顶温度窗为－10～－24 ℃；云底过暖，暖层较厚，大滴多，冰晶自然繁生机制活跃；暖雨过程在雨滴形成中起主导作用。

（2）不适当的播撒部位和不适当的催化剂：暖云部位播撒大量小凝结核，会提高云的胶性稳定度，抑制暖雨过程。

（3）不适当的播撒时机：夏季积云发展早期播云可以提早降水，但时机未成熟的降水会破坏上升气流，云的自然增长会受到抑制或过早发生逆转，缩短云的生命期，有时反而会减少雨量。

（4）不适当的催化剂量（过量播撒）：在对流云（特别是雷暴云或冰雹云）的过冷却区过量播撒成冰核，云迅速冰晶化，释放冻结潜热加强对流的同时也将延缓冰质点淞附增长过程，大量小冰晶被夹带至云砧区在高空向四周流散而削弱局地降水。

（5）动力污染效应：① 由于动力催化，对流加强，相应地在云四周产生范围较广的补偿下沉运动，抑制邻近地区对流活动，从而导致区域雨量减少；② 动力催化一块条件有利的积云使其暴发性增长，其云砧可覆盖数千平方千米，因遮挡太阳辐射而使其下方的层结稳定化导致积云活动消失，抑制区域降水。

从上述分析可以看出，人工增雨作业在催化对象的选择、播撒方案的制定（包括播撒时机和播撒部位的确定）以及催化剂量的控制等方面都要力求符合播云原理假设所限定的可播性条件，方能使作业达到预期的增雨目的，不问天气和云层结构条件就随意作业，难免会出现事与愿违的无效或减雨的反效果。

众多的迹象表明，人工增雨试验中的反效果现象常常出现在云的自然降水条件比较有利的场合，从这里人们得到一个契机，能否利用这种反效应，在多雨季节或地区进行人工减雨特别是人工削弱或抑制局地暴雨以减轻洪涝灾害的试验呢？这是一个新的思路，由于这种试验同样具有重大的潜在社会和经济效益，特别在我国东部暴雨频繁发生的经济发达地区更是如此，所以是一个值得认真探索的研究领域。

3 人工增雨试验中减雨的可能途径

当前人工增雨的基本思想是在合适的天气条件下，用恰当的和适量的催化剂，在最佳的时机播撒到云的最佳部位以获得预期效果。在这种状态下，降水潜力得以最大限度发挥出来，降水效率最高。而人工减雨的指导思想是设法破坏云的良好自然降水状态或过程，使其失调，从而降低降水效率；或使云提早降水，未老先衰；或延缓降水过程使降水在空间再分布。抗旱面临的最大困难是常缺乏合适的播云对象，减雨或降水空间再分布有试验对象，主要问题是能否影响和如何影响。可能的途径包括[34]：

（1）截流效应。在认清暴雨天气前期环流形势基础上，在中尺度水汽和能量供应主通道上提前激发对流过程，使较多的水汽和能量消耗在对流降水活动的过程中，削弱低层向暴雨区输送的水汽、动量和感热通量。这是一种釜底抽薪的策略，方法类似于削弱台风的试验。

（2）竞争场效应。在预期的暴雨区提前进行动力催化，人为促使动力污染效应发挥作用，并对大量较小的积云催化，促使它们"早熟"以削弱其"后劲"，不让水汽和能量在低层充分积聚以达到减雨的目的。

（3）降水空间再分布效应。在沿海暴雨频发地区有利地形条件下对暴雨期（如前汛期和梅雨期）海洋性对流云团的暖云区过量播撒小 CCN，使云大陆化，增加云的胶性稳定度，延缓暖雨形成过程；在冷云区过量播撒人工冰核，增加冻结潜热并延缓冷雨形成过程，使较多云冰质点输入高空云砧区而散布到下游海洋上，从而达到削弱沿海地区局地暴雨强度的目的。现阶段主要从数值模拟试验角度探讨这种方案的可行性。对暖雨过程进行的二维数值模拟试验初步结果表明[35]，改变 CCN 浓度能影响暖雨过程和最终的降雨量及其分布。CCN 从 300 cm^{-3}增至 1 000 cm^{-3}或 2 000 cm^{-3}，最大累积雨量可减少 20%～30%，这项试验尚在继续之中。

参考文献

[1] 游来光.我国人工影响天气的历史现状和面临的某些科学问题.《人工影响天气》(一)，国际气象局科教司，中国气象学会大气物理委员会，1988.23～25.

[2] Battan L J. Relationship between cloud base and initial radar echo. *J. Appl. Meteor.*, 1963, 2: 333～336.

[3] Braham R R Jr and Flueck J A . Some Results of the Whitetop experiment, Preprints of Second National Conference on Weather Modification, 1970. 6～9, 176～179.

[4] Grant L O, Chappell C F and Mielke P W Jr. The Climax experiment for seeding cold orographic clouds. Proceedings of International Conference on Weather Modification. Canberra, Australia, 1971. 78～84.

[5] Mielke P W, Grant L O and Chappell C F. An independent replication of the Climax wintertime orographic cloud seeding experiments. *J. Appl. Meteor.*, 1971, 10(6): 1198～1212.

[6] Grant L O, and Kahan A M. Weather modification for augmenting orographic precipitation. Weather and Climate Modification. Hess, W N ed., Wiley-Interscience, 1974. 282～317.

[7] Hobbs P V and Rangno A L. Comments on the Climax randomized cloud seeding experiments. *J. Appl. Meteor.*, 1979, 18: 1233～1237.

[8] Hobbs P V. Lessons to be learned from the reanalysis of several cloud seeding exptriments. Preprints of Intern. Cloud Phycics Conf., Clermont-Ferrand, France, Amer. Meteor. Soc., 1980. 88～91.

[9] Rangno A L and Hobbs P V. A re-evaluation of the Climax cloud seeding experiments using NOAA published data. *J. Climate Appl. Meteor.*, 1987, **26**: 757～762.

[10] Rangno A L and Hobbs P V. Further analyses of the Climax cloud-seeding experiments. *J. Appl. Meteor.*, 1993, **32**: 1837～1847.

[11] Rangno A L and Hobbs P V. Reply. *J. Appl. Meteor.*, 1995, **34**: 1233～1238.

[12] Mielke P W Jr. Comments on the Climax I and II experiments including replies to Rangno and Hobbs.

J. Appl. Meteor., 1995, **34**: 1228～1232.

[13] Gabriel K R. Climax again? *J. Appl. Meteor.*, 1995, **34**: 1225～1227.

[14] Smith E J. Cloud seeding experiments in Australia. Proceedings of the 5th Berkeley Symposium on Mathematical Statistics and Probability, 1967,**5**: 161.

[15] Smith E J, Adderley E E, Veitch L and Turton E. A cloud-seeding experiment in Tasmania. Proceedings of the International Conference on Weather Modification, Canberra, Australia, 1971. 91.

[16] Gabriel K R. The Israeli Rainmaking experiment 1961～1967. final statistical tables and evaluation. Tech, Rep., Jerusalem, Hebrew University, 1970. 47.

[17] Gagin A and Neumann J. The second Istraeli randomized cloud seeding experiment; Evaluation of the results. *J. Appl. Meteor.*, 1981, **20**: 1301～1331.

[18] Rosenfeld D and Farbstein H. Possible influence of desert dust on seedability of clouds in Israel. *J. Appl. Meteor.*, 1992, **31**: 722～731.

[19] Nirel R and Rosenfeld D. The third Israeli rain enhancement experiment-an intermediate analysis. Proceedings, Sixth WMO Sci. Conf. Wea. Mod., Paestum, Italy, 1994, Ⅱ: 569～572.

[20] Nirel R and Rosenfeld D. Estimation of the effect of operational seeding on rain amounts in Israel. Proceedings, Sixth WMO Scientific Conference on Weather Modification, Paestum, Italy, 1994, Ⅱ: 573～576.

[21] Levi V, Rosenfeld D and Herut B. Relationship between the occurrence of dust, ice nuclei concentrations and rain chemical composition in Israel. Sixth WMO Scientific Conference on Weather Modification, Paestum, Italy, 1994, Ⅱ: 565～568.

[22] Rangno A L and Hobbs P V. A new look at the Israeli cloud seeding experiments. *J. Appl. Meteor.*, 1995, **34**: 1169～1193.

[23] Dennis A S. Changing perceptions of the Israeli weather modification program. J. Wea. Mod., 1996, **28**: 83～85.

[24] Yeh Jiadong, Cheng Kerning and Zeng Guangping. Randomized cloud seeding at Gutian, Fujian, China. WMA, J. Wea. Modif., 1982, **14**: 53～60.

[25] Zeng Guangping, Xiao Feng, Fang Shizeng and Yeh Jiadong. Rainfall results of the Gutian area cloud seeding experiment 1975～1984, China, Reprints, Fourth WMO Scientific Conference on Weather Modification. WMP Report 1985, **1**(2): 513～518.

[26] Dennis A S and Koscielski A. Results of a randomized cloud seeding experiment in South Dakota. *J. Appl. Meteor.*, 1996, **8**: 556～565.

[27] Dennis A S and Schock M R. Evidence of dynamic effects in cloud seeding experiments in South Dakota. *J. Appl. Meteor.*, 1971, **10**: 1180～1184.

[28] Dennis A S and Koscielski A. Height and temperature of first echoes in unseeded and seeded convective clouds in South Dakota. *J. Appl. Meteor.*, 1972, **11**: 994～1000.

[29] Woodley W L. Precipitation results from a pyrotechnic cumulus seeding experiment. *J. Appl. Meteor.*, 1970, **9**: 242～257.

[30] Simpson J and Woodley W L. Seeding cumulus in Florida; new 1970 results. Science, 1971, (172): 117～126.

[31] Rosenfeld D and Woodley W L. Effects of cloud seeding in west Texas. *J. Appl. Meteor.*, 1989, **28**: 1050～1080.

[32] Czys R R, Changnon S A, Westcott N E, et al. Responses of warm-based, mid-western cumulus congestus to dynamic seeding trials. *J. Appl. Meteor.*, 1995, **34**: 1194～1214.

[33] Changnon S A, Gabriel K R, Westcott N E and Czys R R. Exploratory analysis of seeding effects on rainfall; Illinois 1989. *J. Appl. Meteor.*, 1995, **34**: 1215～1224.

[34] 叶家东.关于人工抑制暴雨问题.南京气象学院学报,1993, 16(3): 373～378.

[35] 王春明,叶家东,魏绍远.气溶胶对暖雨过程影响的数值模拟试验.《第12次全国云、降水物理和人工影响天气 科学讨论会》,中国气象学会学术会议文集 CPWM-012号,1996. 79.

2.6 关于人工抑制暴雨问题*

叶家东

(南京大学大气科学系)

摘 要 在阐释人工防雹、人工削台和人工抑制局地暴雨试验的原理假设,及人工增雨试验面临的一些基本问题的基础上,探索人工抑制暴雨的可能性及其发展前景。

关键词 人工抑制暴雨 人工防雹 人工削台

人工影响天气科学试验,一开始就是作为一种抗击气象灾害的科学手段发展起来的,从方法论讲大体上可分为:人工激活自然过程(如人工增雨);人工抑制自然过程,如人工防雹和人工削台等。

1 人工防雹

人工引晶防雹的主要科学依据是:假定在过冷云区引进大量冰晶核争食过冷水滴,使冰雹淞附增长速率减慢,抑制大冰雹的形成。20世纪60年代苏联开展的防雹试验,根据云中上升气流速度随高度呈抛物线分布的现象,提出冰雹生长的累积带理论[1],雹胚的进一步增长主要发生在累积带内。从而提出了在云的过冷却区过量引进人造冰晶,使之能与自然冰粒"争食"过冷水滴,抑制冰雹增长的方法。为使"争食"更有效,提出在云的下部暖区播撒直径为5～10 μm盐粒的辅助方法,使液滴迅速凝结增长,在升到最大上升气流区之前就能增长到克服上升气流而下降的尺度,从而在云的中下部出现一批早熟的降水滴,它们下降并冲刷暖区云水,使进入高层过冷却区的液态水量明显减少,削弱冰雹的淞附增长。这样防雹作业在雹灾面积的减小方面取得了明显的效果。但欧美等地采用类似的方法做了一些较为严谨的验证性试验,结果并不理想,未取得预期的防雹效果。

建立在"累积带"概念基础上的防雹原理,是在某种封闭的、定常的云结构模型基础上发展起来的。当用于复杂的对流风暴系统,考虑云与云、云与环境大气的相互作用以及云发展的时变特征,情况就复杂得多,需要综合考虑冰雹的动力结构及其演变规律、过冷水的含量和部位、雹胚源及其原始冰晶的形成,以及雹块本身的增长轨迹等一系列因素。

关于雹胚的起源,近年来对"馈云单体"的作用重新引起重视。主体雹暴外围的侧向云塔,有可能是雹胚的重要源区,而从主体雹暴中形成的原始霰粒在下降中由于融化和溅散所产生的次生水滴落入馈云单体,能产生再循环增长,可以成为重要的雹胚源[2]。因此,防雹作业时引晶播撒应在主体雹暴外围迎风侧的"馈云单体"顶部进行。为了验证这种构想,

* 收稿日期:1992-07-22

1989年夏美国 Hailswath Ⅱ外场冰雹研究计划[3]，对雹暴的三维流场、冰晶起源、雹胚源及“馈云单体”的小尺度回波区结构、降水和冰雹的形成和发展、雹暴和馈云单体起电等进行综合考察研究，并追踪播云剂和冰雹的轨迹，藉以评价雹胚形成和冰雹增长理论，为防雹提供可信的物理基础。

2 人工抑制台风(飓风)风力的试验[4]

台风(飓风)的主要致灾因素是大风、风暴潮和暴雨造成的洪水泛滥。如果能将台风风速削弱30%，就能使风灾减小一半，因为风力是随风速平方而变化的。据文献[5]的研究，台风的形成和维持，依靠海面向风暴内的空气输送感热和潜热。风暴的能量通过水汽对流过程而释放。这种过程主要位于风暴的眼壁区和主要的中尺度对流雨带里。于是提出了人工影响台风的可能途径：① 在占风暴很小一部分区域的活跃对流区内，人工影响潜热释放率，使加热作用重新分配到较大的范围，导致削弱风暴的目的。这是一种通过影响对流尺度(云体)与天气尺度(台风)两种运动之间相互作用的办法以影响风暴强度，本质上是类似“冷敷”散热的一种方法。② 用人工影响气压梯度的办法改变风速。从流体静压力学角度看，海平面的气压梯度与大气综合形成的温度场有关，故可试用调整温度场的办法影响台风风力，或用减小台风中心与外围温度差的办法影响台风风力。在诸多改变温度场的方法中，比较现实的是人工影响台风眼壁对流区和对流雨带中的云物理过程，把对流释放的热量分配给整个风暴。水汽主要是在对流云里凝结并释放潜热，若能改变对流云的潜热释放形式，就能影响台风中的热量分布。至今设计的一切人工影响台风的试验，都是把台风能量散布到更大区域。

人工影响台风的播云原理假设：

(1) 只在台风眼壁区播云，因这里靠近低层最大气压梯度区。如果云墙里的过冷水通过人工冰晶核化作用而冻结，释放的潜热会使温度升高，从而降低最大气压梯度区附近的气压。如果台风中心气压没有随之减小，就使气压梯度减小，导致风速减小[6]。这种假设的问题在于，不知道眼壁区云内的实际过冷水含量究竟有多少；播云使它们冻结的效率如何；还因风暴的横向环流向眼壁推动，只对眼壁区播撒，就有可能加速这种环流，而使角动量和水汽迅速流向眼壁区，结果加速了系统的发展。

(2) 从眼壁区向外播云，或都在眼壁区外围播云以激发那里的对流和上升气流，让人工激发的对流与眼壁区竞争低层入流空气。若有相当一部分入流空气在播云半径处转向上升，则对原来眼壁的角动量和水汽供应都会减小。结果眼壁对流减弱，台风的最大风速也随之减弱。

雷达观测表明，离台风中心55 km以内，50%以上的积云顶高在610～914 m范围内。由于多数台风的中层几乎没有大尺度的入流或出流，所以在云顶不高的对流云中向上输送的质量必在云体附近下沉，减少了向上的净质量输送。如果人工播云能使这些云产生动力播云效应，有利垂直增长并进入高层出流层，则不仅将释放更多的潜热，而且向上输送的质量将与大尺度出流连接，并从风暴核心区附近抽走，所以这一假设提出了由入流层输送的空气和水汽在中途与其热量(能量)一同抽走的机制。数据试验表明，采用这种方式播云的结果可使台风眼壁向外移动约10 km，使海面最大风速减小3～4 m/s(6%～8%)。

在20世纪60年代进行的台风眼壁附近播云试验STORMFURY计划中，对Debbie飓风的播云结果最令人鼓舞[7]。1969年8月18日，在8小时内对该飓风进行5次播云作业。

飞机在 3.6 km 高度观测的最大风速，由首次作业前的 50 m/s 减小为第 5 次作业后 5 小时的35 m/s[减小了 30%，图 1(a)]；第二天没有作业，最大风速又恢复到最初水平；第三天作业又由首次作业前的 51 m/s 减小到第 5 次作业后 6 小时的 43 m/s[减小了约 15%，图 1(b)]。近年来的飞机观测研究发现，熟台风的对流区 0 ℃层以上液态含水量相当少(<0.59 m^{-3})，而冰质点却相对较多，因而对上述播云原理的假设提出了疑问[8]。

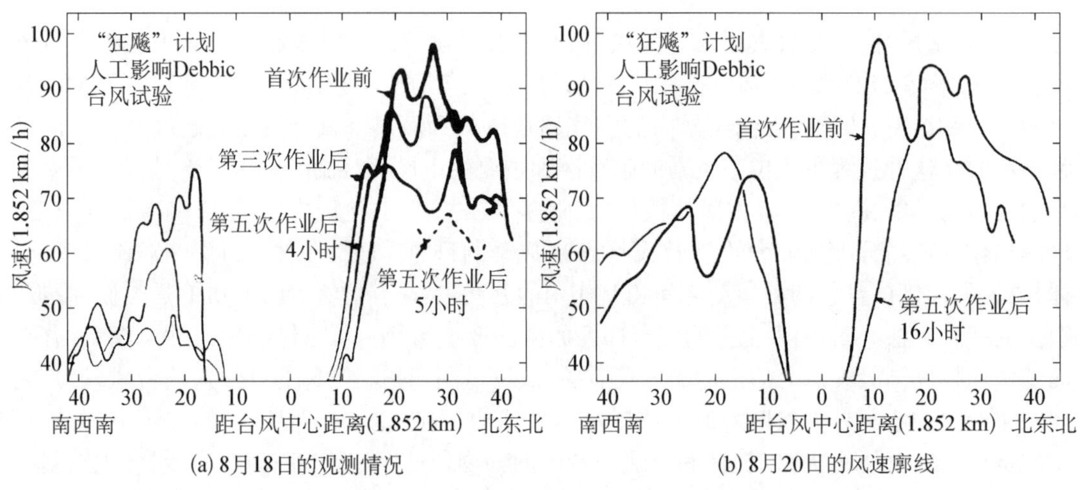

注：曲线已作平滑，风速由飞机从南西南至北东北方向横穿风暴来回飞行时测定[7]

图 1　1969 年 Debbie 飓风 3.6 km 高度风速随时间变化

3　人工抑制局地暴雷——湖效应风暴试验[9]

美国大湖区冬季强降雪常常集中在大湖(如伊利湖和安大略湖)的背风岸 50～100 km 范围内，一次风暴过程局地积雪深度可达 25～75 cm，严重影响沿岸的生产和交通。20 世纪60 年代 9 次雪暴造成纽约州高速公路停止运行累计达 162.3 h(平均每次 18 h，最长达36 h)，造成背风岸强降雪的主要原因是风暴云系在湖泊上空受暖水面影响发展成旺盛的对流云，其霰粒快速而集中降落在湖岸附近。当云系深入内陆后便失去水源和热源而迅速减弱[10]。为此，提出促使降水再分布的播云假设，即在湖泊上空过量播撒(播撒率达 10^6 个/m^3)，促使云在短期内完全冰晶化[11]，使云中水汽大减，抑制冰质点的淞附增长速率，而云中众多的小冰晶通过各自的凝华过程缓缓增长成雪晶，落速较慢从而能离岸入陆，降落在较宽广的区域。使降水在空间再分布，减缓沿湖岸狭窄地带的雪暴灾害。这种假设的主要问题是对降雪物理过程中的聚并效应认识不足。观测表明，自然云和播撒云中都有雪花形成，但播撒云中的雪花密度比自然云的大，所以降落较快。过量播云既能抑制冰质点的淞附增长，也增加了冰晶之间的聚并过程。若产生冰晶聚合体——雪花，上述的假设基础就成问题。因此，播云效应中必须考虑冰晶之间聚并所致的后果，必须搞清聚并过程影响雪花降落速度的问题。只有对播云前后的降雪率及其形态特征进行实际的定量测量，才能真正检验原来的播云假设是否正确。这还有待更深入的试验和研究。

4　人工抑制暴雨的可能性

现在还谈不上设计具体的人工抑制暴雨试验方案。这里只从人工增雨试验中得到的启示，探讨一下人工抑制暴雨的可能性。

4.1 关于反效果问题

在人工增雨试验中，经常出现反效果问题。如采用静力播云原理的白顶试验[12]，统计结果是减少雨量。在动力催化试验中又有动力污染问题[13]。原因如下：① 由于催化，对流加强，相应地产生范围较广的补偿下沉运动，从而有可能导致区域雨量减少；② 动力催化一块云使其爆发性增长，其云砧可覆盖数千平方千米，因遮挡太阳辐射而使其下方的积云活动消失，抑制降水。

此外，云的数值模式试验发现，人工影响冷底大陆性积云会增加总降水量，影响暖底积云则有可能减少雨量。因为较早形成的雪花和霰粒多数被人工催化增强了的上升气流输送进云砧中，而不能形成有效的降水，至少不能形成局地强降水[14]。我们的中-β尺度深对流云系的二维数值模拟试验也表明，改变微物理过程中的参数不仅能显著地影响降水元的形成过程，也能影响云系的动力过程和生命史[15]。可见，人工播云减少雨量的可能性是存在的。

当前人工增雨的基本思想是在合适的天气条件下，用恰当的和适量的催化剂，在最佳的时机播撒到云的最佳部位以获预期效果。人工减雨的指导思想是设法破坏云的良好自然降水状态或过程，使其失调从而降低降水效率。或使云未老先衰，或延缓降水过程使降水在空间再分布。从逻辑上讲，在其他条件相当的情况下，人为建立一种最佳状态会比破坏一种良好的自然状态更困难些。抗旱增雨试验面临的最大困难就是常缺乏合适的播云对象，减雨或降水空间再分布有试验对象，但主要问题是能否影响和如何影响。由于这种试验具有重大的社会和经济效益，是一个很值得认真探索的研究领域。

4.2 暴雨的时空变异性

图 2 是一次突发性暴雨的日雨量图。由图可见，日雨量大于 50 mm 的暴雨区的空间尺度界于 50～200 km 的范围，从雨强变化直方图可见，局地强降水的时间尺度为 1～3 h，具有明显的对流性降水特征[16]。图 3 是 1991 年梅雨期江淮特大暴雨期间(7 月份)的几次暴雨过程日雨量图。尽管整个梅雨锋雨带东西向绵延 1 000～2 000 km，但日雨量大于 100 mm 的特大暴雨区的空间尺度是 50～200 km。局地降水的时间一次过程可持续 3～6 h 或更长(常包含数次强降水峰值，图略)，也具有明显的对流性降水特征。

上述例子说明，不少大暴雨系统的时空尺度还是限于中-β尺度的范畴，大尺度的水汽和能量供应对某一次过程来讲是有一定限度的。对沿海经济发达却又地势低洼的地区，如果通过人工影响使降水形成过程延缓 1～2 h，让较多的雨量至海上，也属造福非浅。

4.3 人工抑制暴雨的可能途径

(1) 截流效应。在认清暴雨天气前期环流形势的基础上，在中尺度水汽和能量供应主通道上提前激发途中的对流过程，使较多的水汽和能量消耗在途中的对流活动中，被人工激活的对流截获变成降水或向高层输送到云砧区，消弱云砧覆盖区的对流活动，同时减少低层向暴雨区输送的水汽、动量和感热通量。这是一种“釜底抽薪”的策略，方法类似于削台试验。

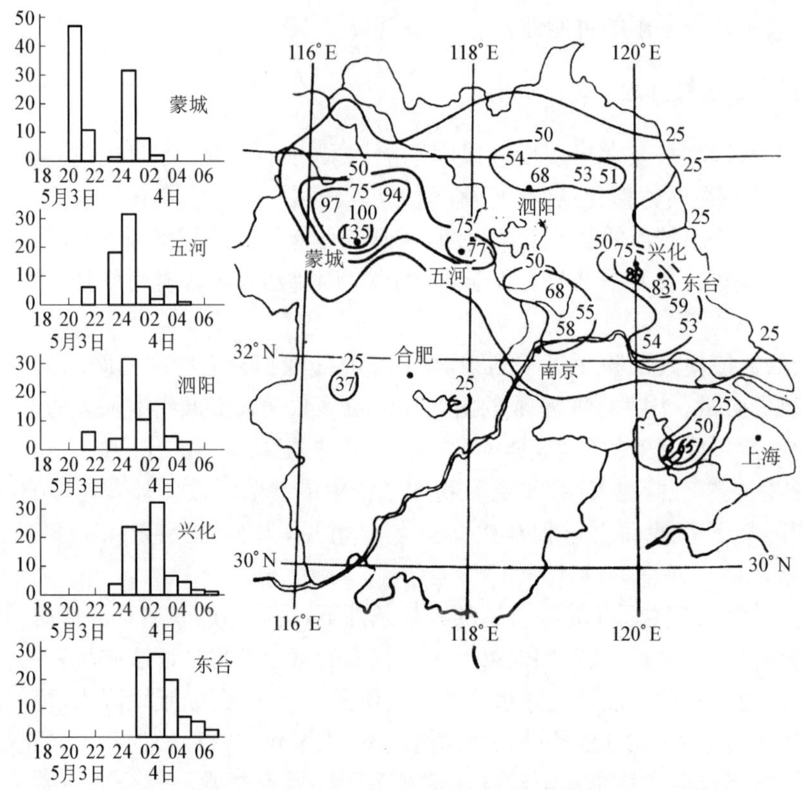

注:左侧是暴雨中心区各站的每小时雨强时变特征(mm/h)

图 2　1988 年 5 月 3 日 05 时—4 月 05 时日雨量(mm)

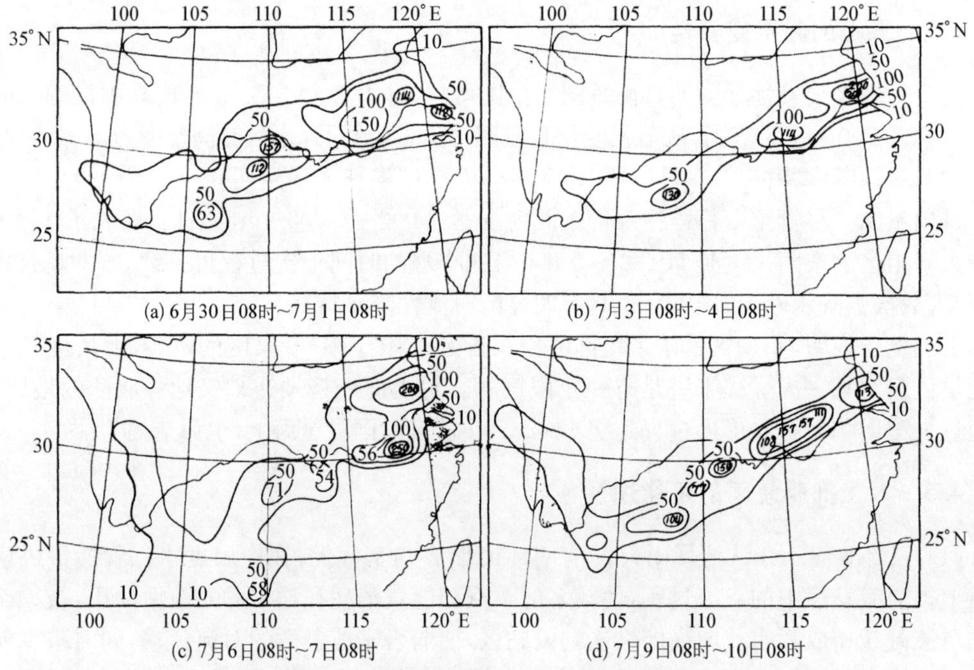

图 3　1991 年梅雨期江淮流域特大暴雨期间(7 月上旬)几次暴雨过程的日雨量图(mm)

(2) 竞争场效应。在预期的暴雨区提前进行动力播云，人为促使动力污染效应发挥作用，并对大量较小的积云进行催化，促使它们“早熟”以削弱其“后劲”，不让水汽和能量在低层充分积聚以致达到不可收拾的程度，还可提出改变降水空间再分布的设想等。但现行的各种人工抑制风暴的原理假设都存在一定的局限性，设想的观测依据不足导致局限性就更大：如“动力污染”效应，在白天可能是对的，夜晚情况就会相反。云顶的长波辐射冷却和云底的地面长波辐射增暖往往加强夜间对流系统的发展。另外，暴雨云系的过冷水储量如何、对暖底云的降水物理过程的认识、冷雨过程在暴雨形成过程中究竟起什么作用也都不十分清楚。

为减少试验的盲目性和可能承受的风险，当前迫切需要对各类型暴雨过程的环流形势、水汽和能量的输送过程、中尺度与对流尺度动力过程及其相互作用，以及云物理过程的实际效应进行深入的研究，提高不同类型的暴雨预报和监测水平，加深对暴雨云系结构及其演变规律的认识。特别需要加强从大气物理角度研究暴雨问题，开展暴雨云系的外场云物理综合考察研究。在现阶段需加强对对流性暴雨云系进行人工影响的数值模拟研究，探索抑制局地突发性暴雨的可能途径。在对暴雨的成因、结构特征及暴雨系统中不同过程的相互作用等方面取得实质性进展的基础上，人工抑制暴雨试验才有可能做出超水平的研究成果。

参考文献

[1] Hess WN，王昂生等译.人工影响天气和气候.北京：科学出版社，1985：284 - 298

[2] Foote GB, Miller LJ, Tuttle JD. *Cloud Physics and weather Modification Research Programme*. Beijing, China: WMO/TD, 1989; (269): 313 - 316

[3] “Hailswath Ⅱ” Preliminary Experiment Design. Bismarck, North Dakota, 1989: 105PP

[4] Hess WN，王昂生等译.人工影响天气和气候.北京：科学出版社，1985：345 - 363, 364 - 386

[5] Charney JG, Eliassen A. *J Atmos Sci*, 1964; 21(Ⅰ): 68 - 74

[6] Simpson RH, Malkus JS. *Sci Am*, 1964; 211(6): 27 - 37

[7] Gentry RC. *Science*, 1970; 168: 473 - 475

[8] Willoughby HE. Jorgensen DD, Black RA, Rosenthal. *Bull Amer Met Soc*, 1985; 66(5) :505 - 514

[9] Hess WN，王昂生等译.人工影响天气和气候.北京：科学出版社，1985:225 - 252

[10] Lavoie RL. *J Atmos Sci*, 1972; 29: 1025 - 1040

[11] Jiusto JE. *J Res Atmos*, 1971; 5:69 - 85

[12] Braham RR Jr, Battan LJ, Byers HR. *Met Monogr*, 1957; 2(11) : 47 - 85

[13] Dennis AS, Koscielski A. *J Appl Met*, 1969; 8: 556 - 565

[14] Orville HD, Kopp FJ, Farley RD, Hoffman RB. *Cloud Physics and weather Modification Research Programme*. Beijin, China: WMO/TD, 1989; (269): 203 - 207

[15] Fan Bei Fen, Yeh Jia Dong, Cotton WR, Tripoli GJ. *Adv Atmos Sci*, 1990; 7(2): 154 - 170

[16] Yeh Jia Dong, Fan Bei Fen, Tang Xunchang, Du Jingchao, Song Hang. *Proceedings*, 11*th International Conference on clouds and Precipitation*, Montreal, Canada: ICCP/IAMAP, 1992; 2: 711 - 714

2.7 人类影响气溶胶和云的气候效应*

叶家东　范蓓芬
(南京大学大气科学系)

提要　本文对人类排放气溶胶的直接气候效应和通过改变云的微结构影响云的反射率而引起的间接气候效应的证据、机理和研究现状做一概括的论述，指出 CCN-云反照率假设(Twomey 效应)是一种重要的影响气候的潜在因素，它的总体作用是降低地面温度，一定程度上可与温室气体增暖效应相抵，所以是一个值得深入研究的领域。

一、引言

人类排放的温室气体如 CO_2 等正在引起全球地面平均温度上升，这一假设近年来正受到社会公众和政府决策阶层以及科学界的普遍关注。鉴于可用的气候资料相对于自然气候变化时间尺度而言历史太短，要用观测分析的方法将温室增暖效应从气候自然变率中鉴别出来是相当困难的，加上全球气候时空尺度太大，进行物理试验来验证温室增暖效应更不可能。所以人们通常都求助于各种气候模式(如 EBM，RCM，ZAM 和 GCM)进行数值试验以"预报"温室气体浓度增加或其他人类影响(如气溶胶排放物，下垫面属性改变等)的气候效应。采用简单的 EBM 和 RCM 估算的 CO_2 浓度倍增引起的全球地面平均温度增幅在 0.48～4.2 ℃之间(Schlesinger and Mitchell，1987)[1]。这类估算值还比较粗糙，主要是未能恰当地处理由 CO_2 浓度增加引起的辐射响应，这些响应有正反馈也有负反馈，包括水汽、云、地面反照率、海洋环流、植被呼吸和光合作用以及有关的生物化学过程等方面的反馈效应。即使是最复杂的三维 GCM，要定量地、恰当地处理这些反馈效应，也还有很多方面需要不断改善。所以现有的温室增暖预测，还只是一种粗糙的一级近似，况且人类影响气候不只限于温室气体增暖效应。本文主要讨论人类影响气候的另一个侧面——人类排放气溶胶并影响云的气候效应。在一定的前提条件下，这类效应与温室气体增暖效应作用相反。所以在气候变化预测研究中也是重要的环节。

为了认识人类影响气候的潜力和基本机理，不妨回顾一下地球系统的辐射能量平衡关系。表 1 是按 MacCracken(1985)[2] 概括的地球系统能量平衡示意图整理的能量平衡表。表中 100 单位就全球平均而言相当于 340 W/m^2。由表可见，大气系统收、支各为 160 单位，而地面系统收、支各为 146 单位，是各自平衡的。只要这种全球能量收支的分配比例保持不变，地球气候就将保持平稳常定，一旦其中一种或多种分量发生系统性改变，如太阳发光度和地球轨道参数、云量和云微结构、气溶胶浓度、温室气体浓度、下垫面属性等发生系统性变化时，就会导致全球能量收支不平衡，从而导致气候发生变化。例如，若将大气中所有的温室气体除去，那么放射到空间的长波辐射将大为增强，结果会使现有的地面平均温度降低 30 ℃。这就是将水汽、CO_2 和 O_3 等吸收地球长波辐射所引起的保暖称为"温室效应"的缘故，虽然"温室"这个术语可能有些用词不当，Bahren(1989)曾指出温室中起主要作用的实际

* 本文获得自然科学基金(49575243 号)资助。

上是温室里湍流散热受到阻滞所致。

表 1 地球系统能量平衡表

太阳短波辐射		地球系统长波辐射	
入射太阳辐射能	100	地面放射长波辐射能至大气	115
水汽、O_3、气溶胶吸收 云吸收 地面吸收	19 4 46	水汽、CO_2、O_3 等温室气体吸收 温室气体放射长波辐射回地面	106 100
云反射回空间 地面反射回空间 空气分子散射回空间	17 6 8	地面长波辐射透过大气层散逸至空间 温室气体向空间放射长波辐射能 云向空间放射长波辐射能	9 40 20
		地面感热通量输送至大气 地面潜热通量输送至大气	7 24

气溶胶粒子的辐射特性依赖于它们的化学成分、外形和尺度谱，如果气溶胶粒子是吸湿性的，则还依赖于大气湿度。相对湿度大于 70%时吸湿性粒子吸取水汽变成霾粒子，其尺度和折射指数都会改变，是相当复杂的，所以通常只能粗糙地估计其辐射特性。尺度为数微米的气溶胶粒子在大气中只能存留数日，而 0.1 μm 和更小的粒子则可在大气中存留数周之久，它们可以远离源区影响区域尺度（10^3 km 量级）或半球尺度的气溶胶浓度分布。气溶胶粒子可以直接影响大气辐射，也可以通过影响云微结构间接影响大气辐射。云的辐射特性也相当复杂，云滴、雨滴和冰晶对辐射的吸收和散射性能取决于入射辐射波长、粒子尺度和相态。小云滴和小冰晶是太阳辐射的强散射体，而雨滴则是强吸收体，其散射贡献很小，云对长波辐射则是强吸收体。冰晶对太阳辐射的反射性能还依赖于形状和在空间的取向。云的另一个重要辐射参量是含水量或液态水路径。所以，如果人类排放的气溶胶和气体能影响云的微结构和含水量，就可以通过云间接影响区域或全球气候。下面我们对气溶胶的直接气候效应和间接气候效应的证据、机理和研究现状做一概要的归纳和论述。

二、气溶胶影响局地和区域天气气候的证据

有许多人类活动会成为气溶胶和气体排放源，包括都市和工业燃烧产物、汽车废气排放物以及植物燃烧产物等，它们可以影响大气辐射、云的微结构和降水过程以及云雾的发生等其他天气现象。

(1) 在一些甘蔗种植地区常在收获前燃烧甘蔗地以除去叶子等废渣，结果下风方凝结核(CCN)浓度大量增加，积云中云滴浓度也有所增加。Warner 和 Twomey(1967)[3]对下风方降水分析表明，与上风方相比，下风方降水减少 25%。他们假设燃烧产生的 CCN 及由其形成的云滴尺度较小，碰并效率低，因而会延缓暖雨过程而导致雨量减少。不过这种现象尚未在其他地区得到验证。

(2) Hobbs 等(1970)[4]则发现纸浆厂和造纸厂及其他一些工厂排放的气溶胶使 CCN 谱变宽，雨量分析表明这类工厂投产后的下风方雨量要比投产前大 50%，观测发现这类工厂下风方大核($d>0.2$ μm)和巨核($d>2$ μm)浓度增加，而小核浓度并无显著改变。所形成的

云滴谱较宽，提高了云滴的碰并效率，因而使降水增加(Hindman，1977)[5]。另外一个可能的影响因素是工厂排放的热量和水汽也有利于下风方雨量增加。观测发现造纸厂下风方云中含水量比周围的云大 30%～50%，不过这个结果统计上并不显著。

(3) 船舶航迹云。从卫星图像上特别是 3.7 μm 波段的红外卫星图像上有时可看到清晰的船舶航迹，像是一条亮度增强的云线(图略)。形成这种船舶航迹的流行假设是由于船上排放出大量 CCN，使云滴浓度增加，因而比周围的云反射更多的太阳辐射而显得更为明亮些。穿云观测表明，船舶航迹云比周围云的云滴浓度大些，尺度小些，液态含水量也高些(Radke et al.，1989)[6]。云滴浓度大、尺度小是与上述假设相一致的。至于较高的含水量，Albrecht(1989)[7]假设云滴浓度大，尺度小，碰并效率低，减缓了毛毛雨的形成速率，结果导致航迹云中有较高的液态含水量。Porch 等(1990)[8]则假设是由于船舶排放的热量和水汽使航迹云中扰动更趋活跃，从而形成更厚、更湿因而更明亮的云。这从航迹云两侧有明显的晴空带也可得到佐证。

(4) 飞机凝结尾迹。在繁忙的高空喷气式飞机航线上经常出现飞机凝结尾迹云，有时多条尾迹云会合并成连片的卷云覆盖在广阔的天空中。其成因主要是飞机排放的水汽。一架波音 727 巡航速度飞行时每小时可排放 3.7 吨水汽。在高空很低的气温下，饱和水汽压很低，加入少量水汽就能导致冰面过饱和而形成凝结尾迹云。这种高空冰晶云能改变辐射收支从而影响地面温度。Kuhn(1970)[9]在飞机观测时发现一 500 m 厚的凝结尾迹云可削弱 15%的入射太阳辐射，而使向下的长波辐射增加 21%，不过仍不足以弥补削弱太阳辐射的损失。结果白天地面温度会降低，而夜间会引起净增暖，减少了地面温度日较差，使局地气候趋向温和。Changnon(1981)[10]对美国中西部近 80 年的云况、日照和地面温度做了一次气候分析，发现自 20 世纪 60 年代以来美国中西部特别是繁忙的航空通道地区云量增加、日照减少以及地面极值温度趋于缓和。至于地面平均温度，凝结尾迹云的影响是净增暖还是净冷却，则取决于云厚和持续性，以及纬度和季节。

其他如汽车含铅汽油燃烧排放物中含有丰富的冰核(IN)，也可影响云雾的微结构。都市和工业区空气污染可直接削弱太阳辐射达 15%，加上某些工厂如发电厂等同时排放大量水汽，在某些高纬度(阿拉斯加和西伯利亚)工业城市冬季可形成持续的冰雾或层云，影响交通和局地气候。至于中低纬度的都市地区，城市热岛效应是影响局地气候的主要因素。

三、气溶胶的直接气候效应

如上所述，在污染的都市和工业区气溶胶可削弱直接太阳辐射约 15%。在偏远的站点也普遍观测到硫酸盐类的气溶胶，北半球的浓度远比南半球高。大气中硫酸盐气溶胶主要由 SO_2 通过云化学或光化学过程形成。虽然 SO_2 可以由自然过程如动植物腐烂、荒原野火及火山喷发等过程产生，但人类活动显然是最重要的因素。Schwartz(1988)[11]指出北半球是主要的 SO_2 排放源，大气中 SO_2 浓度增加主要是近百年之内的事(Cullis and Hirschler，1980)[12]，另外在北极极冰中也发现近百年内硫酸盐浓度有较大增加(Barrie et al.，1985[13]；Mayewski et al.，1986)[14]，而在南极洲却没有类似的增加(Herron，1982)[15]。空气电导率的测量记录也提供了类似的间接证据(Cobb and Wells，1970)[16]。虽然要对硫酸盐气溶胶的浓度及其时、空变率进行精确的测定或估计相当困难，但其浓度正在增加这一点是基本可信的，特别在工业区附近。美国东部自 1948 年以来大气能见度急剧降低，主要是由硫酸盐粒子造成的(Malm,1989)[17]。

一般来讲,气溶胶在吸收太阳辐射从而削弱地面辐射增热的同时,也使大气层自身增暖。但气溶胶对地表面的影响则取决于下垫面的反照率,洋面上反照率低。气溶胶可使地表反照率增加,而在反照率较高的沙漠上,气溶胶可削弱沙漠的反照率。全球大部分地区气溶胶对反照率的影响要超过其吸收效应,但在高纬度冰雪覆盖面上气溶胶的吸收效应是主要的(Charlson et al., 1992)[18]。至于大气对于气溶胶加热的响应则主要取决于气溶胶层的高度和厚度以及该层的稳定度。如在潮湿的中性稳定层里气溶胶加热引起的层结稳定化可压制深对流活动和降水,这种效应通过水分循环具有重要的气候内涵。考虑到云量和地面反照率的地理分布,Charlson 等(1992)估计大气中硫酸盐气溶胶引起的辐射冷却效应在数值上可与 CO_2 增加 25%所引起的增热效应相抵销。当然,这些估计是相当粗糙的。迄今为止,人类活动排放的气溶胶的总体直接气候效应尚不甚明了。

四、气溶胶的间接气候效应——Twomey 效应

地球表面将近 60%为各种类型的云覆盖。云对太阳辐射是一种强反射体,全球平均有 17%的入射太阳辐射被云反射掉(表 1)。气溶胶对云和气候的一种更为重要的影响可能是它们充当云的凝结核(CCN)源而改变云滴浓度。Twomey(1974)[19]最早提出云滴浓度因气溶胶污染而增加,结果使云的反射率增加。他随后指出(Twomey,1977)[20]这种效应对光学厚度薄的云(卷云、层积云)影响最为重要,他认为气溶胶对光学厚度大的云的效应主要是增加云的吸收率。不过,对某些积云来说也不尽然。在人工影响积云的动力催化试验中,人们发现过量播撒使旺盛的对流云中产生大量小冰晶,在激发过冷却部位快速获得额外冻结潜热而加强上升气流的同时,也延缓了冷雨形成过程,大量小冰晶输向高空进入云砧区,云砧扩大覆盖更大的天空而产生"动力污染"效应。这将导致云团的反照率增加。这种机制在人类污染影响云的微结构过程中也会起作用。人类污染增加 CCN 和 IN 使云的微结构稳定化,大量云冰粒子输向云砧也会增大云团的反照率。所以,就其全球气候效应来说,气溶胶对云的吸收效应是小的,而使云反射率增强的效应则起着主要作用(Twomey et al., 1984)[21]。他们论证气溶胶使云反射率增强的效应在量值上可与温室气体增加引起的增暖效应相当,近来 Kaufman 等(1991)[22]指出,虽然燃烧矿物燃料排放的 CO_2 分子要比 SO_2 分子多 120 倍,但每个 SO_2 分子通过所形成的 CCN 增加云的反射率使大气冷却的作用却要比每个 CO_2 分子使大气增暖的作用大 50~1 100 倍。Twomey 还认为,如果海洋性云的 CCN 浓度增加至大陆性云中的 CCN 浓度值,那么就会有 10%以上额外的太阳辐射能被诸如海洋性层积云和卷云这类较薄的云层反射到太空中。全球云的反射率增加 10%所造成的后果要比全球云量增加 10%的效应更大些,这是因为云量增加时虽能减少入射的太阳辐射,但同时也减少了地球放射的红外辐射损失,云量增加的冷却效应和增暖效应是同时起作用的。相反,CCN 增加引起云反射率增加的结果,并不对红外辐射有太大的影响。

全世界的海洋面上有 34%以上的面积为浅薄的层积云覆盖着。这类云的 CCN 浓度一般只有 50~100 cm^{-3} 数量级,而大陆性云的 CCN 浓度达 500~1 000 cm^{-3},严重污染地区更可达每立方厘米数千之多。据说这类浅薄的海洋性层积云对 Twomey 效应最为敏感。因此,任何一种能使海洋上空 CCN 浓度增加的倾向都具有使大气反照率增加从而使全球冷却的潜力。

DMS-反照率假设:一般认为海洋上 CCN 的自然源主要是二甲基硫化物(DMS),是由

浮游生物排泄出来释放到大气中，通过光化学氧化作用形成硫酸盐气溶胶（Bigg et al.，1984[23]；Charlson et al.，1987[24]）。DMS 的主要源区是热带和赤道海洋，它的产生速率与温度有关。因此，如果温室气体增暖效应使海洋面温度增加，就可能使 CCN 及云滴浓度增加，从而增加云的反照率，这就产生了一种与温室气体增暖效应相反的冷却反馈效应。目前这种 DMS -反照率假设尚处于一种信仰性阶段，其中的每一个环节都有待科学研究加以证实（或否定）。不过，它确实促进了关于 CCN -反照率假设的兴趣和争论，使这一论题引向深入。

Schwartz(1988)[11]指出有证据表明过去一个世纪 SO_2 和硫酸盐气溶胶的排放已有很大增加。这种排放主要在北半球。因此，应能鉴别出南北半球云反照率有差异，并且应能检测出由于 CCN 及云滴浓度增加所引起的南北半球平均温度之间有差异。但目前北半球的云既没有显示出更亮些，也没有明显得冷些。不过 Slingo(1988)[25]指出，云对总反照率的贡献不只受云亮度制约，也受云量制约。有证据表明，自 1930 年以来北半球海洋上云量已增加了 3%（Warner et al.，1988）[26]。云量和液态含水量方面的差异有可能掩盖云滴浓度差异所引起的效应。所有这一切表明全球或半球平均反照率或平均温度对人类活动排放的 SO_2 的响应有关的控制因素相当复杂，不能只从 CCN -反照率假设的单一角度去剖释。重要的是需进行系统的 CCN、云滴浓度和云反照率更直接的观测。前面所述的船舶航迹云被认为是连接海洋上空气溶胶浓度改变与云反照率变化的气候效应之间的一种直接观测证据，不过这里仍然还有一些复杂的因素需要通过更多的观测研究和模拟研究加以深入的论证。

不管怎么说，CCN -云反照率假设在物理上是连贯相容的，是一种影响气候的潜在因素，其影响作用与温室气体增暖假设相反。Charlson 等(1992)粗略估算表明全球海洋性层云和层积云平均云滴浓度增加 15%所造成的辐射冷却效应可与当前估计的温室气体增暖效应在量值上相当、符号相反。要使这些估计建立在合理的科学基础之上，需回答：(1) 全球特别是海洋地区 CCN 浓度是否在增加？(2) 浅薄海洋性层积云的云滴浓度是否随时间在变化？(3) 何类云对 CCN 浓度改变响应最敏感？(4) 这类敏感云的覆盖区域是否大到足以引起全球反照率和全球平均温度发生显著改变？对模拟研究来说，这是一个重大的挑战，因为模式应具有真实地模拟毛毛雨形成过程及其对 CCN 浓度和巨型气溶胶粒子响应的敏感性的能力。还需要复杂的三维辐射传输计算，包括辐射与云微物理结构及云的外形结构的相互作用。在人们确信云辐射模式能真实地模拟由于人类活动排放的气溶胶会引起云反射率改变和地面温度变化以前，云辐射模式必须要能够证明它们确实能真实地模拟出观测到的云反照率。

五、后记

本文主要取材于 Cotton 和 Pielke 前不久寄送的新著《Human Impacts on Weather and Climate》一书[27]，作为我们拓宽视野的尝试并期望引起人们对这一论题的关注。文中以偏概全和浅薄之见在所难免，由本文作者负责并恳请行家教正。

1997 年 10 月么枕生先生送来一篇发表在奥地利《理论和应用气候学》杂志上的论文单行本，标题是“气候数值分类的一种聚类分析新方法”。[28]么先生时年 88 岁高龄，犹笔耕不辍，读着却令人神往，感佩之情油然而生，遂作一小诗“赠么枕生先生”曰：

勤培沃土育桃李，
苍松表山蓄正气。
淡功薄利志高远，
八八老骥犹奋蹄。

谨以此小诗作为向么枕生先生九十华诞的献辞，衷心祝愿我们的老师健康、乐观、长寿，奋力攻克百岁大关！为南京大学大气科学系更高地树起一面坦诚正直、勤奋进取的旗帜，作后辈表率，倡气象新风。

参考文献

[1] Schlesinger, M. E., and J. F. B. Mitchell, Climate Model Simulations of the Equilibrium Climatic Response to Increased Carbon Dioxide. *Rev. Geophys.*, **25**, 760～798(1987).

[2] MacCracken, M. C., Carbon Dioxide and Climate Change: Background and Overview, The Potential Climate Effects of Increasing Carbon Dioxide, U. S. Dept, of Energy, Washington, DC, 381 (1985).

[3] Warner. J. and S. Twomey, The Production of Cloud Nuclei by Cane Fires and the Effects on Cloud Droplet Concentration. *J. Atmos. Sci.*, **24**, 704～706(1967).

[4] Hobbs, P. V., L. F. Radke and S. E. Shumway, Cloud Condensation Nuclei from Industrial Sources and Their Apparent Influence on Precipitation in Washington State. *J. Atmos. Sci.*, **27**, 81～89 (1970).

[5] Hindman, E. E., P. V. Hobbs and L. F. Radke, Cloud Condensation Nuclei from a Paper MilL Part I: Measured Effects on Clouds. *J.Appl. Meteeor.*, **16**, 745～752(1977).

[6] Radke, L. F., J. A. Koakely, Jr. and M. D. King, Direct and Remote Sensing Observations of the Effects of Ships on Cloud, *Science*, **246**, 1146～1148(1989).

[7] Albrecht, B. A., Aerosols, Cloud Microphysics, and Fractional Cloudiness, *Science*, **245**, 1227—1230 (1989).

[8] Porch, W.M., C.-Y. J. Kao, and R. G. Kelley, Jr. Ship Trails and Ship Induced Cloud Dynamics, *Atmos. Environ.*, **24**A, 1051～1059(1990).

[9] Kuhn, P. M., Airborne Observations of Contrail Effects on the Thermal Radiation Budget, *J. Atmos. Sci.*, **27**, 937～942(1970).

[10] Changnon, S. A., Jr., Midwestern Cloud, Sunshine and Temperature Trends since 1901: Possible Evidence of Jet Contrail Effects, *J. Appl. Meteer.*, **20**, 496～508(1981).

[11] Schwartz, S. E., Are Global Cloud Albedo and Climate Controlled by Marine Phytoplankton? *Nature*, **336**, 441～445(1988).

[12] Cullis, C. F., and M. M. Hirschler, Atmospheric Sulphur: Natural and Man-made Sources, *Atmos. Environ.*, **14**, 1263～1278(1980).

[13] Barrie, L. A., D. Fisher and R. M. Koemer, Twentieth Century Trends in Arctic Air Pollution Revealed by Conductivity and Acidity Observations in Snow and Ice in the Canadian High Arctic, *Atmos. Environ.*, **19**, 2055～2063(1985).

[14] Mayewski, P. A., et al., Sulfate and Nitrate Concentrations from a South Greenland Ice Core, *Science*, **232**, 975～977(1986).

[15] Herron, M. M., Impurity Sources of F^-, Cl^-, NO_3^- and SO_4^{2-} in Greenland and Antarctic Precipitation, *J. Geophys. Res.*, **87**, 3052～3060(1982).

[16] Cobb, W. E. and H. J. Wells, The Electrical Conductivity of Oceanic Air and its Correlation to

Global Atmospheric Pollution, *J. Atmos. Sci.*, **27**, 814～819(1970).

[17] Malm, W. C., Atmospheric Haze: Its Sources and Effects on Visibility in Rural Areas of the Continental United States. Environ, *Monitoring Assessment*, **12**, 203～225(1989).

[18] Charlson, R. J., et al., Climate Forcing by Anthropogenic Aerosols, *Science*, **255**, 423～430(1992).

[19] Twomey, S., Pollution and the Planetary Albedo, *Atmos. Environ.*, **8**, 1251～1256(1974).

[20] Twomey, S., The Influence of Pollution on the Shortwave Albedo of Clouds, *J. Atmos. Sci.*, **34**, 1149～1152(1977).

[21] Twomey, S., M. Piepgrass and T. L. Walfe, An Assessment of the Impact of Pollution on Global Albedo, *Tellus*, **36**B, 356～366(1984).

[22] Kaufman, Y. J., R. S. Fraser and R. L. Mahoney, Fossil Fuel and Biomass Burning Effect on Climate, *J. Climate*, **4**, 578～588(1984).

[23] Bigg, E. K., et al., Origin of Aitken Particles in Remote Regions of the Southern Hemisphere, *J. Atmos.Chem.*, **1**, 203～214(1984).

[24] Charlson, R. J., et al., Oceanic Phytoplankton, Atmospheric Sulfur, Cloud Albedo and Climate, *Nature*, **326**, 655～661(1987).

[25] Slingo, T., Can Plankton Control Climate? *Nature*, **336**, 421(1988).

[26] Warner, S. G., et al., Global Distribution of Total Cloud Cover and Cloud Type Amounts over the Ocean. Prepared for the U. S. Dept, of Energy, Washington, DC and the NCAR, Boulder, CO. (1988).

[27] Cotton, W. R. and R. A. Pielke, Human Impacts on Weather and Climate, Gambridge University Press, 288(1995).

[29] Yao, C. S., A New Method of Cluster Analysis for Numerical Classification of climate, *Theor. Appl. Climatol*, **57**, 111～118(1997).

2.8 高山仰止　景行行止

——怀念徐尔灏教授

叶家东

徐尔灏先生离开我们 29 年了！但先生的音容笑貌，包括 40 年前讲台上授课时的神韵风采，依然深深地嵌印在我的脑海里。他讲授动力气象课时安祥沉着、胸有成竹、出口成章。我们只需逐句记下，无须多大修饰便是一份很好的讲义。这与徐先生渊博的学识、扎实的数理功底、明快的逻辑推理以及丰厚的语言文学底蕴是分不开的。同学们都乐于听先生讲课，除了授业解惑，还感到是一种讲授艺术的享受。

徐先生平日不苟言笑，人们戏言这可能与他留学英国，多少染上点英国绅士风度有关。但他内心是一个热情正直、刚正不阿的人，对青年教师更是关爱备至、热情引导。记得我的第一篇论文“人工凝结核的实验研究”(气象学报，1962)初稿完成后交徐先生审阅，他逐字逐句细加修改，第三天就交给我。我一看文稿上的细杠、红圈一大串，心里凉了半截：糟了！把我一些自感得意的段落也删掉了。正茫然发愣，徐先生却面带难得的笑容指着文稿说出一句我始料不及的话：这才是真正的科学研究！随即指出最好再做些实验，让论据更充分些、全面些。我心头一热，一扫前期由艰难实验带来的疲惫心理，精神为之一振，兴致勃勃地耐心补做了两个星期实验。近日看报纸上有人说写文章的要领是:删! 感触颇深。38 年前徐先生就指点过,只是实践中未必都能得此要领。使我更加刻骨铭心的是,如果不是徐先生顶风干预,我可能早就舍弃所从事的专业,随风飘至他乡。对于徐先生鼎力扶持之恩,我终生难忘。

徐尔灏教授是著名的气象学家、气象教育家,也是早期我国人工降水科学事业的主要开拓者之一。1958 年他任南京大学气象系系主任,创建了我国第一个大气物理专业并兼任教研室主任。在云雾物理专业方向采取了两个重要举措:教学上举办云雾物理讲习班,聘请苏联专家基留辛讲授云雾降水物理学,学员中有来自中科院、北大、中央气象局等全国主要气象系统,成为当时我国云雾降水物理教学中心;科研上与安徽省科委等协作,于 1959 年开展皖南地区地面暖云人工降水试验研究。他按照历史区域回归方案,亲自设计并组织领导了外场试验和观测研究。这是我国第一个有科学设计的人工降水试验,试验成果受到国家科委重视和嘉奖。云雾组那台价值 2 万美元的德国进口高速照相机,就是由国家科委奖励给南京大学气象系的。鉴于历史回归方案,受历史资料的制约,局限性较大,客观性也较差,徐先生于 1962 年综合国际人工影响天气科学的发展动态提出随机试验的设想“论人工降水随机试验的效果检查问题”(南京大学学报(气象学),1962),这类方案在当时国际人工影响天气领域也才兴起,是国际先进的方案。但在国内实现起来,客观条件的限制和思想认识上的阻力仍是很大。随后徐先生的主要精力转向大气湍流和边界层气象等领域。不久爆发“文化大革命”,一切正常的教学科研秩序均遭破坏,徐先生也蒙受迫害,不幸于 1970 年 7 月 12 日英年早逝,留下许多冤屈与遗憾。1972 年我开始从事人工降水试验设计和效果检验研究,除了任务需要,也怀有一种承继徐先生开拓的思路,发展这一领域研究的心理。1974—1986 年期间我设计了区域回归随机试验方案并与福建省气象局合作开展了我国第一个人工降水随机化试验——古田试验,成为我国有代表性的先进人工降水科学试验,也是我国首次被世界气象组织认可备案的人工降水科学试验计划,并得到 WMO 主管副秘书长和国际同行专家的广泛好评,在国内也获得过全国科学大会奖、江苏省重要科技成果三等奖、中国气象局科技进步二等奖等。我总是想以此来告慰徐先生在天之灵。在我和范蓓芬合著的《人工影响天气的统计数学方法》(科学出版社,1982)一书的前言中指出:“我国研究人工影响天气的效果统计检验工作首先是由徐尔灏教授开创的,谨以此书表示纪念”。

徐先生担任系主任时曾提出过一个“气象物理化”的学科建设方针,我体会这个方针的要点是基于气象科学原本脱胎于地理科学,大体上是从描述性科学发展起来的。早期气象科学重心是“知其然”,为要“知其所以然”,必须从气象演变规律的动力、物理机理上进行剖释。大气科学既是地球科学的一分支,也是应用物理学的一个重要分支,特别是作为大气科学基础学科之一的大气物理学更是如此。气象科学向理科方向发展,这在当时的学科建设方针中,无疑是高屋建瓴之举。当时在专业设置上增设大气物理学和高层大气物理学等新兴专业,教学的课程设置上加强数理基础教学。但是,这并不意味着轻视气象观测和实验性课程的教学。在徐先生任系主任期间,我系的实验室建设得到长足发展。就大气物理专业而论,云雾物理实验室、大气湍流和边界层物理实验室相继建立,雷达气象学也方兴未艾,同时还开展了诸如皖南人工降水试验、沪宁地区飞机人工降水试验和云结构考察研究、东北地区边界层大气扩散试验以及土山试验等一系列现代化的综合考察试验研究,形成了一套教学与科研相结合、以科研任务带动学科建设的生气勃勃的发展体制。徐先生亲自组织领导了其中大部分外场试验。这在当时风行一张纸、一支笔做理论研究的氛围中,特别是对徐先生这样一位承担系行政领导的成名科学家来说，实在难能可贵，是值得后来人继承发扬的一种优良治学精神。从徐先生去世后近三十年大气科学发展动态看来，当时的"气象物理化"方针是符合当时大气科学发展趋势的。发展大气物理学的教学与科研是这一方针的重要内容之一。在大气科学中大气物理学是涉及面最广的基础学科，它综合应用流体力学、热力学、统计物理学、电磁辐射学及量子力学等基本物理学原理，结合地球大气的特点，研究大气结构、大气运动特征、云雨相变、湍流交换、辐射传输和大气热力学过程，以及大气中发生的光、电、声等物理现象，也包括人工影响天气、大气遥感物理学和中层大气物理学等现代大气物理学分支。大气物理过程是气候演变、大气环境和各种灾害性天气发展变化的重要机理，在现代大气科学研究中日益显示出它的重要性。20 世纪 80 年代大气物理学论文发表率占大气科学论文总数的 26%，居大气科学三大核心学科之首（大气物理学 26%，大气动力学 16.1%，大气探测学 12.1%）。随着气候变化，大气和生态环境以及灾害性天气预报和防治等当代热门专题领域的研究趋向深入，对大气物理学基本规律的认知需求愈益提高，推动着大气物理学向更广泛的领域和更深层次拓展。35 年前徐先生高瞻远瞩，在这一领域所做的大量开拓性工作，为后来人开辟了广阔的研究领地。

1978 年徐尔灏教授的冤案得到平反昭雪，但当时我的心情依然沉重，毕竟损失已无可挽回，就在这种心境下写了一首"悼念徐尔灏先生"的诗。我不会写什么诗，只是把当时的心情记录下来作为纪念：

悼念徐尔灏先生

叶家东

（南京大学大气科学系）

嗟乎先师徐，惨惨从何语？
本系寒窗客，面壁西海渡。
赖有红日照，雄鸡唱天晓。
万民得其所，国威震寰宇。
勤恳事业计，革心树宏图。
百步光明顶，耕云惊神女。
"小小足球队"，环保驰先驱。
气象物理化，登高指前路。
育人仰师表，严谨莫含糊。
风云变难测，骇浪卷萍浮。
松柏本梗直，桃李无所措。
明哲不保身，世道忒坎坷。
天地路漫漫，离骚投汨罗。
死生去就轻，空余鹦鹉赋。
忽念归去来，桃源无觅处。
魂飞东南隅，夫子冤千古。
千古有遗恨，失足无人扶。
掩弟暗叹息，炎凉一何殊？
浮尘蔽日昏，鬼蜮掀妖雾。

贼林四人帮，祸国殃民蠹。
除害谢天地，百花得甘露。
冤屈申明日，痛定泪难阻。
四化忆良才，星殒余几许？
春雨催春笋，夜阑闻鹧鸪。[①]

1978 年 5 月作于南京大学
徐尔灏教授追悼会后

① 南秀村一带树高林茂，近日春夏之交夜深人静之际，时闻杜鹃啼鸣林间，声声委婉苍凉，扣人心扉，似在告诫世人：哥哥何苦，世路坎坷，闻之怆然。

3 实验研究

3.1 人工凝结核的实验研究*

叶家东

（南京大学气象系）

提要 本文对几种可溶性的物质微粒[NaCl, $MgCl_2$, NH_4Cl, $CaCl_2$, P_2O_5, NH_4NO_3, $(NH_4)_2SO_4$ 等]和樟脑、矽胶、高岭土、炭黑、生石灰等物质进行了实验研究，测定了它们在不同的空气湿度条件下的凝结增长速度。发现所有上述可溶性核以及樟脑粉末都是活跃的人工凝结核，而炭黑、矽胶、高岭土等在未饱和湿空气中没有发生显著的凝结增长现象，证明这些物质并不是活跃的人工凝结核。

一、仪器装置及实验方法

本文根据实验所测定的各种人工凝结核的凝结增长速度，试图求得有关核的凝结效能的知识。

1. 仪器装置

仪器主要由两部分组成，即湿空气发生器和观测系统。

湿空气发生器要求能够产生一定湿度的气流，其示意图如图 1 所示。

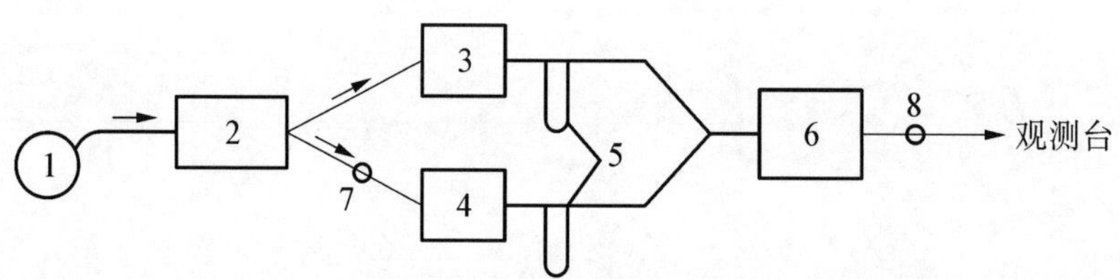

1—鼓风机；2—过滤器；3—湿润器；4—干燥器；5—流量表；
6—混合室，内装干湿表测量空气湿度；7、8—流量控制开关

图 1 湿空气发生器示意图

为了反复观测在不同条件下同一个凝结核的增长情况，除了上述可以调节的湿气流以外，还有一股干空气流直接通向显微镜台，当一次凝结增长过程完毕后，打开干空气流，就可以使液滴重新蒸发获得原来的核。

---

* 本文 1961 年 11 月 6 日收到。

观测系统主要是一架配备有移动游标尺和目镜测微尺的显微镜。放大倍数取450倍或675倍。在镜台上放一小架子，其上绕一根细蜘蛛丝(直径为1～2微米)，将人工凝结核悬挂在丝上，调节架子位置和显微镜焦距，使蜘蛛丝上的核暴露在视野中，打开湿气流即可进行实验。

2. 实验步骤

把新鲜蜘蛛丝绕在小架子上，再设法将要研究的人工凝结核悬挂到蜘蛛丝上，若核是可溶性物质，则可将该物质事先溶解于水中，再将其溶液喷成雾滴悬挂在蜘蛛丝上；若核是不可溶的，则可把它研成细粉再撒上去。此外，我们还用过其他方法。

调节显微镜找到蜘蛛丝和所要研究的人工凝结核，把事先调好湿度的湿气流对准核吹，就可以观测核的凝结增长情况，一边记时间，同时读出液滴的大小，这样就可以得到凝结增长曲线。利用干空气流可以反复对同一个核进行多次观测，以获得足够的数据进行比较。为使测量可靠起见，我们在各种条件固定不变的情况下重复观测五次，发现每次的增长过程基本一致。

二、结果

利用上述仪器和方法，我们对各种物质进行了多次实验，兹将所得资料列于表1。为了比较，部分核的凝结增长曲线绘于图2,3,4,5。

表1　不同相对湿度下各种核的增长率

物质	核半径 $r_0/\mu m$	相对湿度 $f/\%$	气温 $T/℃$	环境湿度 $f/\%$	流速/$(cm \cdot s^{-1})$	平衡半径 $r_m/\mu m$	达到平衡的时间 t_m/s	相对增长率 r_m/r_0
食盐(NaCl)	5.1	～100	17.9	74	33	16	25	3.1
	5.1	98	17.9	80	25	15.5	30	3.0
	10.6	98	16.7	69	29	32.5	70	3.0
	11.7	98	24.5		25	33.3	70	2.9
	3.1	98			25	9		3.0
	27.9	92	17.6	70	22	60	180	2.2
	12.2	99	20.7	82	7	27	150	2.2
	12.2	87	19.6	77	7	24	90	2.0
	27.9	79	18.0	76	37	53.3	280	1.9
	12.2	77	20.9	78	11	22	120	1.8
樟脑($C_9H_{16}CO$)	21	98	20.7	85	11	32	150	1.5
	21	96	20.9	82	22	31	51	1.4
	21	86	20.5	56	15	28	40	1.3
	7	99	17.3	76	20	10	20	1.4
	7	78	18.0	76	16	9	60	1.3
	12	98	20.9	70	22	17	32	1.4

续表

物质	核半径 $r_0/\mu m$	相对湿度 $f/\%$	气温 $T/℃$	环境湿度 $f/\%$	流速/$(cm \cdot s^{-1})$	平衡半径 $r_m/\mu m$	达到平衡的时间 t_m/s	相对增长率 r_m/r_0
硝酸铵(NH_4NO_3)	12	99	31	63	38	22	40	1.6
	12	97	32	63	36	18	60	1.5
	12	85	31	63	38	17	70	1.4
	12	74	31	63	36	15	30	1.2
	8	98	29	65	32	16	60	2.0
	5	98	29	57	34	9	40	1.8
	5	91	29	57	36	9	80	1.8
	5	87	29	57	36	8	30	1.6
	5	78	29	57	42	8	60	1.6
	5	73	29	57	34	7	30	1.4
	9	88	30	41	46	15	15	1.6
	9	84	30	41	59	14	20	1.5
	9	79	30	41	62	13	7	1.4
	9	75	30	41	63	12	10	1.3
	9	73	30	41	73	12	10	1.3
硫酸铵($(NH_4)_2SO_4$)	23	99	30	46	29	45	150	2.0
	23	87	30	46	35	33	90	1.4
	23	81	30	46	35	31	40	1.3
	23	79	30	46	39	30	60	1.3
	9	97	29		29	14	20	1.5
	9	90	29		31	13	30	1.4
	9	86	29		32	13	30	1.4
	9	78	29		35	11	30	1.1
	6	98	31		32	10	100	1.7
氯化铵(NH_4Cl)	18	98			11	33	60	1.8
	18	96	21.0	68	22	33	60	1.8
	18	87	21.3	74	11	29	63	1.6
	18	77	21.1	74	6	27	55	1.5
氯化镁($MgCl_2$)	15	～100				23	80	1.5
	3.5	92				8	20	2.3
	12	90				17	30	1.5

续表

物质	核半径 $r_0/\mu m$	相对湿度 $f/\%$	气温 $T/℃$	环境湿度 $f/\%$	流速/ $(cm \cdot s^{-1})$	平衡半径 $r_m/\mu m$	达到平衡的时间 t_m/s	相对增长率 r_m/r_0
氯化镁 ($MgCl_2$)	28	89				42	160	1.5
	12	67				15	16	1.3
	3.5	65				5	5	1.4
	28	64				33	60	1.2
	15	61				18	20	1.2
氯化钙 ($CaCl_2$)	5	96				8	20	1.6
	7.5	95				17	60	2.3
	7.5	50				13	10	1.6
五氧化二磷 (P_2O_5)	1.5	98				2	3	1.3

对于半径大于1微米的液滴，表面曲率的影响可以忽略，这样，凝结增长公式可以写成

$$\frac{dr}{dt}=\frac{D}{\rho_K R_{\text{Ⅱ}} rT}\left[(e_B-E_B)+c_n E_B\left(\frac{r_0'}{r}\right)^3\right]。\tag{1}$$

由式(1)可以求得在一定湿度下达到平衡时的增长率为

$$\frac{r_m}{r_0'}=\left(\frac{c}{1-f}\right)^{1/3}。$$

式中 r_0'是饱和液滴的半径，对于 NaCl 溶液，饱和浓度可取 $c-0.35$ 克/厘米^3。以 r_0 表示盐核的等值半径，则有

$$\frac{r_m}{r_0}=\frac{r_m}{r_0'}\cdot\frac{r_0'}{r_0}\approx 1.83\left(\frac{c_n}{1-f}\right)^{1/3}。\tag{2}$$

式中 $c_n=0.22$。根据式(2)，可以计算在不同温度下盐核的增长率(表2)。

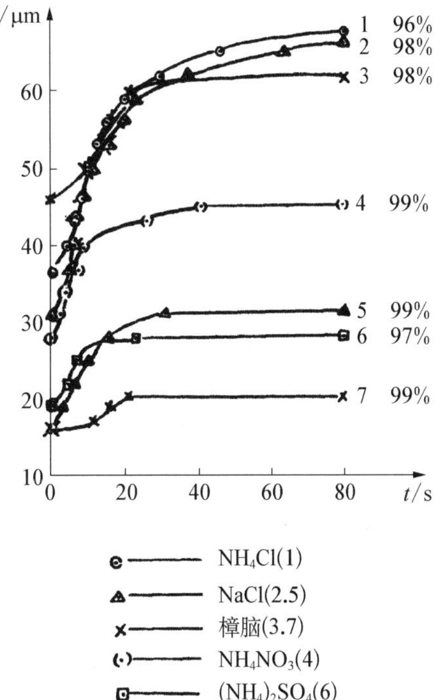

图2　各种人工凝结核的凝结增长曲线

Keith 和 Arons[1]根据溶液表面平衡水汽压的经验定律得到一个计算平衡半径的半经验公式，换算成本文所采用的符号，可以写成

$$\frac{r_m}{r_0}=\left(\frac{4}{3}\pi\rho_{核}\right)^{1/3}\cdot\left(\frac{0.146}{1-f}\right)^{0.299}\approx 2\cdot\left(\frac{0.146}{1-f}\right)^{0.299}。\tag{3}$$

为了比较，将实验结果和根据式(2)和式(3)计算的结果列于表2。

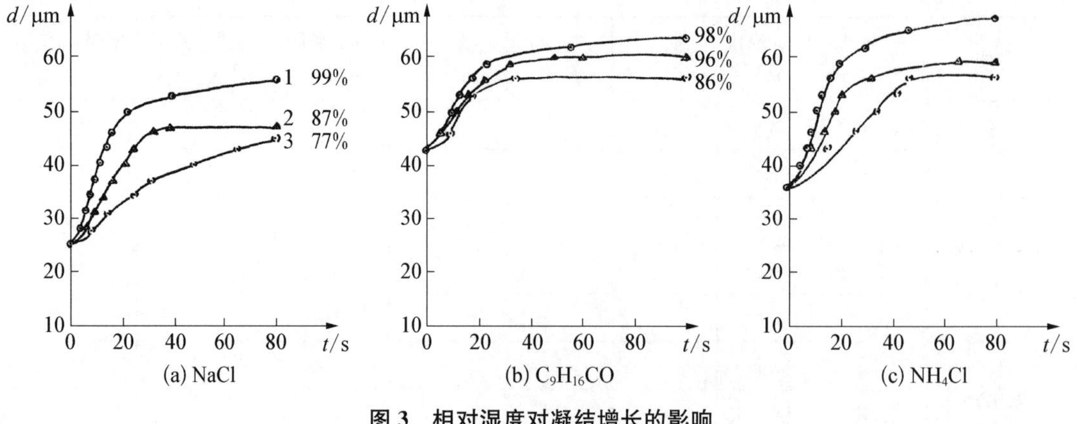

图 3　相对湿度对凝结增长的影响

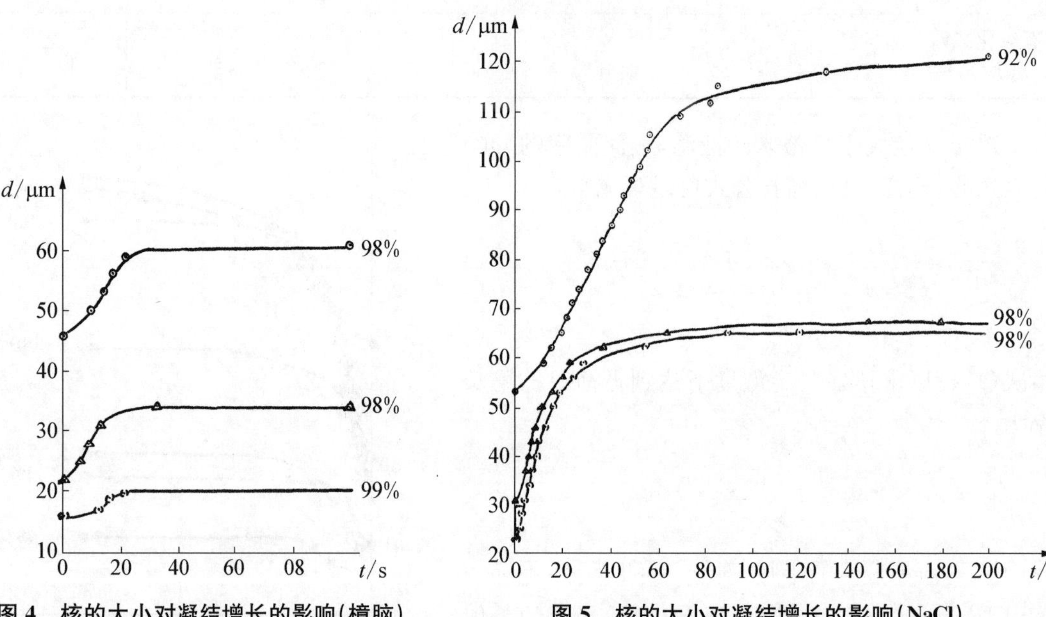

图 4　核的大小对凝结增长的影响(樟脑)

图 5　核的大小对凝结增长的影响(NaCl)

表 2　食盐液滴的增长率

相对湿度 $f/\%$ ＼ 增长率 r_m/r_0	77	78	79	87	92	98	～100
按式(2)计算	1.8	1.83	1.86	2.20	2.52	4.80	∞
按式(3)计算	1.7	1.77	1.78	2.07	2.44	3.53	∞
实验值	1.8		1.9	2.0	2.2	3.0	3.1

由表 2 可见，当相对湿度较小时($f<95\%$)，实验值和计算结果比较符合，偏差介于$+6\%\sim-12\%$之间。公式(2)没有考虑离解系数随溶液浓度的变化，所以数值偏大了。

当湿度接近饱和时，实验值和计算值偏差较大，主要是由于这时环境空气对液滴的增长过程发生较大的影响，致使实际的湿度变小了，这在吹气速度小时尤其明显。如表 1 中

NaCl 核 $r_0=12.2$ 微米，$f=99\%$一例，当时吹气速度很小，仅为 7 厘米/秒，而此时环境空气的湿度是 82%，少量空气卷入气流中就会使实际湿度减小，从而影响核的凝结增长率。这个误差在吹气速度较大、气流的相对湿度较小时，其影响是不大的。

三、可溶性人工凝结核的凝结增长

影响凝结增长的因子很多，主要有凝结核的物理化学性质、核的大小、空气湿度、温度和风速。

根据实验资料(表 1)，可以看出下面几点：

1. 各种人工凝结核的起始凝结增长的相对湿度

在观测核的凝结增长时，可以调节相对湿度到某一数值，使核刚好发生凝结，从而得到各种物质的起始凝结湿度。现把测量结果列于表 3。

表 3　各种核的起始凝结湿度

人工凝结核	红磷烧烟(P_2O_5)	硝酸铵 NH_4NO_3	氯化钙 $CaCl_2$	硫酸铵 $(NH_4)_2SO_4$	氯化镁 $MgCl_2$	氯化钠 NaCl	氯化铵 NH_4Cl
起始凝结湿度	<40%	40%	~50%	50%	60%	~80%	~80%

由表 3 可见，红磷烧烟的产物(主要是 P_2O_5)吸湿性很强，由于它的烟粒小，所形成的水滴也很小，能长期悬浮在空中，因此是一种很好的造雾药剂。

2. 凝结增长速度

吸湿性强弱的一个很重要的判断依据是凝结增长速度，因为只有人工凝结核的凝结增长速度远大于自然云滴的凝结增长速度时，它才能很快地增长成特大的水滴以满足碰并过程充分发挥作用的要求。由表 1 可见，各种人工凝结核的增长速度，一般来说都是很快的，特别是起始阶段更是迅速(图 2)。从开始凝结到平衡所需的时间视相对湿度、核的大小、风速等因子而异。就氯化钠而言，一般不超过 5 分钟(对特别大的核，增长过程会更长一些)，大多数情况是在 1 分钟到 1 分半钟就达到平衡了(见表 1 t_m 栏)。其他各种核的增长时间一般都不超过 3 分钟。这些核在未饱和湿空气中就能凝结增长，所以，当在云底撒药时，可以认为它们在进入云底时就已经具有平衡半径的大小了，当然，在饱和及过饱和空气中，液滴是要继续增长的。

3. 核的凝结增长率

从表 1 最后一行可以得到各种核在相对湿度 $f=98\%$的空气中的平均增长率(见表 4)：

表 4　各种核在相对湿度 $f=98\%$的空气中的平均增长率

凝结核	氯化钠	氯化铵	硝酸铵	硫酸铵	氯化镁	樟脑
增长率$\frac{r_m}{r_0}$	3.0	1.8	1.8	1.7	1.5	1.4

除了氯化钠可以比较精确地测量其大小外，其他物质由于形状不规则，只能测出平均视直径。因此，表 1 中个别例子的增长率与上述平均值会有些出入。

上面所列的平均增长率告诉我们，食盐核的凝结效能是很高的，半径为 10 微米的食盐核撒在云中，单靠凝结作用，一二分钟内就能增长成半径为 30 微米的大云滴，这种水滴的碰并作用已经很显著了。其他几种人工凝结核，只要适当加大颗粒，也是能够起到有效的催化作用的。从增长率的数值看，盐粉的催化效能是最好的。图 2 列出各种物质凝结增长曲线以资比较。

其他如相对湿度、风速等因子的影响与过去对盐核生长所得结果相同。

四、不可溶的人工凝结核的凝结增长

在实验中，发现只有樟脑粉末是不溶于水（溶解度很小）而能吸收大量水汽凝结增长成大水滴的（表 1），其他几种不可溶的物质如炭黑、高岭土、硅胶粉、生石灰等都无此现象。

我们曾用不同的方法获得樟脑微粒的样品，把樟脑研成细粉，设法悬在蜘蛛丝上，发现在湿气流中能够凝结增长成大水滴，水滴中并能发现不可溶的樟脑核心仍然存在。表 1 中所列的关于樟脑的凝结增长资料，都是用这种方法获得的。把樟脑加热，蒸发后复又冷却，樟脑蒸汽就会凝聚在蜘蛛丝上，呈白色结晶。当它暴露在湿气流中时，蒸发很快，一般不超过 5 分钟。蒸发后在蜘蛛丝上会留下几颗很小的液滴，2～3 微米。在无风的空气中，上述樟脑的蒸发速度要慢得多。樟脑蒸发后所残留下来的小液滴在干空气中并无显著的蒸发现象，在湿气流中的凝结现象也不明显。这种小液滴可能是由某种杂质混在樟脑中，或者蜘蛛丝本身含有某种吸湿性的杂质所造成的，这种现象在其他场合也曾发现过。若把樟脑的酒精溶液喷入高热的电炉上，樟脑蒸发，用玻璃片取样可得少量结晶，它在湿气流中并不增长。直接燃烧樟脑酒精溶液时，产生大量黑烟，这是溶液燃烧变成炭粒之故，它也没有凝结增长现象。因此，我们以为，将樟脑用机械法研细所得到的樟脑微粒（$d=5\sim100$ 微米）是活跃的人工凝结核，可作为暖云催化剂，而用其他方法产生的粒子，在未饱和湿空气中并未发现有明显的凝结增长现象。而且，很小烟粒或樟脑晶体，在空气中容易被蒸发掉，而其蒸汽分子作为固定的凝结核的可能性是不大的。不过，较大的粒子是很难一下子被蒸发掉的。

樟脑粉的凝结性能是个很有意义的问题，由于实验条件限制，对于樟脑粉末的凝结机制，一时尚难完全解释清楚。在开始凝结时，由于樟脑表面很不光滑，存在大量毛细孔，因而，此时存在着毛细凝结现象，当核上已包满一层水膜时，毛细凝结过程终止，其进一步的凝结增长可能是由下面的原因引起的：由于核是用机械法研细的，因此，很可能在核上附有某些杂质，或者这种樟脑表面能吸附空气中某种成分的气体，它一方面保护樟脑不致蒸发（由实验得知，由樟脑蒸汽凝聚成的樟脑结晶，由于表面新鲜纯净，很易在空气中蒸发掉），另一方面也能吸收水汽使核增长起来；此外，也可能还有极化分子的吸附等过程在起作用。为了究其根源，需做进一步的研究。但无论如何，樟脑粉末能作有效的人工凝结核，这是可以肯定的。

至于炭黑、硅胶粉、高岭土、生石灰等固体粒子，虽然也都有吸湿性，具有毛细凝结作用，但如上所述，单是这种作用是不能产生明显的凝结增长的，它最多只能形成一层不可见的水膜罢了。实验也证明了这一点。所以，用它们来作为吸湿性的暖云催化剂，其效能是不好的。

五、暖云室的试验结果

上面我们对个别核的凝结增长进行了实验，为了进一步说明人工凝结核对于云雾集体的影响，在这里我们引用了张世丰等同志所做的暖云室催化试验结果[2]。

他们所用的暖云室结构如图 6 所示。其试验方法大致如下：

在整个云室中充雾，当透明度下降到最低时，停止通雾，将中间的布幕放下，云室就隔成两半，一边通入催化剂，一边任其自然消散，比较两边在消散过程中的差别以确定催化的效果。测量项目包括透明度、温度、滴谱等。

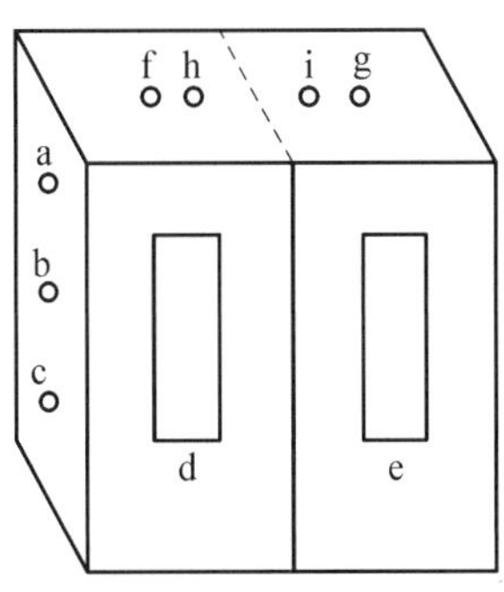

a，b，c—取样孔；d，e—观测窗；f，g—撒药孔；h，i—光源

图 6　暖云室示意图

我们在这里选用一例来具体分析一下。

盐粉用量 1.8 克，催化结果发现：

(1) 催化的那一边，当盐粉通入后透明度迅速好转，而不通盐粉的一边，由于雾的自然消散，其透明度好转的速度远较前者为慢(图 7)。显然，这是由于盐粉在云室里沉降过程中吸收水分致使多数小水滴蒸发的缘故。

盐核在饱和空气中的初始凝结方程可取

$$\frac{\mathrm{d}r}{\mathrm{d}t}=\frac{D}{\rho_{\mathrm{K}}R_{\mathrm{II}}Tr}\cdot c_nE_{\mathrm{B}}(T)\cdot\left(\frac{r_0'}{r}\right)^3。\tag{4}$$

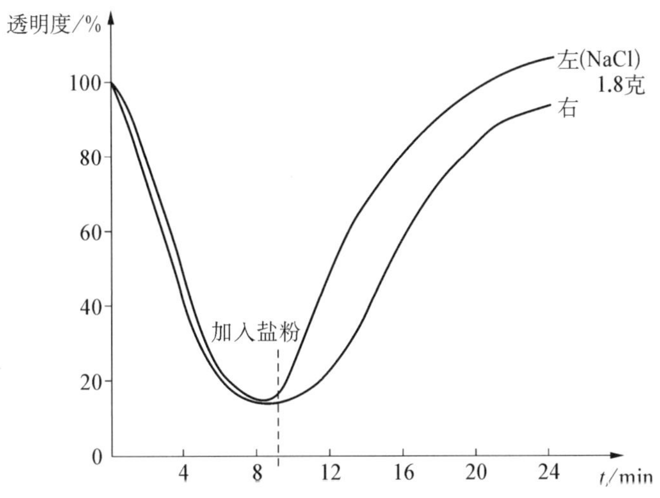

图 7　暖云室中透明度变化曲线

液滴的降落速度 v 取斯托克斯公式，则有

$$\mathrm{d}z=v\mathrm{d}t=c_sr^2\mathrm{d}t=\frac{c_s\rho_{\mathrm{K}}R_{\mathrm{II}}T}{c_nDE_{\mathrm{B}}\cdot r_0'^3}\cdot r^6\mathrm{d}r。\tag{5}$$

设云室高度为 H，积分式(5)可得

$$R=\left[\frac{H}{c}+r_0^7\right]^{1/7}。\tag{6}$$

式中：r_0 为引入云室的盐核初始半径；R 为沉降在云室底部时液滴的半径；c 是常数。显然，在这一过程中，m 克半径为 r_0 的盐核所吸收的水分为

$$\Delta M = m \cdot \frac{\rho_K}{\rho_0}\left(\frac{R}{r_0}\right)^3。$$

令 $r_0=30$ 微米，$T=27$ ℃，取 $\rho_K\approx1.43$ 克/厘米^3，云室高 $H=2.6$ 米。由式(6)可得 $R\approx$ 58 微米，于是

$$\Delta M = m \cdot \frac{1.43}{2}\left(\frac{5.8}{3}\right)^3 \approx 5.1m。$$

由此可见，1 克盐粉就足以把雾迅速消除。当然，这里的计算是在假定空气维持饱和的情况下进行的。

(2) 图 8 表示云室两边的温度情况。由图可见，在停止通雾后温度是要下降的，这是由于雾的温度较高的缘故。但喷盐粉的一边温度下降较慢，不通盐粉的一边下降较快，在雾消散时两边温差可达 1 ℃左右，这可以用盐粉吸收水汽释放凝结潜热来解释。要知道雾滴的蒸发是要耗热的，但是，在做对比试验中，这一项作用不大，因不通盐粉的一边，其中大部分雾滴也是要蒸发掉的。

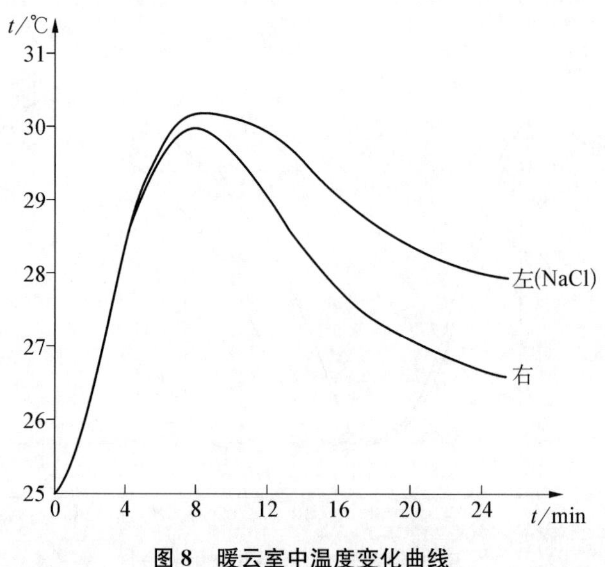

图 8　暖云室中温度变化曲线

暖云室催化试验所显示出来的凝结增温效应很值得注意，因为在暖云催化时，它很可能是起作用的。

本实验是在 Б. В. Кирюхин 专家指导下进行的，参加实验工作的有南京大学气象系大气物理专业各年级的部分同学。在本文内容的组织方面，得到了徐尔灏教授的指导。

参考文献

[1] Keith C. H. and Arons A. B., The growth of sea-salt particules by condensation of atmospheric water vapor, *J. of Met.* **11**, (1954) p.173.

[2] 张世丰等(1960):"暖云室催化试验总结"南京大学气象系.

3.2 斑点法观测大气盐核(Cl^-核)的放大因子问题

叶家东

(南京大学气象系)

提要 本文利用沉降法和显微镜个别对比法测定了盐核(Cl^-核)观测中反应斑点的放大因子。结果发现，放大因子与核的大小有关，可以表示为$K=a+bd_0^{1/2}$，d_0是盐核的等值直径。在我们的实验条件下，系数a，b的数值分别为1.28和1.156($\mu^{-1/2}$)。此外，还讨论了胶膜的厚薄、胶的不同配方以及空气湿度等因素对放大因子的影响，指出所有这些因素都会影响放大因子的数值。因此，观测条件的确定对于取得正确的资料是十分重要的。

一、引言

大气盐核测定的斑点法由来已久，1931年Winckelmann[1]首先利用Liessegang环进行微化学分析，Seely[2]进而用之于气溶胶化学成分的分析，他以氟硅酸亚汞(Hg_2SiF_6)作为试剂，明胶作介质，观测大气盐核(Cl^-核)。Pidgeon[3]对这种方法做了详细的实验研究，肯定了反应斑点(由HgCl沉淀形成)的放大因子为8.7。用这种方法可以观测大核和巨核。但是，正如Lodge等[4]所指出，这种反应斑点极易褪色，所以样品不能保存过久，需要迅速处理。Rau[5]利用硝酸银作试剂，明胶作介质，观测大气盐核(Cl^-核)。将明胶的水溶液和硝酸银($AgNO_3$)溶液混合，制成试剂胶膜。当盐核被捕获在胶膜上时，核中的Cl^-就和胶膜中的Ag^+起化学反应，形成AgCl沉淀，然后在阳光或灯光下曝晒几分钟，就形成红棕色的反应斑点。这种斑点比较稳定，便于保存。斑点的直径和盐核的等值直径之比值就是放大因子，通常采用6倍，这主要是Rau[5]和Podzimek[6]的实验结果。他们分别利用沉降法测量了放大因子，Rau采用逐级分离法，而Podzimek则应用了不同沉降高度上的斑点谱比较法，结果都为6左右，并且与核的大小无关。但是，我们利用沉降法和显微镜个别对比法分别测定了反应斑点的放大因子，得到了与他们不同的结果。

二、自由沉降法测量放大因子

实验方法与Podzimek的相似，主要区别在于我们的实验中可以控制沉降管内比较大的空气湿度，沉降粒子以平衡液滴的状态自由降落，因而就避免了核滴的非球形和在下降途中的蒸发(凝结)影响。

仪器外形如图1所示。沉降管高150厘米，直径7.5厘米，管的上端有开关。在沉降管的不同高度开有取样孔，管内的空气湿度用干湿球温度表测量。为了避免在取样时扰乱管内空气，取样蕊制成圆柱形。沉降管的上部有一段长为30厘米的润湿管，其中湿度和沉降管内部一致，粒子事先在润湿管内充分润湿达到平衡。实验时，将浓度一定的食盐溶液用空气压缩机(约2个大气压)喷成雾滴落入润湿管中，经1～2分钟打开开关(图2(a))。暴露Δt_1时间复又闭上，经t_1时间，A取样。再暴露Δt_1时间，到t_2时刻，B取样(同样暴露Δt_1)。利用装在沉降管内的干湿球温度表测出空气湿度f。实验后取出A，B样品在阳光(或灯光)下曝晒5～15分钟，当斑点明显时就在显微镜(450倍)下分别读出反应斑点的谱分

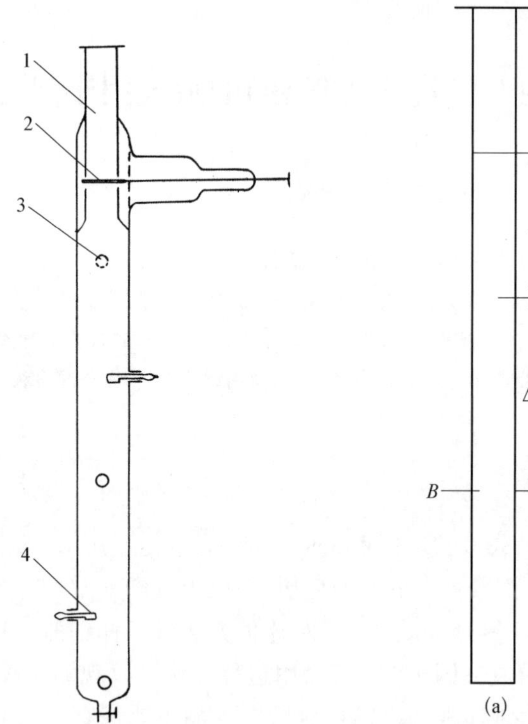

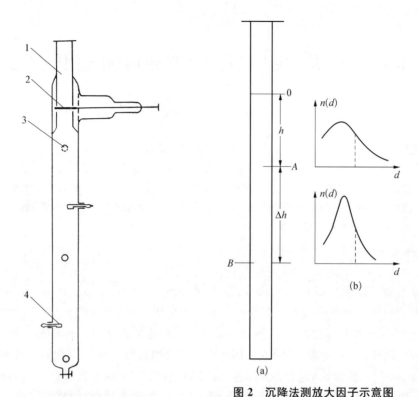

1—润湿管，2—开关，
3—测湿孔，4—取样蕊

图 1　沉降管外形图

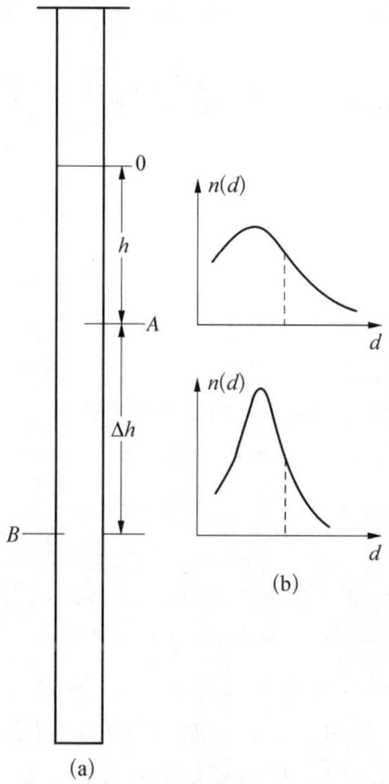

图 2　沉降法测放大因子示意图

布（这里指的是绝对个数的谱分布）。从得到的 A,B 两条谱线（图 2(b)）上可以找到一个交点（有时有两个），这个交点所对应的直径为 d 的斑点是由下降速度为 $v(r)=\dfrac{\Delta h}{\Delta t_2}$ 的液滴形成的（$\Delta t_2=t_1-t_1$）。因为在初始浓度分布均匀的假定下，取样体积相等时其中的液滴个数应该相等（即谱线的交点）。由于 A,B 相距 Δh，其取样时间相隔 Δt_2，所以这种液滴的下降速度应该是 $v(r)=\dfrac{\Delta h}{\Delta t_2}$。至于谱线的另一个交点（不一定有）是不难与之区别的，它总是比上述交点所对应的 d 小。为了得到预期的结果，对于 $\Delta t_1,\Delta t_2$ 以及 t_1 必须适当选择，使之与 $h_1,\Delta h$ 等相配合，如果事先没有这种设计，就会得不到合适的交点。

根据 Δh 和 Δt_2 求出的下降速度 $v(r)$，利用 Stokes 公式

$$v_s(r)=\frac{2}{9}\frac{\rho_k g}{\eta}r^2, \tag{1}$$

可以计算这种液滴的半径 r。式中 ρ_k 是液滴密度，$\rho_k\neq1$，而是与溶液浓度有关的，例如浓度为 26%和 14%时，密度分别为 1.20 和 1.10[7]。

由于沉降管内空气湿度 f 已知，所以液滴的盐分浓度 c 可以根据实验资料[8]或下列经验公式（Keith 等）[9]求出：

$$c=0.24\left(\frac{1-f}{0.146}\right)^{0.897}。 \tag{2}$$

由浓度 c 和半径 r，利用关系式

$$r_0 = r\left(\frac{c}{\rho_{核}}\right)^{\frac{1}{3}} \quad (\rho_{核} = 2.14 \text{ 克 / 厘米}^3), \tag{3}$$

就可以求出相应的干核等值半径 r_0 和等值直径 $d_0 = 2r_0$。最后，放大因子 $K = \frac{d}{d_0}$ 结果列于表 1。①

表 1　沉降法测得的放大因子

次序	沉降高度 Δh/cm	沉降时间 Δt_2/s	沉降液滴半径 r/μm	管内空气湿度 f/%	液滴的盐分浓度 c/(g·cm^{-3})		干核的等值直径 d_0/μm		相应的斑点直径 d/μm	放大因子 $K = \frac{d}{d_0}$		斑点谱中盐核总数 $N = \sum n$	
					据文献[8]资料 c_1	据公式(2) c_2	据 c_1 计算 d_{01}	据 c_2 计算 d_{02}		据 d_{01} 计算 K_1	据 d_{02} 计算 K_2	上片 A	下片 B
1	32	261	2.9	84	0.250	0.261	2.8	2.9	6.9	2.5	2.4	887	731
2	32	195	3.4	90	0.143	0.170	2.8	2.9	6.0	2.2	2.1	1 064	729
3	32	86	5.2	89	0.150	0.186	4.3	4.6	13.6	3.2	3.0	549	473
4	32	86	5.3	94	0.085	0.107	3.6	3.9	10.5	2.9	2.7	1 811	1 578
5	32	33	8.1	86	0.208	0.231	7.4	7.7	22.5	3.0	2.9	861	1 091
6	32	33	8.1	88	0.175	0.200	7.0	7.4	24.0	3.4	3.2	437	447
7	32	32	8.2	86	0.208	0.231	7.5	7.8	25.8	3.4	3.3	1 057	2 394

由表 1 可见，放大因子大致为 2.5～3.5。核大时放大因子也略有增加。

在计算沉降速度时，应用了 Stokes 公式，这公式没有考虑介质分子在沉降粒子表面的滑动效应[10]，当粒子较小时，这种误差比较大，但对于 $r_0 > 3$ μm 的粒子，应用 Stokes 公式所引起的放大因子的误差不超过 1.3%可以不加考虑。

沉降管内的空气湿度是用干湿球温度表测量的，它所引起的误差曾用测量浓度一定的液面平衡湿度的方法检定过，发现有 1%～2%的误差，相应的放大因子绝对误差为 0.1～0.2。

读数误差为±1 μm 所引起的放大因子误差为 0.1～0.3，对于小核，这种误差较大。为了避免沉降管内空气的扰动影响，管子应该放在无热源不通风的室内，操作时不要振动。

从上面的实验结果初步可以看出，放大因子似与盐核的大小有一定的关系，为了进一步了解这个问题，我们又进行了个别对比法来测定放大因子。

三、显微镜个别对比法测量放大因子

由于影响放大因子的因素是很多的，诸如胶膜的厚薄、试剂胶的不同配方，以及取样时

① 表中一部分数据取自金晓钟等的毕业论文(1962 年)。

空气湿度等，都有可能改变放大因子的数值。而且，从上一节的讨论可见，放大因子还随盐核的大小而变。因此，需要做大量的实验才能得出比较确切的结论。由于沉降法实验工作量较大，难以取得大量的数据，因此，我们利用显微镜个别对比法进行实验。测量方法简述如下。

配胶涂片 将 10 克蒸馏水和 1 克明胶混合，微热，待明胶溶化后冷却。将 1 克硝酸银($AgNO_3$)溶于 3 克蒸馏水中，然后与温度为 30～40 ℃的明胶溶液均匀地混合，即得试剂胶。观测时用玻璃棒将试剂胶均匀地涂在干净的取样片上，然后将它放在暗盒中晾干备用。

测量 用喷雾器将食盐溶液喷成雾滴，悬挂在细蜘蛛丝上(丝的直径小于 1 μm)，液滴蒸发后即得待测的盐核。蜘蛛丝绕在一特制的小架子上，可以放在显微镜镜台上进行观测，选择形状规则的单晶体(六面体)，用吹气管轻轻吹动盐核使之转动，测出六面体三边的长度，由此可以求出它的体积和等值直径 d_0，然后把事先涂好晾干的胶片放在显微镜的集光器上，转动集光器升降螺旋，抬升胶片，当盐核与胶片接触以后，把胶片放到装有清水的恒湿皿中，立即调节显微镜找到待测的盐核，此时盐核就吸收水汽迅速扩展，并和胶膜中的 Ag^+ 起反应，形成 AgCl，曝光 10 分钟左右，即形成红棕色的反应斑点(图 3)。测出反应斑点的直径 d 就可以得到放大因子 K(表 2)。我们测量了 136 个盐核的放大因子，结果列于图 4。

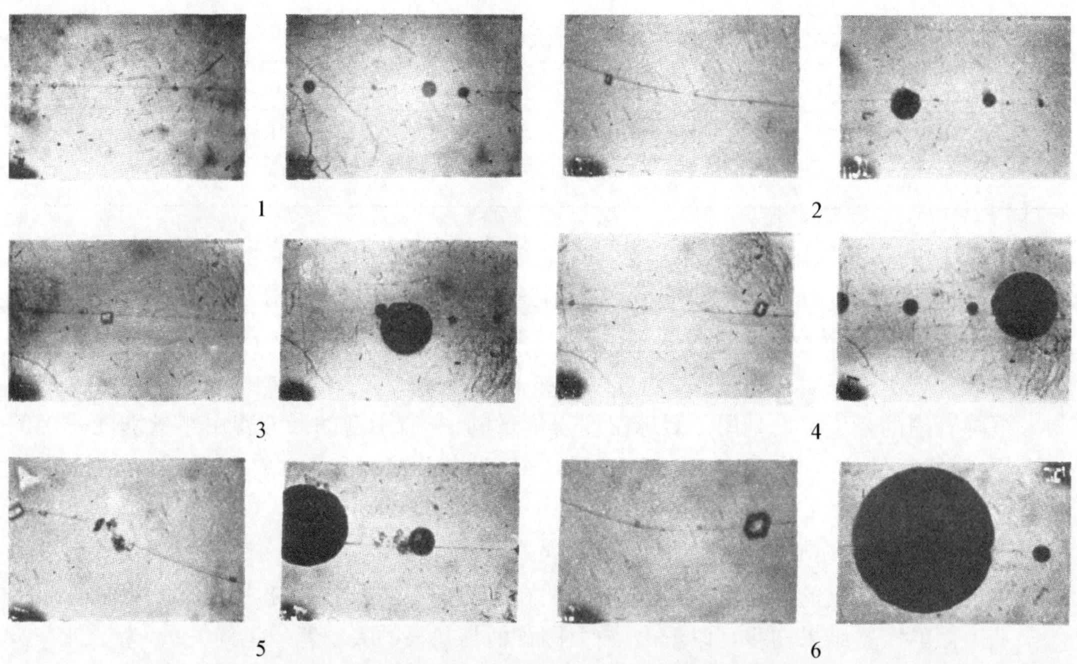

图 3　盐核(1,3,5)和其反应斑点(2,4,6)的显微照相(放大 450 倍)

由图 4 可见，放大因子 K 与核的大小有关，并且遵循着 $K \sim d_0^{\frac{1}{2}}$ 的规律(图 5)，图中的直线是相应的回归直线，利用最小二乘法计算的结果，它可表示为

$$K = 1.28 + 1.156 d_0^{\frac{1}{2}}, \tag{4}$$

表 2 图 3 各照片中盐核的大小及其放大因子 K

相片编号	1					2			3			
盐核号码(自左向右)	1	2	3	4	5	1	2	3	1	2	3	4
$d_0/\mu m$	3.9	1.5	4.8	3.9	1.2	6.6	3.6	2.1	3.3	11.4	1.8	2.1
$d/\mu m$	10.5	4.5	14.4	9.9	3.0	24.6	10.5	4.2	11.4	45.6	6.0	7.5
$K=\frac{d}{d_0}$	2.7	3.0	3.0	2.5	2.5	3.7	2.9	2.0	3.5	4.0	3.3	3.6

相片编号	4				5			6	
盐核号码(自左向右)	1	2	3	4	1	2	3	1	2
$d_0/\mu m$	5.1	3.9	3.0	12.0	12.9	6.0	3.9	4.5	15.9
$d/\mu m$	18.0	12.6	11.1	58.5	66.0	22.5	11.4	13.8	121.5
$K=\frac{d}{d_0}$	3.5	3.2	3.7	4.9	5.1	3.8	2.9	3.1	7.6

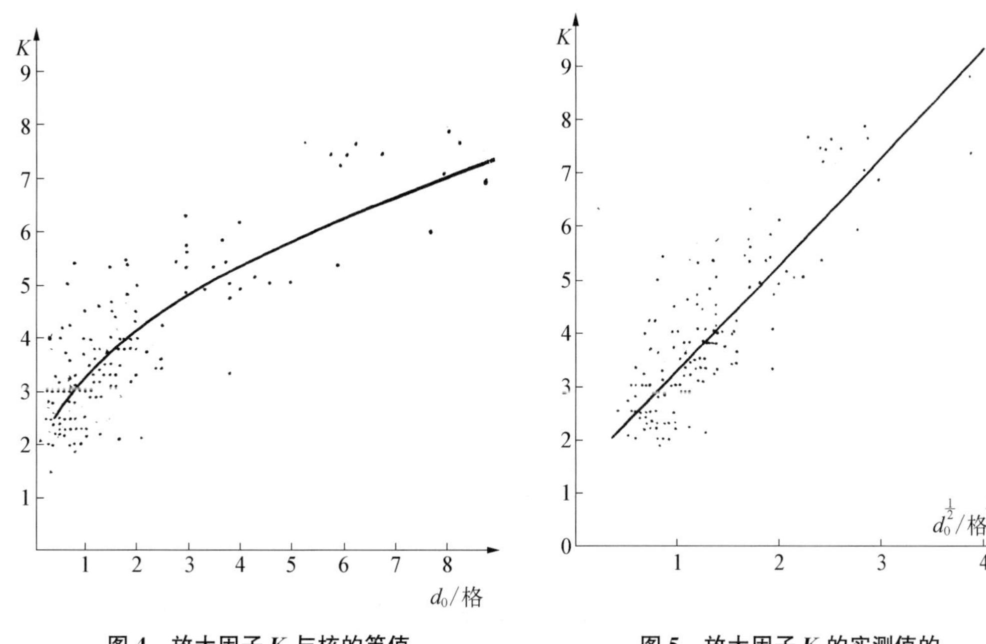

图 4 放大因子 K 与核的等值直径 d 之关系(1 格=3 μm)

图 5 放大因子 K 的实测值的散布图初相应的回归线

式中 d_0 以微米为单位。看来,这种关系是合理的,因为 Cl^- 在胶膜中扩散的体积应该与核的体积成比例。① 假定胶膜的厚度为 h,则有

① 当 Cl^- 和胶中之 Ag^+ 起反应,并形成 AgCl 沉淀时,Cl^- 的扩散便终止了。因此,在胶中 Ag^+ 均匀分布的条件下,Cl^- 扩散的体积和 Cl^- 的质量成比例,也即与核的质量成比例。

$$hd^2 \sim d_0^3,$$

所以

$$K^2 = \frac{d^2}{d_0^2} \sim \frac{d_0}{h}, \text{即} K \sim \left(\frac{d_0}{h}\right)^{\frac{1}{2}}。$$

图 4 和图 5 中点子的离散比较大一些，这主要是由于测量干核厚度时的误差引起的，当核较小时这种误差较大。①

由此看来，放大因子还与胶膜的厚薄有关。在上述一系列实验中，我们曾用 100 倍的物镜镜头测量胶膜的厚度，测得厚度为 2～4 μm。由于胶膜的厚度难以精确测量，因此我们分别用薄胶膜和厚胶膜定性地做了实验，测量结果列于图 6。从图 6 可以看出，胶膜的厚薄对放大因子的影响是很大的，对于大的盐核，尤其明显。当核直径 $d_0 \leqslant 3$ μm 时，这种影响就不大了。

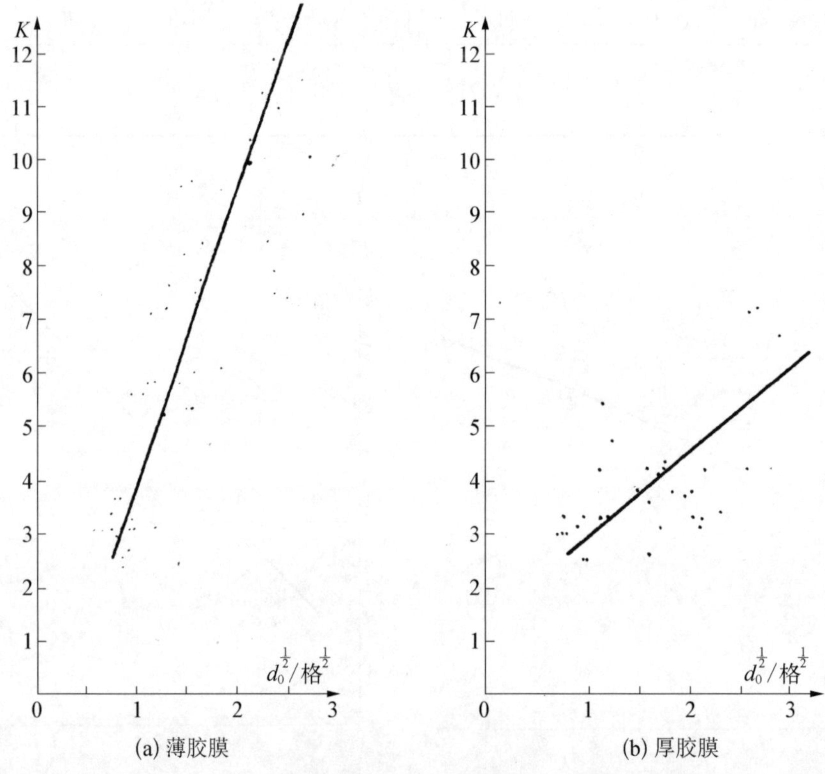

图 6　胶膜的厚度对放大因子 K 的影响

为了检验试剂胶膜的不同配方对放大因子的影响，我们分别用两种不同的配方进行对比试验，结果列于图 7。

由图 7 可见，胶的稠稀程度对放大因子是有显著影响的，这种影响实际上是与胶膜的厚薄的影响类似的。

① 实验数据较多时，这种误差可以适当平滑。

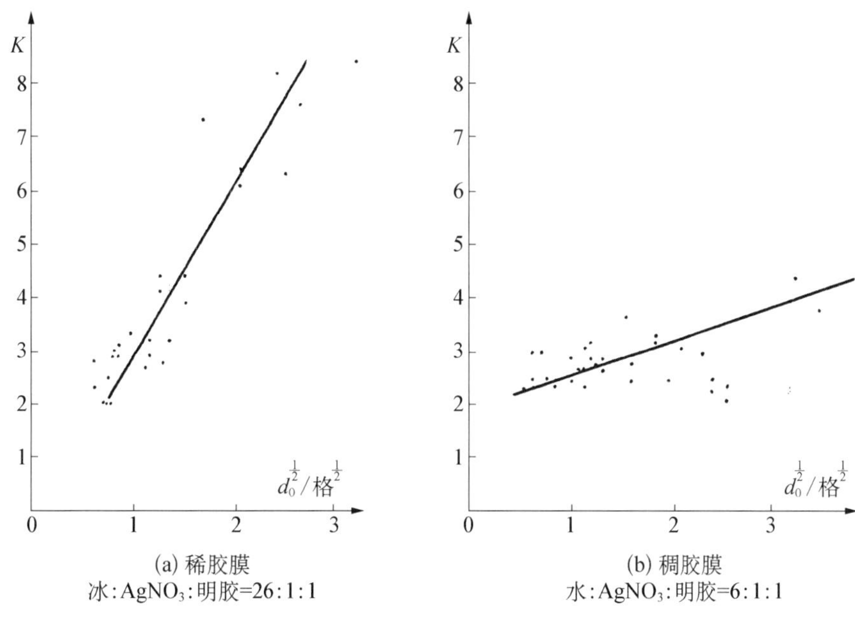

(a) 稀胶膜
冰:$AgNO_3$:明胶=26:1:1

(b) 稠胶膜
水:$AgNO_3$:明胶=6:1:1

图 7　胶膜不同配方对放大因子 K 的影响

观测时如果空气湿度很大,则空气中的盐核实际上已经是液滴了,这在高山云雾站和飞机观测时是常见的,所以需要研究一下空气湿度对放大因子的影响,为此,我们曾在云室内进行试验,此时空气湿度是 92%,实验结果列于图 8。结果发现相对湿度大时放大因子略有增加,当 $f=92\%$ 时可以增大 10%～15%。当相对湿度小于 90%时这种影响不大,因为取样后胶片是曾经放在恒湿皿中充分润湿过的,所以盐核在各种情况下都有可能吸收充分多的水分而扩散开来。

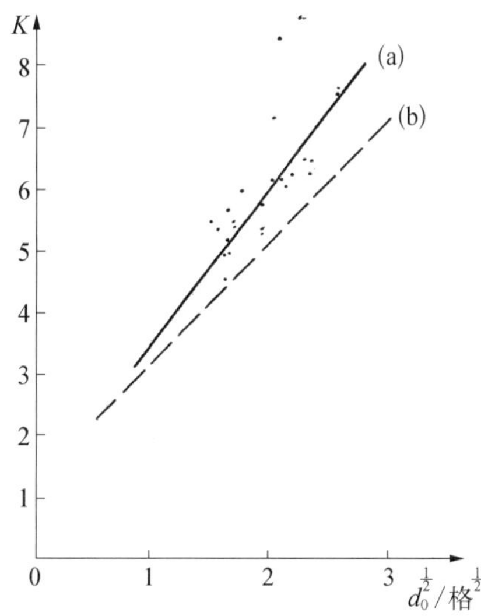

图 8　空气相对湿度对 K 的影响

[(a) $f=92\%$时的 $K(d_0^{1/2})$;
(b) 图 5 的回归线 $K(d_0^{1/2})$]

综上所述,影响放大因子的因素是很多的,而其中最主要的是胶膜的厚度和盐核的大小。为了应用统一的放大因子订正值,观测条件必须固定下来,例如胶膜不能涂得太厚,保持在 2～4 μm。胶膜太厚了,不仅会影响放大因子,而且会使反应斑点边缘不光滑,不便于测量,斑点与背景的对比也不清楚,有碍于正确地辨认。取样时要事先将胶膜晾干,这样反应斑点比较圆,边缘光滑。不然,常常会出现梭状的斑点,或者出现带有尾巴的痕迹。当胶膜部分干了而部分未干时,更容易出现不规则的斑点。胶是有黏性的,晾干不久的胶不会影响捕获性能。取样后将样片放在恒湿皿中(用培养皿即可代用)充分润湿,使所有的核都反应完全,这样可以减小不同空气湿度对放大因子的影响。当相对湿度小于 90%时,这样处理以后,相对湿度的影响就不大了。

四、结论

1. 在以往的观测实践中，斑点法测盐核的放大因子一般都采用 6 倍，这主要是 Rau 和 Podzimek 的实验结果。从前面的分析可见，这是不合适的，这不仅因为他们的实验次数很少（两人分别做了 5 次和 8 次实验），所测的盐核大小范围很窄（$r_0<3.5\ \mu m$），而更重要的是他们的实验条件有不确定性，处理方法上也有不足之处。根据我们的实验结果发现，在一定的观测条件下①，反应斑点的放大因子是与盐核的大小有关的，并可用下列关系式来表示。

$$K=1.28+1.156d_0^{\frac{1}{2}} \tag{4}$$

这关系适用于 $d\geqslant h$ 的情况，当反应斑点的直径 $d<h$ 时，可以取 $K=2.5$。

为了应用方便起见，特做放大因子查用表，如表 3 所示。②

表 3　盐核反应斑点的放大因子查用表

斑点直径 $d/\mu m$	1	2	3	4	5	6	7	8	9	10	11	12	13	14	15	16
放大因子 $K=\frac{d}{d_0}$	2.5	2.5	2.5	2.7	2.8	2.9	3.0	3.1	3.2	3.3	3.3	3.4	3.5	3.6	3.7	3.7
斑点直径 $d/\mu m$	17	18	19	20	21	22	23	24	25	26	27	28	29	30	35	40
放大因子 $K=\frac{d}{d_0}$	3.8	3.8	3.9	3.9	4.0	4.0	4.1	4.1	4.1	4.2	4.2	4.2	4.3	4.3	4.5	4.7
斑点直径 $d/\mu m$	45	50	55	60	65	70	75	80	85	90	95	100	110	120	130	140
放大因子 $K=\frac{d}{d_0}$	4.8	5.0	5.1	5.2	5.3	5.4	5.5	5.6	5.7	5.8	5.9	6.0	6.1	6.3	6.4	6.6

2. 观测条件不同，放大因子会有很大的差异，影响放大因子的主要因素是胶膜的厚度和空气湿度。要严格地测定胶膜的厚度是比较困难的，一般可以从充分曝光后胶膜的颜色变化和透光度大致估计。取样后把胶片放在恒湿皿中充分润湿，这样可以减小空气湿度不同的影响。一般在相对湿度小于 90% 的空气里，可以应用上述表格。当湿度更大时，放大因子就略有增加。

致谢：本实验得到徐尔灏先生的鼓励，他还对本文提出了宝贵的意见。部分实验是在雷连科同志协助下完成的，本文插图由叶品华同志绘制。特此一并致谢。

参考文献

[1] Winckelmann, J., *Mickrocliemie*, **10**(1931), 437—439.
[2] Seely, B. K., *Anal*, *chem.*, **24** (1952), 576—579.
[3] Pidgeon, F.D., *Anal*, *chem.*, **26** (1954), 1832—1835.

① 所指的观测条件包括：① 胶膜的配方为（质量比）：水∶明胶∶$AgNO_3$=13∶1∶1；② 胶膜厚度 $h\approx 2\sim 4\ \mu m$；③ 取样时胶片是事先晾干的，取样后将胶片放在恒湿皿中充分润湿（用装清水的培养皿即可代用），达 15 分钟以上，然后曝光。

② $d_0>30\ \mu m$ 时，实验数据不多，不便外推，但这样大的盐核自然界是不多见的。

[4] Lodge, J. P., Tufts, B. J., *Tellus*, **8** (1956), 184—189.

[5] Rau, W., *Arch.Met. Geophys. Bioklim.*, A, **9** (1956), 224—231.

[6] Podzimek, J., *Studia geoph. et geod.*, **3** (1959), 256—280.

[7] Бачинский, А. И.,物理学手册,商务印书馆,115 页.

[8] Тверской, П. Н.,气象学教程,第二分册,高教出版社,472 页.

[9] 叶家东,气象学报,**32**(1962),232 页.

[10] Фукс, Н. А.,气溶胶力学(中译本),科学出版社,1960 年,52—53,35—37 页.

(本文于 1965 年 3 月 19 日收到,1965 年 12 月 5 日收到修改稿)

3.3 尿素核和盐核的凝结性能*

叶家东　雷连科　谢文彰

(南京大学大气科学系)

尿素作为人工降水暖云催化剂,已有一些地区进行过外场作业和试验。但是关于尿素核的凝结性能,详细的研究甚少。本文对尿素核的凝结增长性能进行了实验研究。① 作为比较,对常用的暖云催化剂盐核的凝结性能也做了相应的研究。

1. 实验方法

实验装置如图 1 所示,主要包括湿空气发生系统和观测系统两部分。湿空气发生系统基本上同文献[1],本实验主要对观测系统做了改进。湿空气引进安装在显微镜台上的凝结盒内。凝结盒为一圆筒形有机玻璃盒(图 1(b)),上方开一圆形口,供物镜伸入盒内进行观测,物镜筒上嵌套一有机玻璃盖,实验时隔离环境空气和观测者呼吸等因素的影响。凝结盒四壁开有四个圆形小孔,分别作为干、湿空气的进气孔和出气孔,以及凝结核托架和干湿球热电偶观测孔。这样,通入凝结盒内的湿空气较少受环境空气的干扰,使核滴的凝结增长过程较为稳定。

(a) 湿空气发生系统

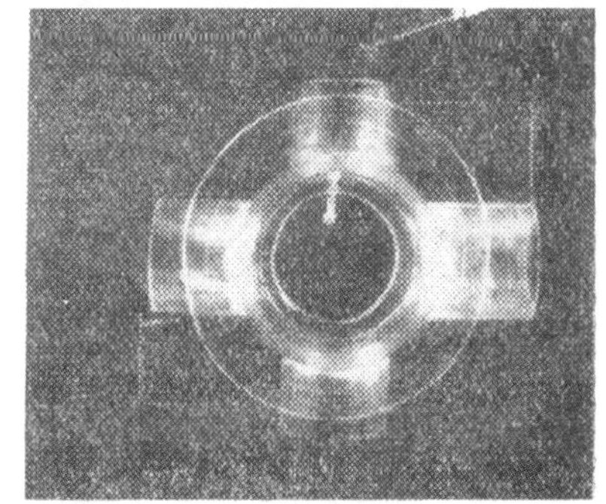

(b) 凝结盒

图 1　凝结实验装置

* 本文于 1983 年 9 月 17 日收到,1984 年 7 月 9 日收到修改稿。

① 尿素核样品由四川省人工降雨办公室提供。

盐核大小的测量，当核滴凝结增长其中干核完全融解时，液滴处于饱和溶液状态，测量此时液滴直径，即可根据溶解度换算出干核等效直径 d_0。尿素核凝结增长后再通干空气使其蒸发，所得干核一般均为球形或近似球形，本实验中一概以此球形干核直径 d_0 作为尿素核的等效直径。

实验时将凝结核悬挂在蜘蛛丝上进行观测，通湿空气后用显微摄影定时拍摄凝结增长过程。为了获得一定数量的凝结增长率资料，我们还用显微镜目测核滴的初始等效直径 d_0 和平衡液滴 d_m，其间不测量增长过程，这样易于获得较多的凝结增长率 $K=\dfrac{d_m}{d_0}$ 的数据。

2. 实验结果

(1) 尿素核的凝结增长性能

尿素核的凝结增长率见表 1。

表 1　尿素核的凝结增长率

核号	d_0/μm	d_m/μm	$K=\dfrac{d_m}{d_0}$	$K_{15}=\dfrac{d_{15}}{d_0}$	f/%
1	59	107	1.80	1.41	89
2	52	97	1.85	1.42	89
3	52	85	1.65	1.37	93
4	53	107	2.00	1.48	89
5	23	45	1.98	1.62	92
6	37	63	1.67	1.39	92
7	49	73	1.47	1.30	90
8	51	73	1.43	1.30	90
9	44	84	1.90	1.58	89
10	41	73	1.78	1.51	89
11	41	73	1.77	1.43	92
12	52	101	1.95	1.45	92
平均	46.2	81.5	1.77	1.44	90.5

注：d_{15} 和 K_{15} 分别为凝结增长 15 秒时的液滴直径和相应的增长率。

由表中可见：

① 尿素核的起始吸湿性增长是十分迅速的，在相对湿度为 90%左右的湿空气中，开始 15 秒的平均增长率为 1.44，约占 4 分钟时增长率的 81%。这一阶段主要是溶液达到饱和以前的阶段，可称为吸湿性增长阶段。

② 从尿素核的凝结增长曲线看(图略)，直至凝结增长时间达 4 分钟，尿素核滴仍有缓慢增长的趋势，表明在 90%左右的湿空气里，尿素核凝结增长达到平衡的时间是相当长的。我们来估算一下这个时间。

不计曲率影响①，溶液滴凝结增长公式可取[2]

$$r\frac{\mathrm{d}r}{\mathrm{d}t}=\frac{f-1+i\,\dfrac{\rho_s r_o^3\mu_l}{\rho_l r^3 u_s}}{\left[\dfrac{L_v\rho_l}{K_rT}\left(\dfrac{L_v}{R_vT}-1\right)+\dfrac{\rho_l R_v T}{D_f e_s(T)}\right]} \tag{1}$$

式中：ρ_s、μ_s 和 ρ_l、μ_l 分别为溶质和水的密度和分子量；r_o 是干核的等效半径；D_f 是水汽扩散系数；i 是溶质的离解系数；f 为空气相对湿度；$e_s(T)$是空气饱和水汽压；R_v 为水汽比气体常数；L_v 为水汽凝结潜热；K_r 是空气导热系数。若按表 1 中的 $r=r_m=\dfrac{d_m}{2}$时的增长速率估算，在 90%的相对湿度下为达到真正的平衡半径，需要 1 小时以上。当然后期的增长速率是十分缓慢的，所以我们仍取起始增长的前 3～4 分钟作为实验凝结增长的时限。

为了获得较多的凝结增长率数据，我们用显微镜目测法测量了尿素核的凝结增长率 K_m。图 2 为尿素核平衡液滴直径 d_m 与干核等效直径 d_0 的实验散布图及相应的回归线，有关的统计量列于表 2。由表 2 可见，实验增长率的回归值是随干核增大而减小的，干核 d_0 从 10 μm 增至 100 μm，相应的增长率从 2.16 减小到 1.32，平均增长率 1.42。如果只计 d_0 = 20～60 μm 的范围，则相应的平均增长率约为 1.50。与表 1 相比较，由于测量方法不同，目测平均增长率大约要偏小 0.30。

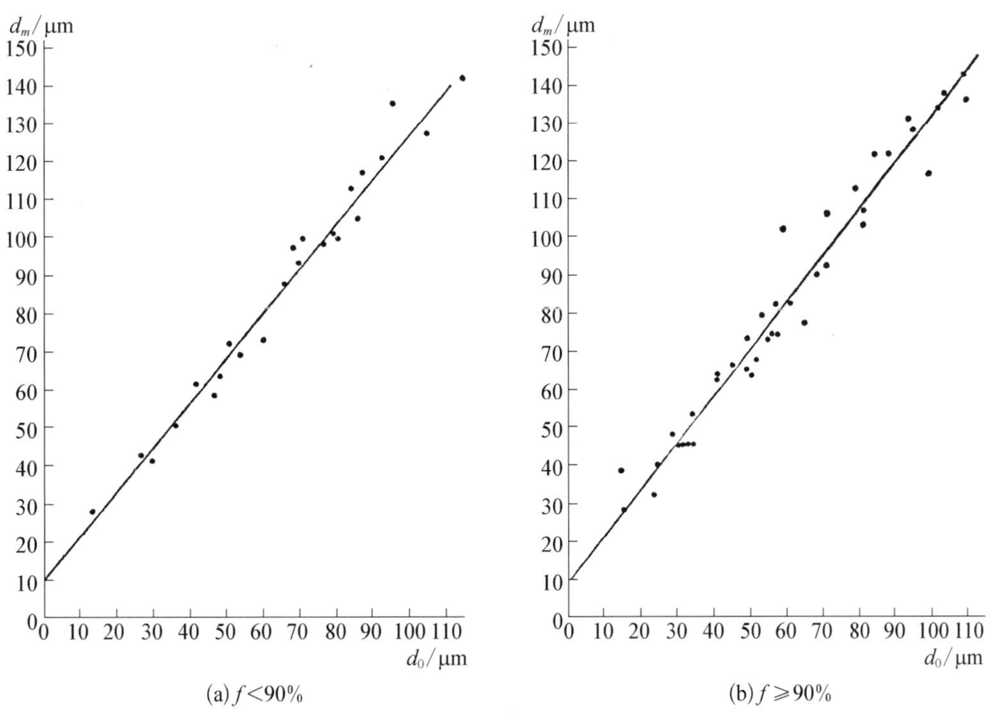

(a) f<90%　　(b) $f\geqslant$90%

图 2　尿素核平衡液滴散布图及其回归线

① 在本实验条件下，曲率项要比湿度项和溶液项小 2 个量级或以上，故略。

表 2　尿素核凝结增长率和有关的统计量

K	相对湿度/%		n	回归方程	r	$\bar{d}_m/\mu m$	$\bar{d}_0/\mu m$	统计平均 $K=\frac{\bar{d}_m}{\bar{d}_0}$	$\bar{K}=\frac{1}{n}\sum_{i=1}^{n}K_i$
	f	$\bar{f}$		$\hat{d}_m=a+bd_0$					
1	<90	85.7	24	$\hat{d}_m=9.22+1.19d_0$	0.99	88.7	66.5	1.33	1.37
2	≥90	93.5	41	$\hat{d}_m=8.91+1.26d_0$	0.99	85.5	60.7	1.41	1.45
3	合计	90.6	65	$\hat{d}_m=9.30+1.23d_0$	0.98	86.7	62.9	1.38	1.42

K	$d_0/\mu m$	10	20	30	40	50	60	70	80	90	100
1	$\hat{K}=\frac{\hat{d}_m}{d_0}$	2.11	1.65	1.50	1.42	1.37	1.34	1.32	1.31	1.29	1.28
2	$\hat{K}=\frac{\hat{d}_m}{d_0}$	2.15	1.71	1.56	1.48	1.44	1.41	1.39	1.37	1.36	1.35
3	$\hat{K}=\frac{\hat{d}_m}{d_0}$	2.16	1.70	1.54	1.46	1.42	1.39	1.36	1.35	1.33	1.32

(2) 盐核的凝结增长性能

盐核的凝结增长率见表 3。

表 3　盐核的凝结增长率

相片		$d_0/\mu m$	$d_m/\mu m$	$K=\frac{d_m}{d_0}$	$K_{15}=\frac{d_{15}}{d_0}$	$f/\%$
张号	核号					
(a)	1	27	76	2.82	1.96	90
	2	28	55	2.84	2.17	90
	3	8	25	2.99	1.60	90
	4	26	69	2.68	1.79	90
	5	22	65	2.99	2.06	90
	6	8	23	2.96	1.71	90
(b)	7	10	34	3.32	2.97	91
	8	6	19	3.28	2.68	91
	9	10	33	3.23	2.73	91
	10	9	29	3.18	2.66	91
	11	12	33	2.78	2.31	91
	12	12	39	3.31	2.57	92

续表

相片		d_0/μm	d_m/μm	$K=\frac{d_m}{d_0}$	$K_{15}=\frac{d_{15}}{d_0}$	f/%
张号	核号					
(c)	13	44	120	2.74	1.92	90
	14	39	106	2.69	2.03	90
	15	8	26	3.28	2.65	90
	16	15	45	2.99	2.54	90
(d)	17	14	43	3.03	2.45	92
	18	12	37	3.05	2.40	92
	19	8	27	3.41	2.94	92
	20	43	116	2.70	1.74	92
(e)	21	14	48	3.42	2.55	92
	22	12	37	3.20	2.49	92
	23	25	69	2.73	2.05	92
(f)	24	10	33	3.23	2.75	92
	25	14	46	3.34	2.75	92
	26	12	37	3.17	2.54	92
	27	15	54	3.48	2.71	92
(g)	28	29	92	3.21	2.03	93
(h)	29	36	88	2.43	1.86	90
	30	42	110	2.63	1.70	90
(i)	31	46	118	2.55	1.55	90
平均				3.02	2.29	91.0

由所列表可见：

① 盐核的凝结增长性能明显优于尿素核，在表 3 的实验湿度下核滴直径的平均增长率达 3 倍左右。起始吸湿增长速率也更为迅速，15 秒时的平均增长率 $\overline{K}_{15}=2.29$，约占实验平衡增长率的 75%。

② 核小时实验平衡增长率一般较大，随着核增大，实验平衡增长率有减小的趋势，这是与增长时间限于 4 分钟之内有关的。从表 3 看出，$d_0 \leqslant 15$ μm 的小核，实验平衡增长率几乎都超过 3 倍，而 $d_0 > 15$ μm 的大核，只有 28 号核的增长率达到 3.21 倍，但此时湿空气相对湿度较大，达到 93%。

为了获得较多的平衡增长率数据，用显微镜目测法测量了盐核的凝结增长率 K。图 3 为盐核平衡液滴直径 d_m 与干核等效直径 d_0 的实验散布图及相应的回归线，有关统计量列于表 4。由表 4 可见，与尿素核类似，盐核的实验平衡增长率的回归值也是随干核增大而减小的，干核 d_0 从 10 μm 增至 100 μm，相应的增长率从 3.01 减小到 2.33，平均增长率为

2.49。如果只计 $d_0=10\sim40\ \mu m$ 的尺度范围(与表 3 的核尺度相仿),则相应的平均增长率约为 2.65。与表 3 相比较,主要由于测量方法不同,目测平均增长率大约要偏小 0.37。

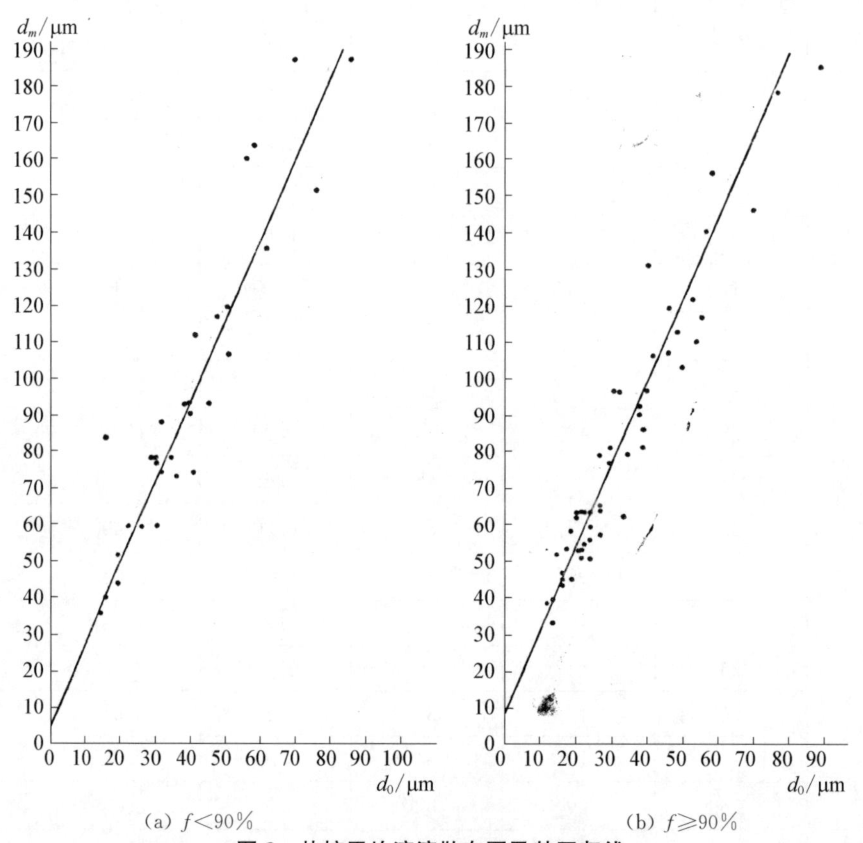

图 3　盐核平均液滴散布图及其回归线

表 4　盐核凝结增长率和有关的统计量

K	相对湿度/%		n	回归方程	r	$\bar{d}_0/\mu m$	$\bar{d}_m/\mu m$	统计平衡	$\bar{K}=$
	f	$\bar{f}$		$\hat{d}_m=a+bd_0$				$K_s=\frac{\bar{d}_m}{\bar{d}_0}$	$\frac{1}{n}\sum_{i=1}^{n}K_i$
1	<90	86.8	29	$\hat{d}_m=4.53+2.23d_0$	0.95	39.7	93.1	2.34	2.41
2	⩾90	93.8	53	$\hat{d}_m=8.11+2.30d_0$	0.95	34.4	87.2	2.53	2.60
3	合计	91.3	82	$\hat{d}_m=7.61+2.25d_0$	0.95	36.3	89.3	2.46	2.49

K	$d_0/\mu m$	10	20	30	40	50	60	70	80	90	100
1	$\hat{K}=\frac{\hat{d}_m}{d_0}$	2.68	2.46	2.38	2.34	2.32	2.31	2.29	2.29	2.28	2.28
2	$\hat{K}=\frac{\hat{d}_m}{d_0}$	3.11	2.71	2.57	2.50	2.46	2.44	2.42	2.40	2.39	2.38
3	$\hat{K}=\frac{\hat{d}_m}{d_0}$	3.01	2.63	2.50	2.44	2.40	2.38	2.36	2.35	2.33	2.33

3. 讨论

(1) 实验平衡增长率与理论平衡增长率

由式(1)可得核滴凝结增长平衡半径 r_m 的理论平衡增长率 K_m

$$K_m = \frac{r_m}{r_o} = \left[\frac{i\rho_s\mu_l}{\rho_l\mu_s(1-f)}\right]^{\frac{1}{3}} \tag{2}$$

对尿素核和盐核分别按式(2)计算 K_m,结果列于表5。计算中对尿素取 $i=1$,$\rho_s=1.33$ 克/厘米^3,$\mu_s=60.6$,对盐核取 $i=2$,$\rho_s=2.16$ 克/厘米^3,$\mu_s=58.5$。表5中还分别列出由表1和表3归纳的实验平衡增长率。由表5可见,尿素核和盐核的平均实验平衡增长率分别比理论计算值大0.10~0.17和0.44~0.62,盐核的偏差尤为明显。究其原因,大致有三个方面:

表5 尿素核和盐核的平衡增长率

	f/%	80	85	90	92	95	98
尿素核	(2)式计算值 K_m 表1的实验平均值 $\bar{K}$	1.25	1.38	1.58 1.75	1.70 1.80	1.99	2.70
盐核	(2)式计算值 K_m 表3的实验平均值 $\bar{K}$	1.88	2.07	2.37 2.81	2.55 3.17	2.98	4.05

① 核滴大小测量误差:主要是干核大小测量比较困难,方法不同会产生系统偏差。由前面的分析知,显微摄影测量的增长率平均要比显微镜目测法偏大0.30左右。

② 湿度测量误差:表6是在湿度检定箱中用阿斯曼干湿表鉴定热电偶干湿表的鉴定参数。由表可见,在凝结实验的湿度范围内(80%~98%),热电偶干湿表测湿度有0.2%~1.3%的系统正偏差。测量偶然误差,回归值的剩余标准差 $S_{y/x}=\pm1.4\%$,合计湿度测量有 $-1.2\%\sim+2.7\%$ 的误差。

表6 阿斯曼干湿表对热电偶干湿表的检定参数 %

n	$f_热-a+bf_阿$	$\bar{f}_热$	$\bar{f}_阿$	s_r	s_s	$s^2_{y/x}$	$s_{y/x}$	$f_阿$	80	85	90	95	100
15	$f_热=7.18+0.93f_阿$	92.7	92.4	5.62	5.89	1.91	1.38	$\hat{f}_热$	81.3	85.9	90.5	95.2	99.8

另外,由于凝结盒容积比较小,只有20立方厘米左右,内装干湿球热电偶,而且离核滴一般仅1厘米以内,湿球上水分不断蒸发,这对核滴凝结增长不无影响。这种影响一般总是使核滴附近空气湿度变大,在空气较干燥时这种润湿作用尤为明显。

③ 理论平衡增长率的计算误差:凝结平衡增长率公式(2)是根据拉乌尔定律导出的,它适用于理想溶液。当溶液浓度较大时,与实际溶液偏差较大,反映在离解系数上,盐核的离解系数 i 是溶液浓度的函数。① 考虑 i 随浓度的变化,对不同湿度下用迭代法求得相应的离解系数 i 并据以计算平衡增长率,结果列于表7。可见,当相对湿度小于94%时,表5中盐核的计算平衡增长率比表7中相应的计算值偏小0.04~0.16。

① 南京大学气象系,云物理学(讲义),1978。

表 7　考虑 i 是浓度的函数时盐核的平衡增长率

f/%	80	85	86.5	90	92	93.8	95	98
i	2.57	2.35	2.30	2.15	2.10	2.02	1.95	1.85
K_m	2.04	2.18	2.26	2.43	2.59	2.79	2.96	3.95

综合上述讨论，尿素核和盐核的实验平衡增长率分别有 0.2～0.3 和 0.2～0.6 的偏差。

(2) 尿素核和盐核作为暖云催化剂的效能

尽管在核滴大小和空气湿度的测量上还存在一些误差，但从显微摄影(图略)和有关数表可以看出，在 90%以上的湿空气里，尿素核直径可凝结增长 1.5～1.8 倍，起始吸湿增长 15 秒即可达到 1.4 倍左右，而盐核的相应增长率则可达到 2.6～3.0 和 2.3 倍左右。在云内饱和或接近饱和的环境空气条件下，按公式(2)的增长率与空气湿度的关系推断，尿素核直径增长率可达 2.0～2.5 倍，而盐核则可增长 3.0～3.5 倍。作为暖云催化剂，与播撒水滴相比，1 吨尿素核相当于 6～12 吨水，而 1 吨盐核则相当于 12～20 吨水的效能。从凝结性能看，1 吨盐核的效能大约相当于 2 吨尿素核的效能。

参考文献

[1]　叶家东，人工凝结核的实验研究，气象学报，32 卷 3 期，232—239，1962。

[2]　B.J.梅森，云物理学，中国科学院大气物理研究所译，科学出版社，1978。

4 人工降水试验设计和效果统计检验研究

4.1 从皖南到古田——人工增水记事

叶家东

（南京大学大气科学系）

1958年12月，在气象系系主任徐尔灏教授领导下，大气物理专业正式成立。我作为一名预备教师，参加了大气物理专业成立的第一次教研室会议。当时的专业方向是以云雾物理为中心，边界层气象和雷达气象为两翼。在云雾物理领域，从1959年开始有两项重要举措：一是教学上聘请苏联专家来华讲授云雾降水物理学，进修班中包括北大、中科院地球物理研究所和中央气象局等6个单位的10多名进修教师，一时间成为我国云雾降水物理教学的中心。结合教学开始筹建我国第一个云雾物理实验室。二是与安徽省科委及中科院协作开展飞机暖云人工降雨和皖南地区地面暖云人工降雨试验研究工作。针对当时人工降水作业缺乏科学设计、催化效果难以评估的情况，徐尔灏教授按照历史区域回归的方案设计皖南人工降雨试验，成为我国第一个有科学设计的人工降雨试验。这两项活动确立了我系云雾物理专业方向在全国同行中的相应地位。1964—1965年，我系与上海民航局及中科院地物所协作，在沪宁地区进行了一系列飞机人工降雨和云的考察研究，取得了十分珍贵的夏季积云和冬季层状云的宏微观结构资料。结合教学在庐山、衡山参与了一系列高山云雾考察。这期间我系云雾物理的科研和教学活动都是十分活跃的。1972年全国人工降水会议在长沙召开，为了推动人工影响天气科学事业健康地发展，减少试验的盲目性，加强科学性，会议开设了一次全国性的人工影响天气讲习班。在八个专题中我系教师主讲三个专题。当时由副系主任陆渝蓉率代表团参加会议。我承担的任务是主讲"人工降水的效果检验"，接受任务时，几经思虑才下的决心。因为这个专题在人工降水实践中是难度较大且比较敏感的问题，它本质上是属于"质量检验"范畴的科学，容易在某些方面与实际作业单位或人员产生认识上的脱节，在系领导的理解和支持下，我在该专题中主要论述了客观评价试验效果的必要性和难点，并提出在现有条件下，随机试验是一种可行的客观评价试验效果的科学方法，当然要付出代价。讲授结果获得了不少业务科技人员的热烈反响，南方至少有三个省的同行表示协作搞科学试验的意向。这些单位的业务科技人员做了巨大的努力，与我们开展了协作，并在此基础上共同筹组了南方人工降水协作片，定期开展学术交流和研讨活动，对促进我国南方的人工降水事业起了重要的推动和协调作用。福建省气象局的科技人员愿与我系协作搞随机试验，于1974年开始开展了我国第一个人工降水随机试验——古田试验，试验方案由我提出，称为区域回归随机试验，目标是雨季增加水库蓄水发电。设计思想是在水库流域设一目标区，在上风方选一地形条件类似的对比区，用常规天气预报和雷达观测判断有利的天气条件以确定试验单元，随后按随机抽样原则确定目标区是催化或不催化，对比区则一直不催化。这样获得四组资料，即催化单元的目标区雨量和对比区雨量，对比单元的目标区雨量和对比区雨量，其中只有催化单元的目标区雨量是实际催化的，其他三组都是自然雨量。由于催化单元与对比单元的划分是随机确定的，所以有理由假定催化单元目标区的自然雨量与对比区雨量之间的回归关系与对比单元的相应关系是一致的，于是可以从对比单元两区雨量的区域回归方程，根据催化单元对比区的实测雨量去推断目标区的自然雨量，它与该区实测雨量之差就是催化效果。这样的效果称为统计效果，它比较客观。当试验次数较多时，可用统计检验的置信水平来估计其可信程度。直到现在，这种观点仍然是正确的。世界气象组织 1992 年 7 月关于人工影响天气现状的声明中指出："在目前可以接受的评价实践中，各种随机化方法被认为是最可靠的检测云播撒效果的方法。像这样的随机化检验要求有大量的个例，每个个例根据水的自然变率和预期效果都容易计算。在信噪比非常低的情况下，可能需要 5～10 年以上的试验期限。"

古田试验期间，福建省气象局参与外场试验的业务科技人员在试验区雨量网的配置、雷达站的筹建、试验点的设置以及日常作业中，克服了许多意想不到的困难，坚持在极其恶劣的自然条件下尽其所能按设计方案，实施高炮播云试验，为试验成果的累积奠定了基础。当然，思想认识上的波动也是难免的。当试验的前期成果获得 1978 年全国科学大会奖后，随着外场主要业务科技人员的更迭，曾提出动议，去闽南另辟试验基地。当时我们基于随机试验需要较长的试验期才能获得较为可靠的结果的特点，坚持在古田按原方案继续试验，并告诫：如果搬到闽南，再搞三年也达不到古田现在这样的水平。终于排除了南迁闽南另辟蹊径的动议。结合试验，我们提出了一整套统计检验方法，实践中我们发现，在古田雨季的气象条件下，催化效果是随着自然雨量的变化而改变的，从统计上讲这意味着试验样本的雨量方差会因催化而改变。数理统计中现有的回归分析方法不能处理这种情况，于是我们发展了方差不相等的双样本回归分析方法。这项成果获江苏省重要科技成果三等奖。

古田试验历时 12 年，试验单元总计达 244 个，统计效果平均增雨 23.8%，统计显著度达 0.01 以上。这项试验是我国第一个被世界气象组织认可并备案的有科学设计的人工降水随机试验计划。前期成果在世界气象组织第三次人工影响天气科学会议上报告交流，获得广泛好评，美、欧、澳、加等国的学者先后来函进行交流或索取详细材料。世界气象组织主管人工影响天气事业的副秘书长利斯特教授在来函中说："你们的研究工作给相当一部分人留下深刻的印象，我们全都希望你们将这项研究继续进行下去。"古田试验的成果"人工降雨效果及其检验方法的研究"获 1989 年国家气象局科技进步二等奖。

4.2 《人工影响天气的统计数学方法》前言

二十多年来，我国人工影响天气试验的规模日趋扩大，其中人工降水和防雹，不少地区曾经或正在作为群众性的抗灾斗争的一种重要措施加以推广。社会生产的发展迫切要求提高人工影响天气的科学水平和试验效果。从根本上讲，只有当人们透彻地认识成云致雨的物理过程，并且能够人为地改变自然过程中某些关键环节的时候，人工影响天气才能真正成为“控制天气”的科学实验。问题在于客观实际的需要常常不能等待科学家把一切都研究清楚了再从事人工影响天气的实践。事物往往是相互牵制而又相互促进的，云和降水物理过程的研究是人工影响天气的科学基础，而人工影响天气的实践又是云物理学发展的动力，在我国尤其如此。因此，如何在现有的科学基础上，科学地组织和客观地评价人工影响天气试验，是一个不能回避的、在实践中且是很重要的问题。由于目前人们对云雨自然过程的认识以及干预自然过程的能力都相当有限，风云变幻又是无奇不有、错综复杂，致使人工影响天气试验的对象和结果都存在着一定的不确定性。因此，在组织人工影响天气的外场试验以及评价试验的效果中，普遍采用统计设计和统计分析的方法，它与试验的物理设计和物理分析是相辅相成、互为补充的。

本书是将近三十年来国内外人工影响天气试验中的统计设计和统计分析的方法作一力所能及的概括，以期对国内人工影响天气试验的进一步开展有所裨益。为了使读者掌握有关的统计分析方法，本书较系统地介绍了线性代数和统计检验中有关的基本概念和方法；回归分析是效果检验的一种重要方法，所以专列一章加以讨论。所举的例子基本上都是国内外人工影响天气试验中实际采用的，目的是给初学者一些具体的实例，便于加深理解。

我国研究人工影响天气的效果统计检验工作首先是徐尔灏教授做的，谨以此书表示纪念。

由于我们水平和经验有限，错误与不当之处在所难免，请读者批评指正。

4.3 人工降水的试验设计和效果检验

叶家东

（南京大学大气科学系）

一、前言

人工降水的试验设计和效果检验是一项重要和复杂的科学研究工作。长期以来，在人工降水的科学试验中，围绕着试验设计和效果检验，有过许多争论。可以说，人工降水科学试验的水平，在某种意义上讲，是取决于效果检验的可信程度。在我国，自从1958年广泛开展人工降水试验以来，在试验设计和效果检验方面做了不少工作，但总的说来，试验设计和效果检验至今还是一个薄弱环节，需要认真研究，逐步解决。本文主要讨论人工降水的试验设计、效果检验以及资料分析方法方面的有关问题。

二、试验设计

试验设计指的是一个详细的研究计划，包括如何搜集资料、分析资料，使得预订的设计能恰当地加以检验和证实，从而做出推断。从国内外现有的效果检验有关的设计方案看，大体上可以分成如下几种：

1. 序列试验

采用单个试验区，一般用月雨量等气候量作为统计变量，利用历史平均值作为试验期自然雨量的估计值。这种方案检验效果的功效是很低的，要在可接受的显著性水平（例如 $\alpha=0.05$）下检验出试验效果来，则效果必须超过一倍以上，而人工降水增加月雨量，常常达不到这么大的程度。只有试验许多年，才有可能较确切地检验出相应的效果。这种方案很少采用，我国只有在防雹试验中才用它。

2. 上风方、下风方的对比试验

以作业点(或作业线)的下风方为目标区，上风方相应的区域为对比区，假定两区自然雨量相等，以试验期上风方的雨量作为目标区的自然降水量估计值，再和实测雨量比较确定效果。这种设计方案只对平原地区层状云降水较为有用，对于地形复杂的地区或积状云降水不宜采用。因为两区自然雨量相等的假定缺乏依据。一般总是选降水条件比较有利的下风方作为试验单元进行作业，而上风方是否有利则不去管它，上风方有利而下风方不利的情况一般是不会去作业的。为了使这种方案的依据比较客观，必须找一些对比单元。例如，在侧风方相应地选一个对比上风方和对比下风方，看看它们之间的雨量是否相等，如果不等，就应采用比率分析法或回归分析法进行检验。

3. 区域回归试验

这是我国目前用得较多的一种试验设计方案，国外一些抗旱的作业性试验，至今仍采用这种方法。这种试验需要选择一个(或几个)和目标区的天气条件、地理条件相似，区域雨量密切相关的对比，对比区的雨量不受作业的影响。根据两区的历史雨量资料，建立区域雨量历史回归方程，然后将试验期对比区的雨量代入回归方程，即可求出试验期目标区的自然降水量的估计值(期待值)，再和实测雨量比较以确定效果。这种试验方案实际上是把对比区雨量作为预报因子来用的，如果对比区选择得当，可用的历史资料较长，分析功效还是比较高的。例如福建古田县1974年8月份，用高炮发射AgI(碘化银)试验一年，获得增加月雨量24.7%、显著性水平$\alpha=0.05$的结果。在结合抗旱开展的作业性试验中，回归试验仍是一种可行的办法，需要充分挖掘潜力，以提高分析结果的客观性和准确性。

历史回归试验的依据是气候相似，如果大气环流形势有长期的系统性变化，这种依据就不充分，这个问题当历史资料较短时尤为突出。历史回归分析不能预测历史上所未曾经历过的现象。

正因为区域回归分析是根据历史气候相似性原则建立起来的，它所用的统计变量一般都是气候量，如季雨量或月雨量。但这样取的结果，势必会冲淡试验效果以致一时鉴别不出来。能否根据天气相似的原则来建立历史回归方程呢？湖南凤凰县的试验对此做了尝试。取统计变量为区域面积平均日雨量，选取历史相似天气的判据有两个：天气形势相似和对比区雨量相似，据此找出历史上和试验日相似的天气日，建立历史相似天气日雨量的区域回归方程，再由试验日对比区日雨量代入回归方程求出试验日目标区自然日雨量的期待值，和实测雨量比较确定效果。如果相似天气的判据选择得更恰当，区域日雨量的相关性有所改善，这种根据天气相似原理建立起来的回归分析方法的功效会有所提高。

4. 随机试验

鉴于回归试验存在一些固有的问题，从20世纪50年代后期开始，国外大多数研究性试验设计都采用随机方案。其基本思想是将宜于催化的云(或日子，或降水过程)随机地分成两组，一组播云，一组不播云留作对比。如果在这两组机会中雨量出现显著的差异，就可以归因于作业；如果只用目标区本身的资料，则这种试验叫作单区随机试验。

随机试验的优点是在较大程度上避免了主观推测，如果试验对象能限制在物理过程比较单一的范畴之内(例如层状云降水或积状云降水，冷云降水或暖云降水)，则在两组机会中把与人工影响无关的因子作为随机因子就有了依据。只是在这种前提下，随机试验的结果才真正符合随机抽样的原则，统计检验的方法才能在它本来的含义下恰当地加以运用。

单个目标区的随机试验功效太低，难以在较短试验期内检验出一定的效果，所以常常附以一些控制变量，以提高其功效。常用的有下面几种：

(1) 区域回归随机试验：福建古田县的试验就采用这种设计方案。在目标区附近选一对比区，目标区的试验单元按随机规则决定是否播云，对比区一直不播云，也不受目标区播云作业的影响。因为目标区是否作业是根据随机规则确定的，所以可以假定作业单元目标区和对比区的自然降水量的区域相关关系是和对比单元一样的。于是可采用和历史回归分析相仿的程序，根据由对比单元目标区自然降水量的期待值，进而和实测雨量比较确定效果。由于这里利用了对比区雨量作为控制变量(协变量)，作业单元目标区的自然降水量的估计值就比较准确，从而可以提高试验的检验功效。古田县用小火箭发射冷云催化剂播云的方法试验两年，62 个单元，就得出三小时区域面积平均增雨量 40%～78.6%的结论，统计显著性水平达到 0.005 以上。

(2) 随机交叉试验：选择两个雨量相关性好而彼此又不污染的试验区，每个试验单元都按随机规则决定在其中一区作业，另一区对比，然后比较作业日雨量和非作业日雨量以评定效果。由于每一个试验单元都在某一区进行了作业，可望试验功效会有所提高。Moran 假定 n 是单区随机试验为在一定的显著性水平检验出一定的效果所需要的试验单元数目，则在一定的假定条件下，区域回归随机试验所需要的试验单元数目 $n\dfrac{(1-r)}{(1+r)}$，随机交叉试验相应的试验单元数目为 $n\dfrac{1-r}{2}$，这里 r 是两个试验区的雨量相关系数。徐尔灏也曾独立地试验过，指出随机交叉试验的功效比单区随机试验高。以色列人工降雨试验采用随机交叉试验方案，试验 7 年，得出日雨量的平均增量为 15.3%(全区)和 21.8%(内地)的结论，显著性水平分别达到 0.009 和 0.002。如果把催化剂对对比区的污染因素除去，则相应的增雨量是 22%和 31%。

上述随机试验的设计方案中一个难以完全避免的问题是催化的污染问题，有两种污染，区域之间的污染和试验单元之间的污染，后者叫作“持续性效应”，有证据表明，播云后数天内冰核数仍维持相当高。

(3) 对云块的随机试验：统计变量是一块云或一群云的降水量，而不是区域雨量，这要用经过仔细校准的雷达进行观测。这种试验针对具体的云进行比较，针对性较强，分析功效也较高。美国佛罗里达州的积云动力催化试验就采用这种设计方案，选择一对符合试验条件的云(或一块云)随机地决定其中一块作业，另一块留作对比。可用比率分析法比较这两组云的降水量以确定效果。比较好的分析方法是利用播云前一段时间内云的自然降水量作为控制变量，再利用和区域回归试验相类似的方法分析效果，这种分析方法功效较高。

随机试验要放弃一半左右的播云机会，这与抗旱斗争的要求是相矛盾的，所以一般在作业性试验中都不大采用它。但做试验应该点面结合，在一些地区建几个试验基地，侧重于研究性的试验，利用较客观的方法来检定某种播云技术和方法是否有效以及在什么条件下有效，是很必要的，只有通过比较细致的研究性试验，才能不断提高人工降水的科学水平和实验效果。

三、效果检验和分析方法

1. 效果的统计检验

统计检验的主要内容是显著性检验，即如果实测雨量和估计雨量之间有差异，就要对这个差值进行统计检验，指出由于降水的自然变差引起这么大差异的可能性有多大。如果这个可能性很大，我们就没有理由认为人工影响显著地改变了雨量，即所谓效果不显著；如果这个可能性很小，例如小于5%，我们就有较大的把握说人工影响有效，即所谓效果显著。上述可能性的大小通常就叫作显著性水平，用 α 表示。

显著性检验在数理统计中常叫假设检验，可分为参量性检验和非参量性检验。在人工降水试验中应用较多的参量性检验法是 t -检验法。t -检验要求统计变量服从正态分布，且要求人工影响的措施只改变总体平均值，不改变总体方差，也就是说人工影响前后的两个样本方差不应该有显著的差异。这两个条件不满足，t -检验的依据就不成立。因此，在决定用 t -检验法以前，要对统计变量的分布进行分布函数检验，常用的方法有 χ^2 -检验和柯尔莫哥洛夫拟合度检验法。这些检验法都要求大样本，一般样本容量应大于50和30。如果检验的结果与正态分布偏离较大(例如拟合度小于50%)，就应该设法使变量正态化，通常用变数变换的方法，例如取雨量的对数，方根、立方根或四次方根比对数拟合得更好。国外一般认为雨量是服从 Γ -分布的，但考虑到用等概率变换法将 Γ -分布正态化，手续较烦，所以一般宁愿采用简单变换的方法。如果拟合度还可以接受，取对数的变换方法是比较好的，因为它将雨量的比值变换成对数的差值。考虑到近年来关于非零降水量作业单元的人工影响的效果，倾向于认为是可乘的而不是可加的，而一般经典的统计检验方法则常常是检验样本平均值之差值的显著性，这是一个矛盾。采用对数正态分布，则雨量对数的差值正好是雨量真数的比值，检验对数的样本平均值之差值的显著性，就相当于检验效果的可乘因子的显著性。这就自然地将效果的可乘性和通常关于差值的统计检验方法一致起来了。如果取 Γ -分布，检验效果比例因子(反映在 Γ -分布上是尺度参数的改变)的显著性就要用Neymann等人建立的 $C(\alpha)$ 检验法。

拟合度多大才是可接受的呢？有人把拟合度大于0.05作为可接受的界限，这不妥当，因为0.05是拒绝原假设的判据，而要接受原假设，0.05是太小了。我们取0.50或0.80作为接受原假设的判据，这本来是一种约定，对于雨量来说，我们认为不小了。

作业样本和非作业样本方差均匀性的检验可用 F -检验法，检验结果如果 $F>F_{0.05}$，则人工影响仍不能应用 t -检验法，而应该用Welch检验法进行显著性检验。

在不知道统计变量属于何种分布的情况下，显著性检验一般采用非参量性检验法，如符号检验法、秩和检验法以及斯米尔诺夫检验法。秩和检验也要求两个样本所代表的总体方差(如果存在)相等，所以也不是无条件的。

2. 效果的物理检验

所谓物理检验，指的是根据云和降水的形成及其人工影响的物理机制，找出相应的物理效应，如微物理效应或宏观动力效应，作为效果的指标，进而通过试验来检验人工影响是否显著地改变这些指标。需要指出的是，所有这些指标，如同雨量一样，也是受许多因素制约的，它们在试验中也不是唯一地确定于播云措施的，存在着相当大的自然变差。要从中鉴别出人工影响的效果，除了效果特别显著或云的结构的时空变差特别小的情况以外，一般仍需采用统计检验的方法。所以，这种物理检验确切地说可以叫作物理效应的统计检验。在微物理效应方面，目前国内除了云中冰晶浓度在人工影响以后曾观测到有显著增加以外，其他物理量指标还没有见到有显著差异的例子，这与观测仪器、探测工作等一系列技术问题有关，是需要大力加强的领域。下面举一个不用统计学的物理检验例子。美国在阿里桑那州曾用两个 AgI 焰弹(每弹含 AgI 25 克)从积云的主体云顶垂直投下进行播云，播云后 7 分钟，观测到云顶上升 1.35 千米，主体云内温度上升 0.75 ℃，在 −5 ℃层的冰晶浓度达到 10^2～10^3 个/升，核心部位几乎完全冰晶化了，比原先增加 3 个量级，液态含水量小于 0.004 g/m^3。上述动力效应和 Weinstein 等人的模式中计算比较甚为一致。为了进行这种检验，需要两架性能良好的飞机和一整套微物理观测仪器配套观测。近年来有一种日益明显的趋势，就是把一系列物理效应的检验和最终的降水效果联系起来，如对佛罗里达州动力播云试验的结果进行分析表明，物理效应和降水效果的相关甚为密切，一般讲，降水效果只有在物理效应上也能得到恰当的验证或解释时，才真正令人信服。

3. 效果的分层统计

效果检验的一个很重要的问题，是要判断在什么条件下有效或效果较好，也就是说要根据试验结果确定有利的天气条件和试验对象，这就要对试验结果进行分层统计，把试验资料按某种物理属性的取值大小分成几个层次，然后分别对各个次级层次的资料进行效果检验。分层的方法很多，福建古田县试验按对比区雨量的大小分层统计，结果发现对比区雨量强度小于1 mm/h，播云效果比较好，相对增加率超过 100%，显著性水平 $\alpha<0.01$，而当对比区雨量强度大于 3 mm/h 时，播云效果就很不显著，特别对积状云降水，效果是负的，当然这个负效果也是不显著的，属于自然变差的可能性大。Grant 等根据云顶温度对 7 个随机试验分别进行分层统计，发现云顶温度介于−10～−25 ℃的层次，播云效果最好，于是把这个范围称为播云有效的“温度窗”。Gabriel 对以色列的试验结果按大气层结特性分层统计，发现各个气象因子的取值范围为下述数值时效果较好：(1) 700 mb 的温度为 −5～−7 ℃；(2) 500 mb风向介于 300°～210°；(3) 500 mb 风速介于 0～35 knot；(4) Showarter 稳定度指数介于 4～7；(5) 从地面到 500 mb 的可降水量为 0～11 mm。他还发现上述各指标对效果的影响彼此并不相关，于是提出一个综合指标：各物理量取值位于上述范围之内者，其指标数为 1，在其外者指标数为 0，然后将试验结果按综合指标分层，结果发现综合指标为 2 和 3、4 时，增雨量分别达到 36.4 和 52.8%，相应的显著性水平达 0.051 和 0.016。

分层统计的重要性还在于它有时能提供新的证据，有些是我们暂时还不理解的甚至是与成见不相容的证据。例如在瑞士的地面烧 AgI 烟的防雹随机试验中，对降水的效果进行分层统计，原先认为大气存在稳定层时效果一定不好，所以想通过按有无逆温层分层统计，剔除这种“不利”的因子。谁知分层的结果发现事实与设想恰恰相反，当低空存在稳定层时效果很显著，而不存在稳定层时，效果大为减小。只是在事后，人们才对此做出气象上的解释：阿尔卑斯山南坡白天增热，破坏了稳定层，原先积聚在稳定层下的催化剂随着云的发展而集中释放出来，从而提高了播云效果。

4. 数值模拟试验在效果检验中的作用

Scott 提出应该对播云条件有利程度不同的云进行试验，然后通过分层统计找出真正有利的播云条件，这样才会不为错误的先入之见所迷惑。但这种观点在我们看来，无异于因噎废食。由于害怕先入之见不正确而一概拒绝气象和物理知识的引导，是不可取的。而且，这种泛泛的试验，效果检验的功效很低。Neymann 和 Scott 自己就曾经指出，在一定的假定条件下(例如雨量服从 Γ-分布，其密度函数的形状参数为 0.70，尺度参数为 0.20)，单区随机试验的效率是很低的，如果显著性水平取 0.10，并且取试验单元 $n=100$，则要鉴别出 20%的播云增雨效果的概率最多只有 0.20，当增雨量为 50%时，这个功效为 0.60，仍不很大，要知道在人工影响天气的试验中，$n=100$ 已是很大一个数了。功效不高是单区随机试验的主要弱点。为了提高统计检验的功效，需要选择适当的预报因子(回归分析中对比区的雨量就是一种预报因子)，在一定的预报因子取值条件下，研究对象(统计总体)的条件变差，即自然降水量的条件变差就可以大大减小，从而提高了统计检验的功效。云和降水物理过程的研究，特别是数值模拟试验可以提供这种预报能力，不仅能够预报自然降水过程，而且能够预报人工影响后的降水过程，至少大致上能做到这一点。美国佛罗里达州的积云动力播云试验表明，利用数值模拟的这种预报能力，使试验的盲目性大为减小，试验效果也更为显著，增加雨量 3 倍之多！中央气象局人工影响天气研究所以及有关单位结合人工降水试验，开展了数值模拟研究。这类模拟试验不论是对人工降水试验，还是对积云发展和降水机制的研究都是有意义的，是需要大力发展和研究的领域。

四、结束语

我们是在对云和降水的物理过程有一定的认识，但认识还是不十分确切的情况下从事试验的。而且，我们是在对人工影响的措施对云和降水的物理过程的作用有一定的了解，但了解是十分不全面的情况下从事试验的。正因为这样，试验效果不是确定无疑的，所以效果的客观评价就显得格外重要。也正因为我们目前的认识能力有限，所以效果的检验就成为一个相当困难的问题。尽管有一些统计检验方法可供客观评价之用，但这种统计效果所提供的只是一些证据，而不是证明，只有统计效果能获得物理上的解释，并为观测到的物理效应所证实的时候，结论才是完整的、真正令人信服的。这是效果检验工作的目标，实际上这也是人工影响天气科学实验发展的一个重要方面。

要通过试验了解人工影响的效果和作用原理，这有赖于对试验的效果进行客观的评价，反之，要对试验效果进行客观的评价，又有赖于对自然的和人工影响的物理过程的了解，两者之间相互依赖、相互制约。所以说，效果检验必须和对云物理过程的研究结合起来，效果的统计检验必须和物理检验结合起来，这是效果检验的发展方向。

目前有两个主要的困难阻碍着效果检验研究工作的深入开展，一个是缺乏必要的观测设备和观测手段，另一个是试验往往缺乏合理的周密的设计。前者是条件问题，后者是实事求是的科学态度问题。首先是观测设备和观测仪器问题，这个问题的解决不仅对效果检验是重要的，对整个云物理和人工影响天气的研究都是十分重要的，而这方面我们和国际水平的差距也最大，所以是一个特别突出的迫切需要解决的问题。效果检验没有可供分析的完整的资料，结论只能是含糊不清的。有了必要的观测手段，进一步的问题是要研究试验的方法。试验无计划，就不能合理地取得资料，有了一些资料也不能恰当地加以运用，从中获取最大的信息供判断之用。从目前国内对人工降雨的效果分析来看，必须强调用科学的态度来对待这项科学试验，试验必须有相应的设计，分析必须力求客观。效果检验本来是一项比较困难的研究工作，要达到客观性好、分析效率又高的检验水平，还有许多工作要做，这要有各方面研究工作的相互配合。“科学有险阻，苦战能过关”。效果检验这一关，不能回避，只能攻克。

4.4　福建省一九七四年八九月份人工降水效果的统计分析

福建省革命委员会气象局　　南京大学气象系

摘要　一九七四年八九月份，在福建古田水库地区设置影响区和对比区进行了高炮人工降水试验。采用随机回归试验方案，利用统计方法检验效果。对系统天气(主要是台风)下的雨层云，以三小时为试验单元，进行了八次随机试验，其中催化四次，平均增加雨量 9.1 毫米/3 小时，相对增加率达 99%，用多个事件 t-检验法，检验平均增值的显著性，信度 $\alpha<0.05$(单侧检验)，雨量增值的 90%置信区间为 $\Delta y_k'>2.51$(毫米/3 小时)〔单侧〕。同时，结合抗旱，八月份共催化 14 次，对月雨量进行了回归分析，经人工催化增加的月雨量达 53.8(毫米)，相对增加率为 24.7%。利用 t-检验法检验其显著性，信度 α 接近 0.025(单侧检验)，雨量增值的 90%置信区间为 $\Delta y'>23.11$(毫米)〔单侧〕。

福建全年雨量主要靠雨季(4～6 月)和台风降水，但 1974 年雨季未能降较多的雨，到了夏秋，旱情严重，古田水库的水位急剧下降。而古田水库是全省最大的水库，福州、三明和南平等地的城乡工、农业用电都依赖古田水电站，若水库缺水、发电量不足，将严重影响全省工业生产，农业生产也颇受影响。在此情况下，我们分析了旱情，根据历史资料，夏秋之交古田地区主要降水系统是地方性对流云(积雨云)和台风(雨层云、积雨云)，其云顶都超过零度线，属混合云。所以在八九月份可以指望通过冷云催化的原理进行人工降水试验。只要云体过冷却层较厚，云顶温度又不太低(>－20 ℃)[1]，向云的过冷却部分播撒冷云催化剂——碘化银或四聚乙醛后，利用它们的成冰性能，在云的过冷却部分发生贝吉龙过程，就可使原来缺少冰晶核的云体提前或者增加降水。

这次试验从 7 月下旬开始到 9 月下旬结束，历时两个月零五天。参加试验的单位有福建省气象局、南京大学、西安化工研究所、古田县革委会、人武部、省古田水电站和宁德、三明、建阳等地区气象站，并组织了古田、宁德、尤溪、建瓯、屏南、南平和闽清等县(市)近 60 个气象、水文雨量站。在省委、县委一元化领导下，组成试验小组担任了抗旱、科研的任务。试验以“三七”高炮发射 AgI 弹催化降雨为主，结合进行小火箭发射四聚乙醛催化降雨试验。

一、作业方案的设计

目前人工降水的一个关键问题是所催化的云能否在影响区增加雨量？若增加，能增加多少？由于降水的自然变差很大，人工催化增加的雨量又常常和自然降水混在一起，用一般的观测方法检验效果，人为的因素较多，很难回答所降的雨究竟是人工催化的结果，还是自然过程本来会下的。统计方法在一定程度上能够回答这个问题。所以，这次试验主要采用统计方法进行效果检验。为此，对大范围降水性云系(主要是台风系统)进行随机回归试验[2]。设计思想是这样的：选择两个试验区，A 区为影响区(目标区)，B 区为对比区(控制区)。根据天气预报、探空和雷达观测资料以及当地的天气实况观测，当宜于催化的天气条件出现时，以三小时为试验单元，按照事先制定的随机程序，随机地决定 A 区催化或不催化，得到两组雨量资料 $y_{A催}$ 和 $y_{A不催}$。B 区一直不催化，作为对比。与 A 区相对应，B 区也

可得两组雨量资料 $X_{A催}$ 和 $X_{A不催}$。这四组雨量资料中只有 $y_{A催}$ 是人工催化后的雨量，其余均系自然降水量。根据 $y_{A不催}$ 和 $X_{A不催}$ 建立区域回归方程，再将 $X_{A催}$ 代入回归方程求出 A 区催化单元的自然降水量的期待值 $\hat{y}$，比较 $y_{A催}$ 和 $\hat{y}$，即可评定人工降水增加的雨量。另外，结合抗旱，八月份大多数有利的天气条件，在 A 区都进行了人工催化作业，因此对八月份的月雨量又进行了回归分析，即根据历史上 A 区和 B 区八月份的月雨量资料建立回归方程，再以今年八月份 B 区的实测雨量代入回归方程，求出 A 区自然降水量的期待值，再和 A 区实测雨量比较以评定人工催化的效果。

试验区的设置：选择影响区和对比区的条件是两区天气条件、地理条件相似，两区雨量密切相关，这个条件满足了，回归分析的效率就高[3]。为此，在福建省古田溪水库上游选了一个东西向 30 千米、南北向 40 千米的面积为 1 200 平方千米的影响区 A，在它的正西向另选一个面积相等的对比区 B(图 1)，两区都是山区，平均高度 400 米左右，两区内又各有一条自北向南的溪流，地理条件大体相似，又因为两区间隔仅 20 千米，所受的天气系统影响基本上是相同的。我们又统计了历史上五年八月份的月雨量资料，结果发现两区面积平均月雨量(对数值)的相关系数 $r=0.997$。用 t 分布检验其信度 $\alpha<0.001$。由此可见，两区八月份的月雨量相关很好。

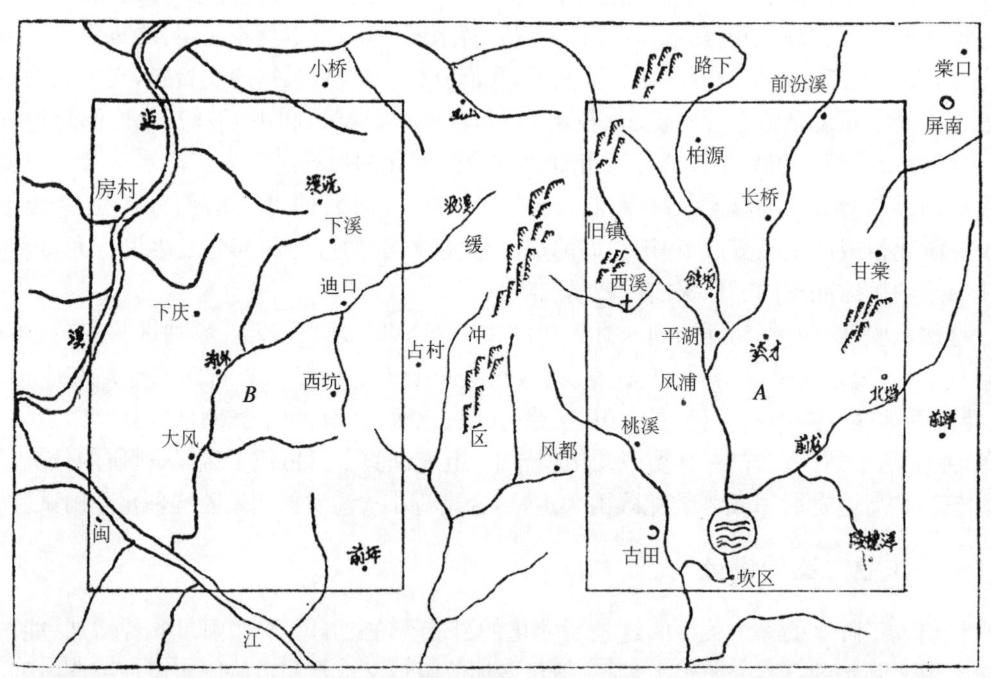

图 1　试验区的地理位置

为了防止影响区 A 催化作业时，催化剂对 B 区的污染，我们统计了历史上四年福州地区 07 时 700 mb 风向，八月份偏东风只占 36%，而九月份只占 17%。由此可见，八九月份福州地区高空吹偏东风的频数是少数。从天气图上看，主要天气系统和偏西风相吻合。所以，把对比区设在影响区的西边是合宜的。此外，在 A 区和 B 区中间又设了一条宽 20 千米的缓冲区。如此设置，对比区基本上不受影响区催化作业时催化剂的污染影响。

雨量站的配置：雨量是这次试验的主要资料。我们在 A 区和 B 区内外组织了 56 个雨量站，其中雨量自计的有 28 个，人工观测八段制（三小时观测一次）有 16 个，人工观测二段制（十二小时观测一次）有 12 个。A 区和 B 区内的雨量站分别有 16 个和 9 个（图 1）。

随机程序的制定：前面讲过，A 区催化或不催化是按随机程序决定的。为此，我们从随机数字表[3]上抽取若干随机数码，顺序编号作为签码，密封备用。事先规定，单数为催化单元，双数为非催化单元（或反之）。为了避免试验人员在判断作业条件时掺杂主观偏见（即如果催化，就选好一点的天气条件；如果不催化，就选差一点的天气条件），上述随机签码是由不参加现场试验人员制定的，现场试验人员事先不知道它的内容，他只是根据天气预报、探空和雷达等观测资料，凭借现有的关于宜于人工催化的天气条件的知识，做出适宜性的判断。如合宜，就作为试验单元，然后顺序抽取一个随机签码，如为单数，就催化；如为双数，就不催化，作为对比，但观测工作照样进行。

有利的天气条件和试验单元：试验时，作业点根据福州、南昌和杭州三个气象台的天气预报，当有利于降水的天气条件出现时，即和雷达联系，并根据福州当天的探空资料和雷达 RHI 回波，定出影响区上空的云层高度、厚度、过冷却层厚度以及高空风向风速。若云厚超过 6 000 米，过冷却层超过 1 500 米，700 mb 高空又偏西风，就作为试验单元。

试验单元的时间我们取三小时，这是因为这次随机试验的对象是雨层云，当时 700 mb 高空风速达 16 米/秒左右（8 月 20 日 07 时），这样，作业云仅在半小时就移出 A 区，但实际上云体移动速度比高空风要慢，从雷达 PPI 回波来看，作业云实际移动速度为 20 千米/小时，所以作业云移过 A 区要 1.5 小时。实际打炮催化作业的持续时间一般为半小时左右，这样从开始作业到作业云移出 A 区，历时约 2 小时。又因为不少雨量站原来就是人工八段制观测，所以就定为三小时一个试验单元。这样前一催化单元所播撒的催化剂基本上不会残留到下一试验单元，从而在一定程度上避免了试验单元之间的污染问题，并且这样从同一总体中抽取的随机样本的容量相应地可以增多。

雨量资料的整理：参加统计分析的雨量资料，我们取区域面积平均雨量。三小时雨量采用加权平均法求取，即根据试验区及其周围的雨量站三小时的雨量资料，作出试验单元试验区的等雨量线图，利用求积仪分区求出平均雨量值，然后求出全区的三小时面积平均雨量。在对八月份月雨量进行回归分析时，则是取各站的简单平均求取区域的面积平均月雨量。

二、发射工具及其性能

这次人工降水试验以“三七”高炮发射碘化银炮弹催化降雨为主，同时也搞了三次自制的“FJ－50－5”型小火箭[4]发射四聚乙醛弹头的降雨试验。

“三七”高炮降雨弹，每发可装 4 克或 6 克碘化银。引信是 9～12 秒、13～17 秒二种。在整个试验中，采用高角度打法，一般都在 65°～75°之间，爆炸时炮弹的垂直高度一般在4 500 米，加上作业点拔海高度（650 米），所以一般炮弹爆炸时的拔海高度为 5 000 米左右。一次催化单元要打多少炮弹，视当时云状、云体大小和过冷却层厚度及云的移动速度而定，各次作业的催化剂量列于表 1。作业时采取打打停停，一次催化单元内的打炮时间集中在半小时左右，有时也集中时间大剂量撒播。

自制“FJ－50－5”型小火箭，外壳由 0.5 毫米厚的铁皮卷制而成，发射的垂直高度为（2 900±100）米，最大飞行速度为 511 米/秒。每枚小火箭弹头内装有 4 克四聚乙醛。四聚乙醛是集中在药柱（TNT）内部压制而成。使用三轨发射架，可同时齐射三枚，也可单枚发射。发射导轨可在仰角 0°～90°范围和水平方位角 0°～360°范围内灵活转动。打法和高炮一样，就是催化剂的用量比高炮要多些，这是因为小火箭的发射高度较低（与高炮比较），但要注意四聚乙醛的成冰阈温要比碘化银高 3～4 ℃[5]。

三、雨量资料的整理和催化效果的统计分析

1. 关于雨量的分布函数的检验

我们知道，雨量的分布是偏态的，因此在作统计检验时就不能直接用 t -分布检验法。关于以一次降水过程为单元的雨量的分布形式，Thom 等[6]用 Γ -分布来拟合，Schickedanz 等[7]用对数正态分布来拟合。我们现在用对数正态分布来拟合月雨量和台风系统影响下三小时雨量的分布。

假设月雨量和三小时雨量服从对数正态分布，利用柯尔莫哥洛夫的“吻合度”检验法，对福州台的雨量资料进行统计检验。检验方法略述如下：设从任一分布为 $F(x)$ 的总体中随机抽取容量为 n 的样本，$F_n(x)$ 为其经验分布函数。若 $F(x)$ 为连续的，则成立柯尔莫哥洛夫定理[8]

$$\lim_{n\to\infty} P\{\sqrt{n}\ \sup_{n}\ |\ F(x)-F_n(x)\ |< y\}=$$

$$K(y)=\begin{cases}\sum\limits_{K=-\infty}^{\infty}(-1)^K \mathrm{e}^{-2K^2y^2} & (y>0),\\ 0 & (y\leqslant 0)。\end{cases}$$

利用柯尔莫哥洛夫定理检验总体分布的大致步骤是这样的，作一原假设 H_0：总体分布为对数正态分布 $F(x)$，首先由样本的经验分布 $F_n(x)$ 算出 $D_n=\sup P\,|\,F_n(x)-F(x)\,|$，当 n 相当大时，可以认为$\sqrt{x}D_n$ 的分布近似于 $K(y)$。这样就可以根据置信水平 $1-\alpha$，由 $K(y_\alpha)=1-\alpha$ 找出 y_α。然后比较$\sqrt{n}D_n$ 和 y_α，若$\sqrt{n}D_n<y_\alpha$，则接受 H_α；而若$\sqrt{n}D_n\geqslant y_\alpha$ 时，则拒绝 H_0。

我们利用上述方法对福州台 30 年的八月份月雨量资料进行统计检验。经计算得[9]

$$D_n^0(x)=\sup_{x}\ |\ F_n(x)-F(x)\ |=0.105\ 4,$$

$$\sqrt{n}D_n^0=0.577\ 3。$$

查 $K(y)$分布表[8]，对应于置信水平 0.95 的 $y_{0.05}=1.36$，所以$\sqrt{n}D_n^{(0)}<y_{0.05}$，于是我们接受 H_0：总体分布为对数正态分布。

实际上，从 $K(y)$分布表可得，对应于 $y_0=\sqrt{n}D_n^{(0)}=0.577\ 3$ 的 $K(y_0)=0.110$，于是由柯尔莫哥洛夫定理知

$$P\{\sqrt{n}D_n\geqslant y_0\}\sim 1-K(y_0)=0.89。$$

这就是说，我们的样本观测值与对数正态分布之间的差异是很小的，由于抽样变差引起的差异比 y_0 大的概率很大，达到 0.89，这个概率就是“吻合度”的指标，所以我们接受假设：八月份月雨量服从对数正态分布。

用同样的方法对台风系统下的三小时雨量分布进行检验：从历史上选取 6 个登陆台风

(6513,6611,6614,6615,7207 和 7209 号台风),登陆的位置在福建福清以北、浙江平阳以南沿海一带,与今年 13 号台风相似。从这 6 个台风降水过程中获得福州台 82 个三小时雨量的样本资料,经计算[9]得

$$D_n^0(x)=\sup_x |F_n(x)-F(x)|=0.078,$$

$$\sqrt{n}D_n^0=0.706<y_{0.05}=1.36。$$

于是接受 H_0:总体分布服从对数正态分布。实际上我们从 $K(y)$分布表可得,对应于 $y_0=\sqrt{n}D_n^0=0.706$ 的 $K(y_0)=0.305$,于是

$$P\{\sqrt{n}D_n^{(0)}\geqslant y_0\}\sim 1-K(y_0)=0.695。$$

这就是说,样本观测值和对数正态分布之间的差异可以归因于样本的抽样变差的概率是 0.695,意即用对数正态分布来拟合观测资料,吻合度是良好的,所以我们接受假设:在闽北、浙南登陆的台风降水过程的三小时雨量服从对数正态分布。

因此,在下面进行催化效果的统计分析时,雨量均取常用对数值,然后采用 t -分布检验法进行统计检验。

2. 月雨量催化效果的回归分析

这次人工降水试验共催化作业 16 次(附表 1),其中 14 次是在八月份进行的。八月份的催化对象主要是地方性对流云和台风影响下的降水性层状云系。我们对八月份的月雨量进行区域回归分析,看看人工催化能否增加月雨量。

为此,先根据历史资料分析一下影响区 A 和对比区 B 的月雨量相关情况。两区历史雨量资料比较完整的有五年。这五年,两区八月份都没有进行人工降水试验。A 区参加统计的雨量站有 11 个,B 区 9 个。采用各站简单平均法求取各区的面积平均月雨量(不计露水量),数据列于表 1。

表 1 历年的月雨量资料

类别 \ 雨量 \ 年份	1965 年	1966 年	1967 年	1968 年	1972 年	1974 年
A 区 y'/毫米	188.2	123.1	44.5	118.3	246.6	271.76
B 区 x'/毫米	182.3	92.6	27.5	103.7	270.7	222.46
$y=\lg y'$	2.274 7	2.090 2	1.648 4	2.073 0	2.392 0	2.434 2
$x=\lg x'$	2.260 8	1.966 6	1.439 3	2.015 8	2.432 5	2.347 3

表 1 还列出了 1974 年的实测月雨量。

根据表 1 的资料,计算了 A 区和 B 区八月份月雨量之间的相关系数[9],其值为

$$r=\frac{S_{xy}}{S_x\cdot S_y}=0.996\ 6。$$

式中

$$S_{xy}=\frac{1}{n-1}\sum_{i=1}^{n}(x_i-\bar{x})(y_i-\bar{y}),$$

$$S_x=\sqrt{\frac{1}{n-1}\sum_{i=1}^{n}(x_i-\bar{x})^2},$$

$$S_y=\sqrt{\frac{1}{n-1}\sum_{i=1}^{n}(y_i-\bar{y})^2},$$

$$n=5。$$

样本相关系数 r 的显著性：

$$t=\frac{r}{\sqrt{\frac{1-r^2}{n-2}}}=20.93,$$

自由度 $\gamma=n-2=3$。

由 t-分布表[10]可得 $t_{0.001}=12.941$，所以 $t>t_{0.001}$。这就是说，A 区和 B 区八月份月雨量之间的相关性很好，相关系数的信度也很高，超过 0.001 的显著水准。

这样就可以建立历史回归方程：$\hat{y}=a+bx$。经计算[9]得

$$回归系数：b_{yx}=\frac{S_{xy}}{S_x^2}=0.748\ 7,$$

$$截距：a_{yx}=\bar{y}-b_{yx}\bar{x}=0.581\ 1。$$

所以

$$\hat{y}=0.581\ 1+0.748\ 7x,$$

其回归线列于图 2。

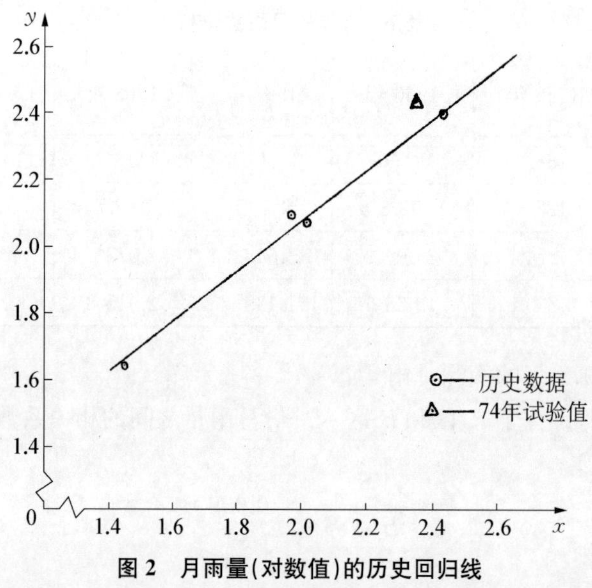

图 2　月雨量(对数值)的历史回归线

B 区是一直不催化的，以 1974 年 B 区八月份的月雨量(对数值)代入上式，求得今年八月份 A 区的自然降水量(对数值)的期待值 $\hat{y}=2.338\ 5$，即 $\hat{y}'=218.0$ 毫米。而 A 区经过人工催化后的实测雨量 $y'=271.8$ 毫米，所以雨量的增加值为 $\Delta y'=y'-\hat{y}'=53.8$ 毫米，相对增加率为 24.7%。

雨量增值的显著性：用 t -检验法，雨量增值(对数值)的 t 值为[6]

$$t=\frac{y_{74}-\hat{y}_{74}}{\sqrt{\frac{1-r^2}{n-2}\cdot\sum_{i=1}^{n}(y_i-\bar{y})^2\cdot\left[1+\frac{1}{n}+\frac{(x_{74}-\bar{x})^2}{\sum_{i=1}^{n}(x_i-\bar{x})^2}\right]}}=3.019,$$

自由度 $\gamma=n-2=3$。

我们只考虑雨量的增加，所以显著性检验时可用单侧检验法，查单侧检验 t -分布表[10]知 $t_{0.025}=3.182$，$t_{0.05}=2.353$，所以 $t\approx t_{0.025}$。这就是说，今年八月份的月雨量(271.76 毫米)是显著地偏多了，自然降雨要达到这么大的可能性很小(约 2.5%)，于是我们可以说：今年八月份人工降水试验有显著效果。

雨量增值(对数值)的 90%置信区间为 $\Delta>0.043\ 8$(单侧检验)，换算为真数值就得雨量增值的范围[9] $\Delta y'>23.11$ 毫米。这就是说，八月份人工催化的结果，平均增加月雨量 53.8 毫米，而且雨量的增值大于 23.11 毫米的概率是 90%。

3. 随机回归试验的统计分析

我们于 8 月 20 日、8 月 21 日和 9 月 18 日对系统天气下的雨层云进行了 8 次随机回归试验。其中不实际催化(空白试验)4 次，实际催化作业 4 次。前面已经指出，A 区催化或不催化是按随机程序决定的，B 区则一直不催化，作为对比。这样，我们就从同一个总体中抽取了两个随机样本，一个催化，一个不催化，看它们的平均雨量是否有显著的差异，引入 B 区的雨量作为控制变量，当 A、B 两区的雨量相关性好时，可以大大缩小所抽样的总体的自然变差，从而提高了检验效率。

试验单元取三小时。统计分析时，雨量取对数值。8 次试验单元的雨量资料列于表 2。

表 2　随机试验的雨量资料

实际催化作业					空白试验				
次数 k	1	2	3	4	次数 k	1	2	3	4
A 区 y'/毫米	12.84	28.10	25.90	6.35	A 区 y'/毫米	10.68	8.87	4.39	0.90
B 区 x'/毫米	11.51	8.08	7.93	13.58	B 区 x'/毫米	9.31	10.32	3.07	0.06
$y=\lg y'$	1.108 6	1.448 7	1.413 3	0.802 8	$y=\lg y'$	1.028 6	0.947 9	0.642 5	1.954 2
$x=\lg x'$	1.061 1	0.907 4	0.899 3	1.132 9	$x=\lg x'$	0.968 9	1.013 7	0.487 1	2.778 2
$\hat{y}=0.500+0.461x$	0.989 2	0.918 3	0.914 6	1.022 3					
$\hat{y}'$/毫米	9.754	8.285	8.215	10.530					
$\Delta y'=y'-y$/毫米	+3.09	+19.81	+17.68	−4.18					

四次空白试验 A 区和 B 区三小时面积平均雨量之间的相关系数[9]为 $\gamma=0.9901$，信度 $\alpha<0.01$。根据这四次空白试验单元的雨量资料(对数值)，建立 A 区和 B 区的三小时自然降水量之间的回归方程

$$\hat{y}=0.500+0.461x。$$

样本回归线和雨量资料列于图 3。

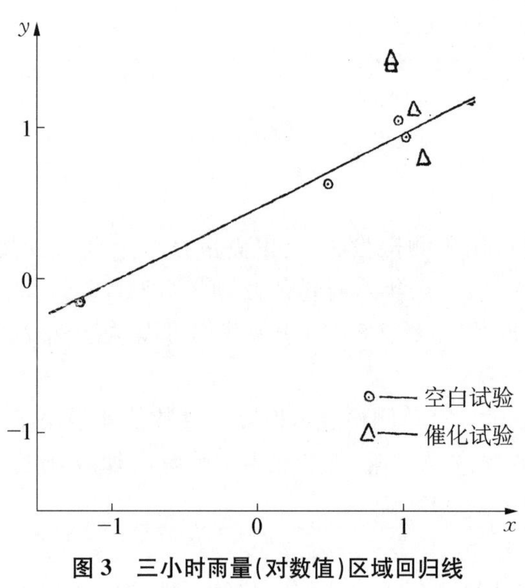

图 3　三小时雨量(对数值)区域回归线

将催化单元的 B 区雨量(对数值)代入回归方程，得到催化单元如果不催化，A 区的自然降水量(对数值)的期待值 $\hat{y}$，再换算成雨量值，列于表 2 左边倒数第二行 $\hat{y}'$；实际催化单元 A 区的实测雨量 y' 减去相应的 A 区自然降水量的期待值 $\hat{y}'$，就得到催化后增加的雨量 $\Delta y'$(表 2 左边最后一行)。

四次试验的平均增雨量[9]为 $\Delta\bar{y}_k'=\frac{\Sigma\Delta y'}{4}=9.1$ 毫米，现在来讨论这个平均增值的统计显著性。

利用多个事件检验法[6]，平均增雨量的 t 值为

$$t=\frac{\Delta\bar{y}_k}{\sqrt{V(\Delta y_k)}},$$

其中

$$V(\Delta\bar{y}_k)=\frac{1}{n-2}(1-\gamma^2)\sum_{i=1}^{n}(y_i-\bar{y})\left[\frac{1}{k}+\frac{1}{n}+\frac{(\bar{x}_k-\bar{x}_n)^2}{\sum_{i=1}^{n}(x_i-\bar{x}_n)^2}\right]。$$

计算结果[9]得 $t=3.45$，查 t -分布表(单侧检验)知 $t_{0.05}=2.92$，所以 $t>t_{0.05}$，这意即在 0.05 的显著水平下，人工降水有显著效果。

雨量增值(对数值)的 90％置信区间为

$$\Delta_k>0.1054(\text{单侧检验}),$$

换算为真数值即得雨量的增值范围[9]，$\Delta y_k'>2.51$ 毫米，意即人工催化的结果，使三小时雨量显著地增加了，平均增加雨量 9.10 毫米/3 小时，雨量增值大于 2.51 毫米/3 小时的概率是 90％。

四、讨论和结论

1974 年 8 月共催化作业 14 次，其中 5 次是台风影响下的层状云，9 次是地方性积状云。对月雨量进行历史回归分析表明，这个月的人工降水试验是有显著效果的，平均增加月雨量 53.8 毫米，相对增加率达 24.7％，雨量增值的统计显著性达到 0.025 的水准(单侧 t -检验法)，月雨量增值大于 23.1 毫米的概率为 90％。

附表 1　1974 年八九月份人工降水试验一览表

编号	作业时间			作业时段	天气形势	试验方法	作业云					发射工具				催化剂		效果摘要
	月	日	时分				云状	云量	回波顶高/km	零度线高/km	−5 ℃高度/km	名称	仰角	催化部位	枚数	名称	剂量/克	
1	8	3	16^{41}～17^{48}	17～20	在副高右侧 高空风: 700 mb,SW,3 米/秒 500 mb,SW,3 米/秒	非随机	Cb	7	9.6	4.6	5.6	"三七"高炮	65°	Cb 前进方向左侧边沿中部	86	碘化银	344	作业前(16^{28}),Cb 在古田,云顶高 9.6 千米,并降雷雨,向 NE 方移动,回波减弱。Cb 边沿经过作业点西溪,经催化后,雷达回波重新出现,并增强,17^{50}云顶高度发展,达 11 千米。在作业点下风方出现一个 14 毫米雨量中心。在 Cb 途经风浦、平湖滴雨无降。作业点下了 13 毫米。
2	8	4	11^{20}～12^{09}	11～14	在副高右侧 高空风: 700 mb,NE,3 米/秒 500 mb,SW,3 米/秒	非随机	Cb	5	7.8	4.9	5.5	"三七"高炮	70°	对西南方 Cb 前沿中部催化	100	碘化银	400	作业前,在作业点西南和西分别有两块 Cb。西南方的 Cb 缓慢地向作业点移动,西方的 Cb 基本上不动,该云顶高在 7.8 千米以下。对西南方向的 Cb 进行了作业,11^{30}目睹两块云接近,并合并,云体变黑,翻滚厉害,并有雷电降雨。作业点11^{28}开始降雨,12^{27}雨止,雨量为 22.4 毫米。12^{31}云顶 10 千米。作业后又分裂成两块 Cb,并向西收缩,作业点正西方 2 千米外,降雷雨。而作业点以东至南 1.5 千米外,滴雨无降。
3	8	5	12^{13}～12^{19}	11～14	副热带高压南侧 高空风: 700 mb,ENE,1 米/秒	非随机	CuCong	2		5.0	5.5	"三七"高炮	50°		15	碘化银	90	CuCong 在作业点北边,向作业点移来,作业开始后 5 分钟,弹着部位,有翻滚出现,随即消失,平缓。又过 10 分钟,作业云略有抬升,随后 8 分钟,逐渐消失。作业点作业开始后降雨数滴。

续表

编号	作业时间			作业时段	天气形势	试验方法	作业云					发射工具				催化剂		效果摘要
	月	日	时分				云状	云量	回波顶高/km	零度线高/km	−5 ℃高度/km	名称	仰角	催化部位	枚数	名称	剂量/克	
4	8	5	15^{19}～15^{40}	14～17	副高南侧 高空风： 700 mb，E，5 米/秒 500 mb，SE，2 米/秒	非随机	Cb	10		5.1	6.1	"三七"高炮	55°	Cb 云的中、上部	54	碘化银	216	Cb 云位于作业点东南方，云体没有移动，仅是 Cb 云帽扩散近作业点。作业后，炮弹爆炸烟团被吸入云内。东南方 Cb 云层加厚变黑，雷声次数增多，作业点下了一些雨滴。在作业点下风方的桃溪、古田，三小时雨量出现了 14 毫米和 18 毫米雨量中心。因 Cb 云体没有移动，又缺少雷达资料，北边雨量中心是否催化所致，还待讨论。
5	8	9	16^{39}～16^{46}	17～20	热带低压北侧 高空风： 700 mb，ESE，12 米/秒 500 mb，E，11 米/秒	非随机	CuCong	2		4.9	6.2	50－5 型小火箭			7	四聚乙醛	28	作业点上空有 CuMin，CuCong，仅发 7 枚小火箭，云体变化不大，发展不起来，即停止作业。作业点仅下了一阵阵雨，雨量为 0.7 毫米。16^{50} 作业点雨止。
6	8	11	10^{37}～11^{06} 12^{03}～12^{31}	11～14	12 号台风影响 高空风： 700 mb，SE，20 米/秒 500 mb，E，10 米/秒	非随机	As Ac Sc		6	5.2	6.0	50－5 型小火箭			20 53	四聚乙醛	252	作业开始前已有连续性雨，作业开始后，仍为连续性雨。15^{51} 云顶高为 6.6 千米，11^{17} 云顶高为 5.5 千米，云顶略有下降，云底下降为 0.7 千米。第二次作业后，有一段时间里，雨带阵性。第一、二两次作业之间，雨转间歇性。从三小时雨量图看，作业点下风方出现 10 毫米雨量中心。

续表

编号	作业时间			作业时段	天气形势	试验方法	作业云					发射工具				催化剂		效果摘要
	月	日	时分				云状	云量	回波顶高/km	零度线高/km	−5 ℃高度/km	名称	仰角	催化部位	枚数	名称	剂量/克	
7	8	11	16^{42}～17^{38}	17～20	12 号台风影响 高空风： 700 mb,ESE,14 米/秒 500 mb,SE,14 米/秒	非随机	Cb		7～7.8	5.4	6.5	“三七”高炮	65°		119	碘化银	666	作业开始后 5 分钟测得回波顶高 7.8 千米，16^{55} 为 7 千米。从三小时雨量图看，作业后，在下风方的柏源出现 10.4 毫米雨量中心，但雨量图上，在古田、迪口、棠口和芝南也同样有 6.6 毫米、5.1 毫米、7.5 毫米和 9.4 毫米雨量中心存在。因 Cb 个体较小，柏源虽然在下风方，这个雨量中心是否是其他地方性对流的自然降水的结果，因缺少雷达资料故不清楚。
8	8	13	11^{12}～11^{40}	11～14	副热带高压影响 高空风： 700 mb,S,6 米/秒 500 mb,S,7 米/秒	非随机	地方性 Cb	10		5.4	6.1	“三七”高炮	55°		43	碘化银	258	Cb 由东向北移动，其左侧边沿经过作业点上空，三小时雨量图上，在下风方的柏源出现一个 20 毫米雨量中心。
9	8	13	17^{26}～17^{39}	17～20	副热带高压影响 高空风： 700 mb,ESE,8 米/秒 500 mb,SSE,7 米/秒	逆随机（动力催化）	地方性 Cb	10	5.7	5.3	6.4	“三七”高炮	60°	Cb 云前进方向中、上部	114	碘化银	684	Cb 由作业点的东南方向移来，三分钟内发射 114 炮弹，Cb 经过作业点向西北移去，发展旺盛，并维持在旧镇，18^{30} 雷达观测到 Cb 回波顶高 5.7 千米、18^{36} 发展为 7.5 千米。18^{40} 发展为 8.1 千米，一直到 18^{51} 才开始下降为 6 千米。从三小时雨量图看，从风浦（东南）经过作业点到旧镇（西北），出现了由中到小、再到大的雨量带，显然是经过催化后发展而致的。

续表

编号	作业时间			作业时段	天气形势	试验方法	作业云					发射工具				催化剂		效果摘要
	月	日	时分				云状	云量	回波顶高/km	零度线高/km	−5 ℃高度/km	名称	仰角	催化部位	枚数	名称	剂量/克	
10	8	20	—	17～20	13号台风外围影响 高空风: 700 mb,WWN,13米/秒 500 mb,NW,19米/秒	随机(空白试验)	N_s	10	6	5.4	6.6							作业前后,云顶高都在6.0千米左右,云底高约在0.7千米。从三小时雨量图看,B区雨量加权平均值9.31毫米,小于A区雨量加权平均值10.68毫米,但雨量中心在B区,是连续性雨。
11	8	20	20^{44}～21^{26}	20～23	13号台风外围影响 高空风: 700 mb,WWN,13米/秒 500 mb,NW,19米/秒	随机试验(实际催化)	N_s	10	7	5.4	6.6	“三七”高炮	75°		105	碘化银	630	连续性雨。从三小时雨量图看,B区雨量加权平均值11.5毫米,小于A区雨量加权平均值12.84毫米。
12	8	21	02^{18}～02^{42}	02～05	13号台风外围影响 高空风: 700 mb,NW,10米/秒 500 mb,WWN,14米/秒	随机试验(实际催化)	N_s	10	7	5.3	6.6	“三七”高炮	75°		100	碘化银	600	14分钟内100发,连续暴雨。从三小时雨量图看,B区雨量加权平均值8.08毫米,小于A区雨量加权平均值28.10毫米,雨量中心在A区。
13	8	21	05^{55}～06^{24}	05～08	13号台风外围影响 高空风: 700 mb,NW,10米/秒 500 mb,WWN,14米/秒	随机试验(实际催化)	N_s	10	7	5.3	6.6	“三七”高炮	75°		112	碘化银	672	连续暴雨。从三小时雨量图看,B区雨量加权平均值7.9毫米,小于A区雨量加权平均值25.90毫米,雨量中心在A区。作业点西溪发大水,老农反映是二三十年未遇。
14	8	21		11～14	13号台风外围影响 高空风: 700 mb,NW,10米/秒 500 mb,WWN,14米/秒	随机试验(空白)	N_s	10	7	5.3	6.6							连续性雨,12^{00}～13^{00}雨势较大。从三小时雨量图看,B区雨量加权平均值10.32毫米,大于A区雨量加权平均值8.87毫米,雨量中心在B区。

续表

编号	作业时间			作业时段	天气形势	试验方法	作业云					发射工具				催化剂		效果摘要
	月	日	时分				云状	云量	回波顶高/km	零度线高/km	−5 ℃高度/km	名称	仰角	催化部位	枚数	名称	剂量/克	
15	8	21	15^{00}～15^{25}	14～17	13 号台风外围影响 高空风： 700 mb，NW，10 米/秒 500 mb，WWN，14 米/秒	随机试验（实际催化）	N_s	10	6.3	5.3	6.6	“三七”高炮	75°		50	碘化银	300	14^{00}～15^{00}和16^{00}～17^{00}雨势较大。从三小时雨量图看，B 区雨量加权平均值 13.58 毫米，大于 A 区雨量加权平均值 6.35 毫米。这是由于过冷水层太薄。作业催化后，下了一阵雨，随后云消散。
16	8	21		17～20	13 号台风影响 高空风： 700 mb，WWN，10 米/秒 500 mb，WWN，6 米/秒	随机试验（空白）	N_s	10	6.5	5.4	6.6							间歇性雨。从三小时雨量图看，B 区雨量加权平均值 3.07 毫米，小于 A 区雨量加权平均值 4.39 毫米。
17	9	18		05～08	东风波影响 高空风： 700 mb，NNW，3 米/秒 500 mb，WSW，8 米/秒	随机试验（空白）	N_s	10		4.9	6.1							从三小时雨量图看，B 区雨量加权平均值 0.06 毫米，小于 A 区雨量加权平均值 0.9 毫米。
18	8	23	23^{17}～23^{40}	23～02	冷锋后降雨 高空风： 700 mb，NNW，5 米/秒 500 mb，NNW，5 米/秒	非随机	Cb	10		5.4	6.1	“三七”高炮	75°	Cb 尾部	38	碘化银	228	从作业前，作业点西溪出现短时暴雨，随后转中阵雨，作业后，又转暴雨。 Cb 由西北移来，炮击部位在 Cb 尾部，即 Cb 主体已移过作业点，向南移去。从三小时雨量图看，作业催化后，作业点下风方的坝区出现一个 31.4 毫米的雨量中心。23^{43}目睹作业点西北方（Cb 移来方向）出现星星，24^{00}作业点雨止，而东南方天空乌黑，作业点西溪23^{00}～24^{00}雨量为 16.1 毫米，坝区23^{00}～24^{00}雨量为 19.3 毫米，24^{00}～01^{00}为 12.1 毫米。

续表

编号	作业时间			作业时段	天气形势	试验方法	作业云					发射工具				催化剂		效果摘要
	月	日	时分				云状	云量	回波顶高/km	零度线高/km	−5 ℃高度/km	名称	仰角	催化部位	枚数	名称	剂量/克	
19	9	1	18^{09}～18^{34}	17～20	高空有一切变线，受西风气流控制 高空风： 700 mb，WWS，4 米/秒 500 mb，WWN，7 米/秒	非随机	CuCong	10	5.0	5.1	6.7	“三七”高炮	70°	Cu-Cong 中、上部	100	碘化银	508	作业前，CuCong 从西北方向向作业点西南方缓慢移动，作业后，天顶云层翻滚明显，18^{16} 翻滚加剧，并在天顶炮弹爆炸后扩展。作业点 18^{00}～20^{00} 的雨量为 2.6 毫米，因作业点西边的缓冲区雨量站很稀疏，估计雨量较大。
20	9	17	14^{29}～15^{32}	14～17	高压入海 高空风： 700 mb，SSE，4 米/秒 500 mb，SSW，9 米/秒	非随机	Cb	10	5.0	5.0	5.8	“三七”高炮	70°	Cb 边沿	76	碘化银	382	作业前，作业点东南方闻雷声，作业后，即 15^{53} 云顶高为 6.0 千米，西北方雷电厉害。从作业开始至 17^{00} 作业点雨量为 7.1 毫米，从三小时雨量图看，作业点下风方的柏源出现 62.5 毫米的雨量极大中心。因雷达和雨量资料不全，待后分析。

从这次回归分析看，只要恰当地选择影响区和对比区，使两区的月雨量有好的相关性，那么回归分析的效率还是相当高的。当然，应该看到，历史回归分析有一个基本的缺点，那就是大气环流形势有长期变化，历史回归分析不能预测历史资料上未曾出现过的天气形势，因此，要求历史资料长一些好。但在抗旱情况下，此种设计还是比较好的一种效果分析方法。

随机试验可以避免依赖历史资料。今年我们对系统天气（主要是台风）影响下的层状云以三小时为一个试验单元，进行了 8 次随机回归试验，其中 4 次催化，4 次不催化。统计分析结果，平均增加雨量 9.1 毫米/3 小时，相对增值达到 99%，雨量增值的显著水平达 0.05（单侧检验），三小时的雨量增值大于 2.51 毫米的概率为 90%。

如果上述结果能在以后的试验中继续得到验证，则对这一类系统天气进行人工影响的意义是很大的。特别是台风系统，它是夏秋季节我国南方的主要降水系统，它从海洋上带来大量的水分和能量，降水的潜力是巨大的。有的台风登陆后又转向出海了，在干旱季节如能对它进行人工影响，尽可能地使更多的水降落在陆地上，受益匪浅。以今年 13 号台风为例：8 月 20、21 日对其影响下的雨层云催化作业 4 次，使影响区（40×30＝1 200 平方千米）的降雨量平均增加了 9.1×4＝36.4（毫米）。也就是说，在水库流域 1 200 平方千米的面积上，由于催化作业，增加的降水量是 4 368 万立方米。按照保守的估计（但影响区植被很好），所增加的降水如有三分之一进入水库，则可使水库增加 1 456 万立方米的水。由于古田水库是 4 级发电，每发 1 度电，用水 1 立方米，则可增加发电 1 400 万度以上。这在配合水库抗旱蓄水发电方面起到了较显著的作用。

随机回归试验的效率比较高，灵敏度也高[11]。今年仅搞了 8 次随机回归试验，虽然通过 t -分布检验效果是显著的，但这一工作还需要继续进行，以便得到进一步的验证。

至于催化剂的扩散问题，今年没有做专门研究。今年仅是从探空和雷达回波来判断三小时内催化云体已移出 A 区，关于催化剂对 B 区的污染和在 A 区对下一作业时段的残留问题，有待今后进行研究探讨，但清楚的是催化剂的污染影响，只会低估效果。关于对不同的作业云用多少剂量为好、有利作业的天气条件和作业时机的选择等，现在还处于摸索阶段，有待今后进一步研究解决。

参考文献

[1] Smith, E. J. (1970): Effects of Cloud-top temperture on the Results of cloud Seeding With Silver Iodide in Australia. J. of Appl., Meteo. vol.9, No 5, p.800—804.

[2] Neyman, J., Scott, E. and Vasilevskis M., (1960): Statistical evaluation of the santa Barbara randomized Cloud—Seeding experiment, Bull. Amer. Meteo. Soc., vol.41, p.531—547.

[3] G.W.斯奈迪格等："应用于农学和生物学实验的数理统计方法".科学出版社，1964，12 页，20 页.

[4] 福建省气象局、南京大学气象系："福建省 1974 年八、九月份人工降水的效果评"定附件 1、2、3，1974.12.

[5] Schickedanz P. T. and F. A. Huff (1971): The Disign and Evaluation of Rainfall Modification Experiments, J. of Appl. Meteo. vol.10, No 3, p.502 - 514.[6] 福建省气象局、古田县炸药厂："FJ—50—5"型小火箭技术总结.1974.

[7] 福建省气象局、南京大学气象系、西安化工研究所："用 TNT 分散四聚乙醛的初 步试验及四聚乙醛室内成冰性能的实验测定。"1974.

[8] 杨纪珂:“数理统计方法在医学科学中的应用”。上海科学技术出版社,1963,241页.
[9] Thom, H. C. S. (1957): A Statistical Method of Evaluating Augmentation of Precipi tation by cloud Seeding. Find Report of the Advisory Committee on Weather Control vol. Ⅱ, p.5 - 25.
[10] 徐尔灏:“论人工降水随机试验的效果检查问题”.南京大学学报(气象学)1962年第2期 p.69 - 83.
[11] 复旦大学数学系:“概率论和数理统计”.上海科学技术出版社,1961.

STATISTICAL EVALUATION OF THE ARTIFICIAL PRECIPITATION STIMULATION EXPERIMENT IN FUKIEN PROVINCE 1974

Weather Bureau of Fujian Province and
Department of Metevrology, Nanjing University

Abstract During the August and September 1974, the artificial precipitation stimulation experiments with silver iodide from the ground by antiaircraft gun were conducted in Gutian reservoir region in Fujian province. And two different kinds of procedures were used for evaluation. For the nimbo-stratus clouds of typhoon, the randomized regression technique was used. In this experiment, 3 h was taken as an experimental time unit. Statistical analysis shows that the amount of precipitation after seeding is about 9.1 mm/3 h more than that without seeding with the level of significance 0.05. Moreover, most of the situation as suitable for claud seeding in August were seeded in the target area. The target-control regression analysis for the rainfall of month was made. The result of analysis shows that the rainfall of month of augmentation under seeding in August is about 53.8 mm with the level of significance 0.025.

4.5 闽中雨季区域雨量统计特性及人工影响的效果*

叶家东　　　程克明　曾光平
(南京大学气象系)　　(福建省气象局科研所)

提要 本文对福建古田水库地区三年人工降水随机试验的效果,应用双样本回归分析方法做了统计检验。结果发现:(1) 区域雨量的四次方根变换正态性很好,拟合度达0.94。(2) 区域雨量回归分析中,催化单元与对比单元的余方差不一定相等。(3) 三年试验目标区平均增雨量20.3%,平均绝对增雨量1.11(毫米/3小时),显著度$\alpha=0.05$。(4) 在各种天气条件下,以锋面天气的催化效果最好,平均增雨量达70.6%($\alpha<0.05$),平均绝对增雨量2.82(毫米/3小时)。其他天气条件下的平均效果都不显著。(5) 自然雨量小时,相对增雨量大,它随自然雨量增大而减小,当对比区自然雨强大于3(毫米/小时)时,催化效果就不显著。(6) 催化效果大于20%的区域位于作业点下风方10~50千米,宽约20千米的范围。(7) 介乙醛与碘化银的催化效果,差异不显著。

1975、1977和1978年4~6月,福建省气象局科研所和南京大学气象系等①在古田水库

* 本文于1980年4月5日收到,1980年10月20日收到修改稿。
① 参加试验的还有古田县气象站、古田县人武部、西安化工研究所等单位。

地区进行了三年人工降水随机试验。试验的目的是要了解该地区雨季人工增雨的可能性、有利的催化条件以及人工影响有效的区域范围。

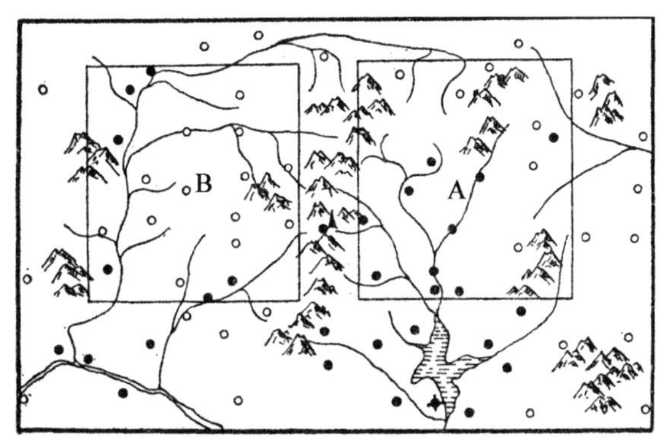

○ 人工观测雨量点 ● 自记雨量计

图 1 试验区的地理位置

试验区试验期的平均月雨量 217.8 毫米，月平均降雨日数 22.1，自然降水条件比较好。但是，到 1974 年为止，建库 15 年中，古田水库平均蓄水量只有 64.6%，其中 73%的年份不足 90%，影响到水力发电。所以从解决水库蓄水发电着眼，进行人工降水是有实际需要的。

试验采用区域回归随机试验方案。有关的试验对象、催化方法、试验区设置以及试验的统计设计在前两年的报告中已有阐述[1]。本文根据三年的试验资料，采用双样本回归分析方法对试验效果进行检验。

一、区域雨量的统计特性和效果分析方法

正态性检验 参加统计分析的雨量资料取三小时区域面积平均雨量。为了使统计变量正态化，我们将对比区三小时区域面积平均雨量进行变数变换，然后利用柯尔莫哥洛夫定理进行正态分布的“拟合度”检验，结果列于表 1。

表 1 三小时区域面积平均雨量 x' 及其变换值的正态性“拟合度”检验表

统计变量 x	x'	$\sqrt{x'}$	$\sqrt[3]{x'}$	$\sqrt[4]{x'}$	$\sqrt[5]{x'}$	$\lg x'$
拟合概率$[1-k(y_0)]$	0.004 1	0.457 8	0.744 2	0.937 3	0.943 0	0.537 7
区域相关系数 r	0.667 6	0.775 4	0.810 2	0.824 5	0.835 8	0.8435
相关系数显著度 α	0.001	0.001	0.001	0.001	0.001	0.001

由表 1 可见，变量$\sqrt[4]{x'}$和$\sqrt[5]{x'}$的正态性很好，拟合概率达 0.94，区域相关系数也高，超过 0.82。为方便计，统计分析时统计变量取 $x=\sqrt[4]{x'}$。

方差相等性检验 通常的效果检验方法，要求供比较的样本所代表的总体方差相等。但是，从古田试验的回归分析发现，这个条件并不总是满足的。由表 2 可见，合并统计与积状云，催化样本与对比样本余方差有显著差异，显著度达到 0.05。层状云差异不显著。所以，对试验效果进行回归分析时，除了采用通常的多个事件检验法[2]检验平均效果以外，主要采用方差不相等的双样本回归分析方法检验效果[3]。

表 2　回归分析中催化样本和对比样本的余方差相等性检验

	k	s_{1r}^2	n	s_{2r}^2	F	α
合并统计	50	0.027 0	51	0.044 7	1.655 6	<0.05
层状云	29	0.030 7	30	0.027 4	1.120 4	>0.05
积状云	21	0.030 8	21	0.070 5	2.289 0	<0.05

注：k，n 分别表示催化样本和对比样本的容量；s_{1r}^2，s_{2r}^2 分别表示催化样本和对比样本的样本余方差。

双样本回归分析　令催化单元目标区雨量指标(本文均指雨量的四次方根)y 倚对比区雨量指标 x 的区域回归方程是

$$\hat{y}_1 = a_1 + b_1 x, \tag{1}$$

对比单元的区域雨量回归方程为

$$\hat{y}_2 = a_2 + b_2 x。 \tag{2}$$

相应的一元线性正态回归模型是

$$y_1 = \alpha_1 + \beta_1 x + \varepsilon_1, \tag*{(1)$'$}$$

$$y_2 = \alpha_2 + \beta_2 x + \varepsilon_2。 \tag*{(2)$'$}$$

其中 $\alpha_1 + \beta_1 x = y_{10}$，$\alpha_2 + \beta_2 x = y_{20}$ 是相应的总体回归值，而 $\varepsilon_1 \sim N(0, \sigma_1)$，$\varepsilon_2 \sim N(0, \sigma_2)$，这里 $\sigma_1 \neq \sigma_2$。

待检验的假设是 $H_0: y_{10} = y_{20}$。由文献[3]知，在原假设 H_0 成立的前提下，统计量

$$Z_{\hat{y}} = \frac{\hat{y}_1 - \hat{y}_2}{\sqrt{\left[\frac{1}{k} + \frac{(x - \bar{x}_1)^2}{\sum_{i=1}^{k}(x_i - \bar{x}_1)^2}\right] s_{1r}^2 + \left[\frac{1}{n} + \frac{(x - \bar{x}_2)^2}{\sum_{j=1}^{n}(x_j - \bar{x}_2)^2}\right] s_{2r}^2}}, \tag{3}$$

近似地服从自由度为

$$\gamma_{\hat{y}} = \frac{\left\{\left[\frac{1}{k} + \frac{(x - \bar{x}_1)^2}{\sum (x_i - \bar{x}_1)^2}\right] s_{1r}^2 + \left[\frac{1}{n} + \frac{(x - \bar{x}_2)^2}{\sum (x_j - \bar{x}_2)^2}\right] s_{2r}^2\right\}^2}{\frac{1}{k-2}\left\{\left[\frac{1}{k} + \frac{(x - \bar{x}_1)^2}{\sum (x_i - \bar{x}_1)^2}\right] s_{1r}^2\right\}^2 + \frac{1}{n-2}\left\{\left[\frac{1}{n} + \frac{(x - \bar{x}_2)^2}{\sum (x_j - \bar{x}_2)^2}\right] s_{2r}^2\right\}^2} \tag{4}$$

的 t-变量，于是，可以利用 t-分布对差值 $(\hat{y}_1 - \hat{y}_2)$ 进行显著性检验。式中 $\bar{x}_1 = \frac{1}{k}\sum_{i=1}^{k} x_i$，$\bar{x}_2 = \frac{1}{n}\sum_{j=1}^{i} x_j$ 是样本平均数；$s_{1r}^2 = \frac{1}{k-2}\sum_{i=1}^{k}(y_i - \hat{y}_i)^2$，$s_{2r}^2 = \frac{1}{n-2}\sum_{j=1}^{n}(y_j - \hat{y}_j)^2$ 分别是 σ_1^2，σ_2^2 的样本估计量。

多个事件检验法所用的统计量是

$$t=\frac{\hat{y}_k-\bar{\hat{y}}_k}{\sqrt{\left[\frac{1}{k}+\frac{1}{n}\frac{(\bar{x}_1-\bar{x}_2)^2}{\sum(x_j-\bar{x}_2)^2}\right]s_{2r}^2}},\tag{5}$$

t 是自由度为$(n-2)$的 t -变量。式中$\hat{y}_k$ 是k 次催化试验目标区多测平均雨量指标，$\bar{\hat{y}}_k$ 是相应的平均自然雨量期待值。易知当 $x=\bar{x}$ 而时，$\hat{y}_1=\hat{y}_k$，$\hat{y}_2=\bar{\hat{y}}_k$。

二、催化效果统计分析

1. 区域面积平均雨量增雨效果的回归分析

计算结果列于表 3。合并统计平均增加雨量 20.3%($\alpha=0.05$)，平均绝对增雨量 1.11(毫米/3 小时)。按云形分层统计，层状云和积状云的平均相对增雨量分别为 21.7%和 17.5%，平均绝对增雨量也是 1 毫米/3 小时左右。但统计显著度不高，层状云是 0.06，积状云 $\alpha<0.20$。

比较有意思的是按天气形势分类统计的结果，我们根据地面锋和副高位置将试验单元分成锋前(锋面未过武夷山)、锋面(冷锋或静止锋位于武夷山和闽南之间，或入海之前)、锋后(锋面入海或进入闽南)和副高影响四类天气。除了 1 个个例不属于上述天气范围，其他 100 个试验单元全都包括在内。分类统计结果发现，只有在锋面天气平均催化效果是显著的，其他三种场合的平均催化效果都不显著。这表明在锋面附近人工降水是最有利的，平均增雨量可达70%以上。三年试验的平均效果要比头两年的平均效果差[1]，这一方面是由于样本容量较小，自然随机起伏较大引起的，同时在一定程度上也反映了有利天气条件选择上的偏差。分析表明，1978 年 19 次催化试验中，锋面天气只有 1 次，锋前和副高天气有 13 次，所以效果较差。由此也可看出，选择有利天气条件和有利催化时机对于提高催化效果是十分重要的。

表 3　区域面积平均雨量增雨效果的回归分析表

			合并统计	按云形分层统计		按天气条件分层统计			
				层状云	积状云	锋前	锋面	锋后	副高
试验总次数			101	59	42	33	23	30	14
催化试验	次数 k		50	29	21	18	10	16	6
	区域雨量相关系数 r_1		0.877 1	0.863 3	0.866 7	0.883 0	0.521 0	0.916 9	0.905 9
	回归系数	a_1	0.347 7	0.294 8	0.455 0	0.370 0	0.418 7	0.329 3	0.832 7
		b_1	0.808 5	0.837 4	0.747 0	0.779 6	0.822 3	0.834 0	0.375 8
对比试验	次数 n		51	30	21	15	13	14	8
	区域雨量相关系数 r_2		0.824 5	0.900 1	0.713 8	0.770 8	0.828 2	0.879 4	0.937 2
	回归系数	a_2	0.140 9	0.224 4	−0.276 3	−0.492 0	0.250 1	0.306 4	0.165 4
		b_2	0.895 2	0.833 6	1.177 7	1.325 6	0.799 4	0.820 0	0.890 0
试验效果	绝对增雨量/(毫米/3 小时)		1.11	1.14	1.01	−0.79	2.82	0.72	−0.72
	相对增雨量/%		20.3	21.4	17.5	−9.4	70.6	11.8	−17.0

续表

		合并统计	按云形分层统计		按天气条件分层统计			
			层状云	积状云	锋前	锋面	锋后	副高
显著度	α_1	≈0.05	≈0.06	≈0.225	>0.25	<0.025	≈0.213	<0.25
	α_2	<0.05	<0.10	<0.20	>0.25	<0.05	≈0.20	<0.20
90%置信区间	绝对增雨量/(毫米/3小时)	>0.25	>0.22			>1.28		
	相对增雨量/%	>4.0	>3.6			>23.1		

注：α_1 是按多个事件检验法检验的单边显著度；α_2 是按双样本回归分析法检验的单边显著度。

2. 不同自然雨量强度下的增雨效果

催化单元和对比单元的区域雨量回归线与实测雨量散布点示于图 2 和 3。我们还按对比区雨量大小分级，根据公式(3)和(4)对催化效果进行双样本回归分析。计算的增雨效果随对比区自然雨量 x 而变化的关系曲线示于图 4。

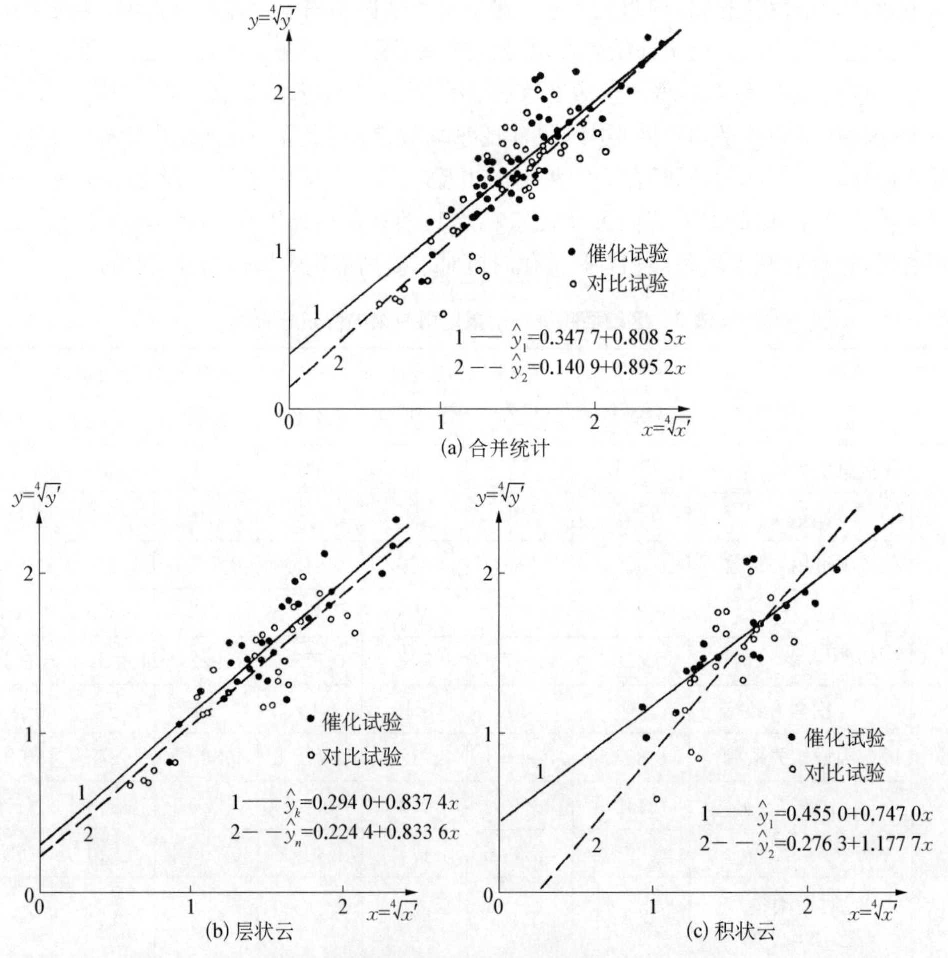

图 2　区域雨量回归线和实测雨量散布图

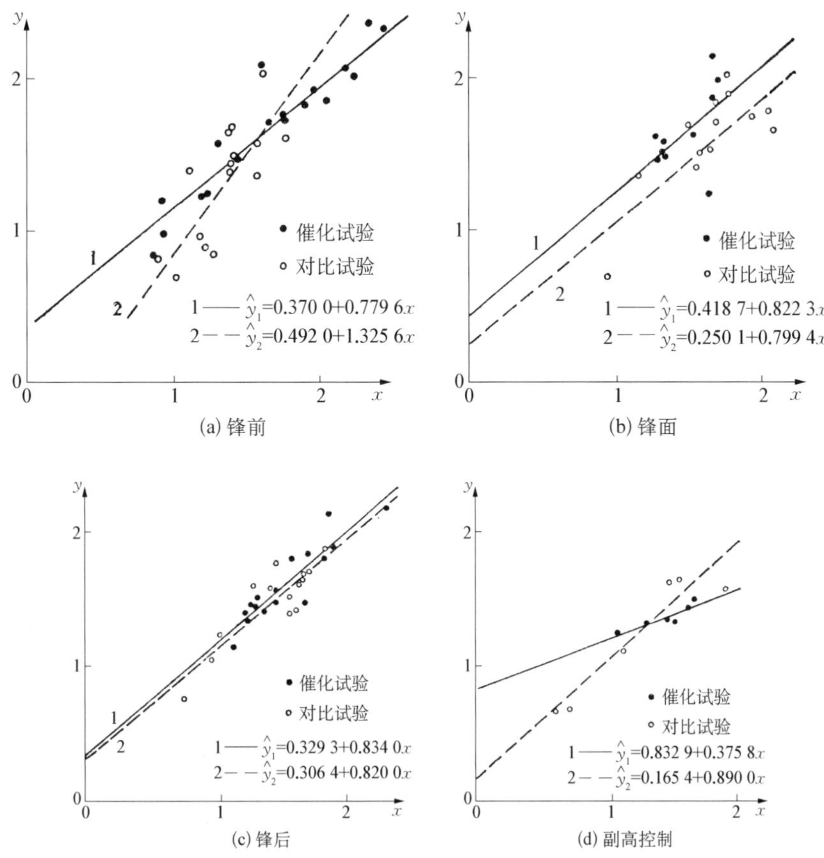

图 3　不同天气条件下区域雨量回归线和实测雨量散布图

合并统计　由图 2 和图 4 可见，全部试验单元合并统计的相对增雨量介于 0.121～0.814，随着对比区自然雨量增大而减小，但绝对增雨量却是随之而增加的，从 0.44～1.14（毫米/3 小时）。当 $x<1.55$（相当于 $x'<5.79$ 毫米/3 小时）时，增雨效果是显著的（$\alpha<0.05$），而当 $x\geqslant1.8$（相当于 $x'\geqslant10.50$ 毫米/3 小时）时，效果就不显著了。

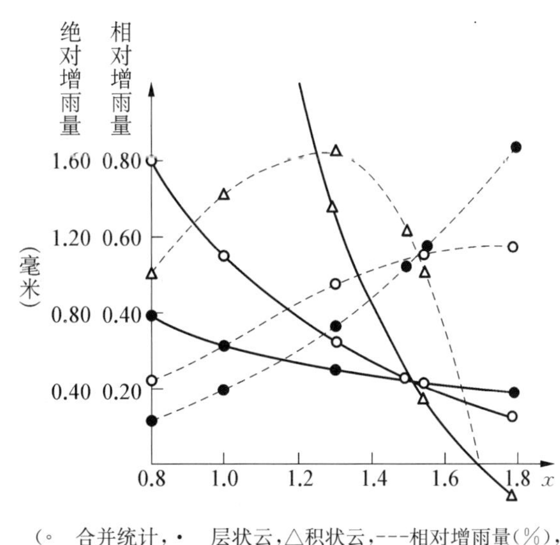

（◦ 合并统计，• 层状云，△积状云，---相对增雨量（%），
----绝对增雨量（毫米/3 小时））

图 4　增雨效果随对比区自然雨量而变化的关系曲线

分层统计　按云形分类的结果，发现层状云和积状云的催化效果各有特点。虽然相对增雨量都是随对比区自然雨量指标 x 增加而减小的，但层状云的绝对增雨量是一直随 x 增加而增加的，效果显著的区间在 $1.30\leqslant x\leqslant1.55$（相当于 2.86 毫米/3 小时$\leqslant x'\leqslant$5.80 毫米/

3 小时)，相对增雨量达 22.3%～38.4%，绝对增雨量 0.75～0.76 毫米/3 小时；积状云效果显著的区间 $x \leqslant 1.3$，相对增雨量超过 66.9%，$x=1.0$ 时达 2.16 倍(积状云试验 x 都大于 0.8)，绝对增雨量为 1.03～1.66 毫米/3 小时。当 x 小时，积状云的绝对增雨量随 x 增大而增加，$x=1.3$ 时最大，达到 1.66 毫米/3 小时，而后随 x 增加而减小。说明存在一个催化有效的雨强区间和最优的催化条件。层状云是中雨时(对比区雨强约为 2 毫米/小时)最优，可增加雨量 22%；积状云是小雨时(对比区雨强约为 1 毫米/小时)最优，可增加雨量 67%。

按天气条件分类的结果，各种天气条件相对增雨量都是随 x 增加而减小的。锋前天气，试验单元对比区雨量指标 x 都大于 0.8，只有当 $x=1.0$ 时效果是显著的($\alpha=0.05$)，相对增雨量达 2.62 倍，绝对增雨量约 1.26 毫米/3 小时；锋面天气，$1.46 \leqslant x \leqslant 1.5$(相当于 4.50 毫米/3 小时 $\leqslant x' \leqslant$ 5.66 毫米/3 小时)时效果显著($\alpha=0.05$)，相对增雨量达 70%左右，绝对增雨量约 2.82～3.04 毫米/3 小时，是各种天气条件下效果最好的；锋后天气和副高天气效果都不显著。

看来人工降水效果很可能既不是定常的，也不是与自然雨量成比例的，而以往常常认为催化效果是可乘的[4,5]。较为可能的是存在一种催化有效的最优条件。就古田试验而言，从自然雨强看，层状云中雨时最好，积状云小雨时最好。从天气条件看是锋面天气最优。

3. 目标区单站雨量增雨效果的回归分析

古田试验的前两年，我们曾用双比分析法分析试验区效果较明显的区域范围[1]。但是，应该指出，双比分析中试验区的各雨量站的单站自然雨量与对比区的面积平均雨量成比例的假定是需要验证的。而且，双比分析假定雨量的区域回归线通过坐标原点，这也常常不能满足。

现在我们不用双比分析，而采用双样本回归分析检验目标区单站雨量的催化效果。控制变量仍取对比区区域面积平均雨量，倚变量则取目标区各单站雨量。为了使回归分析有意义，先计算目标区单站自然雨量与对比区区域面积平均雨量的相关系数，结果示于图 5。对合并统计、层状云和积状云，$\alpha=0.01$ 的相关系数分别为 $r=0.358$、0.457 和 0.549。由图 5 可见，合并统计和层状云，整个目标区的单站雨量与对比区面积平均雨量的相关系数都是显著的，表明目标区单站雨量的区域回归分析是有意义的。至于积状云，目标区单站雨量与对比区面积平均雨量的相关性就很差，整个目标区只有西南角靠近作业点的两个站的相关系数接近显著。所以，积状云的单站雨量区域回归分析大部分地区都没有意义。同样，在这

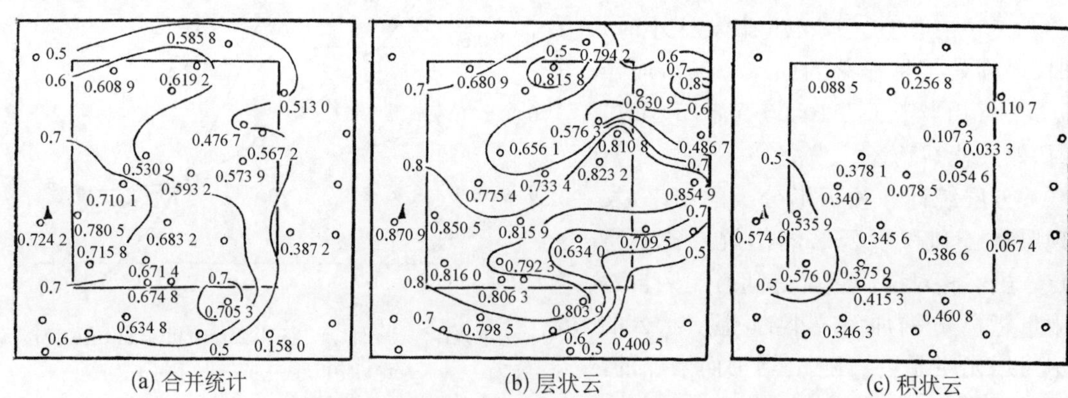

(a) 合并统计　　(b) 层状云　　(c) 积状云

图 5　目标区单站雨量与对比区区域面积平均雨量的相关系数分布图

种情形下做双比分析也只是形式上的，无统计意义。

单站雨量的双样本回归分析：取各单站雨量作为因变量 y，对比区面积平均雨量 x 作为控制变量。仿照前面区域面积平均雨量的双样本回归分析方法，计算目标区各站的增雨效果及其统计显著度，结果示于图 6。可见，合并统计人工催化增雨效果超过 20%的区域（图 6(a)）位于作业点下风方 10～50 千米、宽约 20 千米的范围，其中大多数雨量站的增雨效果统计上都是显著的（$\alpha=0.05$）。层状云相对增雨量超过 20%的区域与合并统计的相仿，位于作业点下风方 15～50 千米、宽约 20 千米的范围，其中约有一半雨量站的增雨效果统计上显著（$\alpha=0.05$）。

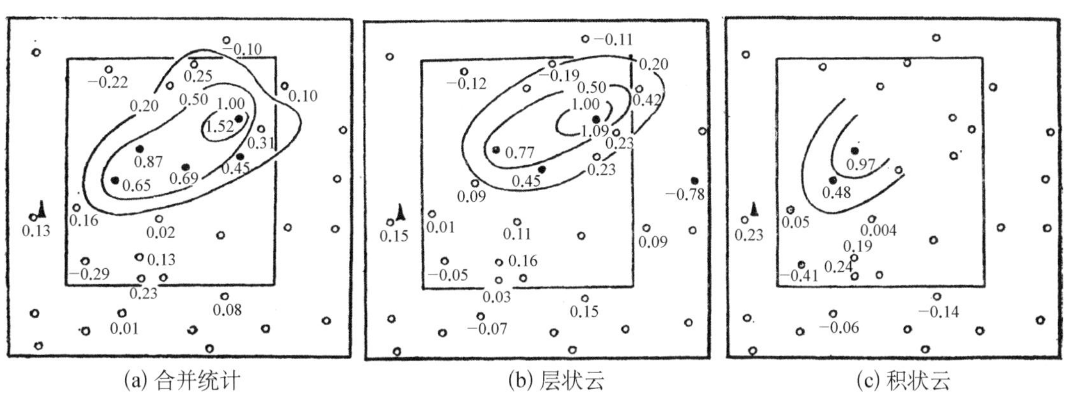

(a) 合并统计　　(b) 层状云　　(c) 积状云

◦ 统计不显著　• 统计显著（$\alpha=0.05$）

图 6　目标区单站增雨效果（相对增雨量）的分布

4. 不同催化剂增雨效果的差异显著性检验

三年试验，采用介乙醛（MA）29 次，碘化银（AgI）21 次。这两种冷云催化剂的增雨效果是否有显著差异呢？对此，用方差分析法进行检验，结果列于表 4。

表 4　MA 和 AgI 增雨效果差异的显著性检验——方差分析表

	离差来源	自由度	离差平方和	F	$F_{0.05}$
合并统计	组间	$r-1=1$	0.052 7	1.939 9	4.04
	组内	$n-r=48$	1.302 7		
	总和	$n-1=49$	1.355 4		
层状云	组间	$r-1=1$	0.024 0	0.940 3	4.20
	组内	$n-r=27$	0.687 6		
	总和	$n-1=28$	0.711 6		
积状云	组间	$r-1=1$	0.034 3	1.088 5	4.41
	组内	$n-r=19$	0.598 0		
	总和	$n-1=20$	0.632 3		

由表 4 可见，两种催化剂的增雨效果差异都是不显著的，合并统计和层状云、积状云的 F 值分别是 1.939 9、0.940 3 和 1.088 5，远小于 $F_{0.05}=4.04$、4.20 和 4.41 值。

参考文献

［1］ 福建省气象局科研所、南京大学气象系，古田水库地区人工降水试验效果分析，大气科学，3 卷，2 期，131—140，1979.

［2］ Thom，H. C. S.，A statistical method of evaluating augmentation of precipitation by cloud seeding，Final Report of the Advisory Committee on Weather Control，Vol. Ⅱ，5—25，1957，preprints，n.d.

［3］ 叶家东、范蓓芬，方差不相等的双样本回归分析，大气科学，5 卷，2 期，214—224，1981.

［4］ Biondini，R.，Cloud motion and rainfall statistics，*J. of Appl. Meteor.* **15**，205—224，1976.

［5］ Scott，E. L.，Problems in the design and analysis of weather modification studies，Third Conf. on Probab. and statis. in Atmos. Sci.，65—72，1973，Boulder，Colorado；preprints，n. p. n. d.

THE STATISTICAL CHARACTERISTICS OF AREAL RAINFALL AND THE EFFECTS OF RANDOMIZED CLOUD SEEDING EXPERIMENT IN GUTIAN FUJIAN

Ye Jia-dong

(*Department of Meteorology Nanjing University*)

Cheng Ke-ming Zhen Guang-ping

(*Institute of Meteorology，Fujian Province*)

Abstract From April to June，1975，1977 and 1978，a randomized cloud seeding experiment with metaldahyde (MA) and silver iodide (AgI) from the ground by the small rockets was carried out in Gutian reservoir area，Fujian province. In this experiment，the design of randomized regression of area was used and the experimental effect was evaluated with the two-sample regression analysis with unequal residusl variances. The experimental objects were mainly the stratiform and cumuliform clouds under the influence of precipitating weather system，and the period of three hours was taken as experimental time unit. The experimental units were selected totaled 101，50 of them were seeded and the others unseeded as controlled by randomized procedure. The statistical analysis indicated that：(1) The normality of the transformation of areal rainfall was very good，which the goodness of fit reached 0. 94，and the correlation coefficient of this rainfall index between the two experimental areas exceeded 0. 82；(2) the difference between the residual variances of seeding and of control samples in the regression analysis might be significant，hence it was necessary to use the two-sample regression analysis method with unequal residual variances for testing the seeding effect；(3) the average rainfall in the target area (1 500 km^2) was increased aboud 20. 3% by seeding，with a significant level of 0. 05 (one side test)；(4) the seeding effect under frontal weather was 70. 6% in average ($\alpha <$ 0. 05)，which was most favourable for seeding in the various condtions of weather；(5) the relative seeding effect was larger when natural rainfall was small，and decreased with the increasing of natural rainfall in the control area，when it was more than 3 mm/h，the effect was non-significant；(6) the area of effect exceeding 20% extended 10—50 km downwind of the operation set，and its width was about 20 km；(7) the difference between seeding effects of MA and AgI was non-significant，although the amount released once of MA was about ten times larger than AgI.

4.6 古田水库地区人工降水试验效果统计分析①

福建省气象局气科所** 南京大学气象系

提要 1975 年和 1977 年 4～7 月,在福建古田水库地区进行了由小火箭携带介乙醛(MA)和碘化银(AgI)播云的人工降水试验,采用区域回归随机试验方案,利用统计方法检验效果。试验对象主要是系统天气影响下的降水性层状云和积状云,以三小时时段为试验单元,总共进行了 62 次随机试验,其中实际作业 31 次,对比试验 31 次。试验结果平均增加目标区雨量达 40%～78.7%,显著性水平 $\alpha<0.0025$。有效区域的范围可延伸到作业点下风方 40～60 千米远,层状云和积状云的有效范围区是不同的。对比区自然雨量强度小时,试验效果好,随着自然雨强的增大,效果变差。MA 和 AgI 两种催化剂的播云效果,差异不显著。

一、引言

近年来,我国各地广泛开展了以高炮和小火箭发射冷云催化剂播云的人工降水试验。这类试验的效果究竟如何? 在什么条件下效果较好? 效果较明显的区域有多大? 所有这些是实际工作中迫切需要解决的问题。但是,由于自然降水的时空变差很大,影响降水量的因素又多,降水的定量预报问题尚未解决,所以人工降水效果的客观评价一直是一个相当复杂的问题。迄今为止,差不多所有的人工降水试验效果的定量估价,都是依靠或结合统计分析的方法获得,而要进行这种分析,试验必须有严谨的设计。在各种试验设计方案中,较客观的是随机试验。1974 年在福建古田水库流域利用"三七"高炮发射 AgI 炮弹播云的方法,进行了人工降水试验。区域回归分析结果发现可以增加日雨量 24.7%,统计显著度接近 0.05,对台风天气影响下的降水性云系进行区域回归随机试验,结果表明三小时时段区域面积平均增雨量接近 1 倍,显著度 $\alpha<0.05$[1]。在 1974 年试验的基础上,1975 年和 1977 年 4～7 月份,利用机动性更大的小火箭发射介乙醛和碘化银播云的方法,开展人工降水试验。试验的目的是结合增加水库蓄水发电的具体任务,检验这种播云方法的实际效果,判别有利的天气条件和有效区域的范围。

二、试验设计

试验设计的主要目的之一是使样本资料符合随机抽样的原则,为此,我们采用区域回归随机试验方案。选择两个试验区,A 区为目标区,B 区为对比区。在有利的天气条件下,以三小时时段为试验单元,按照事先制定的随机程序,随机地决定 A 区播云或不播云,得到两组雨量资料 Y_{AS} 和 Y_{An}。B 区一直不播云,作为对比,与 A 区相对应,B 区也可得到两组雨量资料 X_{AS} 和 X_{An}。这四组雨量资料中只有 Y_{AS} 是播云后的雨量,其余均系自然降水量。根据 Y_{An} 和 X_{An} 建立区域回归方程,再将 X_{AS} 代入回归方程求出 A 区播云单元的自然降水量的期待值 $\hat{Y}$,比较 Y_{AS} 和 $\hat{Y}$ 可以确定试验效果。由于 A 区两组降雨机会中的试验单元是否播云是随机决定的,所以在这样两组机会中,除了播云这一人为因素有着系

① 1978 年 7 月 8 日收到修改稿。

** 本文由叶家东、程克明两位同志执笔。

统差异以外，其他因子对降水的贡献，彼此是相当的。因此，那些对降水有影响而与人工播云无关的因子，在这里是作为偶然因素出现的，从而当试验次数足够多时，它们将作为“随机误差”而从两组机会的平均结果中排除出去。因此，如果在这两组机会中观测到雨量有系统的差异，或者两组的平均雨量之间存在显著的差异，就可以归因于播云作业，据此即可评定人工降水的效果。利用对比区雨量作为控制变量，可以减小目标区估计雨量的误差，提高效果分析的准确率。

作业点设在古田西北 32 千米处的石塔山。试验期(4～7 月)的主要降水性云系是自西向东的。目标区 A 设在作业点的下风方，面积约 1 500 平方千米。对比区 B 选在作业点上风方，面积与目标区相等，两区间隔 11 千米以防止催化剂对对比区的污染(图 1)。两个试验区都是山区，地理条件大体相似。根据两年试验期内 31 次随机决定的对比试验，计算两区雨量的相关系数 $r=0.8270$($\alpha<0.001$)，可见两区雨量的相关性很好，回归分析比较有效。

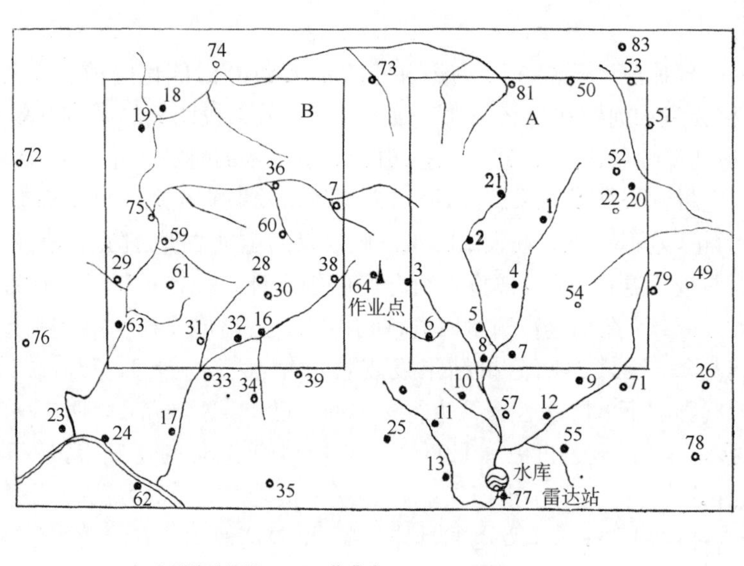

图 1　试验区的地理位置

两个试验区的雨量站，A 区 15 个，B 区 16 个，连同试验区周围地区共有 81 个雨量站的雨量资料供制作雨量图之用。

为了增加试验期随机样本的容量，试验单元时间取 3 小时时段。与日雨量相仿，3 小时雨量也是一种具有雨强性质的统计变量，但它与降水过程的平均雨强不同，还叠加了一个该时段在降水过程中所处部位这样一个随机因素，由于比较是在随机选取的播云单元与对比单元之间进行的，因此，自然条件的有利程度，在两组雨量中在统计上仍是彼此相当的。所以仍可将它们作为随机变量处理。此外，据雷达观测，试验云的移动速度一般在 20 千米/小时左右，它移过目标区只需 2 小时，而作业时间一般不超过半小时，所以前一播云单元所播撒的催化剂基本上不会残留到下一试验单元，从而在一定程度上避免了试验单元之间的污染问题。

有利天气条件主要根据天气预报、雷达观测和探空资料判断，当天气预报有降水系统过境时，即由雷达监视，若在试验单元前半小时两试验区的云的回波顶高在 5 000(层状云)和 5 500(积状云)米以上，该天 500 mb 高空风向介于 217°～290°，且估计在该时段云层条件仍能维持或发展的，即选作试验单元。2 年的试验期共选了 62 个试验单元，按随机规则决定其中 31 次播云、31 次对比。按云形分类，层状云播云 13 次，不播云 15 次；积状云播云 18 次，不播云 16 次。

31 次播云试验中，19 次用介乙醛，平均每次用量 2 450 克；12 次 AgI 播云试验，平均每次用量 224 克。发射工具采用自制小火箭，发射高度 2 600～4 000 米，加上作业点高度 1 625 米，催化剂可以发射到 4 200～5 600 米高度上爆炸播撒，试验期福州地区 0 ℃层高度为 4 400～5 200 米，所以催化剂一般是在 0 ℃层附近播撒，而后随上升气流带到云的过冷却区的。

三、效果分析

参加统计分析的雨量资料取 3 小时区域面积平均雨量。由于雨量的概率分布通常是偏态的，因此在做统计检验时就不能直接用 t -检验法。关于一次降水过程为单元的雨量的分布函数，Thom 用 Γ -分布来拟合[2]，Schickedanz 等则用对数正态分布来拟合[3]。现在我们假设试验期 3 小时面积平均雨量服从对数正态分布，利用柯尔莫哥洛夫定理进行“拟合度”检验。检验结果拟合概率达到 0.544，说明对数正态分布和观测资料的拟合度还是较好的，所以统计分析时，统计变量取雨量的对数，变换成正态变量，然后采用回归分析求取人工降水的增雨量，并用 t -检验法对它进行显著性检验。

对试验效果进行双比分析时，取雨量本身，所以效果的显著性检验采用非参量性的秩和检验法。

1. 双比分析

假定目标区的自然降水量和对比区的雨量成比例，作双比值

$$\bar{R}=\frac{\bar{Y}_{AS}/\bar{X}_{AS}}{\bar{Y}_{An}/\bar{X}_{An}} \tag{1}$$

式中：$\bar{Y}_{AS}$ 和 $\bar{X}_{AS}$ 是历次播云单元目标区和对比区的平均面积平均雨量；$\bar{Y}_{An}$ 和 $\bar{X}_{An}$ 是历次非播云单元目标区和对比区的平均面积平均雨量。如果人工降水无效，则双比值 $\bar{R}$ 的期待值为 1；$\bar{R}>1$ 表示有正效果，$\bar{R}<1$ 表示负效果，计算结果列于表 1。

表 1 双比分析表

类别	n_{AS}	n_{An}	$\bar{Y}_{AS}$	$\bar{X}_{AS}$	$\bar{Y}_{An}$	$\bar{X}_{An}$	$\bar{R}$	u	α(秩和检验)
合并统计	31	31	8.61	7.11	5.42	6.25	1.40	−3.19	0.000 7
层状云	13	15	9.20	7.04	4.64	6.58	1.86	2.46	0.006 9
积状云	18	16	8.18	7.16	6.15	5.95	1.11	−2.00	0.022 7

由表 1 可见，合并统计的结果，双比值 $\bar{R}=1.40$，人工播云平均增加雨量 40%，统计显著度达 0.000 7。按云形分层统计，层状云和积状云的增雨量分别为 86%($\alpha=0.006\ 9$)和

11%(α=0.022 7)。

我们进一步假定整个试验区中各雨量站的单站自然雨量是和对比区的面积平均雨量成比例的,据此可以分析效果比较明显的区域与作业点的位置及高空风的关系。为此,作双比值[4]

$$R=\frac{Y_{AS}/\overline{X}_{AS}}{Y_{An}/\overline{X}_{An}} \quad (2)$$

式中 Y_{AS} 和 Y_{An} 分别是试验区中某个雨量站的播云单元和对比单元的平均雨量,其他符号如前述。这样对整个试验区内外的雨量站分别求出双比值 R,并用秩和检验法检验其显著性,结果列于图 2。由图可见,雨量增值达到 50%以上的区域位于从作业点到下风方约 50 千米处,而 $R\geqslant2.0$ 的范围在作业点下风方 15 千米到 40 千米、宽约 20 千米的区域。$R\geqslant2.0$ 区内 5 个雨量站的双比值统计上都是显著的。

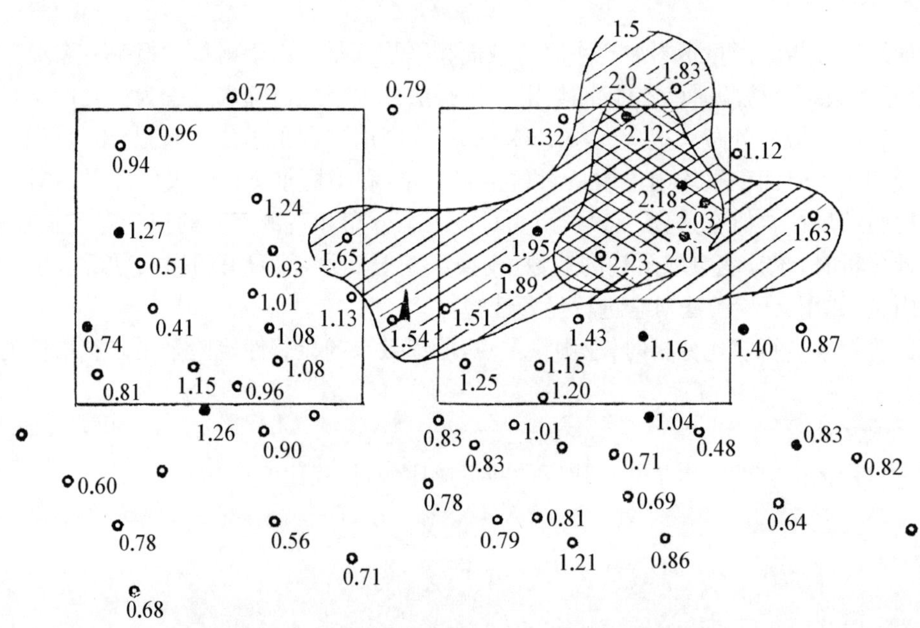

图 2　试验区各雨量站的双比值($R\geqslant1.5$ 的区域用影线表示,$\alpha\leqslant0.05$ 的站用·表示)

层状云和积状云相应的双比值分布见图 3 和图 4。层状云主要是对系统天气的降水性层状云系试验的,500 mb 高空风向主要是西偏北或西偏南风;积状云主要也是系统天气影响下的 Cb 云,有的是层状云幕中的对流云团,高空风以西偏南风为主。所以图 3 和图 4 $R\geqslant1.5$ 和 $R\geqslant2.0$ 的区域都落在作业点的下风方,这与雷达"PPI"回波的移向也是一致的。层状云 $R\geqslant1.5$ 的区域最远可延伸到离作业点 60 千米远处,而 $R\geqslant2.0$ 的区域在作业点下风方 15~50 千米范围、宽度 20 千米左右;积状云 $R\geqslant1.5$ 的范围限于作业点下风方 10~40 千米之间、宽度约 20 千米的区域,基本上都在预定的目标区内。积状云效果较显著的区域较小,这与积状云生命期较短有关。看来对不同的降水过程,目标区的设置应该是不同的。

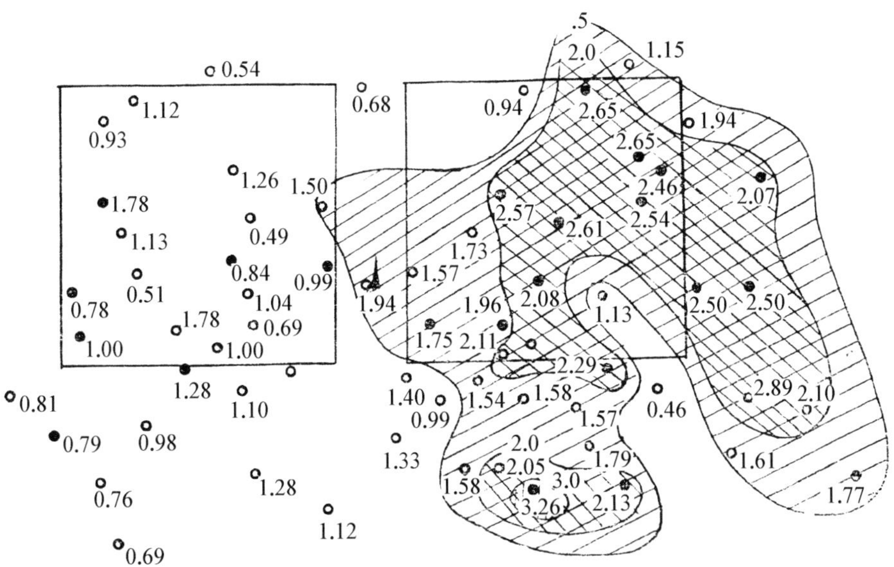

图 3　层状云试验各雨量站的双比值（$R \geqslant 1.5$ 的区域用影线表示，$\alpha \leqslant 0.05$ 的站用·表示）

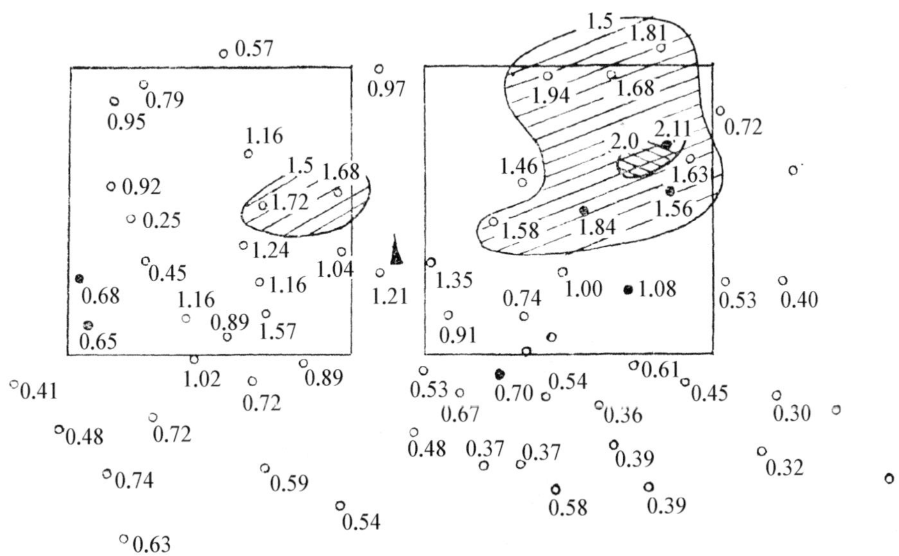

图 4　积状云试验各雨量站的双比值（$R \geqslant 1.5$ 的区域用影线表示，$\alpha \leqslant 0.05$ 的站用·表示）

2. 回归分析

31 次非播云单元 A、B 两区的 3 小时面积平均雨量(对数值，下同)之间的相关系数 $r=0.8270$，其显著度 $\alpha<0.001$，根据这 31 次自然雨量数据建立区域回归方程为

$$\hat{Y}=-0.1235+1.0122X \tag{3}$$

样本回归线和实测雨量散布点列于图 5。将第 j 次播云单元的 B 区雨量 X_j 代入回归方程，

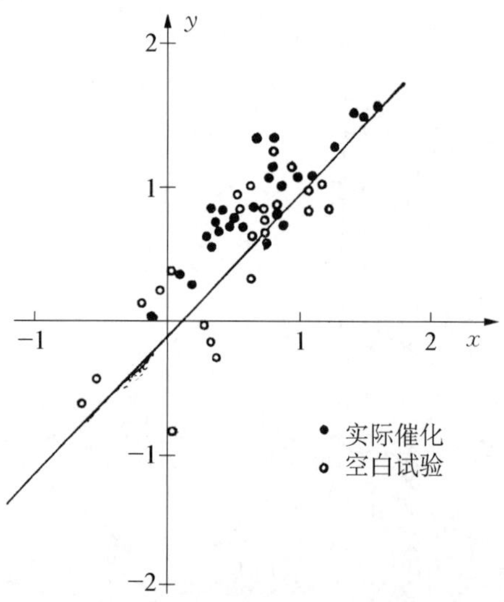

图 5　3 小时雨量(对数值)区域回归线和实测雨量散布图

得到 A 区自然降水量的期待值$\hat{Y}_j$，并求出雨量增值

$$\Delta Y_j = Y_j - \hat{Y}_j (j = 1, 2, \cdots, k)$$

31 次播云单元的平均值分别为 $\bar{Y}_k = 0.8062$，$\bar{\hat{Y}}_k = 0.5542$ 和 $\Delta\bar{Y}_k = 0.2520 (k = 31)$，相应的雨量真数值为 $\bar{Y}_k' = 6.40$(毫米 /3 小时)，$\bar{\hat{Y}}_k' = 3.58$(毫米 /3 小时) 和 $\Delta\bar{Y}_k' = 1.787$。$\Delta\bar{Y}_k'$ 是实测平均雨量 $\bar{Y}_k'$(几何平均，下同)对期待平均雨量$\bar{\hat{Y}}_k'$的比数，所以人工降水的平均相对增雨量为 0.787，而平均绝对增雨量为

$$E' = 0.787\, \bar{\hat{Y}}_k' = 2.82 (\text{毫米}/3\ \text{小时})$$

利用多个事件检验法对这个增雨量进行显著性检验，平均增雨量的 t 值为

$$t = \frac{\Delta \bar{Y}_k}{\sqrt{\dfrac{1-r^2}{n-2}\sum_{i=1}^{n}(Y_i - \bar{Y}_n)^2\left[\dfrac{1}{k} + \dfrac{1}{n} + \dfrac{(\bar{X}_k - \bar{X}_n)^2}{\sum_{i=1}^{n}(X_i - \bar{X}_n)^2}\right]}} = 3.15 \tag{4}$$

式中 $n=31$ 和 $k=31$ 是对比单元和播云单元数目。由 t -分布表知 $t > t_{0.005}$，意即人工降水效果显著，单边检验的显著度 $\alpha < 0.0025$。

雨量增值的 90%单边置信区间为 $\Delta Y_k' > 1.44$(毫米/3 小时)，表示人工降水增加雨量超过 1.44(毫米/3 小时)的概率达到 90%。

3. 分层统计

按天气条件对播云效果进行分层统计，是判断有利天气条件的一种有用的方法。分层统计不仅是缩小抽样变差提高统计分析效率的一种手段，而且也是提高样本的随机性所需要的。它把统计总体限制在一个物理过程比较单一的统计实体内。这样，在各个层次的次级总体内，对所欲检验的效应或属性有重要影响的那些物理因子就不存在系统性的差异，于是把与人工影响无关的各种因素当作随机因子就比较合理。下面我们就现有的资料，按云形催化剂和对比区自然雨量强度对试验效果进行分层统计。

(1) 按云形分层统计

分别对层状云和积状云的试验效果进行区域回归分析，结果列于表 2、图 6 和图 7。

层状云降水 A、B 两区的自然雨量相关系数 $r=0.9288$，其显著度 $\alpha<0.001$。人工影响的结果，目标区平均增加雨量 3.65 毫米/3 小时，相对增加率达 91.6%($\alpha<0.005$)；雨量增值的 90%单边置信区间为 $\Delta Y_k'>2.07$ 毫米/3 小时。

积状云降水 A、B 两区的自然雨量相关系数 $r=0.8657$，显著度 $\alpha<0.001$。人工影响使得目标区平均增加雨量 2.85 毫米/3 小时，相对增值达 102.5%，其显著度 $\alpha<0.01$；雨量增值的 90%单边置信区间为 $\Delta Y_k'>1.24$ 毫米/3 小时。

表 2　区域回归随机试验效果分析表

项目		合并该计			分层统计					
					积状云			层状云		
		75 年+77 年	75 年	77 年	75 年+77 年	75 年	77 年	75 年+77 年	75 年	77 年
随机试验次数	播云试验次数 k	31	19	12	18	11	7	13	8	5
	对比试验次数 n	31	15	12	16	8	7	15	7	5
	合并次数	62	34	24	34	19	14	28	15	10
A、B 两区雨量相关系数 r		0.827 0	0.854	0.723 1	0.865 7	0.922	0.750 0	0.928 8	0.955	0.964 4
相关系数 r 的显著性水平 α		<0.001	<0.001	<0.01	<0.001	<0.01	≈ 0.05	<0.001	<0.001	<0.01
A、B 两区雨量的回归方程		$\hat{Y}=-0.1235+1.0122X$	$\hat{Y}=-0.1716+1.086X$	$\hat{Y}=0.0005+0.7956X$	$\hat{Y}=-0.7042+1.8094X$	$\hat{Y}=-0.8885+2.1071X$	$\hat{Y}=-0.4113+1.3519X$	$\hat{Y}=0.0199+0.8093X$	$\hat{Y}=-0.0661+0.8268X$	$\hat{Y}=0.3135+0.4968X$
A 区平均雨量增值/(毫米/3 小时)		2.82	3.16	2.39	2.85	3.01	2.18	3.65	4.22	3.07
A 区雨量增值的相对增长率		78.65%	81.8%	76.16%	102.49%	82.7%	101%	91.6%	126.5%	65.35%
A 区雨量增值的显著性水平 α〈单边〉		<0.005	<0.05	<0.05	<0.01	<0.075	≈ 0.1	<0.005	<0.01	<0.025
A 区雨量增值的置信区间	置信水平〈$1-\alpha$〉	0.90	0.90	0.90	0.90	0.90		0.90	0.90	0.90
	置信下限/(mm/3 小时)	1.44	0.85	0.61	1.24	0.53		2.07	1.95	1.69

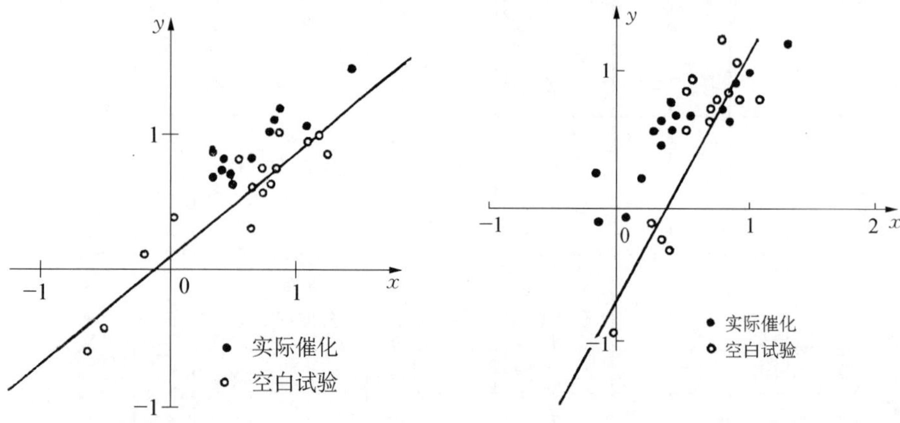

图 6　层状云降水区域回归线和雨量散布图　　图 7　积状云降水区域回归线和雨量散布图

由图 6、图 7 和表 2 可见，不同云形 A、B 两区的雨量回归关系是不一样的，所以合并统计时，两区的雨量相关系数小，回归分析的效率也较低，不过两种降水过程的区域雨量相关系数的差异是不显著的，统计显著度 α 只为 0.21。

(2) 按催化剂分层统计

1975 年采用介乙醛，1977 年采用碘化银。两年所用催化剂是不同的，这两种不同的催化剂的播云效果，从表 2 看是有差异的，介乙醛的效果普遍好一些。我们用方差分析方法对这种差异的显著性进行统计检验，结果列于表 3～5。

表 3　层状云效果的一元方差分析表

	自由度	离差平方和	F	$F_{0.05}$
总和	$n-1=12$	$Q=0.2368$	3.528	4.84
组间	$r-1=1$	$Q_2=0.0575$		
组内	$n-r=11$	$Q_1=0.1793$		

表 4　积状云效果的一元方差分析表

	自由度	离差平方和	F	$F_{0.05}$
总和	$n-1=17$	$Q=5.7842$	0.020	4.49
组间	$r-1=1$	$Q_2=0.0073$		
组内	$n-r=16$	$Q_1=5.7769$		

表 5　合并统计一元方差分析表

	自由度	离差平方和	F	$F_{0.05}$
总和	$n-1=30$	$Q=1.1245$	0.036	4.18
组间	$r-1=1$	$Q_2=0.0014$		
组内	$n-r=29$	$Q_1=1.1232$		

由表可见，两种催化剂的播云效果的差异都是不显著的，合并统计和积状云的 F 值分别是 0.036 和 0.020，远小于 $F_{0.05}=4.18$ 和 4.49；层状云的 $F=3.53$，仍达不到 $\alpha=0.05$ 的水平，所以还没有足够的证据说介乙醛的效果比碘化银好。但介乙醛有较好的效果这个迹象值得在今后的试验中进一步分析，看是否有实质性的意义。

(3) 按对比区自然雨量强度分层统计

为了比较不同的自然降水条件下的播云效果，我们根据播云单元对比区的雨量强度对播云效果进行分层统计，结果列于表 7。

表 6　播云效果按对比区雨量强度分层统计表

类别	对比区面积平均雨量强度/(毫米/小时)								
	$X'\leqslant 1$			$1<X'\leqslant 3$			$X'>3$		
	n	相对增雨量/%	α	n	相对增雨量/%	α	n	相对增雨量/%	α
合并统计	11	116.0	<0.005	15	82.9	≈0.005	5	9.5	≈0.42
层状云	2	139.8	≈0.01	9	91.8	<0.005	2	52.6	<0.1
积状云	9	361.6	<0.005	6	39.6	<0.15	3	−64.0	<0.21

由表 6 可见，对比区雨量强度小于 1 毫米/小时时，播云效果比较好，相对增加率超过 1 倍，显著度都达到 0.01 以上，而当对比区雨量强度大于 3 毫米/小时时，播云效果就很不显著，特别是积状云，效果是负的，当然这个负效果也不显著。播云效果随对比区自然雨量增加而减小的这种趋势，在美国佛罗里达的试验中也相当明显[5]，如果这种趋势能在更多的试验中继续得到证实，那它说明了一个很重要的事实，就是人工降水的效果很可能既不是可加的，也不是可乘的，而以往总认为效果是可乘的[6]。从降水物理的角度看，这也许意味着自然降水条件好时，自然降水效率比较高，人工播云就“无用武之地”，弄不好反而会“帮倒忙”，减少雨量；而当自然雨量小时，人工播云在提高降水效率方面起作用的余地就比较大。从统计上看，这意味着人工影响不仅改变了雨量的平均值，而且也改变了雨量的方差。如果这是真的，则将对效果检验的方法本身提出重大的修正，因为通常的检验方法都是在方差不变的前提下适用的，所以这是一个值得进一步研究的问题。

参考资料

[1]　福建省气象局、南京大学，南京大学学报(自然科学版)，第一期，137—151 页，1975.

[2]　H. C. S. Thom, Final Report of the Advisory Committee on Weather Control, vol II, p. 5 - 25. 1957.

[3]　P. T. Schickedanz and F. A. Huff, *J. of Appl, Meteor.*, vol. 10, No. 3, p. 502 - 524, 1971.

[4]　E. D. Elliott, P. S. Amand, and J. R. Thompson, *J. Appl. Meteor.*, vol. 10, p. 785 - 795. 1971.

[5]　J. Simpson, et al., *J. of Appl. Metor.*, vol. 10, No. 3, p. 526 - 543, 1971.

[6]　R. Biondini, *J. of Appl. Meteor.*, vol. 15, p. 205 - 224, 1976.

STATISTICAL EVALUATION OF THE ARTIFICIAL RAINFALL STIMULATION EXPERIMENT IN GUTIAN RESERVOIR REGION, FUJIAN PROVINCE

Weather Bureau of Fujian Province and Department of Meteorology, Nanjing University

Abstbact From April to July, 1975, and 1977, a series of artificial stimulating precipitation experiments by seeding clouds with metaldehyde (MA) and silver-iodine (AgI) from the ground by the small rockets were carried out in Gutian reservoir region of Fujian province. In these experiments, the design of random regression of area was used and the experimental effect was evaluated with statistical method. The objects in these experiments were mainly the strati- and cumuliform clouds under the influence of precipitating weather system, and the 3 h period was taken as experimental time unit. Of all 62 stochastic experimental units were selected in which, 31 were seeded and the others were unseeded as controls. The statistical analysis indicated that (1) the average rainfall on target area was increased about 40%—78.7% by seeding, with a significance level of 0.0025 (one side), (2) the area of marked effect extended 40—60 km downwind of the operation set, (3) the effect of seeding was larger when the natural rainfall was small, and decreased with increasing natural rainfall, (4) there was not appreciable difference in seeding effectiveness between MA and AgI.

4.7 Randomized Cloud Seeding at Gutian, Fujian, China

Jia-dong Yeh Ke-ming Cheng and Guang-ping Zhen
Institute of Meteorology Weather Bureau of Fujian Province
Department of Meteorology, Nanjing University

Abstract To increase rainfall over the watershed of Gutian reservoir, half-way from Shanghai to Canton and 100 km inland from Taiwan strait, metaldehyde and silver iodide were injected by rocket into clouds during three seasons, April—June 1975, 1977, and 1981. Randomization provided 50 seeded and 51 unseeded three-hour cases, indicating a 20% increase (significant at 5% level) in mean rainfall over the 1 500 km^2 target, mostly in frontal situations. Fourth-root transformations provided adequate transformation to normality of the three-hour rainfall totals. Residual variances of regressions of seeded and unseeded cases differed markedly, requiring special analyses. Contrary to previous assumptions, seeding effects were not proportional to control area rainfall, and regression intercepts were not zero.

1. INTRODUCTION

From April to June of 1975, 1977, and 1978, a series of randomized cloud seeding experiments were conducted over the Gutian (Kutien) reservoir in the coastal mountains of Fujian (Fuchien) province, about 100 km inland from Taiwan strait, almost midway between Shanghai and Guangzhou (Kuangchou, Canton) (Fig. 1). (Throughout this report, place names are in Pinyin, the new official Chinese transliteration system, with the older Wade-Giles and other forms in parentheses on first usage. Personal names are in western style, family name last. —Ed.)

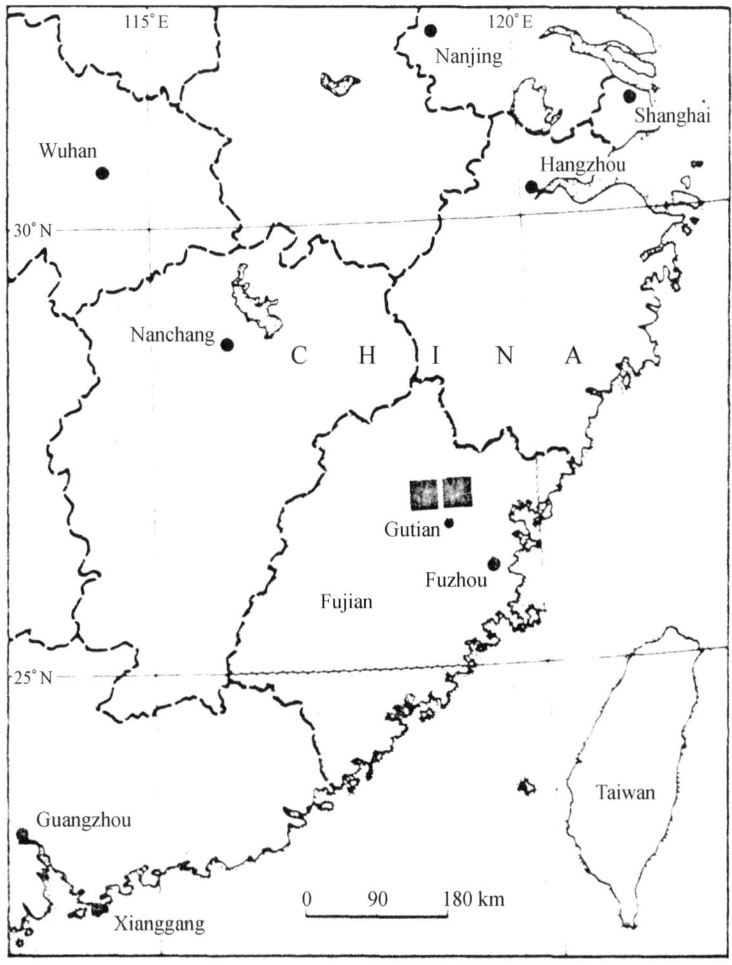

Fig.1 Location of target (right) and control (left) areas, each about 1 500 km² (39 km square), in northern Fujian province, China, of 1975, 1977, and 1978 April—June randomized cloud seeding experiment

For 15 years up to 1974, the mean water storage at the end of June in Gutian reservoir had been only 65% of capacity, insufficient for hydro-electric generation. A research program was organized by Mr. Yuen-yun Sue, Director of the Institute of Meteorology, Weather Bureau, Fujian province, to study synoptic situations suitable for cloud seeding to increase rainfall over the reservoir's watershed, and to estimate the area possibly affected by such seeding.

Long-term mean monthly rainfall in April, May, and June at Gutian is 218 mm on an average of 22 rainy days per month. The chief precipitation in late spring is from frontal cyclones with stratiform and cumuliform clouds. Cloud tops usually are higher than 5 000 m, with the upper parts generally supercooled. Thus a cold cloud seeding technique was adopted for the rain stimulation experiment, using metaldahyde (C_2H_4O or "MA" and silver iodide (AgI).

Small rockets, made in Fujian (Inst. of Met., 1975), each carrying 100 g of MA or 5 g of AgI, could be shot to heights of 2.6 km to 4.0 km, according to rocket type. The operation site, on Mt. Shita 32 km NW of Gutian, was about 1.6 km MSL, so seeding agents could be sent as high as 4.2 km to 5.6 km MSL. During the experimental period the 0 ℃ level over Fuzhow (Fuchou or Fuchow) is 4.4 km to 5.2 km, so that the MA or AgI usually was released near the 0 ℃ level and carried by updrafts to the supercooled regions of the cloud. On the 29 cases using MA, average dose was 3 295 g; for the 21 uses of AgI, average dose was 213 g. Laboratory experiments (Inst. of Met., 1977) showed the nucleation efficiencies and average numbers of activated nuclei given in Table 1.

Table 1 Nucleation efficiencies and number of activated nuclei

	Temperature/℃	Efficiency per gram	Number of activated nuclei
MA	−10 to −15	10^{10} to 5×10^{10}	3.3×10^{13} to 1.7×10^{14}
AgI	−12 to −13.5	10^{11} to 10^{12}	2.2×10^{13} to 2.2×10^{14}

2. EXPERIMENTAL DESIGN

To evaluate objectively the effects of cloud seeding, a randomized program was adopted (Weather Bureau, 1979), using regressions of precipitation in a specified target area on that in a control area. When a synoptic situation suitable for seeding appeared, whether or not the target area would actually be seeded was decided by a randomized procedure. The control area was never seeded. Separate regressions were constructed predicting rainfall in the target area from that in the control area for seeded and unseeded cases. The difference between the two regressions showed whether the seeding had any effect. This design avoids use of historical rainfall data. The randomized procedure gave a firm foundation for the assumption of similarity between the synoptic situations of the seeded and control cases, thereby increasing the efficiency of the test of seeding effect.

The target area was a rectangle of about 1 500 km^2 over the Gutian reservoir watershed (Fig.2). Winds at the 500 mb level blow from WSW 82%, 83%, and 69% of the time in April, May, and June, respectively, and the main precipitating systems move from west to east at about 20 km/h, drifting across the target in about two hours. Thus a three-hour period was adopted as the experimental unit, assuming that a seeding agent produced during one experimental period would not last into the next one, and no sequential contamination was likely.

The control area was another 1 500 km^2 rectangle west (upwind) of the target,

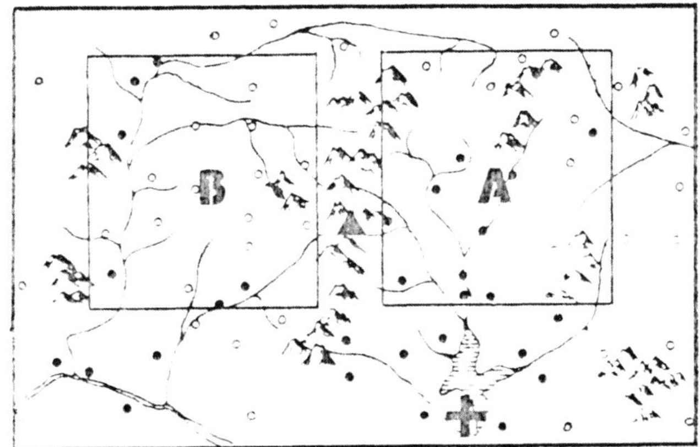

Fig.2 Raingages (solid circle = recording, open circle = ordinary), locations of rocket launch site on Mt. Shita (solid triangle) and of radar on Mt. Turtle at Gutian (cross), and drainage and topography of target (right] and control left areas, separated by 11-km buffer strip

topographically similar, separated from it by a buffer zone 11-km wide containing Mt. Shita, on the watershed between the two areas.

The target area already had 15 to 17 raingages, the control area 16 to 18. For the experiment, additional raingages were places in both area, making a total of 81 to 93, some of them recording. The ordinary gages were read every three hours. Isohyetal maps based on the raingage readings were drawn and mean area rainfall obtained by integrator. For the 51 unseeded units during the three years, the correlation between average areal precipitation in target and control areas was 0.824 5. (The 95% confidence interval, determined from Fisher's z-prime transformation, for a correlation of 0.82 based on 51 pairs extends from 0.70 to 0.90. — Ed.)

Suitability of synoptic conditions for seeding was determined mainly from weather predictions, radar observations, and radiosonde data. When a precipitating system was predicted, the type 711 radar on Mt. Turtle at Gutian began to watch the clouds over the experimental region.

Half an hour before a three-hour experimental unit would begin, if stratiform cloud tops in the two areas were higher than 5 000 m or cumuliform tops higher than 5 500 m, and would probably remain or develop during the period, and the wind direction at the 500 mb level was from 217° to 290°, then the period was declared experimental and a randomized decision made whether seeding should begin.

In this way, during the three years 101 experimental units were selected:

	Seeded	Control	Total
Stratiform	29	30	59
Cumuliform	21	21	42
Total	50	51	101

In addition to this cloud-type stratification, experimental units were classified synoptically into prefrontal, frontal, postfrontal, and high pressure situations.

3. STATI STICAL OPERATIONS

Because the statistical distribution of the three-hour rainfall totals was quite skewed and thus markedly different from that of the normal distribution, on which the usual statistical procedures are based, transformations were needed. Logarithmic and square, cube, fourth, and fifth root transformations were compared to a normal distribution by the Kolmogorov-Smirnov test (Table 2).

Table 2 Goodness of fit and area correlations of various transformations

Transformation exponent	none	1/2	1/3	1/4	1/5	log
Goodness of fit, $1-K(y_o)$	0.004 1	0.457 8	0.744 2	0.937 3	0.943 0	0.553 7
Area correlation coefficient	0.667 6	0.775 4	0.810 2	0.824 5	0.835 8	0.843 5

The logarithmic transformation yielded the strongest correlation between the average rainfalls in the two areas, but a poor fit to normality. Both the fourth and fifth root transformations gave distributions very close to normal, with goodness of fit of 0.94 and correlation greater than 0.82 between transformed rainfall amounts in the two areas. For all comparisons, therefore the fourth root of the 3-hour precipitation totals was used.

The general multiple event regression procedures for cloud seeding evaluation developed by Thom (1957) assume equality of variances of the seeded and control samples. Tests of the residual variance in regression analysis (Table 3) showed that such equality may be acceptable for rain from stratiform clouds but not from cumuliform clouds or for the entire sample including both types. Consequently, a two-sample regression analysis for unequal variances was developed (Yeh and Fan, 1981) for use here.

Table 3 Residual variances in regressions

	n_s	s_{sr}^2	n_c	s_{cr}^2	F	Signif.
Stratiform	29	0.030 7	30	0.027 4	1.120 4	>0.05
Cumuliform	21	0.030 8	21	0.070 5	2.289 0	<0.05
Total	50	0.027 0	51	0.044 7	1.655 6	<0.05

In the randomized cloud seeding experiment with control areas, the regression equations for areally-average rainfall are

$$\hat{Y}_j = a_j + b_j X$$

where $\hat{Y}_j$ is the regression estimate of target area rainfall, X is the observed control area rainfall, and $j = s$ for seeded 3-hour units, $j = c$ for control units. The corresponding linear normal regression models are

$$Y_j = \alpha_j + \beta_j X + e_j.$$

The residuals e_j are normally distributed with zero means and possibly unequal variances σ_s^2 and σ_c^2. These can be estimated from the sample residual variances

$$s_{jr}^2 = \sum_{j=1}^{n_j} (Y_{ji} - \hat{Y}_{ji})^2 / (n_j - 2).$$

For a test of the null hypothesis of equality of the means of the regression-computed target rainfalls for the seeded and unseeded units, a student's test can be used. For it, the sample residual variances must be adjusted to obtain the pooled standard deviation

$$Q_j^2 = s_{jr}^2 \left[\frac{1}{n_j} + (X - \overline{X}_j)^2 / \sum_{n_j} (X_i - X_j)^2 \right].$$

In the denominator of the t-statistic

$$Z_{\hat{y}} = (\hat{Y}_s - \hat{Y}_c)(Q_s^2 + Q_c^2)^{-1/2}.$$

The adjusted residual variances also appear in the expression for the degrees of freedom of the t-statistic for the unequal variance case. When $s_{sr}^2 = s_{cr}^2$, the simpler multiple event test of Thom (1957) is used.

4. CLOUD SEEDING EFFECTS

Results of the cloud seeding were examined through comparisons of the fourth roots of the average precipitation in the target and control areas, stratified by cloud type, synoptic situation, and rainfall amount in the control area. In addition, results were examined for individual target raingages, and differences sought between the MA and AgI treatments.

The mean increase in target rainfall, overall, was 20. 3%, significant at the 0. 05 level in a onesided test; the corresponding absolute increment was 1. 11 mm/3 h (Fig. 3). By cloud type, mean inrement of rainfall from stratiform clouds was 21. 7%, from cumuliform clouds 17. 5%, each corresponding to absolute increments of about 1 mm/3 h, but none of these was significant at the 5% level (Table 4).

More interesting results came from the synoptic stratification: Seeding effects were significant only for frontal weather, where the mean increment was 70. 6%, or 2. 82 mm/3 h, with less than 5% probability of being due to chance (Table 4). Effects under prefrontal, postfrontal, and high pressure conditions were not significant.

Effects of rainfall intensity on the experiment were studied by stratifying all experimental data of seeded and unseeded units according to rain intensity in the control area (Fig. 4). The relative increments ranged from 12% to 81%, but decreased with increasing control area rainfall, while absolute additions increased, from 0. 44 mm/3 h to 1. 14 mm/3 h. For all cases, seeding effects were significant at the 5% level when control area rainfall was between 5. 80 mm/3 h and 10. 50 mm/3 h, not significant at lesser or

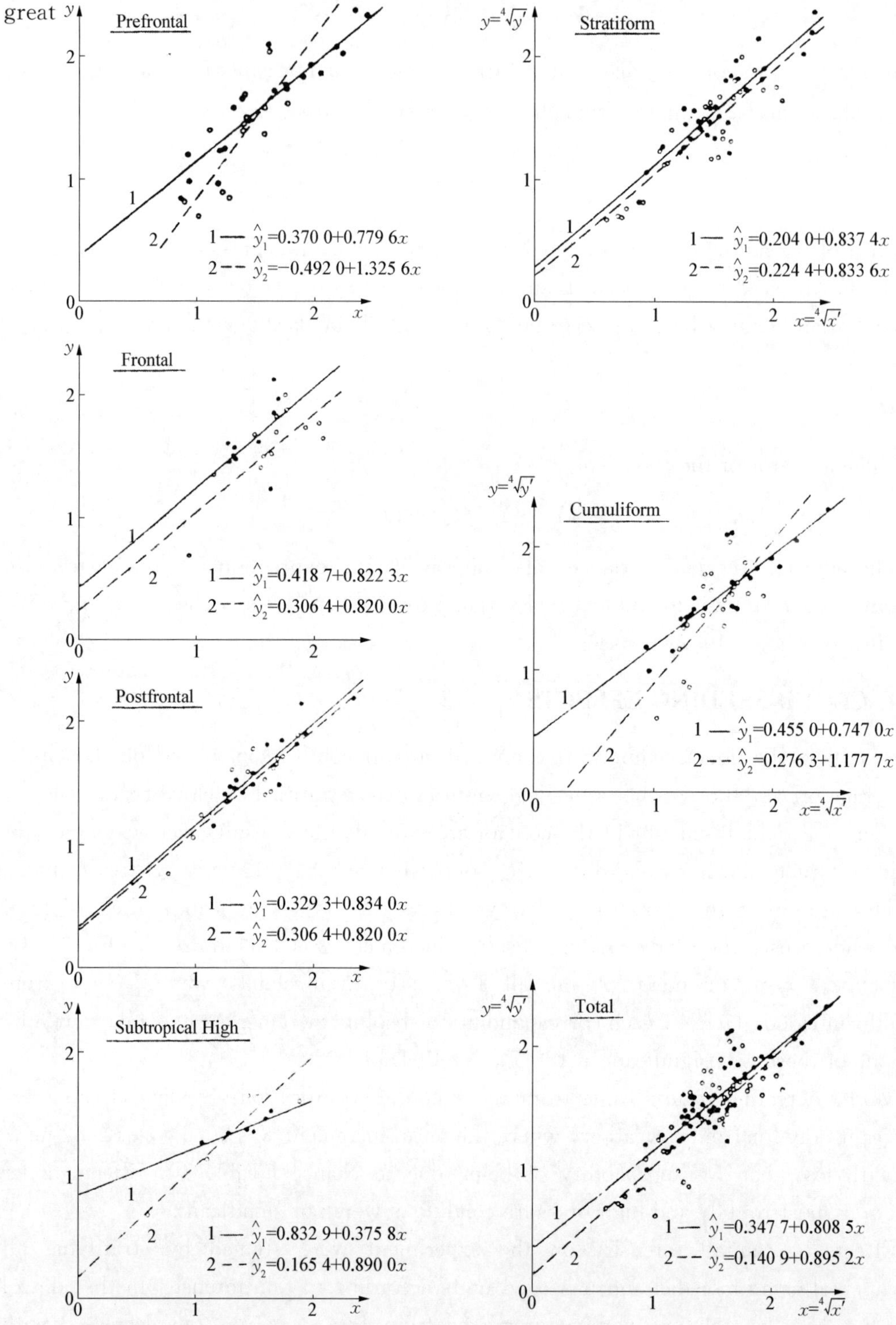

Fig. 3 Target-control regressions of fourth roots of areal precipitation for seeded (dots) and control (circles) units, by cloud-type and synoptic stratifications

Table 4 Correlations, regression parameters, and significance levels of regression comparisons of the fourth root of seeded (S) and control (C) rainfall, under two different stratifications

		By cloud types		Stratified by synoptic situations				
		Strati-form	Cumuli-form	Pre-front	Front	Post-front	Subtrop. High	Total
Numbers of 3-hour units	S	29	21	18	10	16	6	50
	C	30	21	15	13	14	* 8	51
	T	59	42	33	23	30	14	101
Correlations of areal precipitation	S	0. 863 3	0. 866 7	0. 883 0	0. 521 0	0. 916 9	0. 905 9	0. 877 1
	C	0. 901 1	0. 713 8	0. 770 8	0. 828 2	0. 879 4	0. 937 2	0. 824 5
Regression constants, α	S	0. 294 8	0. 455 0	0. 377 0	0. 418 7	0. 329 3	0. 832 7	0. 347 7
	C	0. 224 4	0. 276 3	0. 492 0	0. 250 1	0. 306 4	0. 165 4	0. 140 9
Regression Coeff., β	S	0. 837 4	0. 747 0	0. 779 6	0. 822 3	0. 834 0	0. 375 8	0. 808 5
	C	0. 837 6	1. 177 7	1. 325 6	0. 799 4	0. 820 0	0. 890 0	0. 895 2
Increment estimates	mm/ 3 h	1. 14	1. 01	−0. 79	2. 82	0. 72	−0. 72	1. 11
	%	21. 4	17. 5	−9. 4	70. 6	11. 8	−17. 0	20. 3
Approx. levels of significance	α_1	0. 063	0. 225	0. 250	0. 025	0. 213	0. 250	0. 050
	α_2	0. 100	0. 200	0. 250	0. 05	0. 200	0. 200	0. 050

α_1 is from the multiple event test, α_2 is from the two-sample regression test, both one sided.
* one control case wasn't classified synoptically.

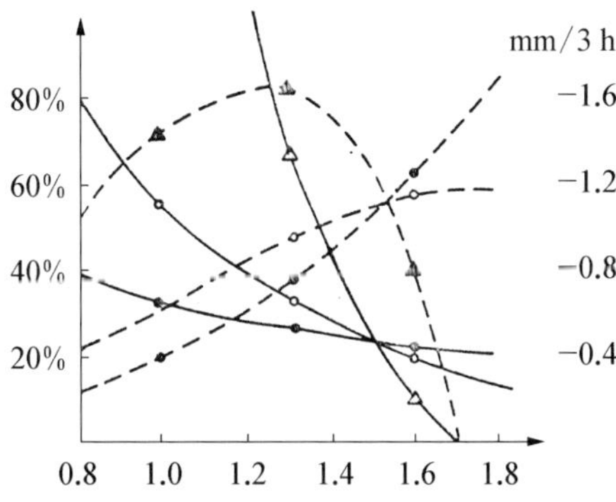

Fig. 4 Increments (relative, %, solid, and absolute, mm/3 h, dashed) in seeding stratiform (dots), cumuliform (triangles), and all cloud typeps (circles) in target area as function of control area precipitation

(Table 5). Stratiform clouds provided the largest increases (about 22%) when control area rainfall was around 2 mm/3 h, cumuliform clouds (up to 67%) when control area rainfall was only about 1 mm/3 h. Absolute increments from stratiform clouds were significant when control area rainfall was between 2. 86 mm/3 h and 5. 80 mm/3 h, from cumuliform clouds only when it was less than 2. 86 mm/3 h.

Table 5 Seeding effects indicated by two-sample regression analyses for two cloud and four synoptic stratifications, and various control area rainfall rates

Category seed/not	Control area rain		Target area increment			Significance test		
	mm/ 3 h	4th rt	$y_s - y_c$	mm/3 h	%	z_y	y	α
Stratiform 29/30	0.41	0.80	0.073 2	0.24	38.42	0.773 8	48.26	0.20
	1.00	1.00	0.074 2	0.39	31.14	1.000 0	48.58	0.10
	2.86	1.30	0.075 3	0.73	25.09	1.475 6	51.58	0.08
	5.06	1.50	0.076 1	1.05	22.29	1.661 6	54.78	0.05
	5.80	1.55	0.076 3	1.15	21.67	1.626 9	55.00	0.06
	10.50	1.80	0.087 2	1.67	18.90	1.473 0	54.76	0.20
Cumuliform 21/21	0.41	0.80	0.386 7	1.03	524.42	1.860 9	25.97	0.05
	1.00	1.00	0.300 6	1.43	216.28	1.928 2	27.01	0.05
	2.86	1.30	0.171 4	1.66	66.89	1.928 0	31.16	0.05
	5.06	1.50	0.085 2	1.23	24.90	1.217 1	33.03	0.20
	5.77	1.55	0.063 8	1.01	17.52	0.884 9	32.23	0.20
	10.50	1.80	−0.044 0	−1.06	−9.21	−0.395 0	26.44	>0.25
Prefrontal 18/15	0.41	0.80	0.425 2	0.87	833.41	2.029 6	20.70	0.05
	1.00	1.00	0.316 0	1.26	261.67	2.053 3	23.07	0.025
	2.86	1.30	0.152 2	1.37	59.39	1.614 0	27.56	0.10
	5.06	1.50	0.043 0	0.60	12.00	0.441 0	22.88	>0.25
	7.50	1.66	−0.040 7	−0.79	−9.45	−0.334 9	17.94	>0.25
	10.50	1.80	−0.120 8	−2.98	−23.17	−0.752 2	16.16	0.25
Frontal 10/13	0.41	0.80	0.186 9	0.72	114.48	0.530 8	11.16	>0.25
	1.00	1.00	0.191 5	1.16	95.51	0.744 8	11.88	0.25
	2.86	1.30	0.198 4	2.14	77.28	1.513 3	15.18	0.10
	4.50	1.46	0.202 0	2.82	70.58	2.050 8	16.61	0.05
	5.06	1.50	0.203 0	3.04	68.94	2.040 2	15.72	0.05
	10.50	1.80	0.209 8	4.86	59.73	1.093 8	10.02	0.20
Postfrontal 16/14	0.41	0.80	0.034 1	0.13	14.93	0.284 2	24.69	>0.25
	1.00	1.00	0.036 9	0.22	13.76	0.402 4	24.80	0.25
	2.86	1.30	0.041 1	0.44	12.53	0.714 8	24.70	0.25
	5.06	1.50	0.043 9	0.66	11.93	0.860 8	22.63	0.20
	5.63	1.54	0.044 5	0.72	11.84	0.855 8	22.34	0.20
	10.50	1.80	0.048 1	1.13	11.24	0.667 1	20.78	>0.25

续表

Category seed/not	Control area rain		Target area increment			Significance test		
	mm/ 3 h	4th rt	$y_s - y_c$	mm/3 h	%	z_y	y	α
Subtropical high 6/8	0.41	0.80	0.255 7	1.06	178.08	2.569 8	9.87	0.025
	1.00	1.00	0.152 9	0.89	71.78	2.006 6	9.55	0.05
	2.86	1.30	−0.001 5	−0.01	−0.45	−0.025 0	7.42	>0.025
	4.12	1.43	−0.065 5	−0.72	−17.06	−1.023 4	6.92	0.20
	5.06	1.50	−0.104 3	−1.27	−25.04	−1.490 0	7.03	0.10
	10.50	1.80	−0.258 7	−4.58	−46.89	−2.523 9	7.71	0.025
Total 50/51	0.41	0.80	0.137 6	0.44	81.38	1.680 1	94.76	0.05
	1.00	1.00	0.120 1	0.63	55.05	1.876 6	95.81	0.05
	2.86	1.30	0.094 1	0.93	32.12	2.219 3	96.03	0.025
	5.06	1.50	0.076 8	1.08	22.37	1.984 5	93.63	0.025
	5.79	1.55	0.072 3	1.11	20.29	1.868 2	87.84	0.05
	10.50	1.80	0.050 7	1.14	12.07	0.958 4	82.56	0.20

In all synoptic classes, relative increments declined with increasing control area rainfall, but were significant at the 5% level only when control area prefrontal rain was around 1 mm/3 h, frontal rain from 4.50 mm/3 h to 5.66 m/3 h, when relative increases were as much as 70% and absolute additions 2.82 mm/3 h to 3.04 mm/3 h. No significant increases were obtained in postfrontal or subtropical high pressure conditions, the latter partly because only 14 cases were so classified.

For the first two years of the experiment, double-ratio analysis had been used (Weather Bureau, etc., 1979). This assumes that rainfall at a single target station is proportional to mean control area rainfall, and with no regression intercept, neither assumption seemed acceptable. Thus, the two-sample regression method was adopted for comparing separately each target station rainfall with control area averages. Correlations between individual target gages and control area averages (Fig. 5) differed significantly (1% level) from zero for all stations in the stratiform cases but, for cumuliform cases at only two stations, both in the southwest corner of the target. Thus both regression analysis and double-ratio tests for single-station rainfall from cumulus clouds gave meaningless results over most of the target area.

Two-sample regressions for single-station rainfall in all cases indicated increments of more than 20% over an area some 20 km wide and 10 km to 50 km downwind of Mt. Shita (Fig.6). At most stations in this swath, effects were significant at the 5% level. For stratiform clouds, the area of effect, in which about half the stations showed increases significant at the 5% level, was similar, about 20 km wide and 15 km to 50 km long. No significant differences in the effects of metaldehyde (29 cases) and silver iodide (21 cases) were shown by analysis of varience (Table 5) for all cases; the F values were only about a quarter as large as the critical values.

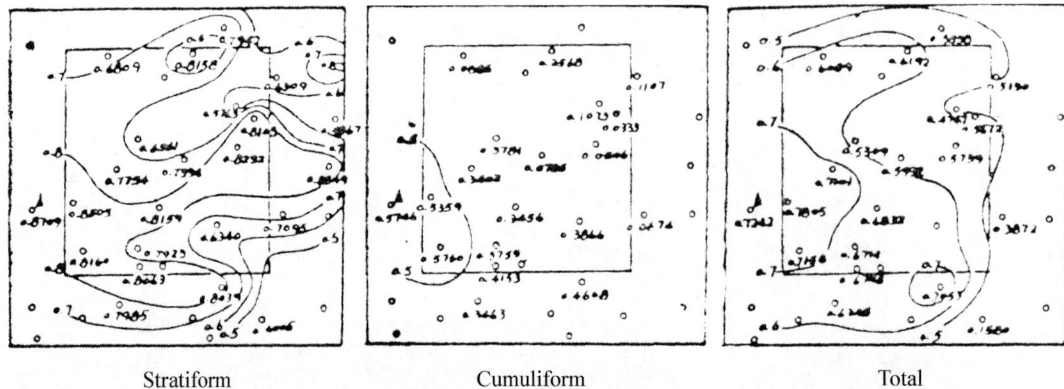

Stratiform Cumuliform Total

Fig. 5 Correlation coefficients between individual station reainfalls in target area and mean rainfall in control area, for cloud type stratification.

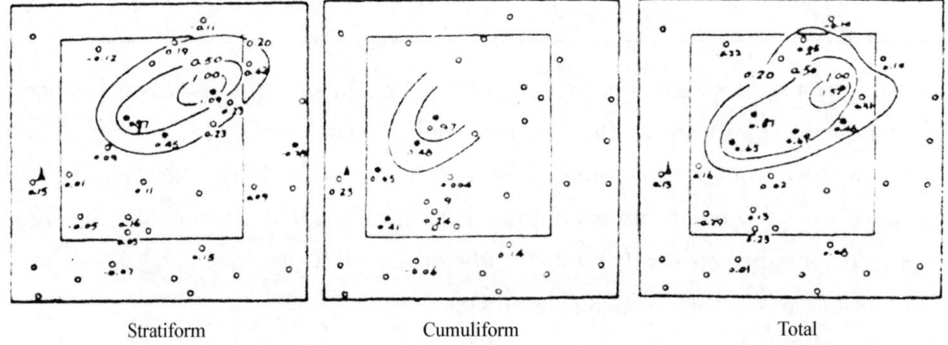

Stratiform Cumuliform Total

Fig. 6 Seeding effects (relative increments) of individual stations in target area. Solid circles significant at 5% level, open circles not.

5. SUMMARY AND CONCLUSIONS

Detailed examination of the 50 seeded and 51 unseeded cases during three eseasons leads to seven conclusions:

(1) Fourth-root and fifth-root transformations of 3-hour rainfall amounts/provide acceptable norm-ality; goodness of fit reached 0.94 and correlation between fourth-root amounts in target and control areas exceeded 0.82.

(2) Residual variances of regressions for seeded and unseeded samples, separately, differed significantly, so that unequal variances were considered in the regression analyses.

(3) Mean target rainfall was increased by about 20.3% by seeding, with one-sided significance of about 5%.

(4) Frontal conditions, with an increase of 70.8% (significant at the 5% level), were

most favorable for rainfall increase; other synoptic conditions showed little or no significant increases.

(5) Relative seeding effects were the greatest when control area rain was light, and were generally insignificant when control area rain exceeded 3 mm/3 h. Absolute amounts, however, increased with control area rainfall, less for cumuliform and prefrontal rain than otherwise. Thus, seeding effects apparently were neither constant not proportional to natural rainfall, as others have supposed (Biondini, 1976; Scott, 1973).

(6) The area in which rainfall increases exceeded 20% was about 20 km wide and 20 km to 50 km long, downwind from the rocket release site.

(7) Although ten times as much metaldehyde as silver iodide was used, no significant differences in effects was found.

6. REFERENCES

Biondini, R., 1976: Cloud motion and rainfall statistics. *J. Appl. Meteor.* *15*, 205 - 224.

Institute of Meteorology, Fujian; Department of Meteorology, Nanjing Univ.; and Institute of Chemistry Engineering, Sian, 1977. Dispersing method of metaldehyde and its active nucleation property. *Data of Meteor. Sci. and Tech., Fujian*, *2*, 51 - 54.

Institute of Meteorology, Fujian, 1975. On the small rocket for rain enhancement and hail suppression of Type FJ - 50. *Data of Meteor. Sci. and Tech., Fujian*, *5*, 29.

Scott, E. L., 1973. Problems in the design and analysis of weather modification studies. *Preprints, Third Conf. on Prob. and Stat. in Atmos. Sci.* (A.M.S.) *6*, 65 - 72.

Thom, H. C. S., 1957. A statistical method of evaluating augmentation of precipitation by cloud seeding. *Final Report of the Advisory Committee on Weather Control*, *2*, 5 - 25.

Weather Bureau of Fujian Province and Department of Meteorology, Nanjing Univ., 1979. Statistical evaluation of the artificial rainfall stimulation experiment in Gutian reservoir region, Fujian Province. *Scientia Atmospherica Sinica* 3(2), 131 - 140.

Yeh, Jian-dong and Pei-fen Fan, 1981. Two-sample regression analysis with unequal variances. Scientia Atmospherica Sinica 5, (2).

4.8 Statistical Power in Randomized Precipitation Enhancing Experiment

Yeh Jia-dong, Lou Xing-pin

(Department of Meteorology, Nanjing University, China)

Zeng Guang-ping, Xiao Feng

(institute of Meteorology, Fujian Province, China)

1. INTRODUCTION

Owing to the considerable natural variability of precipitation both in space and time, the long duration of experiment is necessary for detecting the effect of cloud seeding. The statistical power in precipitation enhancement experiment can give a numerical estimate of detecting probability of seeding effect in special period of time at special significant level (0.05 in general). The numerical computation of statistical power presented here simulates a randomized cloud seeding experiment by superimposing multiplicative seeding effects into the natural rainfall data which have been got from the control units of three hours cases, all of 200 control units during the randomized cloud seeding experiment (1975—1983) at Gutian region, Fujian, China. The statistical power is computed by "naive" method used by Gabriel(1980), Hsu (1979) and Salvam et al. (1979) in the present paper.

2. COMPUTATION PROCEDURE

Target area is a rectangle of about 1 500 km^2 over the Gutian reservoir watershed, and control area is another 1 500 km^2 rectangle west (upwind) of the target, separated from it by a buffer zone 11-km wide. As in Gution cloud seeding experiment, a three-hour period is adopted as experimental unit, and the fourth root of 3-hour areally-average rainfall is used as statistical variables. In power computation, the sequence of rainfall data in target area is randomly divided into seeded and control samples. For simulating a cloud-seeding experiment, the rainfall in seeded sample is modified by superimposing various multiplicative seeding effects $\delta = 1.1, 1.2, 1.3, 1.5, 1.7$ and 2.0, corresponding multiplicative factors of statistical variable by "seeding" $\theta = (1.1)^{\frac{1}{4}}, (1.2)^{\frac{1}{4}}, (1.3)^{\frac{1}{4}}, (1.5)^{\frac{1}{4}}, (1.7)^{\frac{1}{4}}$ and $(2.0)^{\frac{1}{4}}$, respectively. The rainfall data in control area are not modified by "seeding", so both two samples in control area divided corresponding in target area are as control samples.

Several experimental designs are used for power computation, such as one-target randomized, target-control area randomized and cross-over randomized designs, corresponding test statistics of seeding effect are respectively:

Ratio of seeded vs. control sample for one-target area randomized design as

$$R = \theta \frac{\bar{y}_1}{\bar{y}_2}$$

where $\bar{y}_1$ and $\bar{y}_2$ denote target areally-average of fourth root of rainfall correspond to seeded and control unit, respectively, and θ the multiplicative factor of seeding effect.

Double ratio for target-control area randomized design as

$$DR = \theta \frac{\bar{y}_1}{\bar{x}_1} \cdot \frac{\bar{x}_2}{\bar{y}_2}$$

where $\bar{y}_i$ and $\bar{x}_i$ denote average of statistical variable of rainfall in target and control area, subscript $i=1, 2$ denote seeded and control sample, respectively.

Double ratio value for cross-over randomized design as

$$CR = \theta \left(\frac{\bar{y}_1}{\bar{x}_1} \cdot \frac{\bar{x}_2}{\bar{y}_2} \right)^{\frac{1}{2}}$$

where $\bar{y}_i$ and $\bar{x}_i$ are the same as mentioned above, but subscript $i=1$ denotes in y area-seeded, and $i=2$ in x area-seeded unit.

Besides, in the target-control areal randomized regression analysis, several evaluation methods of seeding effect have been adopted (Yeh et al, 1982), such as multiple event test, two-sample regression analysis with equal and unequal residual variances, corresponding test statistics are as follows:

$$t_1 = (\theta\, \hat{y}_1 - \hat{y}_2) \left[s_{2r}^2 \left(\frac{1}{k} + Q_2^2 \right) \right]^{-\frac{1}{2}}$$

$$t = (\theta\, \hat{y}_1 - \hat{y}_2) \cdot \left[\frac{(k-2)s_{1r}^2 + (n-2)s_{2r}^2}{n+k-4} \left(\frac{1}{k} + Q_2^2 \right) \right]^{-\frac{1}{2}}$$

$$z = (\theta\, \hat{y}_1 - \hat{y}_2) \left(\theta^2 \frac{s_{1r}^2}{k} + s_{2r}^2 Q_2^2 \right)^{-\frac{1}{2}}$$

where

$$s_{jr}^2 = \frac{1}{n_j - 1} \sum_{i=1}^{n_j} (y_{ji} - \bar{y}_j)^2 \quad (j=1,2;\ n_1 = n;\ n_2 = k)$$

$$Q_2^2 = \frac{1}{n} + \frac{(\bar{x}_1 - \bar{x}_2)^2}{\sum_{j=1}^{n} (x_j - \bar{x}_2)^2}$$

The powers of various test methods mentioned above are also computed in the present paper.

3. NUMERICAL ANALYSIS OF POWER

Table 1 shows the power computed for various experimental designs and test methods of evaluation of seeding effect by "naive" method. The power analyses are as follows:

Table 1 Statistical power values for various designs and test methods of seeding effects

		N	Seeding effects/%					
			10	20	30	50	70	100
R	Total	31	0.076	0.122	0.169	0.283	0.411	0.555
		51	0.101	0.174	0.279	0.481	0.650	0.845
		73	0.111	0.197	0.331	0.586	0.790	0.949
		88	0.123	0.228	0.356	0.625	0.820	0.971
		118	0.129	0.275	0.435	0.762	0.908	0.985
		150	0.147	0.320	0.536	0.837	0.964	0.999
	S	51	0.104	0.216	0.338	0.562	0.755	0.929
	C	37	0.084	0.124	0.171	0.256	0.369	0.522

续表

		N	Seeding effects/%					
			10	20	30	50	70	100
DR	Total	31	0.134	0.237	0.354	0.616	0.812	0.938
		51	0.150	0.331	0.521	0.834	0.973	0.999
		73	0.166	0.389	0.629	0.913	0.996	1.000
		88	0.209	0.478	0.720	0.968	1.000	1.000
		118	0.262	0.607	0.869	0.998	1.000	1.000
		150	0.226	0.506	0.782	0.988	1.000	1.000
	S	51	0.158	0.347	0.564	0.881	0.986	1.000
	C	37	0.141	0.236	0.374	0.633	0.832	0.965
CR	Total	31	0.241	0.529	0.793	0.989	1.000	1.000
		51	0.354	0.752	0.969	1.000	1.000	1.000
		73	0.412	0.864	0.995	1.000	1.000	1.000
		88	0.513	0.926	1.000	1.000	1.000	1.000
		118	0.633	0.985	1.000	1.000	1.000	1.000
		150	0.538	0.966	1.000	1.000	1.000	1.000
	S	51	0.370	0.820	0.985	1.000	1.000	1.000
	C	37	0.244	0.562	0.822	0.995	1.000	1.000
t_1	Total	31	0.120	0.234	0.348	0.635	0.833	0.967
		51	0.140	0.301	0.516	0.835	0.980	0.998
		73	0.177	0.387	0.643	0.926	0.997	1.000
		88	0.226	0.497	0.775	0.987	1.000	1.000
		118	0.252	0.631	0.893	1.000	1.000	1.000
		150	0.261	0.578	0.860	0.997	1.000	1.000
	S	51	0.182	0.392	0.648	0.951	0.999	1.000
	C	37	0.129	0.225	0.352	0.595	0.796	0.961
t	Total	31	0.115	0.203	0.306	0.543	0.750	0.914
		51	0.151	0.320	0.521	0.845	0.979	0.998
		73	0.162	0.354	0.598	0.915	0.996	1.000
		88	0.224	0.479	0.765	0.981	1.000	1.000
		118	0.262	0.631	0.887	0.999	1.000	1.000
		150	0.261	0.583	0.868	0.998	1.000	1.000
	S	51	0.187	0.413	0.692	0.372	0.998	1.000
	C	37	0.122	0.214	0.324	0.558	0.768	0.934
z	Total	31	0.113	0.203	0.306	0.543	0.750	0.916
		51	0.150	0.320	0.522	0.843	0.979	0.998
		73	0.164	0.371	0.617	0.917	0.996	1.000
		88	0.222	0.477	0.762	0.981	1.000	1.000
		118	0.261	0.630	0.887	0.999	1.000	1.000
		150	0.259	0.581	0.866	0.999	1.000	1.000
	S	51	0.180	0.399	0.678	0.970	0.998	1.000
	C	37	0.125	0.218	0.329	0.563	0.770	0.939

3.1 Statistical powers in various enhancing rainfall effect

As shown in Table 1 and Figure 1, statistical powers are always increase with increasing of seeding effects. The variations in power along with seeding effect are more rapid when the seeding effect is small (less than 50%), and varied slowly with effect when it is more than 50%.

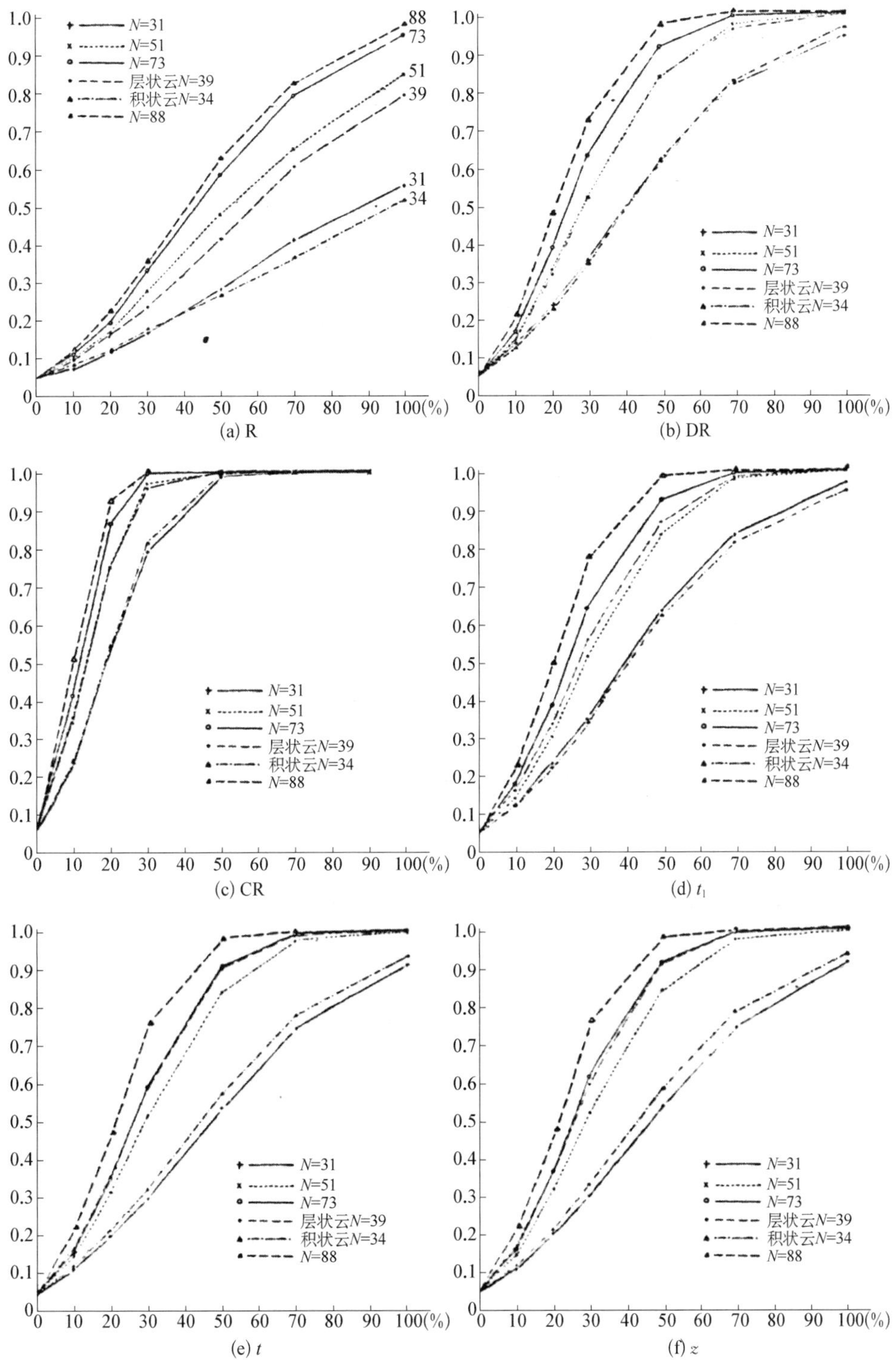

Figure 1 Statistical power values for randomized experiment

3.2 Statistical power vs. experimental period(sample size)

One of the main purpose of power analysis is to determine the period for a cloud seeding experiment in which the seeding effect can be detected by a specific probability. Figure 2 gives the powers with various sample sizes for seeding effects of 20% and 30%, respectively. Obviously, statistical power is increased with increasing of sample sizes in general when the experimental units are permuted randomly in advance ($N \leqslant 88$). But the variation is tend to slow up when sample sizes is more than 50.

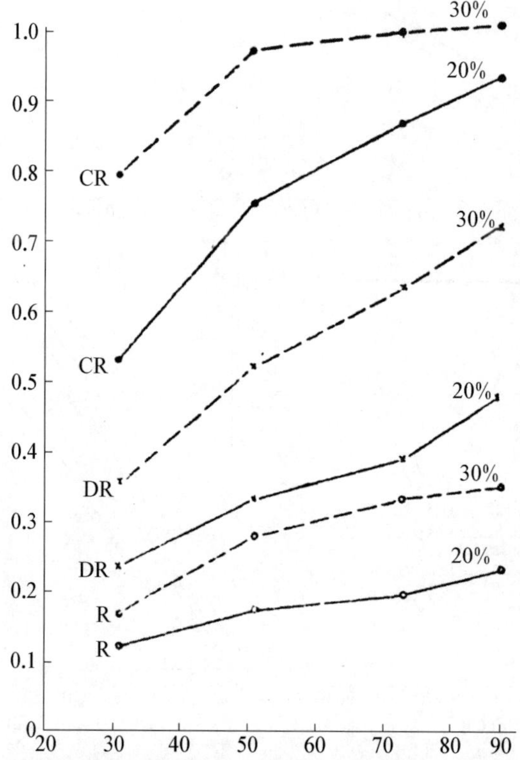

Figure 2 Relation between powers and sample sizes

3.3 Statistical power for various experiment design

The statistical powers for three experiment designs (Table 1) indicated the power for cross-over desigh (CR) is the largest, next one is target-control area randmized design (DR), and the power for one-target area randomized design is the smallest. These are of the same with analysis of Schickedang et al (1971). However, even if for one-target design, the powers for $N=88$, as shown in Table 2, are larger than the powers computed by Scott (1973) used analytic method. Although the values of powers computed in present paper for $N=88$ and by Scott are near each other (Table 2), the sample size in Scott's computation is 100, larger than in present paper. The season probably is that the natural variance of areal average rainfall in Gutian region in raining season is less than the typical case presumed by Scott (the rainfall variable satisfy Γ-distribution, shape parameter is 0.70 and scale parameter is 2.0).

Table 2 Statistical power values for one-target area design

	N	Relative increments of rainfall/%				
		10	20	30	50	70
R	88	0.12	0.23	0.36	0.63	0.82
Scott	100	0.14	0.20	0.30	0.60	0.80

Since the power of one-target area design is small in general, it was suggested that to choose efficient meteorological covariate for enhancing statistical power in the weather modification is necessary. The rainfall in control area, as shown in Table 1 and Figure 1, can be used as a efficient covariate to enhance statistical power if the high area correlation between target and control areas is presented.

If the power value of 0.80 is designed as a acceptable criterion that corresponding seeding effect can be detected at a specific significant level (0.05 in general), for one-target area randomized design, when the sample sizes are 51, 73 and 88, only for seeding effects larger than 69%, 48% and 37% can be detected in significant level 0.05, corresponding seeding effects are 48%, 42% and 37% for target-control area randomized design, and only 22%, 19% and 17% for cross-over randomized design. In general, cloud-seeding effects, as found in Gution experiment are only 20%—30%, so if want to detect the seeding effect with a detecting probability of 0.80 in a period of 2—3 seasons (correspond to sample sizes of 60—90), using cross-over design is necessary. To detect the same effects and detecting probability, as mentioned above using target-control area randomized design, the period of 4—6 seasons (120—180 units) is probably necessary as estimated from the two groups of $N=118$ and 150 in Table 1.

3.4 Influence stratified statistic to power

The sample of $N=88$ which have records of cloud types can be stratified according to cloud type. The results of power analysis for various cloud types indicate the powers for stratiform cloud can be enhanced obviously (Table 1), power values of $N=51$ for stratiform cloud is close to power of $N=73$ for total sample, especially in statistics of areal regression analysis t_1, t and z. The powers for cumuliform cloud sample have not been improved. The reason is higher areal correlation in stratiform cloud category, corresponding correlation coefficient is 0.900 2, and for cumuliform and total samples are 0.799 4 and 0.845 5, respectively.

3.5 Powers of various test methods in areal regression analysis

Multiple events test and two-sample analysis with equal and unequal residual variances have been used in evaluation of seeding effects in Gutian experiment. The statistical powers for these test methods (t_1, t and z) are also presented in Table 1 and Figure 1. Powers of these three test methods are closed each other, deviation have not exceeded 5% in general and is close to powers of double ratio values DR. It is noted that the experimental design is same, difference is only in evaluating method.

In addition, the power values of multiple events test method perhaps is a bit of larger than that of two-sample regression analysis methods, more apparent in $N=31$ and 73 (Table 1). The reason propably is in the assumption of constent multiplication factor of seeding effect, so the variances in seeded samples are enlarged. In these cases, the efficiences of evaluation of seeding effects will be decreased (Yeh and Fan, 1981).

3.6 Influence of areal correlation coefficient to power

Table 3 gives the statistical powers for various areal correlation coefficients when $N=30$, it is found the powers for various statistical variables denoting seeding effects expecting R are more or less increased with increasing of correlation coefficient, especially for the group of $r=0.975\ 3$, corresponding powers are considerably large excepting of R. Evidently the areal correlation coefficient is a quite important factor to determine the statistical power in areal regression analysis or cross-over design.

Table 3 Influence of correlation coefficient to power (N=30)

	r	R	DR	CR	t_1	t	z
1.20	0.783 1	0.119	0.151	0.304	0.146	0.157	0.159
	0.832 9	0.138	0.191	0.396	0.187	0.201	0.198
	0.864 1	0.134	0.218	0.510	0.227	0.187	0.195
	0.890 9	0.138	0.320	0.588	0.297	0.272	0.274
	0.917 8	0.130	0.261	0.629	0.294	0.305	0.309
	0.975 3	0.151	0.618	0.989	0.632	0.635	0.642
1.30	0.783 1	0.162	0.205	0.491	0.212	0.217	0.216
	0.832 9	0.205	0.191		0.187	0.201	0.198
	0.864 1	0.185	0.340	0.757	0.344		0.291
	0.890 9	0.185	0.445	0.857	0.427	0.403	0.405
	0.917 8	0.183	0.409	0.875	0.441	0.478	0.478
	0.975 3	0.211	0.559	1.000	0.903	0.907	0.907

It is also noted from the groups of calculation of $N=118$ and 150 in Table 1, most of the powers in group of $N=118$ are larger than of $N=150$ (except of R). It seems the reason is the area correlation coefficient of variables in control year of 1976 is high coming up to 0.975 2, these rainfall data is included in the group of $N=118$, so in which the powers have been made considerable increases, and the areal correlation coefficient of rainfall variables control year of 1979 is lower, only 0.757 9, these data are additioned to the group of $N=150$, decreasing the power values as well.

4. REFERENCE

1. Schickedang, P. T. and F. A. Huff, The design and evaluation of rainfall modification experiment, J. Appl. Meteor., 10, 504 - 514, 1971
2. Scott, E. L., Problems in the design and analysis of weather modification studies, Third Conf. on Probab. and Statis. in Atmos. Sci., 65 - 72, 1973
3. Gabriel K. R. and Chin-Fei Hsu, Power study of re-rando-mization test, Third WM0 Sci. Conf. on Weather Modification, Clermont-Ferrand, France, July, 21 - 25, 1980
4. Hsu Chin-Foi, Two methods of computing statistical powers with application to weather modification, IMC. Statis. Comp. Section, American Statical Association, Washington, D. C. 6.
5. Salvam, A. M., et al., Numerical simulation of cloud seeding experiments in Maherashtra, State, India, WMA J. Weather Modification, Vol. 11, No. 1, 116 - 140, 1979
6. Yeh Jia-dong, Ke-ming Cheng and Guang-ping Zeng, Randomized cloud seeding at Gutian, Fujian, China, WMA J. Weather Modification, Vol.14, No. 1, 53 - 60, 1982
7. Yeh Jia-dong and Fan Pei-Feng, Two-sample regression analysis with unequal variance, Scientia Atmospherica Sinica, Vol.5, 214 - 224,1981

4.9 湖南凤凰地区高炮人工降水效果分析*

叶家东

（南京大学气象系）

一、试验概况

凤凰地区位于湖南省西部，地处云贵高原东侧，属山区，地形复杂，气候多变。本地区全年雨量主要靠雨季（4～6 月）降水，而到夏末秋初，往往干旱成灾。根据历史资料，凤凰地区夏、秋两季主要是地方性对流云，午后常有强盛的对流云生成，云顶超过 0 ℃层（5 000 m 左右）。因此，在 7～9 月间可根据冷云催化原理进行人工降水试验。

人工降水试验采用“三七”高炮发射碘化银炮弹催化降水。试验中炮弹爆炸高度一般在 2 400～4 200 m，加上作业点海拔高度（400～800 m），碘化银可在 2 800～5 000 m 高度上播撒。

炮点的选择直接影响作业效果，同时也影响安全及统一指挥等。我们考虑以下几个条件选择炮点、炮位：① 位于降水性云的主要源地或移动路径上；② 交通和通信联络方便；③ 四周开阔，视野广，阵地平整坚实，离居民点 5 km 以上，以确保安全。表 1 为三年试验的炮点、用弹量。

---

* 本试验由湘西自治州气象局、凤凰县人工降雨办公室主持。参加总结工作的有张文侠、余夕刚、刘爱琼、焦艾彩、丁代湘、张智群、陈历舒、王道藩、费中运、俞香仁等。本文由叶家东执笔整理。

表 1　试验区炮点设置和催化剂总量

年份	炮点	高炮数	试验期	催化剂总量	
				用弹量/发	AgI 用量/g
1975	阿拉，廖家桥，吉信，木里	4	7～9 月(51 天)	3 059	12 236
1976	阿拉，吉信，禾库，寨阳	4	8～9 月(43 天)	1 007	4 028
1977	廖家桥，吉信，禾库	3	7～9 月(45 天)	666	3 090

二、效果分析

人工降水试验能否增加雨量和增加多少，这是效果检验研究的主要课题。由于降水的自然变差很大，也由于人工降水的催化方法对不同的催化对象会有不同的效果，使得催化效果的检验工作变得十分复杂。我们的试验是一种以抗旱为主的作业性试验，采用非随机试验方案，试验期内凡有适宜催化的天气都争取时机作业。同时，受条件限制，试验中无现代化探测仪器和设备，配合观测试验效果，可供利用的仅限于雨量站的雨量资料。在这种情况下如何分析试验效果，是一个很棘手的问题。但是，它又是一般作业性试验中都面临的迫切问题。我们主要采用区域雨量历史回归分析方法进行试验效果的统计检验，包括月雨量的回归分析、雨量图个例分析和主要天气系统影响下的日雨量回归分析。

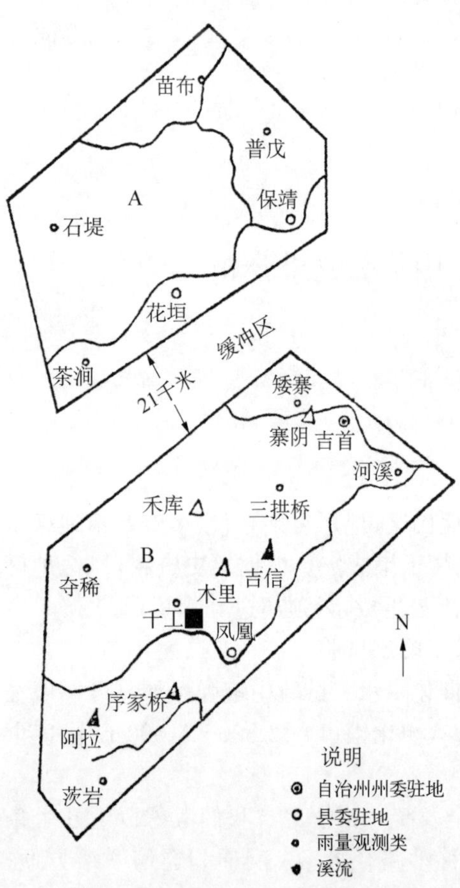

图 1　试验区的地理位置

试验区的设置：区域雨量回归分析需要选择一个和目标区的天气条件、地理条件相似，区城雨量密切相关的对比区。由于我们是事后分析，所以根据凤凰地区三年的炮点分布和芷江高空盛行风向，选择一个面积为 2 686 km^2 的目标区 B，在它的北面偏西另选一个面积为2 135 km^2 的对比区 A，两区间隔约 21 km(图 1)。A 区有历史资料可用的雨量站 6 个，B 区 8 个。我们统计了历史上九年 8 月份的月雨量资料，结果得到两区面积平均月雨量(对数值)的相关系数 $r=0.861(\alpha=0.005)$。可见，两区 8 月份的月雨量相关性较好，区城回归分析较为有效。

为了避免对比区受催化剂的污染，在目标区和对比区之间划出宽为 21 km 的缓冲带。此外，我们统计了芷江的高空风资料(芷江距凤凰 50 km)，发现 8 月份芷江 500 mb* 高空盛行西南

* 1 mb＝100 Pa。

和偏西风，对比区大体处于目标区的侧风方，可以认为目标区播撒的催化剂基本上不会污染对比区。

雨量资料的整理：回归分析所用的雨量都是区域面积平均雨量(平均月、日雨量)，采用各站简单平均的方式求取。在雨量图个例分析中，由于不涉及历史资料，所以将临时增设的雨量点的雨量资料一并包括在内。考虑到雨量站的代表性，凤凰县境内有 5 个常规雨量站，平均每 300 km^2 一个。我们根据三年试验期资料，用两种方法计算了六天区域面积平均日雨量：一个是 5 站资料简单平均，一个是根据临时增设的雨量点的资料(在凤凰县境内有 30 个左右)和雨量站的全部资料求平均。以后者为标准，由 5 个雨量资料站计算的区域面积平均日雨量的平均误差是 4.7%，最大误差是 35.6%。这说明至少在中、大雨情况下，根据现有的雨量站的雨量资料进行统计，代表性还可以。

根据福建古田试验[1]和我们对长沙的 8 月份雨量进行分布函数的检验，假定月雨量服从对数正态分布是可以的，长沙 8 月份雨量的拟合概率达到 0.99 以上。所以，月雨量回归分析中统计变量取区域面积平均月雨量的对数。日雨量的回归分析中，统计变量也取区域面积平均日雨量的对数。对数正态分布在催化效果的统计分析中有一个优点，就是它能将相对增雨效果转换成雨量对数的差值，于是可以方便地应用通常的统计检验方法检验此差值的显著性。

1. 月雨量的历史回归分析

1975—1977 年 8 月份，凡是试验期目标区出现有利于催化的天气条件都尽可能催化。现在我们考虑 8 月份目标区的月雨量是否因为人工催化的作用而显著地增加了。所谓增加，是对自然月雨量而言。但并不知道自然月雨量，我们根据历史资料来估计它。表 2 列入试验区历史上 9 年 8 月份的区域面积平均月雨量(以下简称月雨量)。试验期 8 月份的实测月雨量也列入表中。从中看出，历史上(9 年中)目标区 8 月份月雨量比 1975、1976 和 1977 年大的分别为 3、7 和 3 年，表明这三年目标区的 8 月份月雨量并不十分大，自然降水达到这么大的可能性不一定小。根据这种比较，不能得出人工降水有效的结论。但是，我们的人工降水试验是结合抗旱进行的，试验期常常是干旱期，自然雨量往往偏小，所以光从历史雨量比较是否显著偏大并不合理。需要找一个指标，表示试验期目标区自然雨量的大小。回归分析的目的就是根据附近不受催化影响的对比区的雨量，假定区域相关性历史上稳定不变，利用两区历史雨量资料建立区域回归方程，进而求出试验期目标区自然雨量的期待值，再和实测雨量比较，以确定效果，计算结果也列于表 2。由表 2 可见，三年平均增雨量为 46.16 mm，相对增加率为 55.5%，其统计显著度 $\alpha=0.10$。注意到 1977 年 8 月份增雨效果是负值，其雨量减小值为 9.72 mm，相对减小率 4.8%，但不显著($\alpha>0.45$)。从对比区自然降雨条件看，1977 年 8 月雨水偏多，达 196.3 mm，历史上 9 年中只有 2 年超过它，这是否意味着自然降水条件好时效果差有待于进一步试验证实。根据福建古田地区的随机试验[2]，当自然雨强大于 3 mm/h 时，催化效果就不显著。

表 2 8 月份区域面积平均月雨量增雨效果历史回归分析表

年份	X'/mm	Y'/mm	$X=\lg X'$	$Y=\lg Y'$	$\hat{Y}=1.3517X-0.7935$	增值 $(Y-\hat{Y})$	增雨量/mm	n (单边检验)
1966	144.48	80.83	2.159 9	1.907 4				
1967	160.32	218.86	2.204 9	2.340 3				

续表

年份	X'/mm	Y'/mm	$X=\lg X'$	$Y=\lg Y'$	$\hat{Y}=1.3517X-0.7935$	增值$(Y-\hat{Y})$	增雨量/mm	n(单边检验)
1968	235.17	179.63	2.371 5	2.254 5				
1969	241.00	322.86	2.382 0	2.509 1				
1970	111.97	125.42	2.049 2	2.098 3				
1971	128.38	149.87	2.108 6	2.175 8				
1972	51.00	32.04	1.707 6	1.505 1				
1973	150.20	200.32	2.176 7	2.301 6				
1974	117.07	60.29	2.068 6	1.780 3				
1975	86.98	187.65	1.939 4	2.273 3	1.828 0	0.445 3	120.35	0.025
1976	61.60	59.90	1.789 6	1.777 4	1.625 5	0.151 9	17.70	<0.25
1977	196.28	192.53	2.293 0	2.284 4	2.305 9	−0.021 5	−9.72	>0.45
三年平均					1.919 8	0.191 9		0.10

2. 日雨量图个例分析

为了配合试验，在试验期临时增设一批雨量点，用以分析作业效果。连同原有的气象站、水文站，在凤凰县境内，1975、1976 和 1977 年分别有雨量点 33、37 和 21 个。我们把六次天气系统影响下，试验区全部炮点都作业的试验日雨量绘成日雨量分布图(图 2)，图中还标明芷江站(745 站)500 mb 20 时高空风向、风速以供参考。

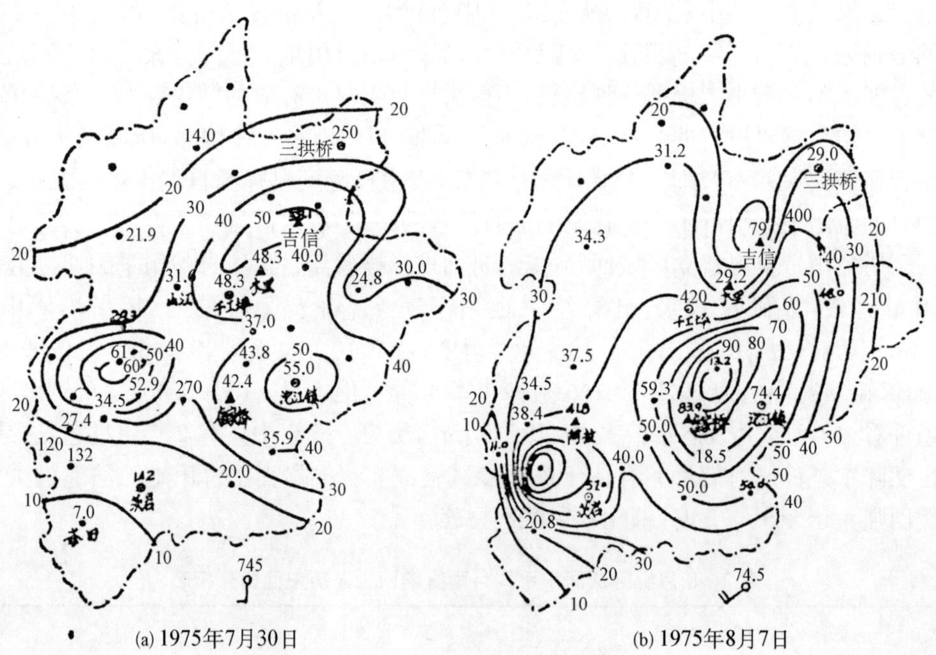

(a) 1975年7月30日　　(b) 1975年8月7日

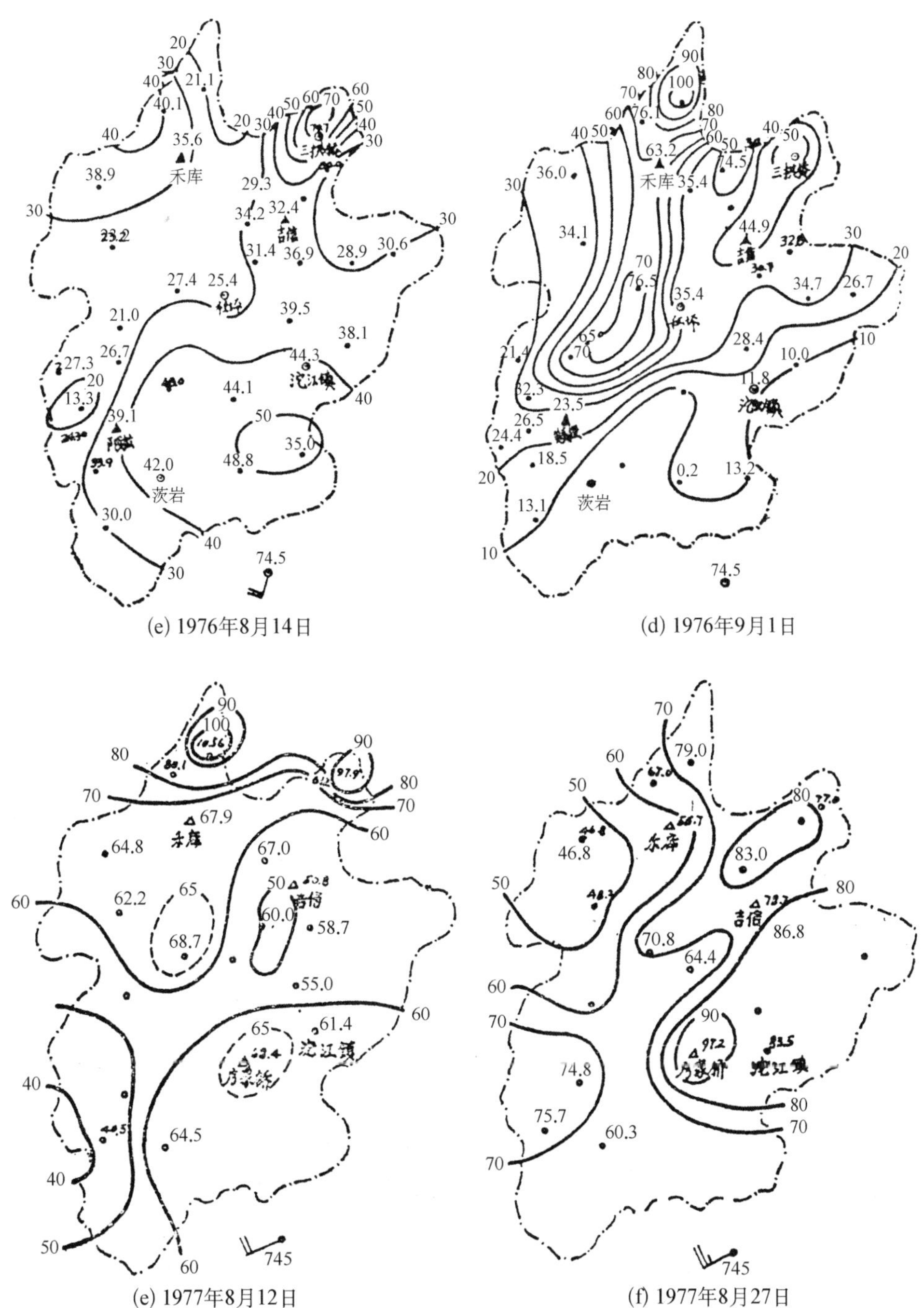

(e) 1976年8月14日　　(d) 1976年9月1日

(e) 1977年8月12日　　(f) 1977年8月27日

图 2　试验区试验日雨量

从图 2 可以看出，试验日在炮点或其下风方都出现了较大的雨量中心。例如：1975 年 7 月 30 日，芷江站 500 mb 高空风是南风 4 m/s。当天 4 个炮点下风方大约 5 km 处(包括吉信炮点本身)出现了三个雨量中心，最大日雨量分别为 61.0、55.0 和 58.0 mm。如果按芷江的高空风速推算，从炮点到雨量中心点约需 20 分钟，这与催化剂起作用的时间较为吻合。

其他各日均有类似的雨量中心。从各炮点现场观测发现，在有利天气条件下，催化后有时会促使对流云合并、云层变厚、雨强增大等现象，可以作为人工催化有效果的迹象。

3. 日雨量的统计分辨

由于自然雨量的空间分布是起伏多变的，特别是夏季山区对流云降水的时空分布本来就很不均匀，在作业点及其下风方出现一些雨量中心，虽然提供了人工催化有效果的迹象，但并不能排除自然的因素也可能在起作用。因此，有必要分析一下：如果试验区试验日的日雨量比较大，那么，由于自然原因引起这么大雨量的可能性有多大，为此，根据历史回归分析对试验期主要天气形势影响下的日雨量进行效果检验。假定在相同的天气形势下，目标区和对比区的区域面积平均日雨量（以下简称日雨量）的相关关系是相同的。因此，从历史上找出试验日的相似天气条件下，目标区和对比区日雨量的相关关系，建立回归方程，并据此检验目标区试验日的催化效果。

为此，将系统天气影响下的各炮点共同作业的试验日选出来作为分析的对象，按天气形势分成三型：① 锋面低槽；② 副高边缘；③ 低压辐合。我们选择相似天气的判据比较简单，一个是天气型相似，另一个是对比区雨量相似，具体定为试验日对比区雨量的±30%区间之内。第一型天气分为小雨和大雨两种情况，第二、三型由于历史相似天气次数少，所以对比区雨量相近这个判据放宽为介于 0.05～50.0 mm 之间。相似天气的资料和有关的计算结果列于表 3。

由表 3 可见，第一、三两型相似天气（除第一型的大雨情况外），目标区和对比区的日雨量相关系数略大于 0.5，统计显著度达 0.05，回归方程勉强可用。第二型的区域日雨量相关性较好，相关系数达 0.83，显著度达 0.01。在第一型的大雨情况下，两区日雨量的相关性很差，故不能用回归分析。

回归分析的结果表明，第一、二、三型（除第一型大雨情况外）的日雨量平均增值分别为 2.51、−0.46 和 13.61 mm，显著度分别为 0.20、0.40 和 0.05。除第三型较显著外，其他两型都不显著。需要指出，由于试验次数较少，历史雨量资料年限较短，历史相似天气的样本容量较小，再加上相似天气的判据比较简单，以致区域日雨量的相关性不高，这都影响统计检验的效率。

三、结果与讨论

区域面积平均月雨量的回归分析表明，凤凰地区 1975—1977 年三年的高炮人工降水试验效果还是比较显著的，三年平均增雨量为 46.16 mm，相对增加率为 55.5%，统计显著度 $\alpha=0.10$。1977 年效果不好，可能与这一年自然雨水偏多有关，需要进一步试验分析，从统计分析的角度看，若恰当地选择目标区和对比区，使两区月雨量有较好的相关性，回归分析方法还是可行的。关于不同雨量站，不同历史期回归分析结果会有所不同，这与雨量站的代表性以及历史月雨量的稳定性等问题有关。有条件的地区当然历史资料长一些好。看来在以抗旱为主的作业性试验中，历史回归分析还是一种可行的定量估计效果的方法。

日雨量图个例分析，从六次试验日的日雨量分布图的分析看出，试验日在作业区及其下风方常常出现大的雨量中心，它们与炮点位置及高空风向、风速的配合相当好，这定性表明高炮催化降雨增大了作业区及其下风方的雨量。当然，由于催化作业总是在作业区有较好的降雨条件时进行，分析时不能完全撇开这种主观的“预报”效应，而认为试验日自然降雨条

表 3　相似天气日雨量回归分析表

天气型	试验日	对比区雨量		目标区雨量		相似天气判据		n	r	α_r	回归方程	$\hat{y}$	$d=y-\hat{y}$	平均雨量增值			α_d	$\alpha_{\bar{d}_k}$
		X'/mm	$X=\lg X'$	Y'/mm	$Y=\lg Y'$	相似天气	对比区雨量							$\bar{d}_k=\frac{\sum d}{k}$	相对增加率	日雨量平均增值/mm		
锋面低槽	75.7.26	2.11	0.324 3	1.30	0.113 9	地面:锋面位于成都、西安以南至本站 高空:槽线位于太原、西安、成都一线以东南至本站	1.47~3.87	13	0.53	0.05	$\hat{Y}=3.083X-1.206\ 1$	−0.206 3	0.320 2	0.443 0	1.312 9	2.51	>0.25	<0.20
	76.9.1	2.98	0.474 2	14.28	1.154 7							0.255 9	0.898 8				<0.20	
	77.8.11	4.93	0.696 8	6.61	0.820 2							0.929 8	−0.109 6				>0.45	
	77.8.22	2.75	0.439 3	6.43	0.808 2							0.148 3	0.659 9				0.20	
	75.7.30	11.60	1.064 5	30.19	1.479 9		8.1~15.1	26	0.2									
	77.8.27	44.90	1.652 2	71.61	1.855 0													
副高边缘	76.8.14	2.95	0.469 8	24.75	1.393 6	地面:受静止锋或切变线影响 高空:538 线位于南昌以西至本站,316 线位于南昌以西至本站		8	0.83	0.01	$\hat{Y}=0.472\ 9X+0.713\ 0$	0.935 2	0.458 4	−0.103 3	−0.067 7	−0.46	0.10	0.40
	76.8.25	0.48	−0.318 8	1.89	0.276 5							0.562 6	−0.286 1				0.20	
	76.8.26	0.10	−1.000 0	3.89	0.589 9							0.240 1	0.349 8				0.20	
	77.8.10	13.77	1.139 0	10.79	1.033 1							1.251 6	−0.218 5				>0.20	
	77.8.23	9.38	0.972 2	2.25	0.352 2							1.172 8	−0.820 6				<0.25	
低压辐合	75.8.7	23.00	1.361 7	25.00	1.397 9	地面:受低压、切变线影响 高空:(1) 受高空低压辐合影响 (2) 受副高控制		21	0.50	0.05	$\hat{Y}=0.597\ 2X-0.032\ 6$	0.780 6	0.617 3	0.671 7	3.833	13.61	0.20	0.05
	75.8.23	0.97	−0.013 2	1.50	0.176 1							−0.024 7	0.200 8				>0.25	
	75.8.26	4.05	0.607 5	27.20	1.434 6							0.330 2	1.104 4				<0.10	
	77.8.12	99.95	1.999 8	27.41	1.438 0							1.161 7	0.276 3				0.25	
	77.8.13	9.57	0.980 9	51.64	1.712 9							0.553 0	1.159 7				0.10	

注:n 为历史上相似天气日数;r 为目标区与对比区日雨量的历史相关系数;α_r 为相关系数 r 的统计显著度;$\hat{y}$ 为目标区试验日自然雨量的期待值;d、$\bar{d}_k$ 为目标区试验日雨量增值及其平均值;α_d、$\alpha_{\bar{d}_k}$ 为 d 和 $\bar{d}_k$ 的统计显著度。

件处处均匀。

日雨量催化效果的统计分析，我们以天气形势和对比区雨量作为控制因子，寻找历史相似天气。历史相似天气日雨量的区域回归分析表明，除第三型天气催化效果比较显著外，其他两型的效果都不显著。如果在选择相似天气方面能找出更多更强的控制因子（相当于预报因子），试验次数继续积累，可望日雨量催化效果回归分析的效率会有所提高。月雨量历史回归分析是根据气候相似的原则进行的，而日雨量历史回归分析是根据天气相似的原则进行的，如果能与单站预报的分析研究结合起来，相似天气的判据会更有效，这种日雨量历史回归分析还是有发展余地的，值得进一步研究改进。1972 年广东新丰江地区人工降水试验，把各类相似天气合在一起建立回归方程，用以分析作业效果。从本文表 3 可以看出，不同天气型日雨量的区域回归方程是很不相同的。因此，回归分析应以按天气型分别进行为宜。

本文所用的水文雨量站雨量资料，是由湖南湘西自治州水文站提供的，谨此致谢。

参考文献

［1］ 福建省气象局、南京大学气象系.福建省 1974 年 8、9 月份人工降水效果的统计分析[J].南京大学学报，1975(1)：137－151.

［2］ 叶家东，程克明，曾光平.闽中雨季区域雨量统计特性及人工影响的效果[J].气象学报，1981，39(4)：474－482.

4.10 区域趋势控制协变量回归分析效果评估方法研究①

叶家东[1]　李铁林[2]

（1 南京大学大气科学系，南京 210093；2 河南省人工影响天气办公室，郑州 450003）

摘要 本文根据区域趋势控制和气象-物理协变量相关设计了三套非随机化人工增雨作业效果评估方案：个例作业区域趋势对比双比分析评估方案、区域趋势相关回归分析评估方案和气象-物理协变量多元回归分析评估方案。对河南省实例评估计算表明，利用物理协变量作为控制因子，可以提高作业区自然降水量估计值的准确度，从而能提高非随机化作业的效果评估效率。

关键词 效果评估　区域趋势控制　协变量回归分析

分类号 P456.8

引　言

冷云人工增雨的静力催化原理假设依然是：冷云降水是由冰晶发动并通过“水-冰”转化的贝吉隆过程完成的；有些云降水效率不高或根本无降水是因为云中缺乏足够的冰晶；通过人工引晶（播撒干冰或碘化银等冷云催化剂）的方法可弥补冷云中自然冰晶不足，致使降水过程得以有效地发动而达到增雨目的。

因此，有利的播云条件至少应该有三条：(1) 冷云部分有较丰富的过冷水含量；(2) 过冷水区自然冰晶较少；(3) 云层较厚，而云中大水滴较少，水滴碰并及冰晶繁生效率均较低。

① 收稿日期：1998－08－30；修改稿日期：1999－12－14

这些“可播性”条件,不仅对研究性试验重要,对抗旱增雨的业务性作业也同样是重要的。由于抗旱增雨作业往往缺少相应的云物理探测设备,对上述“可播性”条件进行实时仪器检测有一定困难,但还是可以根据飞机结冰和冰晶闪烁等目测现象进行定性判断。20 世纪 80 年代我国北方层状云人工降水试验研究项目所开展的综合性外场考察与试验及相应的层状降水性云系的天气、气候背景、人工降水资源、云与降水微物理结构和人工催化的外场试验、效果检验、数值模拟等方面取得了卓有成效的进展。我们需要吸取上述研究的成果,在现有认识水平和设备条件下,在作业区的设置、催化对象的选择以及播撒方案和剂量控制等技术方面力求使作业符合播云原理假设,力求使每次作业有增雨效果。这是抗旱增雨作业的基本目标所在,也是效果评估的前提条件。

河南省开展多年人工增雨试验和作业,积累了不少经验和资料,对本地区的作业对象也有一定的认识,但尚缺乏相应的云微物理结构方面的考察资料。邻近的河北省邯郸、石家庄和衡水地区以及陕西南部地区都有过一些云物理结构考察资料可供参考[1-2]。

我国北方诸省(陕西、新疆、吉林、黑龙江、内蒙古、河北、河南等)飞机和雷达观测表明,在气旋、锋面、低涡等降水性云系中,降水常常是由不同尺度、不同走向和不同伸展高度的雨雪带组成,70%以上的强降水带宽度介于 10～50 km,宽的达 100 km 或以上。局地强降水持续时间 1～4 小时,也有达 6 小时以上的[1-2]。自然强降水带的这种时空分布特征给人工增雨效果评估带来困难,在作业区(作业影响区)和对比区以及作业单元(时段)的设计中需要考虑自然降水的这种由时空分布不均匀性带来的自然“噪声”干扰。

1 人工增雨效果评估中控制变量(协变量)的选择

鉴于现有的定量降水预报的准确率远未达到可以用来评估人工增雨作业效果的程度,而且,业务数值预报模式所预报的主要是大尺度形势和大尺度降水量,难以检别中-β 尺度以下的降水,特别是系统性降水中的中尺度对流性降水特征,所以通常采用其他来源的气象协变量作为降水量的控制变量。

选择协变量的基本条件有二:

(1) 协变量与目标区(作业影响区)的自然降水量密切相关(物理上有内在联系,数值上相关显著)。

(2) 协变量应不受催化作业措施的影响。这两个条件有时会相互抵触,难以同时满足。例如:雷达回波顶高或卫星云图测算的云顶温度,虽然静力催化原理假设中并没有涉及播云的动力学效应,于是上述两种参量是可以用来评估自然降水量的,但实际上人们知道,通过人工引晶引起的“水-冰”转化过程会释放额外的 80 卡/克水的冻结潜热,在云中热力学层结条件接近中性或呈弱不稳定的场合,这附加的加热效应可以增强云的对流性扰动,抬高云顶,也可以改变云的含水量结构和云中垂直气流强度。在这种场合,回波顶高度或云顶温度是受人工引晶的播云措施影响的,用它们来作为自然降水量的控制变量(协变量)是有局限的。只有假定人工引晶引起的附加冻结潜热释放不足以增强云的热力层结不稳定性和云中对流扰动强度的前提下才是有效的控制变量。但是,作业开始前时段的回波顶高、回波强度或云顶温度等物理量并未受播云作业影响,可作为协变量参与效果评估。

利用邻近的对比区雨量和作业影响区作业时段前时段的前期雨量作为控制变量来估计作业区作业时段(作业单元)的自然雨量有一个前提条件:即自然雨量在时空分布的延续性。

对于系统性层状云系，当降水的空间分布比较均匀时间变化较为平缓时，上述假定是能满足的。但正如前面指出的，我国北方系统性层状云系降水中也常嵌有对流降水带，其时空变化都比较大，这就削弱了利用区域雨量或前期雨量作为控制变量的有效性。为了减少这种干扰，作业区和对比区都不能选得太小、作业单元时段也不能选得太短。一般讲，作业区面积以 10^4 km^2 为宜，作业时段以 3～6 小时为宜，具体尺度需通过自然雨量分布特征结合作业技术各方面因素加以痡定。

根据现阶段能收集的资料，控制变量（协变量）可选择：

对比区和作业区作业时段的雨量、雷达回波参量和卫星红外探测云顶温度；

对比区和作业区作业前期的雨量、雷达回波参量和卫星红外探测云顶温度。

2　对比样本选择和效果评估方案

严格地讲，科学试验的条件应是可控制的且试验过程是可以重复的，而人工增雨试验恰恰是在自然条件无法控制甚至难以预测或监测的大自然环境中进行的试验，几乎不可能重复，为了科学地进行这类试验，人们往往求助于随机化试验设计，将适于作业的对象随机地分出一部分不作业，保持其自然状态作为对比样本，并据以检验作业样本的效果。这类方案原则上可以做到符合随机抽样规则，因而可以根据随机抽样理论的一套成熟的统计方法，定量地检验效果并指明其可靠程度。但是，这类方案需要牺牲一部分作业机会，是人们在抗旱作业实践中不易或不能采用的一个重要因素。

抗旱人工增雨作业虽然不是严格意义上的科学试验，但实际上它仍是一种试验，是在不确定因素众多的场合下进行的一种难以预料其结果的试验。因此，不仅作业方案、催化技术需要符合科学原理，作业效果的评估也要有科学性，特别要强调客观地评估作业效果。在目前物理监测手段还相当不齐备而又不宜采用随机化设计的前提下，如何客观地评估作业效果，确实是一个难题。现有的各种非随机化试验所采用的统计评估方法都是在一定的假设前提下利用一定的控制变量（预报变量或协变量）估计作业区自然雨量，再与作业区实测雨量对比评估作业效果：

（1）序列试验以作业区历史平均雨量作为作业期自然雨量估计值，它假定作业区自然雨量在历史上是平稳的随机序列。由于天气形势不同或局地气候条件有变化，这个假定常常不能成立。况且历史雨量变率太大，用它的平均值来估计作业期自然雨量功效很低。

（2）区域对比试验利用同期对比区雨量作为作业区自然雨量的估计值。它假定作业期自然雨量的空间分布统计上是均匀的，由于地形条件差异以及作业单元选择时往往偏向于天气条件有利于作业区等主观偏倚，上述假定也常常难以满足。

（3）区域历史回归试验利用对比区自然雨量作为预报因子（控制变量），对作业区的自然雨量进行统计推断。它假定作业期作业区与对比区雨量的统计相关关系与历史上同类天气条件下的雨量的区域相关性相同。由于样本容量通常较小，加上历史相似天气的选择难免有主观性，所以估计值不稳定，或存在系统性的主观偏倚，估计值随着样本容量改变显示出较大的波动。

（4）近年提出采用区域控制模拟试验评估作业效果[3-4]。这方法的实质是对历史相似天气样本选择中，利用统计数值模拟的方法将“不相似”的样本（实质是含有少数极值个例的样本）从相似天气系列中剔除掉，从而将具有较大离差的历史对比样本人为地修匀，以达到形式上提高效果评估功效的目的。这方法与直接将对比样本中造成大离差的少数极值个例删去本质上是一样的，不符合统计抽样和统计检验的原则。它的立论基础有些问题，例如，从历史相似天气长系列中进行随机抽样，就意味着该长系列是一个次级总体，从中随机抽样所获得的每一个样本，不论它与试验期对比区样本特征值的差异是否“显著”，都是从该次级总体中抽取的一个合格的随机样本，都“有权”参与回归分析或对比分析。只选择这些样本中的某些或某个样本作为相似样本，就引进了人为的主观偏倚，将人为的主观删除效应引入效果分析中。如果要从相似天气长系列中进行随机抽样而进行统计数值模拟试验，就应该采用规范的复随机化或自然复随机化方案进行效果检验或功效分析[5]。这里没有引进任何人为的删除效应，是客观的，但也难以直接提高效果检验的“功效”或灵敏度。要提高历史回归分析检验效果的功效，还是要从物理指标上下工夫，结合播云原理，引进更为有效的物理控制变量，提高对比样本的物理相似性。

从人工增雨作业的抗旱需求出发，作业区和作业云的选择通常总是找最有利于降水的或自然发展最旺盛的云或云区，因此，在进行效果评估时，不论采用时间对比或区域对比，或用改进的回归分析方法，都存在一个如何减少或消除上述主观预报能力给效果评估带来的偏倚。这是非随机化作业效果评估中的一个关键问题。在现有监测设备和作业条件下，相似天气对比样本的选择建议如下方案：

2.1　方案1——个例作业区域趋势对比双比分析评估方案

通常在抗旱增雨作业中需要对每次作业的增雨效果即时进行评估，为适应这种需求，对于系统性降水过程，可以采用区域趋势对比双比分析方法评估作业效果。事先根据雷达观测和卫星云图资料，选择云系结构与作业区相似的邻近地带作为对比区。分别收集作业区和对比区的作业时段雨量 Y_2 和 X_2，以及作业前时段相应的雨量 Y_1 和 X_1，在自然情况下假定对比区作业期前后时段的雨量比$\dfrac{X_2}{X_1}$是与作业区作业期前后时段的对应比值$\dfrac{Y_2}{Y_1}$相同的，这意味着假定对比区与作业区处于降水系统的同一发展阶段，所以对比区以选择雷达回波结构及其演变趋势与作业区大体相似的侧风方为宜。在这样的假定前提下，如果人工增雨作业无效，应有双比值：

$$R=\frac{Y_2/Y_1}{X_2/X_1}=1。$$

如果 $R>1$，意味着有正效果；$R<1$，表示有负效果。上述双比值也可以写成

$$R=\frac{Y_2/X_2}{Y_1/X_1}=1。$$

这意味着假定自然降水情况下作业时段作业区与对比区雨量比值是与作业前期的对应比值相同的。如果作业有增雨效果，应有 $R>1$。这在一定程度上有利于减缓由于作业飞行技术人员往往选择云层结构发展旺盛的云区进行作业所引发的主观预报偏倚。因为是单次个例分析，自然难以检验这评估值的统计可信度。

这个方案是否可行或在何种程度上可行，尚需结合本地区作业季节盛行的降水系统的

时空变化特征，可根据雷达回波结构和卫星云图资料选择相应的作业区和对比区，利用自然降水资料对上述假设($R=1$)进行检验，找出 R 值的变异程度以判断此方案的适用性。关键是云层结构及其发展趋势相似的对比区是否选择得当。此外，当作业区或对比区某一时段无雨时，此方法失效。

2.2 方案2——区域趋势相关回归分析方案

这是一种事后分析方案，是对一系列作业结束后进行的总体效果评估方案。选择与作业区作业时段天气形势和云层条件相似的对比区，相似判据为：

(1) 天气形势相似：指的是对比区对比时段所处的天气系统中的部位(如距低涡中心距离、方位，距地面冷锋的水平距离等等)与作业区作业时段相似(作业时段与对比时段一般取同一时段)。

(2) 自然云系的雷达回波结构相似：作业时段前 1 小时的回波顶高、回波强度、强回波区面积和垂直厚度等参量对比区均与作业区相近。

(3) 自然云(作业时段前 1 小时)的卫星红外云顶温度及水汽含量对比区与作业区也相近。

上述分析均需在雨量资料分析前进行，切忌根据地面雨量资料选择对比区对比时段。

效果评估方法：

(1) 基本资料：

① 作业时段前一时段对比区区域面积平均雨量 X'_1；

② 作业时段前一时段作业区雨量 Y'_1；

③ 作业时段对比区雨量 X'_2；

④ 作业时段作业区雨量 Y'_2。

(2) 雨量资料正态性检验取 $X'^{\frac{1}{2}}$，$X'^{\frac{1}{3}}$，$\lg X'$ 等变数变换(统计变量)，选择其中正态性最佳的作为统计变量 X_1，Y_1，X_2，Y_2。

(3) 分析 X_1 与 X_2 的统计相关性，分析 X_1 与 Y_1 的统计相关性。如果这两种相关均显著，则：

① 建立 Y_1 倚 X_1 的线性回归方程 $Y_1=a+bX_1$，并检验回归系数的显著性。

② 假定作业影响区(简称作业区)作业时段的自然降水量 Y_2 与作业期对比区的雨量 X_2 的相关关系(回归关系)是与作业期前 1 时段两区的雨量对应关系(Y_1 倚 X_1 的回归关系)相似的。于是以 X_2 代入回归方程 $Y_1=a+bX_1$ 求出作业区作业时段自然雨量估计值 $\hat{Y}_2=a+bX_2$，于是人工增雨的效果就可由 $\Delta Y=Y_2-\hat{Y}_2$ 或 $R=\dfrac{\Delta Y}{\hat{Y}_2}$ 评估，并检验其统计显著性。

③ 替代方案：类似地建立 X_2 倚 X_1 的线性回归方程 $X_2=a_1+b_1X_1$ 并检验回归系数的显著性；假定作业区作业时段自然降水量 Y_2 与作业区作业前一时段的雨量 Y_1 的回归关系是与对比区相应关系(X_2 倚 X_1 的回归关系)相似的，于是以 Y_1 代入上述回归方程求出作业区作业时段自然雨量估计值 $\hat{Y}_2=a_1+b_1Y_1$，再和实测雨量 Y_2 比较确定增雨效果。

以上的评估方法主要的特点是避免了人为选择历史相似天气这一烦难而又容易引入众

多主观偏倚和争议的操作程序。当然，对比区的选择依然需谨慎而公正地进行。它的分析方法与区域回归试验的分析程序相似，相应的统计分析和检验方法可参看引文[6]第285～303页。关于回归分析和相关分析可参看引文[6]第191～202页和第216～223页。

算例1：根据河南省的地理条件和历史天气资料，在郑州市东偏南侧选取一面积为 $1.17\times10^4\ km^2$ 的作业影响区（A区），内设有11个自记雨量站，在郑州市南偏西方向约80千米以外选一面积为 $1.26\times10^4\ km^2$ 的对比区（B区），内设12个自记雨量站。B区位于A区西南方（图1）。河南省主要的作业天气系统有冷锋、切变线和低涡云系。选取1996、1997年上述云系影响下的 $n=29$ 个非催化单元（3小时时段雨量），对雨量分布进行正态检验，表明取3小时区域面积平均雨量的立方根值，其分布正态性较好，故作为统计变量。分析 X_1 与 Y_1 及 X_1 与 X_2 的统计相关性，发现 X_1 与 Y_1 的相关系数 $r_{X_1Y_1}=0.6769$，X_1 与 X_2 的相关系数 $r_{X_1X_2}=0.8351$，均大于 $r_{0.01}=0.487$，但 X_1 与 Y_1 的区域相关性仍较差。利用一元线性回归分析方法建立了 X_2 与 X_1 的时间相关一元回归方程：

$$X_2=0.1638+1.0917X_1。\tag{1}$$

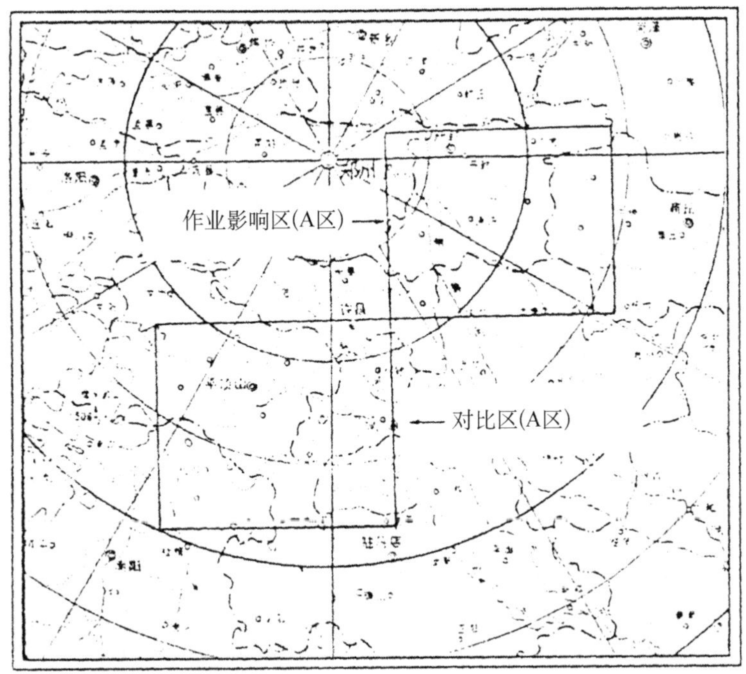

图1　郑州地区飞机人工增雨作业区与对比区的配置

Fig.1　Loction of operational area of artificial enhanced rainfall by aircraft and its contrast area

假设对比区前后时段的时间相关一元回归关系适用于作业区的对应关系，于是作业区作业时段的自然雨量估计值 $\hat{Y}_2$ 可用式(2)表示：

$$\hat{Y}_2=0.1638+1.0917Y_1。\tag{2}$$

式中 Y_1 是作业区作业时段前1小时的区域雨量（立方根值）。

利用该方程对29个作业区实测自然雨量进行回报检验，其平均绝对偏差达1.25 mm/

3小时,平均相对偏差达76%。可见,由于X_1与Y_1的区域相关性较差,将上述对比区的回归关系用于作业区,回归分析的功效较差。

作为一个算例,对河南飞机人工增雨在上述作业区实施作业而B区未作业的7个个例进行效果检验,结果列于表1。

表1 区域相关时间回归分析检验飞机人工增雨效果

Table 1 Enhanced rainfall effects of cloud seeding by aircraft with areal correlation regression analysis

作业序号	Y_2'/mm	$Y_1/(\text{mm})^{\frac{1}{3}}$	$\hat{Y}_2'$/mm	$\Delta Y'$/mm	$\Delta Y'$秩次
1	10.3	1.338 9	4.3	6.0	7
2	3.3	0.965 5	1.9	1.5	3.5
3	9.5	1.442 2	5.3	4.2	6
4	3.7	1.600 5	7.0	−3.3	−5
5	4.2	1.144 7	2.8	1.4	2
6	1.7	1.062 7	2.3	−0.6	−1
7	6.1	1.651 0	7.6	−1.5	−3.5
和	38.8		31.2	7.7	−9.5
平均	5.54		4.46	1.1	

由表可见,7次作业平均绝对增雨量1.1 mm/3 h,相对增雨量为24.7%,按照成对试验符号铁和检验法(文献[6]第162~164页)对这一增雨量进行显著性检验,负铁和为−9.5,查文献[6]附表9得显著度$\alpha=0.262$,不显著。如上所述,由于区域相关性较差,这种方法在本例的计算中,检验效率并不强。

2.3 方案3——区域趋势协变量多元回归分析方案

这也是一种总体效果评估方案,对比区的选择原则同方案2。

效果评估方法:

(1) 基本资料,协变量取:① 作业区作业时段前1小时雨量X_1;② 对比区作业时段雨量X_2;③ 作业区作业时段前1小时雷达回波最大垂直厚度X_3,最大回波强度X_4,强回波区面积X_5;④ 卫星云图红外探测云顶温度X_6。

(2) 自然降水的协变量回归方程。

① 选择相似天气条件下未作业的自然雨量Y(取原始雨量Y'的正态化变量$Y^{1/2}$,$Y^{1/3}$或$\lg Y$中的一个,余同)与X_1,X_2,X_3,X_4,X_5,X_6的统计相关性;其中Y与X_3,X_4,X_5之间的相关性宜取对数变量$\lg Y$与$\lg X_5+\lg X_4$之间的相关。分别检验各相关系数的显著性,剔除不显著的相关变量X_i(参看文献[6]第216~223页)。

② 参照多元回归分析方法(文献[6]第231~233页)建立自然降水的多元回归预报方程(检验方程):

$$Y=a+b_1X_1+b_2X_2+b_3X_3+\cdots$$

并进行回归方程和回归系数的显著性检验(参看文献[6]第233~243页)。最后剔除回归系数不显著的预报变量,将回归方程的自变量个数减至最少几个(每剔除一个变量,需重新计算多元回归方程的系数)。

(3) 利用协变量多元回归方程估计作业区作业时段自然降水量。

将作业区作业时段对应的协变量观测值 $X_1, X_2, X_3, \cdots$ 代入多元回归方程,求出作业区作业时段的自然降水量估计值$\hat{Y}$。这里隐含着一个重要假定,即协变量 $X_1, X_2, X_3, \cdots$ 是不受人工催化措施影响的。

(4) 将作业区作业时段的实测雨量 Y 与$\hat{Y}$全对比,求出增雨量 $\Delta Y = Y - \hat{Y}$ 或 $R = \dfrac{\Delta Y}{\hat{Y}}$;对 ΔY 或 R,或多次作业的平均增雨量$\overline{\Delta Y}$或$\bar{R}$ 进行显著性检验或区间估计(参看文献[6]第 202～216 页)。

(5) 当相似天气的自然雨量样本难以选择,样本太小而不足以建立上述自然降水的多元回归预报方程时,一种替代办法是利用对比区作业时段对应的 $X'_1, X'_2, X'_3, X'_4, X'_5$和 X'_6,再按上述步骤建立对比区的协变量多元回归方程:

$$X_1 = a' + b_2' X_2' + b_3' + b_4' + b' X_4 + b_5' X_5' + b_6' X_6'。$$

从中选择回归系数显著的变量,剔除不显著的变量。以此作为作业区自然降水量的预报方程。将作业区作业时段的协变量 $X'_2, X'_3, X'_4, \cdots$ 代入上式求出 X_1 的估计值 $\hat{X}_1$,这时 $\hat{X}_1$ 已是作业区作业时段的自然降水量估计值,将它与实测雨量比较评估作业效果。这时附加了一个假设条件:对比区的协变量和雨量的统计相关与作业区的相应关系在统计上是一致的。

算例 2:试验区的设置同算例 1。选取的协变量包括:作业区作业时段前 1 小时区域面积平均雨量立方根值 X_1;对比区作业时段(3 小时)区域面积平均雨量的立方根值 X_2;作业区作业时段前 1 小时雷达回波最大强度平均值的自然对数值 X_3 和最大回波顶高度的自然对数值 X_4。待检验(或预报)的量是作业区作业时段(3 小时)区域面积平均雨量的立方根值 Y。所有的变数变换均因统计分布正态性而作。对 1996、1997 年主要作业天气系统影响下的 33 个非催化单元的 Y 与各协变量的相关分析,结果表明 $r_{YX_1} = 0.839\ 2, r_{YX_2} = 0.686\ 3, r_{YX_3} = 0.868\ 3, r_{YX_4} = 0.823\ 9$,均大于 $r_{0.001} = 0.554$,相关显著。利用多元回归分析方法(参看文献[6]第 231～233 页)建立作业区自然降水的协变量多元回归预报方程(检验方程):

$$Y = -1.041\ 3 + 0.392\ 1X_1 + 0.201\ 1X_2 + 0.282\ 5X_3 + 0.631\ 9X_4。 \quad (3)$$

回归方程的复相关系数 $r = 0.927\ 9$,用 F 检验法对回归方程作显著性检验,得 $F = 43.335\ 8 \gg F_{0.01} = 4.07$,回归方程高度显著。利用上述检验方程对河南省飞机人工增雨在作业区 A 实施作业的个例进行效果检验,结果列于表 2。

表 2 协变量多元回归分析检验飞机人工增雨作业效果

Table 2 Enhanced rainfall effects of cloud seeding by aircraft with covariable multiple regression analysis

作业序号	Y'/mm	X_1/(mm)$^{\frac{1}{3}}$	X_2/(mm)$^{\frac{1}{3}}$	X_3 ln(dBZ)	X_4 ln(km)	$\hat{Y}$/mm	$\Delta Y'$/mm	$\Delta Y'$秩次
1	10.3	1.338 9	2.387 0	3.555 3	1.667 7	8.3	2.0	7
2	3.3	0.965 5	0.736 8	3.218 9	1.568 6	2.7	0.6	5

续表

作业序号	Y'/mm	X_1/(mm)$^{\frac{1}{3}}$	X_2/(mm)$^{\frac{1}{3}}$	X_3 ln(dBZ)	X_4 ln(km)	$\hat{Y}$/mm	$\Delta Y'$/mm	$\Delta Y'$秩次
3	9.5	1.442 2	2.432 9	3.465 7	1.568 6	7.8	1.7	6
4	3.7	1.600 5	0.584 8	2.995 7	1.609 4	3.9	−0.2	−2
5	4.2	1.144 7	1.922 0	3.135 5	1.386 3	3.8	0.4	4
6	1.7	1.062 7	1.118 7	2.708 1	1.335 0	1.8	−0.1	−1
7	6.1	1.651 0	1.574 1	3.258 1	1.504 1	5.8	0.3	3
和	38.8					34.1	4.7	−3
平均	5.54					4.87	0.67	

表中：Y'为作业区催化后的3小时实测区域平均雨量；X_1为作业区作业时段前1小时区域平均雨量的立方根值；X_2为对比区作业时段区域平均雨量的立方根值；X_3和X_4分别为作业区作业时段前1小时最大回波强度和最大回波顶高度的自然对数值。

由表可见，上述作业个例催化后3小时平均增雨量0.67 mm，平均相对增雨量约13.8%。按成对试验符号秩和检验法对这一增雨效果进行显著性检验，得$\alpha=0.039$，表明增雨效果显著，显著性水平优于5%。

由此可见，利用物理协变量作为控制因子，可以提高作业区自然降水量估计值的准确度，这表明按照这样的思路，不断改进或引入新的更有效的协变量，包括设置相关性更强的的对比区在内，有可能提高非随机化作业的效果评估效率。

3 结 论

本文提出三套非随机化人工增雨作业效果评估方案：

(1) 方案1根据区域趋势对比分析进行个例作业效果评估，它假定作业时段作业区与对比区自然雨量比值与作业前期的对应比值相同，这比通常的区域对比分析法假定作业区与对比区自然雨量相等合理些，在作业简报中是有用的即时评估方法。

(2) 方案2假定作业时段作业区自然雨量与对比区雨量的回归关系与作业前期的对应关系一致进行区域回归分析评估作业效果，避免了选择历史相似天气这一烦难而又容易引入诸多人为主观偏倚的操作程序。在抗旱人工增雨作业中只有雨量资料可供使用的情况下，这是一种可行的方案。

(3) 方案3除了利用对比区雨量外，还引进物理协变量作为控制因子进行协变量多元回归分析，作业区自然雨量的估计值准确度有明显提高，从而提高了非随机化人工增雨作业的效果评估效率，在有雷达、卫星等观测资料可供利用的场合是一种有效的方案。

参考文献

［1］ 北方层状云人工降水试验课题组.北方层状云人工降水试验研究.国家重点课程，气科院803X-2，1991

［2］ 游景炎，段英，游来光主编.云降水物理和人工增雨技术研究.北京：气象出版社，1994.83～88、155～163

[3] 曾光平,刘峻.人工降水试验效果检验的统计模拟方法研究.气象学报,1993,**51**(2):241～247

[4] 曾光平,郑行照,方仕珍,李顺来.非随机化人工降雨试舱效果评价方法研究.大气科学,1994,**18**(2):233～242

[5] 叶家东,罗幸贫,曾光平.肖锋.随机试验功效的数值分析.气象学报,1984,**42**(1):69～79

[6] 叶家东,范蓓芳.人工影响天气的统计数学方法.北京:科学出版社,1982,162～164、191～223、231～243、285～303

EVALUATION METHODS OF CLOUD SEEDING EFFECT WITH REGIONAL CONTROL AND COVARIABLE REGRESSION ANALYSIS

Ye Jiadong[1] Li Tielin[2]

(1 *Department of Atmospheric Sdences, Nanjing University* 210093;
2 *Weather Modification Office, Henan Province* 450003)

Abstract Three evaluation methods for the nonrandomized precipitation enhancement operation effects have been developed based on the regional rainfall control and meteorological covariable correlation: Dopple ratio analysis evaluation method using regional rainfall tendency control for single cloud seeding operation case; regression analysis evaluation effects on the bases of regional correlation and developing tendency of rainfall; and multiple regression analysis with meteorological and physical covariables. It is shown from the evaluation examples of effects for cloud seeding operation cases carried out in the central region of Henan Province, that the inferential accuracy of natural rainfall on the cloud seeding operational area could be improved by using appropriate physical covariables as control factors, then the evaluational efficiency of efficiency for the nonrandomized cloud seeding operations can be increased.

Key words Effect evaluation Regional control Covariable regression ananlysis

4.11 方差不相等的双样本回归分析

叶家东 范蓓芬

(南京大学气象系)

提 要 在人工降水试验效果的区域雨量回归分析中,当余方差不相等时,通常的多个事件检验法就不适用。本文在Welch检验法的基础上,提出了余方差不相等时适用的双样本回归分析方法,并且指出,我们根据双样本回归分析提出的统计量(19),当余方差相等时要比多个事件检验法的统计量(6)有效。

利用双样本回归分析,对福建古田地区人工降水随机试验的效果进行评价。结果表明,人工降水引起的目标区平均相对增雨量随着对比区自然雨量的增大而减小。当对比区为小雨($x'<2$ 毫米/小时)时,催化效果显著($\alpha=0.05$),相对增雨量达20%～81%,但绝对增雨量不大,只有0.44～1.15毫米/3小时,而当对比区雨量 $x'>2.8$ 毫米/小时时,催化效果就不显著。

人工影响天气试验效果的统计检验,通常都采用 t-检验法或秩和检验法[1]。这类检验方法有一个共同的前提条件,就是假定待检验的两个样本所抽取的总体方差是相等的。但是，实际上这个条件并不总是满足的。在福建古田地区的人工降水试验中[2]，我们发现人工影响后的相对增雨量有随自然雨量强度增大而减小的趋势（表 1）。这种趋势在美国佛罗里达[3]、苏联乌克兰[4]以及以色列[5]等地的试验中均有所发现。从统计上看，这意味着人工影响不仅改变了雨量的平均值，而且也改变了雨量的方差。如果这是真的，则通常的检验方法就不适用。同样，在区域雨量的回归分析中，利用多个事件检验法[6]检验平均增雨量的显著性时，也是假定了供分析的两组样本（称催化样本和对比样本）所抽取的总体余方差相等，如果这个条件不满足，多个事件检验法也不适用。

表 1　增雨效果与自然雨量的关系

对比区雨量强度/(毫米/小时)	$x'\leqslant 1$	$1<x'\leqslant 3$	$x'>3$
样本容量	18	22	12
目标区相对增雨量/%	125.2	24.4	−1.0

方差不相等条件下的双样本检验问题，至今尚无一般性的解法，只有一些近似方法可供利用[7]，Behrens 曾提出过一个检验法，所以这种问题常称为 Behrens-Fisher 问题。对于人工影响天气试验，常采用 Welch 检验法[8]。至于余方差不相等条件下的双样本回归分析，尚未见到有人讨论过。本文在 Welch 检验法的基础上，提出一种余方差不相等条件下的双样本回归分析方法，并对余方差相等时适用的多个事件检验法做出修正。

一、Welch 检验法

设二独立样本 $x_{11},x_{12},\cdots,x_{1n_1}$ 和 $x_{21},x_{22},\cdots,x_{2n_2}$ 服从正态分布，总体平均值为 μ_1 和 μ_2，方差 $\sigma_1^2\neq\sigma_2^2$，相应的样本统计量为 $\bar{x}_1,\bar{x}_2$ 和 s_1^2,s_2^2。现在要根据上述样本数据检验假设 $H_0:\mu_1=\mu_2$。为此，需要决定统计量

$$z=\frac{(\bar{x}_1-\bar{x}_2)-(\mu_1-\mu_2)}{\sqrt{\dfrac{s_1^2}{n_1}+\dfrac{s_2^2}{n_2}}}\tag{1}$$

的分布。令

$$s^2=\frac{s_1^2}{n_1}+\frac{s_2^2}{n_2}。$$

易知

$$z=\frac{\dfrac{(\bar{x}_1-\bar{x}_2)-(\mu_1-\mu_2)}{\sqrt{\dfrac{\sigma_1^2}{n_1}+\dfrac{\sigma_2^2}{n_2}}}}{\sqrt{\dfrac{s^2}{\dfrac{\sigma_1^2}{n_1}+\dfrac{\sigma_2^2}{n_2}}}}$$

的分子是 $N(0,1)$变量。如果分母是一个 χ -变量的倍数,则 z 是 t -变量。但现在 $\sigma_1^2 \neq \sigma_2^2$,所以分母不是一个 χ -变量的倍数。Welch 检验法的要点是以一个于 χ^2 -变量的倍数 s'^2 来近似代替 s^2,选择 s'^2 的自由度 v',使得 s'^2 与 s^2 的数学期望及方差都相等。这样,z 就近似地成为自由度是 v'的 t -变量。据此可以导出

$$v' = \frac{\left(\frac{\sigma_1^2}{n_1} + \frac{\sigma_2^2}{n_2}\right)^2}{\frac{1}{n_1 - 1}\left(\frac{\sigma_1^2}{n_1}\right)^2 + \frac{1}{n_2 - 1}\left(\frac{\sigma_2^2}{n_2}\right)^2}。 \tag{2}$$

实际上 σ_1^2 与 σ_2^2 是不知道的,以相应的样本估计量 s_1^2 与 s_2^2 近似代替,就得

$$v' \approx \frac{\left(\frac{s_1^2}{n_1} + \frac{s_2^2}{n_2}\right)^2}{\frac{1}{n_1 - 1}\left(\frac{s_1^2}{n_1}\right)^2 + \frac{1}{n_2 - 1}\left(\frac{s_2^2}{n_2}\right)^2}, \tag{3}$$

v'称为有效自由度。于是,在原假设成立的前提下,统计量

$$z = \frac{\bar{x}_1 - \bar{x}_2}{\sqrt{\frac{s_1^2}{n_1} + \frac{s_2^2}{n_2}}} \tag{4}$$

可近似看作是自由度为 v'的 t -变量,从而可利用 t -检验法加以检验。

Welch 检验法与双样本 t -检验的比较,由式(4)和通常的双样本 t -检验的统计量

$$t = \frac{\bar{x}_1 - \bar{x}_2}{\sqrt{\frac{(n_1 - 1)s_1^2 + (n_2 - 1)s_2^2}{n_1 + n_2 - 2}} \cdot \sqrt{\frac{1}{n_1} + \frac{1}{n_2}}}, \tag{5}$$

比较可知,它们的区别在于分母不同。设 $\frac{s_1^2}{s_2^2} = F$,可得

$$\frac{z^2}{t^2} = \frac{\frac{v_1 s_1^2 + v_2 s^2}{v_1 + v_2}\left(\frac{1}{v_1 + 1} + \frac{1}{v_2 + 1}\right)}{\frac{s_1^2}{v_1 + 1} + \frac{s_2^2}{v_2 + 1}} = \frac{(Fv_1 + v_2)(v_1 + v_2 + 2)}{(v_1 + v_2)(1 + F + v_1 + Fv_2)}。$$

如果两个样本容量相等,则 $v_1 = v_2 = v$,这时

$$\frac{z^2}{t^2} = 1。$$

因此,对于容量相等的样本,统计量 z 和 t 数值相同,此时两种检验法的区别仅在于 t 的自由度是 $2v$,而 z 的自由度为

$$v' = v\frac{(F+1)^2}{F^2+1} = v\left(1+\frac{2F}{F^2+1}\right)。$$

当 $F=1$ 时，$v'=2v$，z 还原为 t，这就是方差相等的情形；当 $F\neq 1$ 时，v' 比 $2v$ 小，但对于 $0.333<F<3$，自由度的偏差不超过 20%。所以当样本容量相等而样本方差相差不大时，两种方法的差异是小的。但是，当样本容量不相等时，例如当 $v_1 \gg v_2$ 时，$z^2/t^2 \to F$，此时用 t 统计量代替 z 统计量所产生的误差近似地随着 $\sqrt{F}$ 增大而增加，这种情况下就不宜用双样本 t -检验法。

二、双样本回归分析

通常，在人工降水效果检验的区域雨量回归分析中，利用统计量[6]

$$t = \frac{\bar{y}_k - \bar{\hat{y}}_k}{\sqrt{\dfrac{\sum\limits_{i=1}^{n}(y_j-\hat{y}_j)^2}{n-2}\left[\dfrac{1}{k}+\dfrac{1}{n}+\dfrac{(\bar{x}_k-\bar{x}_n)^2}{\sum\limits_{j=1}^{n}(x_j-\bar{x}_n)^2}\right]}} \tag{6}$$

检验平均差值

$$\bar{d}_k = \bar{y}_k - \bar{\hat{y}}_k$$

的显著性时，也是假定了供分析的两组样本（催化样本和对比样本）所代表的倚变量（目标区雨量）的余方差是相等的。但是，从福建古田试验的回归分析发现，这个条件并不满足（表 2）。由表 2 可见，催化样本与对比样本的余方差有显著差异，显著度达到 0.05，这时就不宜用上述多个事件检验公式(6)进行催化效果的统计检验。为此，我们作如下双样本回归分析。

表 2　回归分析中催化样本与对比样本的余方差相等性检验

	样本容量	$s^2_{余}$	$F=\dfrac{s^2_{2余}}{s^2_{1余}}$	α
催化样本(1)	$k=50$	0.027 0	1.655 6	<0.05
对比样本(2)	$n=51$	0.044 7		

令催化单元目标区雨量 y 倚对比区雨量 x 的区域回归方程为

$$\hat{y}_1 = a_1 + b_1 x, \tag{7}$$

而将对比单元的区域雨量回归方程写成

$$\hat{y}_2 = a_2 + b_2 x。 \tag{8}$$

相应的一元线性正态回归的结构模型是

$$y_1 = \alpha_1 + \beta_1 x + \varepsilon_1, \tag{7'}$$

$$y_2 = \alpha_2 + \beta_2 x + \varepsilon_2。 \tag{8'}$$

其中 $\alpha_1 + \beta_1 x = y_{10}$，$\alpha_2 + \beta_2 x = y_{20}$ 是相应的总体回归值，而 $\varepsilon_1 \sim N(0, \sigma_1)$，$\sigma_2 \sim N(0, \sigma_2)$，$\sigma_1 \neq \sigma_2$。

由回归分析知[3]

$$\hat{y}_1 \sim N\left(\alpha_1 + \beta_1 x, \sigma_1 \sqrt{\frac{1}{k} + \frac{(x - \bar{x}_1)^2}{\sum_{j=1}^{k}(x_i - \bar{x}_1)^2}}\right), \tag{9}$$

$$\hat{y}_2 \sim N\left(\alpha_2 + \beta_2 x, \sigma_2 \sqrt{\frac{1}{n} + \frac{(x - \bar{x}_2)^2}{\sum_{j=1}^{n}(x_j - \bar{x}_2)^2}}\right)。 \tag{10}$$

其中 k，n 分别为催化样本和对比样本的容量，而

$$\bar{x}_1 = \frac{1}{k}\sum_{i=1}^{k} x_i,\quad \bar{x}_2 = \frac{1}{n}\sum_{j=1}^{n} x_j。$$

现在要检验假设 $H_0: y_{10} = y_{20}$。为此，作统计量

$$z_{\hat{y}} = \frac{\hat{y}_1 - \hat{y}_2 - (y_{10} - y_{20})}{\sqrt{\left(\frac{1}{k} + \frac{(x - \bar{x}_1)^2}{\sum_{i=1}^{k}(x_i - \bar{x}_1)^2}\right) s_{1余}^2 + \left(\frac{1}{n} + \frac{(x - \bar{x}_2)^2}{\sum_{j=1}^{n}(x_j - \bar{x}_2)^2}\right) s_{2余}^2}}。 \tag{11}$$

其中

$$s_{1余}^2 = \frac{1}{k-2}\sum_{i=1}^{k}(y_i - \hat{y}_i)^2,\quad s_{2余}^2 = \frac{1}{n-2}\sum_{j=1}^{n}(y_j - \hat{y}_j)^2$$

分别是 σ_1^2，σ_2^2 的样本估计量。

仿照 Welch 的处理方法，在原假设成立的前提下，统计量

$$z_{\hat{y}} = \frac{\hat{y}_1 - \hat{y}_2}{\sqrt{\left(\frac{1}{k} + \frac{(x - \bar{x}_1)^2}{\sum_{i=1}^{k}(x_i - \bar{x}_1)^2}\right) s_{1余}^2 + \left(\frac{1}{n} + \frac{(x - \bar{x}_2)^2}{\sum_{j=1}^{n}(x_j - \bar{x}_2)^2}\right) s_{2余}^2}} \tag{12}$$

可近似地看作是自由度为

$$v_{\hat{y}} = \frac{\left\{\left(\frac{1}{k} + \frac{(x - \bar{x}_1)^2}{\sum (x_i - \bar{x}_1)^2}\right) s_{1余}^2 + \left(\frac{1}{n} + \frac{(x - \bar{x}_2)^2}{\sum (x_j - \bar{x}_2)^2}\right) s_{2余}^2\right\}^2}{\frac{1}{k-2}\left\{\left(\frac{1}{k} + \frac{(x - \bar{x}_1)^2}{\sum (x_i - \bar{x}_1)^2}\right) s_{1余}^2\right\}^2 + \frac{1}{n-2}\left\{\left(\frac{1}{n} + \frac{(x - \bar{x}_2)^2}{\sum (x_j - \bar{x}_2)^2}\right) s_{2余}^2\right\}^2} \tag{13}$$

的 t -变量。

表 3 是福建古田地区三年人工降水试验中目标区和对比区雨量的基本资料。计算表明，变量取雨量的四次方根值的正态性很好，拟合度达 0.94，所以分析时统计变量均取雨量的四次方根值。表 3 还列入有关的统计量。我们根据上述方法对古田地区试验的效果进行双样本回归分析，计算结果列于表 4。催化样本(1)和对比样本(2)的区域雨量回归线及实测雨量散布图见图 1；催化效果随对比区自然雨量而变化的关系见图 2。

表 3　样本数据和基本统计量(x_i'，y_i'与x_j'，y_j'均系三小时区域面积平均雨量)

催化样本(1)		对比样本(2)		催化样本(1)		对比样本(2)		$k=50$
$x_i=\sqrt[4]{x_i'}$	$y_i=\sqrt[4]{y_i'}$	$x_j=\sqrt[4]{x_j'}$	$y_j=\sqrt[4]{y_j'}$	$x_i=\sqrt[4]{x_i'}$	$y_i=\sqrt[4]{y_i'}$	$x_j=\sqrt[4]{x_j'}$	$y_j=\sqrt[4]{y_j'}$	
0.929 9	1.181 5	1.479 1	1.165 9	1.143 5	1.146 9	1.905 4	1.724 8	$\bar{x}=1.551\ 0$
1.605 3	2.077 4	0.758 6	0.764 0	1.278 3	1.433 3	1.675 4	1.680 7	$\bar{y}_1=1.601\ 7$
1.318 0	1.546 4	1.776 1	1.598 3	1.242 1	1.343 7	1.636 4	1.602 2	$\sum_{i=1}^{k}(x_i-\bar{x}_1)^2=6.624\ 1$
1.251 0	1.576 7	1.019 3	0.588 3	1.660 4	1.698 4	1.200 9	0.948 7	$\sum_{i=1}^{k}(y_i-\bar{y}_1)^2=5.630\ 2$
0.941 7	0.969 3	0.725 9	0.678 5	1.648 2	1.503 5	1.897 4	1.579 9	$\sum_{i=1}^{k}(x_i-\bar{x}_1)(y_i-\bar{y}_1)=5.355\ 9$
1.593 0	1.789 2	1.569 4	1.394 5	1.755 6	1.718 4	1.455 7	1.633 5	$a_1=0.347\ 7$
1.260 3	1.452 6	1.608 1	1.418 0	2.239 9	2.014 8	1.732 1	1.987 1	$b_1=0.808\ 5$
1.324 4	1.509 3	0.930 4	1.047 5	1.286 6	1.315 0	1.140 2	1.322 6	$r_1=0.877\ 1$
1.914 5	1.801 2	1.753 8	1.864 7	1.495 4	1.330 1	1.537 6	1.650 5	$\hat{y}_1=0.347\ 7+0.808\ 5x$
2.326 7	2.350 0	1.665 3	1.797 9	0.876 4	0.814 4	1.081 9	1.115 8	$s_{1余}^2=\frac{1-r_1^2}{k-2}\sum_{i=1}^{k}(y_i-\bar{y}_1)^2=0.027\ 0$
1.831 0	1.795 9	1.526 5	1.383 1	1.205 2	1.208 1	0.699 9	0.692 5	
1.366 0	1.411 4	1.466 4	1.772 4	1.450 8	1.361 9	1.095 4	1.122 9	
1.679 5	1.476 6	1.674 2	1.677 6	1.608 9	1.445 9	1.465 3	1.613 0	
1.299 2	1.444 0	1.239 8	0.872 1	1.057 4	1.252 4	1.133 4	1.388 8	
2.172 3	2.033 4	1.623 2	2.023 0	1.621 3	1.209 5	1.416 9	1.487 8	$n=51$
1.227 1	1.397 5	1.564 5	1.483 9	1.472 4	1.474 8	0.904 7	0.809 8	$\bar{x}_2=1.424\ 6$
2.440 5	2.300 1	0.419 5	1.584 9	1.901 7	1.891 5	0.600 5	0.668 7	$\bar{y}_2=1.416\ 2$
1.311 6	1.467 8	1.572 5	1.521 0	2.297 0	2.171 9	1.398 9	1.664 7	$\sum_{j=1}^{n}(x_j-\bar{x}_2)^2=5.835\ 4$

续表

催化样本(1)		对比样本(2)		催化样本(1)		对比样本(2)		$k=50$
$x_i=\sqrt[4]{x_i'}$	$y_i=\sqrt[4]{y_i'}$	$x_j=\sqrt[4]{x_j'}$	$y_j=\sqrt[4]{y_j'}$	$x_i=\sqrt[4]{x_i'}$	$y_i=\sqrt[4]{y_i'}$	$x_j=\sqrt[4]{x_j'}$	$y_j=\sqrt[4]{y_j'}$	
1.627 4	2.098 0	1.578 9	1.555 1	1.855 4	2.125 0	1.659 8	1.644 3	$\sum_{j=1}^{n}(y_j-\bar{y}_2)^2=6.878\ 7$
1.311 7	1.562 5	1.397 1	1.428 1	1.694 8	1.824 1	1.836 7	1.870 1	$\sum_{j=1}^{n}(x_j-\bar{x}_2)(y_j-\bar{y}_2)=5.223\ 7$
1.411 6	1.477 8	1.412 4	1.666 3	2.045 7	1.832 2	1.711 0	1.702 4	$a_2=0.140\ 9$
1.627 1	1.833 1	1.291 2	0.836 7	1.960 7	1.893 3	1.285 4	1.602 2	$b_2=0.895\ 2$
1.499 8	1.579 9	1.028 7	1.215 1	1.775 6	1.734 9	1.571 6	1.254 9	$r_2=0.824\ 5$
1.677 1	1.959 1	1.632 9	1.496 8	1.216 5	1.220 0	1.414 2	1.773 4	$\hat{y}_2=0.140\ 9+0.895\ 2x$
1.351 9	1.458 9	2.023 6	1.745 9			1.388 8	1.377 5	$s_{2余}^2=\frac{1-r_2^2}{n-2}\sum_{j=1}^{n}(y_j-\bar{y}_2)^2=0.044\ 7$
1.458 9	1.569 6	2.070 8	1.626 6					

表 4　试验效果的双样本回归分析

对比区自然雨量		目标区平均增雨量			显著性检验		
x	$x'/\%$	$\hat{y}_1-\hat{y}_2$	绝对增雨量/毫米	相对增雨量/%	$z_{\hat{y}}$	$v_{\hat{y}}$	α(单边)
0.8	0.41	0.137 6	0.438 8	81.38	1.680 1	94.76	<0.05
1.0	1.00	0.120 1	0.634 5	55.05	1.876 6	95.81	<0.05
1.3	2.86	0.094 1	0.930 8	32.12	2.219 3	96.03	<0.025
1.5	5.06	0.076 8	1.084 0	22.37	1.984 5	93.63	0.025
1.551 0	5.79	0.072 3	1.110 1	20.09	1.868 2	87.84	<0.05
1.7	8.35	0.059 4	1.152 4	15.08	1.302 6	84.33	<0.10
1.8	10.50	0.050 7	1.139 3	12.07	0.958 4	82.56	<0.20
2.0	16.00	0.033 4	0.987 7	7.10	0.483 4	82.50	>0.25

由所列图表可见:(1) 平均相对增雨量是随对比区雨量的增加而减小的,当 $x>1.7$(相当于雨量 8.35 毫米/3 小时)时,催化效果就不显著了($\alpha=0.05$);(2) 平均绝对增雨量介于 0.44～1.15 毫米/3 小时之间,对比区雨量 $x\leqslant 1.7$ 时是随着 x 增加而增大的,$x=1.7$ 时达到最大,约 1.15 毫米/3 小时,但此时显著度不高($\alpha<0.10$)。看来这种小火箭催化冷云降雨的方法在小雨时效果较好,可增加 20%～81%,但绝对增雨量并不大,只有 1 毫米/3 小时左右。

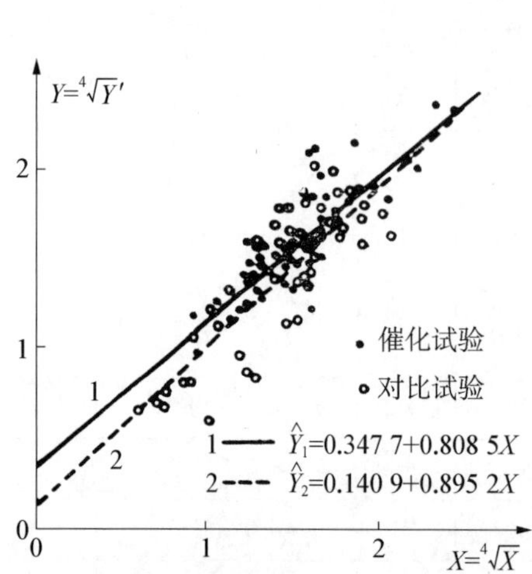

图 1　区域雨量回归线和实测雨量散布图

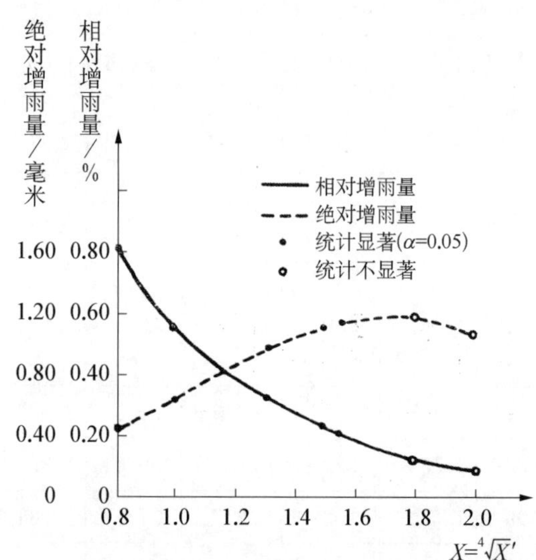

图 2　增雨效果随对比区自然雨量而变化的关系曲线

与多个事件检验法比较，多个事件检验法是检验 k 次试验的平均效果。为了比较起见，在式(12)中取 $x=\bar{x}_1=\bar{x}_k$，于是 $\hat{y}_1=\hat{y}_k$，$\bar{\hat{y}}_k=\hat{y}_2$。令

$$\frac{s_{1余}^2}{s_{2余}^2}=F,$$

且因 $\bar{x}_2=\bar{x}_n$，则由式(12)和式(6)可得

$$\frac{t^2}{z_{\hat{y}}^2}=\frac{\left[\frac{1}{k}\cdot F+\frac{1}{n}+\frac{(\bar{x}_1-\bar{x}_2)^2}{\sum(x_j-\bar{x}_n)^2}\right]}{\left[\frac{1}{k}+\frac{1}{n}+\frac{(\bar{x}_k-\bar{x}_n)^2}{\sum(x_j-\bar{x}_n)^2}\right]}。\tag{14}$$

如果 $F=1$，则 $\frac{t^2}{z_{\hat{y}}^2}=1$，$z_{\hat{y}}$ 还原为 t；如果 $F<1$，则 $t<z_{\hat{y}}$；反之，当 $F>1$ 时 $t>z_{\hat{y}}$，即人工影响后目标区雨量指标的余方差变小时，统计量 $z_{\hat{y}}$ 比 t 有效。反之，则差。

至于自由度，如果 $F=1$，且 $n=k$，则由式(13)可得

$$v_{\hat{y}}=(n-2)\left\{1+\left[1+\frac{\left[\frac{(\bar{x}_1-\bar{x}_2)^2}{\sum(x_j-\bar{x}_2)^2}\right]^2}{\frac{2}{n}\left[\frac{1}{n}+\frac{(\bar{x}_1-\bar{x}_2)^2}{\sum(x_j-\bar{x}_2)^2}\right]}\right]^{-1}\right\}\approx 2(n-2)=2v。$$

因为在一般情况下，

$$\left[\frac{(\bar{x}_1-\bar{x}_2)^2}{\sum(x_j-\bar{x}_2)^2}\right]^2 与 \frac{2}{n}\left[\frac{1}{n}+\frac{(\bar{x}_1-\bar{x}_2)^2}{\sum(x_j-\bar{x}_2)^2}\right]$$

相比，是二阶小项。在这种情况下，$z_{\hat{y}}$的自由度在$n=k$、$F=1$时几乎是多个事件检验法中t的自由度的两倍。

我们认为，问题在于多个事件检验法没有充分利用样本资料所提供的全部信息，所以它不是一个最有效的统计量。对此，我们作如下方差相等时的双样本回归分析。

在$\sigma_1=\sigma_2=\sigma$的情况下，当假设$H_0: y_{10}=y_{20}=y_0$成立时，对任一固定的x值而言，由式(9)和式(10)可知统计量

$$\begin{aligned} u &= \frac{\hat{y}_1-\hat{y}_2}{\sqrt{\sigma^2\left(\frac{1}{k}+\frac{(x-\bar{x}_1)^2}{\sum(x_i-\bar{x}_1)^2}\right)+\sigma^2\left(\frac{1}{n}+\frac{(x-\bar{x}_2)^2}{\sum(x_j-\bar{x}_2)^2}\right)}} \\ &= \frac{\hat{y}_1-\hat{y}_2}{\sigma\cdot\sqrt{\left(\frac{1}{k}+\frac{(x-\bar{x}_1)^2}{\sum(x_i-\bar{x}_1)^2}\right)+\left(\frac{1}{n}+\frac{(x-\bar{x}_2)^2}{\sum(x_j-\bar{x}_2)^2}\right)}} \end{aligned} \tag{15}$$

服从$N(0,1)$分布，于是

$$t=\frac{\hat{y}_1-\hat{y}_2}{s_{余}\cdot\sqrt{\left(\frac{1}{k}+\frac{(x-\bar{x}_1)^2}{\sum(x_i-\bar{x}_1)^2}\right)+\left(\frac{1}{n}+\frac{(x-\bar{x}_2)^2}{\sum(x_j-\bar{x}_2)^2}\right)}} \tag{16}$$

是t-变量，其中$s_{余}^2$是余方差σ^2的样本估计量。我们来考虑$s_{余}^2$。既然样本1和2都是从正态回归模型

$$y=\alpha+\beta x+\varepsilon=y_0+\varepsilon$$

中抽取的独立样本，式中$\varepsilon\sim N(0,\sigma)$，则样本余方差

$$s_{1余}^2=\frac{1}{k-2}\sum_{i=1}^{k}(y_i-\hat{y}_i)^2,\quad s_{2余}^2=\frac{1}{n-2}\sum_{j=1}^{n}(y_i-\hat{y}_i)^2,$$

以及

$$s_{余}^2=\frac{1}{n+k-4}\left[\sum_{i=1}^{k}(y_i-\hat{y}_i)^2+\sum_{j=1}^{n}(y_i-\hat{y}_i)^2\right]$$

都是σ^2的无偏估计量，于是

$$t_1=\frac{\hat{y}_1-\hat{y}_2}{\sqrt{\frac{1}{k-2}\sum(y_i-\hat{y}_i)^2\left\{\left(\frac{1}{k}+\frac{(x-\bar{x}_1)^2}{\sum(x_i-\bar{x}_1)^2}\right)+\left(\frac{1}{n}+\frac{(x-\bar{x}_2)^2}{\sum(x_j-\bar{x}_2)^2}\right)\right\}}}, \tag{17}$$

$$t_2=\frac{\hat{y}_1-\hat{y}_2}{\sqrt{\frac{1}{n-2}\sum(y_j-\hat{y}_j)^2\left\{\left[\frac{1}{k}+\frac{(x-\bar{x}_1)^2}{\sum(x_i-\bar{x}_1)^2}\right]+\left[\frac{1}{n}+\frac{(x-\bar{x}_2)^2}{\sum(x_j-\bar{x}_2)^2}\right]\right\}}}, \tag{18}$$

以及

$$t=\frac{\hat{y}_1-\hat{y}_2}{\sqrt{\frac{\sum(y_i-\hat{y}_i)+\sum(y_j-\hat{y}_j)^2}{n+k-4}\left\{\left[\frac{1}{k}+\frac{(x-\bar{x}_1)^2}{\sum(x_i-\bar{x}_1)^2}\right]+\left[\frac{1}{n}+\frac{(x-\bar{x}_2)^2}{\sum(x_j-\bar{x}_2)^2}\right]\right\}}} \tag{19}$$

分别是自由度为 $v_1=k-2$，$v_2=n-2$ 以及 $v=n+k-4$ 的 t -变量，都可以用来检验假设

$$H_0: y_{10}=y_{20}。$$

但是，以 t 值最有效，因为它的余方差估计量最充分地利用了全部样本数据。

这样，若令 $x=\bar{x}_1=\bar{x}_k$，则式(18)就还原为式(6)，而式(19)所表达的 t 就是式(12)所表达的 $z_{\hat{y}}$ 当 $F=1$ 且 $n=k$ 时的特例。

所以我们认为，要检验假设 $H_0: y_{10}=y_{20}$，在 $\sigma_1^2=\sigma_2^2$ 的情况下，宜用式(19)，它比多个事件检验法中的式(6)更充分地利用了样本资料，且不限于只检验$(\bar{y}_k-\bar{\hat{y}}_k)$的显著性。在$\sigma_1^2\neq\sigma_2^2$ 的情况下，宜用式(12)。当 $F=1$ 且 $k=n$ 时，式(12)还原为式(19)。

三、讨论和结论

(1) 双样本检验中，当方差不相等且两个样本容量相差较大时，宜采用 Welch 检验法；如果两个样本容量接近相等且样本方差相差不大时($0.333<F<3$)，采用 t -检验法误差不大，只在自由度上有不超过 20%的误差。

(2) 回归分析中，当催化单元倚变量的余方差与对比单元的不相等时，可利用双样本回归分析方法，根据统计量(12)检验两条回归线的差异。这种检验方法比多个事件检验法进了一步。首先，它不需要假定两个回归方程的余方差相等；其次，它可以检验自变量 x 取不同值时的倚变量$\hat{y}$的差异显著性。就人工降水试验来说，可以检验对比区雨量不同时的催化效果。即使在总体余方差相等的条件下，多个事件检验法也不是最有效的。采用式(19)作为待检验的统计置，由于它充分利用了样本资料，所以比多个事件检验法的统计量式(6)更有效，它是式(12)当 $F=1$ 且 k 和 n 相差不大时的特例。

(3) 福建古田地区人工降水试验的催化效果。区域回归分析表明，平均相对增雨量是随对比区自然雨量的增加而减小的，小雨时($x'<2$ 毫米/小时)效果比较显著，相对增雨量达 20%～81%，但绝对增雨量并不大，一次试验(3 小时时段)增加 0.44～1.15 毫米。当 $x>1.7$(相当于 $x'>8.35$ 毫米/3 小时)时，效果就不显著了。

由古田试验看来，利用小火箭发射冷云催化剂(碘化银或介乙醛)对古田地区 4～6 月份的降水性云系进行人工催化的方法，一次试验平均在 3 小时时段内可增加目标区雨量 1 毫

米左右。而且人工降水的效果，看来既不是可加的，也不一定像通常认为的那样是可乘的[10]。因为不少地区的试验都曾发现相对增雨量随自然雨量增加而减小，绝对增雨量则在某一自然雨量强度下达到极大。

古田试验平均每次试验的催化剂用量为碘化银（AgI）220 克或介乙醛（MA）2 450 克。室内成冰核率试验表明[11]，在－10～－15 ℃范围内，MA 的成冰核率约 10^{10}～5×10^{10} 个/克；在－12～－13.5 ℃时，AgI 的成冰核率为 10^{11}～10^{12} 个/克。于是，每次试验平均输入云中的有效冰核数为 2.5×10^{13}～1.2×10^{14} 个（MA）或 2.2×10^{13}～2.2×10^{14} 个（AgI）。如果每个有效核在降水性云系中都能长成 $r=1$ 毫米的雨滴且都落在目标区，则在目标区 1 500 平方千米面积上平均每次可增加雨量 0.07～0.34 毫米（MA）或 0.07～0.70 毫米（AgI）。如果冰晶在云中有繁生作用，或者大多数人工冰核都先后在不同温度下活化，则人工冰核产生的冰晶再增加几倍甚至 1 个量级是有可能的。这样，只要自然云中水分是充足的，人工催化增加 1 毫米或更多一些雨量是可能的。大雨时效果不显著可能是自然起伏大，检查不出来；或者自然降水过程本身已充分有效，增加冰核无济于事。

参考文献

[1] 叶家东，人工降水的统计研究方法，江西省气象科学研究所印，1977.

[2] 福建省气象局气科所，南京大学气象系，大气科学，3 卷 2 期，131—140，1979.

[3] Simpson, J., Woodley, W. L., Miller, A. R. and Cotton, G. F. *J. Atmos, Sci.*, 10, 526—544, 1971.

[4] Лесков, Б. Н., Proce. of the WMO/IAMAP Sci. Conf. on weather Modification, Tashkant, 17, 10, 1973, WMO－No. 399, 143—146, 1974.

[5] Gabriel, K. R., The Israeli Rainmaking Experiment 1961—1967 Final Statistical Tables and Evaluation Hebrew University, 1970.

[6] Thom, H. C. S., Final Report of the Advisory Committee on Weather Control, Vol. II., Tech. Rep, No. 1, 1—25, 1951.

[7] Kendall, M. G. and Stuart, A., The Advanced Theory of Statistics, vol. 2, p. 146, 1961.

[8] Brownlee, K. A., Statistical Theory and Methodology in Science and Engineering, p. 299, 1965.

[9] 上海师范大学数学系概率统计教研组，回归分析及其试验设计，上海教育出版社，1978.

[10] Scott, E. L., Third Conf. on Probabi. and Statis. in Atmos. Sci., 65—72, 1973.

[11] 福建省气象局、南京大学、西安化工研究所，福建气象科技，2 期，51—54，1977.

TWO-SAMPLE REGRESSION ANALYSIS WITH UNEQUAL VARIANCES

Yeh Jia-dong Fan Pei-fen

(Department of Meteorology, Nanjing University)

Abstract In the regression analysis with control area of rainfall, the usual method of multiple event test is unsuitable in the case of unequal variances. In this paper, based on Welch test, a method of the two-sample regression analysis was proposed, which is suitable in the case of unequal variances. Futher analysis shows that the statistic (19) with equal variances is more powerful than the statistic of multiple event test (6).

The effect of cloud seeding experiment in the area of Gutian, Fujian province, has been evaluated with the two-sample regression analysis. It is shown that on average in the target area the relative enhancement of rainfall by seeding decreases with increase of the rainfall in control area. When the rainfall in control area is small ($x'<2$ mm/h), the seeding effect is significance ($\alpha=0.05$), the relative enhancement of rainfall reaches 20%—81%, but absolute enhancement is only 0.44—1.15 mm/3 h, and the effect is nonsignificant when $x'>2.8$ mm/h.

4.12 随机试验功效的数值分析*

叶家东　罗幸贫
（南京大气气象系）
曾光平　肖　锋
（福建省气象科学研究所）

提　要　根据自然复随机化法对福建古田地区人工降水试验的统计功效进行数值试验，结果表明，随机交叉设计试验2～3年能以80%的检出概率检验出20%～30%的增雨效果($\alpha=0.05$)，区域控制设计则要试验4～5年才能检验出相应的增雨效果；区域回归随机试验中，不同的统计检验方法：双比分析、多个事件检验和双样本回归分析，其检验功效差别不大，一般不超过5%；分层统计在一定条件下能提高统计功效，如层状云降水，由于其自然变差小，区域相关性高，分层统计其功效显著提高，积状云降水的统计功效则没有改善。区域相关性是影响统计功效的重要因素。

一、引言

人工降水试验的对象总是千差万别、变化多端的。尽管可以通过适当的物理考虑，事先进行有利天气条件或可播度的预报，或事后进行分层统计将统计组群限制在天气、物理条件大体相近的范围之内，但总不能做到试验对象千篇一律，而往往存在较大的自然变差。统计检验的目的就是要从降水的种种自然变差中检验出人工降水的平均效果。原则上讲，显著度 α 就衡量了统计效果的可信程度。但实际上，或者由于统计概型与资料拟合不佳，或者由于个别极值的权重太大，或者是由于试验对象实际上并不属于同一统计总体等等，致使统计效果及其显著度都会有较大的起伏(表1)。由表1可见，随着试验期延长，效果及其显著度渐趋平稳，问题是究竟需要试验多长时期，才能得出可信的统计效果呢？由于上述诸因素无法在参量性统计检验中反映出来，更由于实际效果并不知道，所以这是一个难以直接回答的问题。但是我们可以通过试验功效的分析给以统计上的解答。

表1　古田试验历年的统计效果及其显著度

试验期	1年 (1975)	2年 (1975+1977)	3年 (1975+1977+1978)	4年 (1975+1977+1978+1980)	5年 (1975+1977+1978+1980+1981)
统计效果/%	81.8	78.7	20.3	27.6	26.9
显著度 α	0.005	0.05	0.05	0.05	0.002 5

* 本文于1982年11月8日收到，1983年4月1日收到修改稿。

我们先看一下降水的自然变差有多大。图1是古田试验中样本容量分别为 $N=32$,83 和 124 的自然降水资料,各根据假想的 100 次区域回归随机试验求取的由降水自然起伏引起的目标区相对"增雨效果"的累积频率曲线[1]。可见,相应于上述三种容置的样本,要在 $\alpha=0.05$ 的显著度上获得显著的增雨效果,相对增雨量需分别超过 55.6%,24.8%和 20.1%。显然,如果进行一次实际的催化试验,则要通过 32 个单元的区域回归随机试验,在 0.05 的显著度上检验出一定的比方说 20%的增雨效果可能性是很小的,因为降水自然起伏引起的"噪声"远大于效果这个"信号"。这就是古田试验历年的统计效果有较大起伏的重要原因所在。试验功效的研究在确定试验区、试验期以及选择有效的试验设计方案和统计检验方法等方面起着指导性作用。试验功效是从雨量的自然变差中检别出试验效果的能力的一种统计量度。它的确切提法是:一定的试验期内在一定的显著度上检别出一定的试验效果的概率。与此等价的另一种提法是:要在一定的显著度上以一定的检出概率检别出一定的试验效果,需进行多长的试验期(试验单元)?

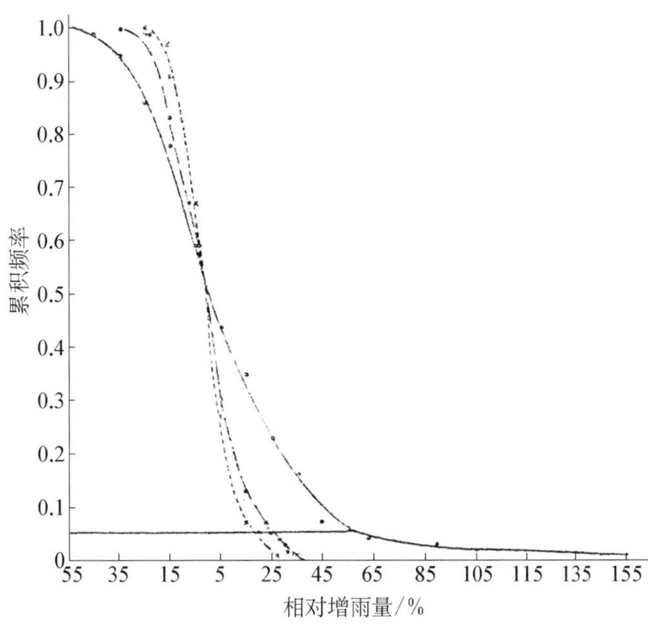

图1　降水自然起伏引起的相对"增雨效果"的累积频率曲线

研究功效的方法主要有两类:经典的统计理论分析方法和统计数值模拟试验方法。统计理论分析方法根据雨量的理论概率分布或渐近概率分布律求出效果检验统计量的概率分布,进而根据此统计量的分布律和假定的试验效果,比较自然的和经过"人工催化"影响后的该统计量的分布律,在一定的显著度 α 上求出犯第二类错误的概率 β,于是功效就是 $P-1-\beta$。它是试验有效时做出有效判断的概率,即在替代假设成立的条件下拒绝原假设的概率。

功效的统计理论分析方法要求自然雨量服从某一已知的概率分布律[2-6],而在实际上这一点常常难以达到,所以近年来发展了统计数值模拟试验方法,即所谓的复随机化试验法(Re-randomization method)。这类方法的基础是 Fisher 提出的排列检验或随机化检验[7]。它不要求雨量以及表征效果的统计量服从已知的概率分布,比通常的参量性检验有更强的稳健性(Robustness)。这一点对气象试验尤其重要,因为气象试验中的试验单元是它一出现就得采用的,事先无法进行标准化或独立性的选择。可是,由于这种非参量性检验法计算量大,长期以来其应用受到限制。随着现代计算技术发展,排列检验法重新复活[8]。1961 年 Addley 首先采用复随机化法检验人工降水效果。Kempthorne 和 Doerfler(1969)提出用复随机化法计算功效[9]。近年来一些地区的功效分析都采用复随机化法[10-11]。为了简化计算量,Gabriel[12] 和 Salvam 等[13] 提出一种简易计算功效的方法,称之为 Naive method,可以叫作自然复随机化法。分析表明,这种方法计算的功效略大于复随机化法计算的精确功效,偏差最大不超过 7%，然而计算量却可减少一个量级。

二、功效计算的统计试验法

1. 复随机化试验法

设有 N 个试验单元，首先进行一次主试验。为此，随机抽取其中 K 个单元按一定的试验效果进行“催化”处理，其余 $n=N-K$ 个单元留作对比。设 $R[E,\theta(E)]$ 为效果统计量，其中 $\theta(E)$ 是抽样方式为 E 时的效果，它在复随机化试验中保持不变。其次，进行复随机化副试验。对上述资料，其中 K 个单元经过“催化”处理的，采取不同的随机程序从中抽 K 个单元作为“催化”样本，其余 $n=N-K$ 个对比，计算相应的表示“效果”的统计量 $R[e,\theta(E)]$，其中 e 表示副试验是按抽样方式 e 实施的。重复这种复随机化试验，比如说进行 100 次，可得到 100 个 $R[e,\theta(E)]$ 的大小分布。设 $R_{\alpha}[e,\theta(E)]$ 是从大到小计数累积频率为 α 的 $R[e,\theta(E)]$ 值，若 $R[E,\theta(E)]\geqslant R_{\alpha}[e,\theta(E)]$，则上述主试验的“催化”效果在 α 水平上显著，否则，不显著。为计算功效，需重复进行上述主-副试验，比如重复 100 次主试验及各自相应的复随机化副试验。于是在 100 次主试验中在 α 水平上“效果”显著的百分比率 P 就是试验功效，它是通过 N 个单元的试验，在 α 显著度上能检验出效果 $\theta(E)$ 的概率的估计值。

2. 自然复随机化试验法

复随机化法计算量很大，如上所述，一种试验样本需进行 10 000 次复随机化试验。自然复随机化试验作为估计功效的近似计算法，大致可减少计算量一个量级。其基本程序为：对 N 个单元的试验样本，事先不做任何“催化”处理，而对原始数据进行复随机化试验，比方说进行 1 000 次，据此可以求出在雨量的自然随机变差影响下，表征“效果”的统计量 R 的大小分布，并求出相应的 R_{α} 值，以此 R_{α} 为判据，设进行一次“催化”试验，其中 K 个单元做了“催化”处理，求出相应的效果统计量 $R(\theta)$，如果 $R(\theta)\geqslant R_{\alpha}$，则“效果”显著。如此进行，比方说 1 000 次“催化”试验，每次试验的随机抽样程序不同，其中“效果”显著的比率 P 就是试验功效的估计值。

本文根据古田试验 1975，1977，1978，1980，1981，1982 年 4～6 月试验期对比单元的自然雨量资料，对该地区不同试验设计方案，不同效果检验方法的功效以及分层统计对功效的影响，采用自然复随机化法进行数值模拟试验。每次复随机化试验都进行 1 000 次。试验中效果是人为假定的，但可以选择不同的效果值，分析其对功效的影响程度，这相当于效果检验的灵敏度分析。

三、计算方案

1. 催化单元与对比单元

将试验单元的雨量资料随机排列，根据随机数字发生器随机产生 0 或 1，顺序抽样，这样把资料随机分成催化单元和对比单元两组样本，为保证两样本有基本相同的容量，抽样是成对进行的①。这一随机抽样过程由计算机执行。

2. 试验效果

为了分析功效对催化效果的响应程度，在模拟试验中选取增雨效果为 0%，10%，20%，

① 一般讲，非成对随机抽样限制性条件较少，但根据 Welch 检验和双样本回归分析知[15]，当供比较的两个样本容量接近相等时，方差不相等的影响小。故本文仍采用成对随机抽样。

30%,50%,70%和100%,各增雨效果在各自的试验中保持定常。试验中雨量资料系三小时区域面积平均雨量,为了在回归分析中使雨量正态化,统计变量取此平均雨量的四次方根[14],于是相应的增雨效果比例因子分别表示为 $\theta=(1.0)^{1/4}$、$(1.1)^{1/4}$、$(1.2)^{1/4}$、$(1.3)^{1/4}$、$(1.5)^{1/4}$、$(1.7)^{1/4}$ 和 $(2.0)^{1/4}$。$\theta=1.0$ 表示催化效果为 0 的情形,此时“效果”统计量即为雨量自然随机起伏所引起的“噪声”。

3. 效果统计量

本试验分析三种试验设计方案的功效:单区随机试验、区域控制随机试验和区域交叉随机试验,相应的效果统计量分别为:

单区随机试验取比率 R:

$$R=\frac{\frac{1}{K}\sum_{i=1}^{K}\theta y_i}{\frac{1}{n}\sum_{j=1}^{n}y_j}=\theta\frac{\bar{y}_1}{\bar{y}_2}。\tag{1}$$

其中:$\bar{y}_1$ 和 $\bar{y}_2$ 分别为目标区催化单元和对比单元的雨量指标平均值,θ 是效果指标。

区域控制随机试验效果统计量取双比值 DR:

$$\mathrm{DR}=\frac{\frac{1}{K}\sum_{i=1}^{K}\theta y_i}{\frac{1}{K}\sum_{i=1}^{K}x_i}\Bigg/\frac{\frac{1}{n}\sum_{j=1}^{n}y_j}{\frac{1}{n}\sum_{j=1}^{n}x_j}=\theta\frac{\bar{y}_1}{\bar{x}_1}\Big/\frac{\bar{y}_2}{\bar{x}_2}。\tag{2}$$

其中:$\bar{y}_1$,$\bar{x}_1$ 分别为催化单元目标区与对比区雨量指标的平均值;$\bar{y}_2$,$\bar{x}_2$ 分别为对比单元的相应值。

区域交叉随机试验效果统计量取双比值 CR:

$$\mathrm{CR}=\left[\frac{\frac{1}{K}\sum_{i=1}^{K}\theta y_i}{\frac{1}{K}\sum_{i=1}^{K}x_i}\Bigg/\frac{\frac{1}{n}\sum_{j=1}^{n}y_j}{\frac{1}{n}\sum_{j=1}^{n}\theta x_j}\right]^{1/2}=\theta\left[\frac{\bar{y}_1}{\bar{x}_1}\Big/\frac{\bar{y}_2}{\bar{x}_2}\right]^{1/2}。\tag{3}$$

其中:$\bar{y}_1$,$\bar{x}_1$ 分别为 y 区催化 x 区对比单元两区雨量指标的平均值;$\bar{y}_2$,$\bar{x}_2$ 分别为 x 区催化 y 区对比单元两区雨量指标的平均值。

此外,在区域回归随机试验中,对试验效果进行回归分析时,我们采用过三种统计检验方法[15]:多个事件检验法、方差相等的双样本回归分析和方差不相等的双样本回归分析。相应的效果检验统计量分别为:

$$t_1=\frac{\theta\hat{y}_1-\hat{y}_2}{\sqrt{\frac{(1-r_2^2)\sum(y_j-\bar{y}_2)^2}{n-2}\left[\frac{1}{K}+\frac{1}{n}+\frac{(\bar{x}_1-\bar{x}_2)^2}{\sum(x_j-\bar{x}_2)^2}\right]}},\tag{4}$$

$$t=\frac{\theta\hat{y}_1-\hat{y}_2}{\sqrt{\frac{\theta^2(1-r_1^2)\sum(y_i-\bar{y}_1)+(1-r_2^2)\sum(y_j-\bar{y}_2)^2}{n+K-4}\left[\frac{1}{K}+\frac{1}{n}+\frac{(\bar{x}_1-\bar{x}_2)^2}{\sum(x_j-\bar{x}_2)^2}\right]}},\tag{5}$$

$$z=\frac{\theta\hat{y}_1-\hat{y}_2}{\sqrt{\frac{\theta^2(1-r_1^2)\sum(y_j-\bar{y}_1)^2}{K(K-2)}+\frac{(1-r_2^2)\sum(y_j-\bar{y}_2)^2}{n-2}\left[\frac{1}{n}+\frac{(\bar{x}_1-\bar{x}_2)^2}{\sum(x_j-\bar{x}_2)^2}\right]}}。\tag{6}$$

其中 r_1 和 r_2 分别为催化单元和对比单元的区域相关系数。为比较不同检验方法的功效，对上述三个统计量也分别做了功效数值模拟试验。

功效值采用图解法求取。图 2 所示为样本容量 $N=73$ 时计算的各效果统计量的累积

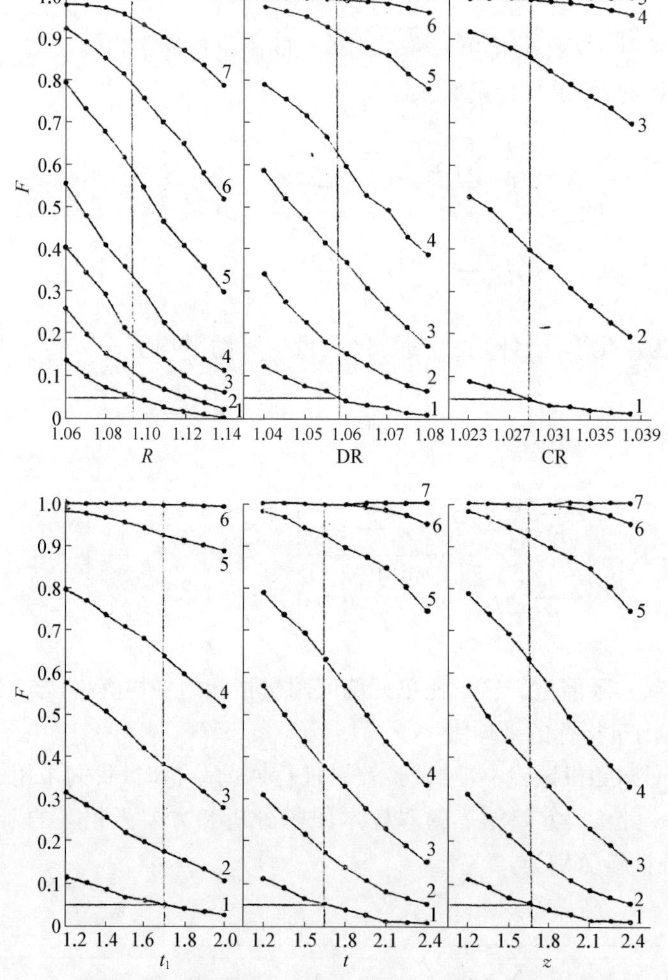

图 2　效果统计量的累积频率曲线(曲线 1～7 分别代表增雨效果 0%,10%,20%,30%,50%,70%和 100%)

频率图。取显著度 $\alpha=0.05$，在 $\theta=1.0$ 曲线(1)的累积频率为5%处作平行于纵坐标的直线，它与各累积频率曲线的交点所对应的纵坐标值即为该效果下的统计功效估计值。

四、功效数值分析

各种试验设计方案和统计检验方法的功效数值计算结果列于表2。据此我们做如下功效分析：

表2 各种试验设计方案及检验方法的统计功效值①

统计量		N	增雨效果/%					
			10	20	30	50	70	100
R	合并统计	31	0.076	0.122	0.169	0.283	0.411	0.555
		51	0.101	0.174	0.279	0.481	0.650	0.845
		73	0.111	0.197	0.331	0.586	0.790	0.949
		88	0.123	0.228	0.356	0.625	0.820	0.971
		118	0.129	0.275	0.435	0.762	0.908	0.985
		150	0.147	0.320	0.536	0.837	0.964	0.999
	层状云	51	0.104	0.216	0.338	0.562	0.755	0.929
	积状云	37	0.084	0.124	0.171	0.256	0.369	0.522
DR	合并统计	31	0.134	0.237	0.354	0.616	0.812	0.938
		51	0.150	0.331	0.521	0.834	0.973	0.999
		73	0.166	0.389	0.629	0.913	0.996	1.000
		88	0.209	0.478	0.720	0.968	1.000	1.000
		118	0.262	0.607	0.869	0.998	1.000	1.000
		150	0.226	0.506	0.782	0.988	1.000	1.000
	层状云	51	0.158	0.347	0.564	0.881	0.986	1.000
	积状云	37	0.141	0.236	0.374	0.633	0.832	0.965
CR	合并统计	31	0.241	0.529	0.793	0.989	1.000	1.000
		51	0.354	0.752	0.969	1.000	1.000	1.000
		73	0.412	0.864	0.995	1.000	1.000	1.000
		88	0.513	0.926	1.000	1.000	1.000	1.000
		118	0.633	0.985	1.000	1.000	1.000	1.000
		150	0.538	0.966	1.000	1.000	1.000	1.000
	层状云	51	0.370	0.820	0.985	1.000	1.000	1.000
	积状云	37	0.244	0.562	0.822	0.995	1.000	1.000

① 表中 $N=118,150$ 的样本是逐年加进1976(30个)和1979年(32个)两个非试验年的雨量资料计算的。由于该两年无雷达观测配合，试验单元的确定随意性大，故将计算结果列入作为参考。

续表

<table>
<tr><th rowspan="2">统计量</th><th rowspan="2"></th><th rowspan="2">N</th><th colspan="6">增雨效果/%</th></tr>
<tr><th>10</th><th>20</th><th>30</th><th>50</th><th>70</th><th>100</th></tr>
<tr><td rowspan="8">t_1</td><td rowspan="6">合并统计</td><td>31</td><td>0.120</td><td>0.234</td><td>0.348</td><td>0.635</td><td>0.833</td><td>0.967</td></tr>
<tr><td>51</td><td>0.140</td><td>0.301</td><td>0.516</td><td>0.835</td><td>0.980</td><td>0.998</td></tr>
<tr><td>73</td><td>0.177</td><td>0.387</td><td>0.643</td><td>0.926</td><td>0.997</td><td>1.000</td></tr>
<tr><td>88</td><td>0.226</td><td>0.497</td><td>0.775</td><td>0.987</td><td>1.000</td><td>1.000</td></tr>
<tr><td>118</td><td>0.252</td><td>0.631</td><td>0.893</td><td>1.000</td><td>1.000</td><td>1.000</td></tr>
<tr><td>150</td><td>0.261</td><td>0.578</td><td>0.860</td><td>0.997</td><td>1.000</td><td>1.000</td></tr>
<tr><td>层状云</td><td>51</td><td>0.182</td><td>0.392</td><td>0.648</td><td>0.951</td><td>0.999</td><td>1.000</td></tr>
<tr><td>积状云</td><td>37</td><td>0.129</td><td>0.225</td><td>0.352</td><td>0.595</td><td>0.796</td><td>0.961</td></tr>
<tr><td rowspan="8">t</td><td rowspan="6">合并统计</td><td>31</td><td>0.115</td><td>0.203</td><td>0.306</td><td>0.543</td><td>0.750</td><td>0.914</td></tr>
<tr><td>51</td><td>0.151</td><td>0.320</td><td>0.521</td><td>0.845</td><td>0.979</td><td>0.998</td></tr>
<tr><td>73</td><td>0.162</td><td>0.354</td><td>0.598</td><td>0.915</td><td>0.996</td><td>1.000</td></tr>
<tr><td>88</td><td>0.224</td><td>0.479</td><td>0.765</td><td>0.981</td><td>1.000</td><td>1.000</td></tr>
<tr><td>118</td><td>0.262</td><td>0.631</td><td>0.887</td><td>0.999</td><td>1.000</td><td>1.000</td></tr>
<tr><td>150</td><td>0.261</td><td>0.583</td><td>0.868</td><td>0.998</td><td>1.000</td><td>1.000</td></tr>
<tr><td>层状云</td><td>51</td><td>0.187</td><td>0.413</td><td>0.692</td><td>0.972</td><td>0.998</td><td>1.000</td></tr>
<tr><td>积状云</td><td>37</td><td>0.122</td><td>0.214</td><td>0.324</td><td>0.558</td><td>0.768</td><td>0.937</td></tr>
<tr><td rowspan="8">z</td><td rowspan="6">合并统计</td><td>31</td><td>0.113</td><td>0.203</td><td>0.306</td><td>0.543</td><td>0.750</td><td>0.916</td></tr>
<tr><td>51</td><td>0.150</td><td>0.320</td><td>0.522</td><td>0.843</td><td>0.979</td><td>0.998</td></tr>
<tr><td>73</td><td>0.164</td><td>0.371</td><td>0.617</td><td>0.917</td><td>0.996</td><td>1.000</td></tr>
<tr><td>88</td><td>0.222</td><td>0.477</td><td>0.762</td><td>0.981</td><td>1.000</td><td>1.000</td></tr>
<tr><td>118</td><td>0.261</td><td>0.630</td><td>0.887</td><td>0.999</td><td>1.000</td><td>1.000</td></tr>
<tr><td>150</td><td>0.259</td><td>0.581</td><td>0.866</td><td>0.999</td><td>1.000</td><td>1.000</td></tr>
<tr><td>层状云</td><td>51</td><td>0.180</td><td>0.399</td><td>0.678</td><td>0.970</td><td>0.998</td><td>1.000</td></tr>
<tr><td>积状云</td><td>37</td><td>0.125</td><td>0.218</td><td>0.329</td><td>0.563</td><td>0.770</td><td>0.939</td></tr>
</table>

(1) 统计功效与增雨效果的关系：由图 3 可见，试验功效总是随增雨效果增加而增大的。当增雨效果小于 50%时，功效随效果增加而迅速增大，效果大于 50%时，功效曲线渐趋平稳。实际上当功效值超过 0.80 时，其值随效果的变化就缓慢了。

(2) 试验功效与试验期的关系：功效分析的主要目的之一是要确定一个试验，究竟需要进行多长时间才能较有把握地检验出一定的试验效果。图 4 示出增雨效果为 20%和 30%时不同容量的统计功效。显见，随着试验期延长(样本容量增多)，功效是增加的，但一般并不是随容量增加而按比例增大的。在试验单元随机排列情况下($N\leqslant 73$)，随着容量增大，功效增加逐渐变缓。$N=88$ 的功效增加较大，是加进 1982 年试验期 15 个对比单元计算的，该年区域相关系数高达 0.936 1，大于前 5 年合计的相关系数 0.845 5，所以功效增加明显。

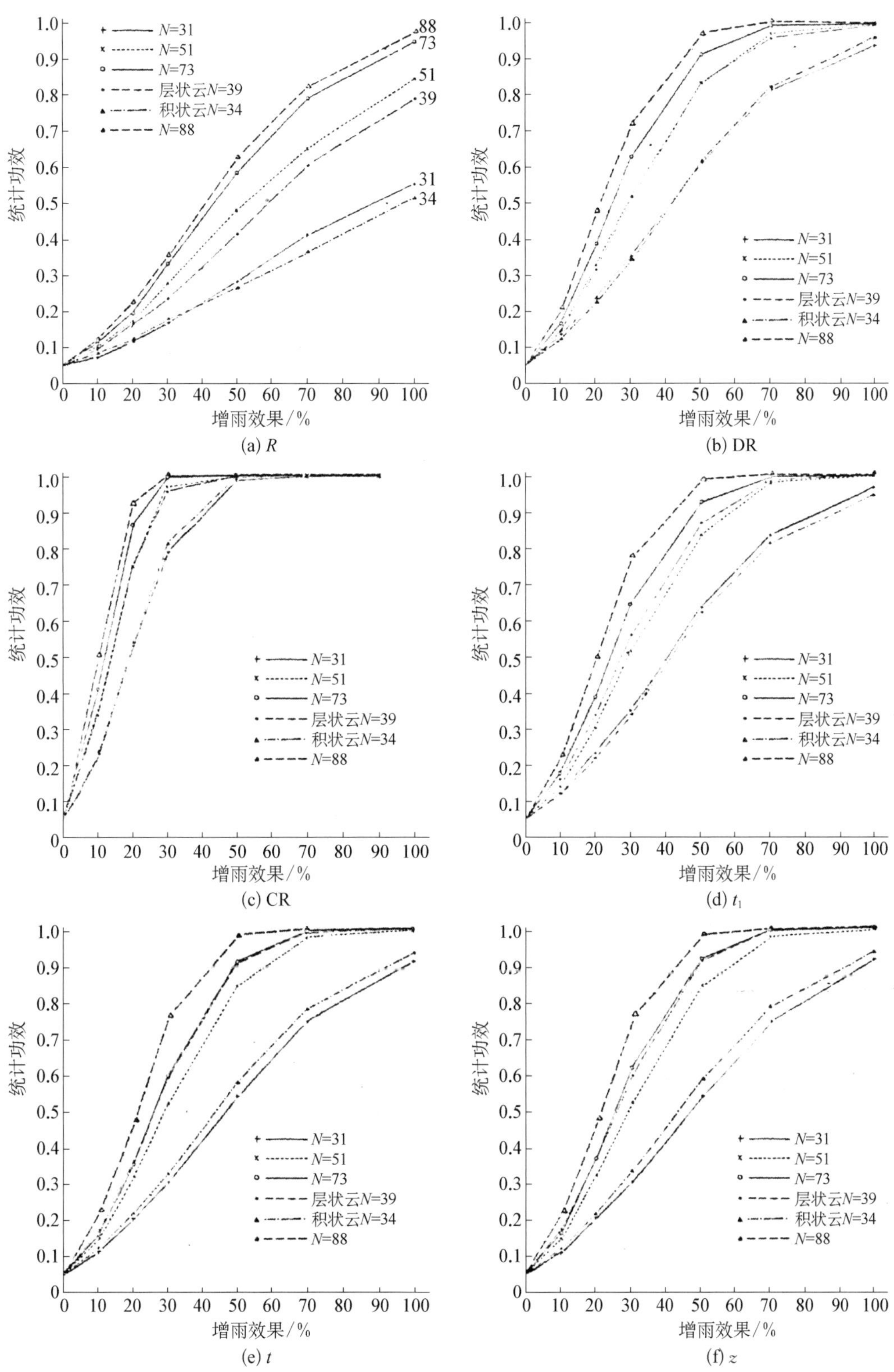

图 3 随机试验的统计功效

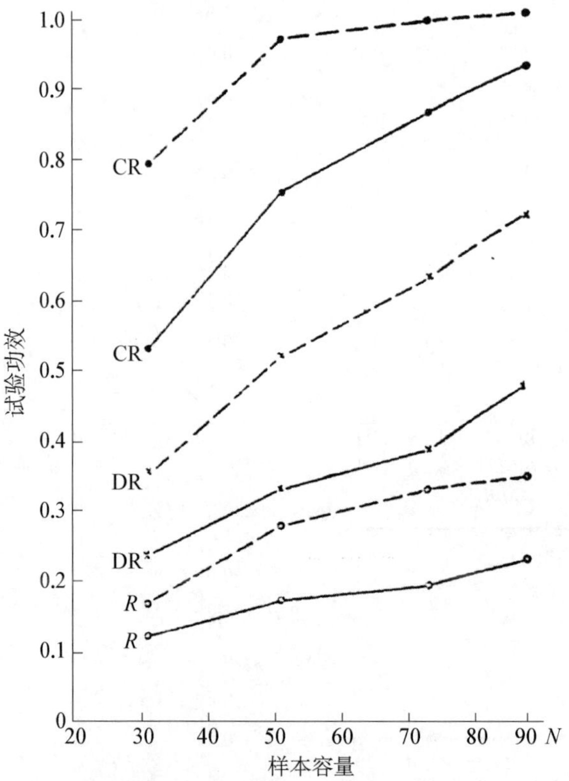

图 4　试验功效与样本容量的关系(实线和虚线增雨效果分别为 20%和 30%)

(3) 不同试验设计下的统计功效,图 5 是三种不同试验设计方案下的功效值。可见,随机交叉试验(CR)的功效最高,区域控制随机试验(DR)次之,单区随机试验的功效(R)最低。这与 Schickedanz 等的结果是一致的[4],但是即使是单区试验,表 2 所列 $N=88$ 的功效也比 Scott 计算的功效大[5]。虽然从表 3 看出,两者的功效数值相近①,但 Scott 的样本容量是 100,大于本试验的容量。这可能是古田地区雨季雨量的自然变差较 Scott 假定的典型情况(雨量服从 Γ-分布,其形状参数取 0.70,尺度参数取 2.0)小的缘故。

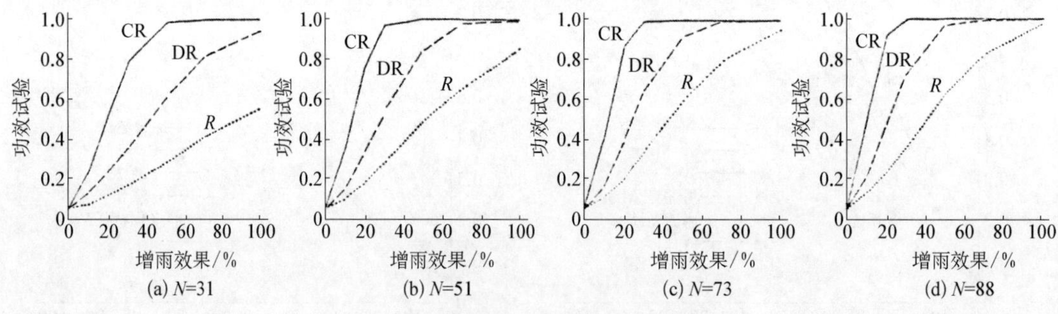

图 5　不同试验设计方案下的统计功效

① Scott 计算功效所用的统计显著度为 0.1,但从其对效果的假设检验定义看,用的可能是双边检验,与本文的单边检验显著度 $\alpha=0.05$ 是一致的,不然他的功效比本文计算得更低。

表 3 单区随机试验的功效值

	N	相对增雨效果/%				
		10	20	30	50	70
本试验 R	88	0.12	0.23	0.36	0.63	0.82
Scott	100	0.14	0.20	0.30	0.60	0.80

由于单区试验功效一般较低，Scott 提出选择有效的气象协变量对于提高统计功效十分必要。由图 5 可见，利用对比区雨量作为协变量，若区域相关性好，是能显著提高试验功效的。

如果我们定义功效值等于 0.80 作为可接受的界限，则对单区随机试验，样本容量为 51，73 和 88 时，只有当增雨效果达 93%，72%和 68%时才能检验出来。对于区域控制随机试验，容量为 31，51，73 和 88 时，所要求的增雨效果分别为 69%，48%，42%和 37%。而对随机交叉试验，相应的增雨效果只需 31%，22%，19%和 17%。从古田地区前 5 年试验的经验看，试验效果一般只有 20%～30%，所以只有采用随机交叉试验，才能在 2～3 年的试验期(平均每年 30 个试验单元计)以 80%的检出概率检验出相应的效果来。区域控制随机试验要达到同样的检出概率(功效)，在同样的增雨效果下，由表 2 的 $N=118$ 和 150 两个参考算例估计，在同样的区域相关性下需试验 4～5 年。

(4) 分层统计对试验功效的影响：将 73 和 88 个试验单元按云形分层统计，计算的功效值分别列于图 3 和表 2。分层的结果，层状云的功效有明显提高，其 $N=39$ 和 51 的功效分别与合并统计 $N=51$ 和 73 的相差不多。积状云的功效则没有改善，与 $N=31$ 时的合并统计结果相近，这主要是层状云降水的自然变率较小、空间相关性较高所致。合并统计($N=73$)的区域雨量相关系数为 0.845 5，按云形分层后，层状云和积状云雨量指标的区域相关系数分别为 0.900 2 和 0.799 4。

(5) 区域回归随机试验的功效：古田试验采用区域回归随机试验设计方案，检验功效除了初期采用过双比分析以外，主要应用多个事件检验法和方差不相等的双样本回归分析法，分别适用于催化样本与对比样本余方差相等和不相等的情况。表 2 和图 3 中分别示出多个事件检验法(t_1)、方差相等的双样本回归分析法(t)以及方差不相等的双样本回归分析(z)的统计功效。可见，这三种统计检验方法的功效彼此相差不大，除个别值外都不超过 5%，且与区域控制随机试验的双比值 DR 的功效相近。这四种统计量的试验设计是相同的，只是统计检验方法不同。从数值试验的结果看，t 与 z 的功效几乎处处相同。这是近乎自然的，因为当催化样本与对比样本容量相同时，方差不相等的双样本回归分析结果与方差相等时的相应结果一致[15]。另外，多个事件检验法 t_1 的功效，在多数情况下似乎比 t 和 z 的功效值略大，$N=31$ 和 73 时尤为明显，看来这不能完全归因于随机抽样或数值试验的误差。原因在于我们做功效分析时，假定了催化效果与自然雨量成比例(θ 是一定常比例因子)，于是催化样本的方差就要比对比样本的方差大。在这种假设条件下，t 以及 z 的检验效率都会降低[15]。这从式(5)和式(6)也可以看出，θ 值使该两统计量分母增大，从而降低其检验功效。双比分析 DR 的功效当 $N=31$ 和 51 时与 t_1 的功效相近，$N=73$ 和 88 时比 t_1 的功效小。

(6) 由表 2 的参考算例可以看出，除 R 以外 $N=150$ 的功效值普遍比 $N=118$ 时的小，这与预想的不同。究其原因，可能是 1979 年所选的 32 个试验单元雨量的区域相关性较差，相关系数只有 0.757 9，而 1976 年 30 个单元的雨量资料区域相关系数又很高，达 0.975 2。所以 $N=118$ 时的功效增加显著，而 $N=150$ 时反而降低了。从这里也可以看出，区域雨量相关性是影响除单区试验外的各种试验功效的重要因素。

参考文献

[1] 福建省气象科学研究所，非催化因子对人工降水效果统计检验的影响，福建"气象科技"，1980，4.

[2] Moran, P. A. P., The power of a cross-over test for the artificial stimulation of rainfall, Australian J. Star., I, 47—52, 1959.

[3] 徐尔灏，论人工降水随机试验的效果检查问题，南京大学学报(气象学)，69—83，2，1962.

[4] Schickedanz, P. T. and F. A. Huff, The design and evaluation of rainfall modification experiment, *J. Appl. Meteor.*, 10, 504—514, 1971.

[5] Scott, E. L., Problems in the design and analysis of weather modification studies. Third Conf. on Prob. and Stat. in Atmos. Sci., 65—72, 1973.

[6] Stinson P. J., Experimental design and statistical evaluation of weather modification experiments, Final Report, PB 186523, 1969.

[7] Gabriel K. R. and Chin-Fei Hsu, Power study of re-randomization test. Third WMO Scient. Conf. on Weather Modification, Clemont-Ferrand, France, July, 21—25, 1980.

[8] Hsu Chin-Fei, Two methods of computing statistical powers with application to weather modification, IMC. Statist, comp, section, American Statistical Association, Washington, D. C. 6.

[9] Kempthorne, O and T. E. Doerfler, The behavior of some significance tests under experimental randomization, Biometrika, 56, 231—248, 1969.

[10] Twomey, S. and I. Robertson, Numerical simulation of cloud seeding experiments in selected Australian areas, *J. Appl. Meteor.*, 12, 473—478, 1973.

[11] Smith E. J. and D. E. Shaw, Some aspects of the assessment of a site for a cloud-seeding experiment, Secend WMO Scient. Conf. on Weather Modification, Boulder, Colorado, 2—9 August, 165—172, 1976.

[12] Gabriel, K. R., Some statistical issues in weather experimentation. Comm. Statist. ——Theor. Meth., A8(10), 975—1015, 1979.

[13] Salvam, A. M., et. al.. Numerical simulation of cloud seeding experiments in Maherashtra state. India, WMA J. Weather Modification, Vol. 11, No. 1, 116—140, 1979.

[14] Ye Jia-dong, Ke-ming Cheng and Guang—ping Zhen, Randomized cloud seeding at Gutian. Fujian. China, WMA J. Weather Modification, Vol. 14, No. 1, 53—60, 1982.

[15] 叶家东，范蓓芬，方差不相等的双样本回归分析，大气科学，5 卷 2 期，214—224，1981.

NUMERICAL ANALYSIS OF STATISTICAL POWER IN RANDOMIZED PRECIPITATION ENHANCEMENT EXPERIMENT

Ye Jiadong Luo Xingpin
(*Department of Meteorology, Nanjing University*)
Zeng Guangping Xiao Feng
(*Institute of Meterorology, Fujian Province*)

Abstract The statistical power of precipitation enhancement experiment in Gutian region is computed numerically with naive method. The results of calculation showed, that the cloud seeding effects of 20%—30% can be detected with power of 80% in 2—3 years duration with cross-over randomized design ($\alpha=0.05$), and the duration of 4—5 years is necessary to detect corresponding increase of rain with target control randomized design. In the randomized regression experiment of area, the deviations of statistical power are less than 5% generally with various evaluation methods: double ratio analysis, multiple event test and two-sample regression analysis. The statistical powers for cloud type stratification are different, which is larger in stratiform category than cumuliform category owing to the strong areal correlation.

5　云对流动力学、中尺度对流系统的结构分析和数值模拟研究

5.1　《积云动力学》概论

一、研究云动力学的意义

云动力学是云和降水物理学的一个重要组成部分。通过长期的云物理研究和人工影响天气试验，人们日益认识到云中降水的形成不能仅仅根据云的微物理过程加以认识。云和降水的微物理过程在相当大的程度上受云中大气运动的制约。云发展和降水形成的宏观动力过程提供了微物理过程进行的背景，决定性地影响了云质点的数密度、初始大小分布及其物理性质，规定了微物理过程进行的速率和持续时间，以及最终降水量的大小，而且只有研究了云动力学过程，才能知道微物理过程中哪些是重要的过程。所以，云动力学在云和降水物理学的研究中具有基本的意义。自然，微物理过程对于宏观动力过程又有重要的反馈作用。云和降水质点的凝结和凝华、蒸发和升华改变了水汽密度，伴随着相变潜热的释放和吸收，又提供了一种重要的热源和热汇，它极大地影响了云（特别是对流云）内外空气的运动，而降水质点的拖曳作用又常常是促使云体消散、崩溃的一个重要因素。近年来，越来越多的人致力于研究云的动力学与微物理学的相互作用。如果说20年以前云动力学主要是指云的宏观动力学条件和机制的研究，那么当今云动力学研究的一个重要特点是将宏观动力过程和微物理过程结合起来，把云的微物理结构和动力学结构联系起来，组成统一的云模式加以研究。云和降水的数值模拟试验已成为云动力学研究的一种重要理论手段。

在云动力学的研究中，积云对流动力学占着突出的地位。首先，很多局地灾害性天气现象，如雷暴、闪电、强风、冰雹、龙卷和暴雨等都与对流云的猛烈发展有关。低纬度地区大约有四分之三的雨是由对流性降水造成的，中纬度地区夏季对流云也是主要的降水源，是夏季人工降水和防雹的主要对象。冬季中纬度一些地区的气旋风暴中中尺度对流雨带有时会产生暴雨，特别是中纬度地区许多强烈的灾害性天气如冰雹、龙卷等常常由有组织的强对流过程引起。因此，积云对流动力学的研究将为监测和预报这些局地灾害性天气提供一定的理论基础。

其次，积云对流在热带大气的热量收支中以及在大尺度热带扰动的形成和生长过程中起着重要作用。这是过去早就知道而现在为人们所广泛强调的。尽管各种物理因子的相互作用使热带扰动问题变得很复杂，热带深厚积雨云云塔中凝结潜热的释放仍是热带扰动动能的主要来源，积雨云在这里起着一种"热塔"或"燃烧室"的作用。卫星云图分析表明，热带海洋云系是高度组织化的，往往组成云簇或云团。台风发展的物理机制，是以水汽凝结潜热

为能源的扰动不稳定自激增长的结果。所谓第二类条件不稳定(CISK),就是中小尺度积雨云系统与天气尺度台风相互作用的过程。此外,积云对流在全球大气环流的能量平衡中起着关键作用。积云对流是大气中最有效的一种能量转换器,是水汽、热量和动量垂直输送的重要机制,它对更大尺度的天气系统或环流发生重要的影响。众所周知,就全球而论,热带海洋地区是能量的源,38°以外的中高纬度是能量的汇。为了平衡中高纬度能量收支的亏损,必须有能量从热带向中高纬度输送。研究表明,在热带地区为数不多的(1 500～5 000 个)巨大积雨云塔,其面积仅约占赤道辐合带面积的千分之一,就承担着维持热带哈特来环流的能量供应。热带积雨云塔是平流层内水汽的一个重要来源,因此也就成了调节全球热量收支的一个重要的辐射“阀门”。这些云塔的数目和集中程度的变化趋势与信风强弱及季风变化有关,它影响平流层的水汽含量并进而可能影响气候。而直到目前为止,在全球热量收支计算中难以肯定的项目仍然是与云的形成、演变和分布以及云的辐射性质有关的热通量。它们还决定海表面的热量平衡,而这对建立有意义的稳态或时变气候模式是必需的。值得强调指出,不同尺度大气过程的相互作用已日益成为全球气候模式、数值天气预报和大气环流研究的注意焦点,而在这一尺度链中,积云对流动力过程及其与大中尺度天气系统的相互作用是至关重要的、曾经一度被忽视的、而研究起来又相当困难的关键环节。

所以,积云对流动力学对于云物理学研究和人工影响天气试验,对于改善对流云降水,特别是局地灾害性天气的监测和短期或临近预报,以及增进热带天气的长期预报能力和完善全球气候模式及大气环流的研究都有重要的意义。这是近年来积云对流动力学得以迅速发展的主要动力。

另外,近 20 年来由于计算机技术迅猛发展,计算速度和容量大大增强;也由于各种现代探测技术,如多普勒雷达、气象卫星和各种类型的飞机探测平台等现代探测技术的发展和应用,使得人们认识和研究云动力学的能力明显提高。这是云动力学能够取得重大进展的基本条件。

云动力学的理论基础仍是把动量、质量、能量和水汽守恒这样一些基本物理原理,应用于牛顿(纳维-斯托克斯)运动方程、质量连续方程、热力学方程和气体状态方程。为了将蒸发、凝结、降水这样一些云物理效应包括进来,方程组中增加了水汽和液水量(或冰水量)连续方程,并将加热项加以修正,以便把潜热释放效应考虑在内,再根据小尺度对流运动的特性对方程组做必要的简化,组成各种类型的云动力学模式。云和降水的物理-数学模式同实际大气的复杂性相比尽管比较简单,但它应能合理地表示出那些控制云体时空尺度发展演变的主要物理过程和动力过程,应能适当地表示出云体(或云系)运动与环境较大尺度天气系统的相关关系及其非线性相互作用,应能恰当地表示出所有小尺度湍流运动和云微物理过程对云体运动以及输送和转换能量和质量的总体效应。后者通常是通过参数化方法来表示它们的统计平均特征而加以考虑的。这是云动力学模式研究的基本任务。

二、云动力学研究的主要进展

现代关于积云对流性质的知识主要源自 20 世纪 40 年代。由于战争迫切需要了解雷暴天气的飞行条件,雷暴的研究引起了广泛的重视。一系列气象雷达和飞机气象观测揭示了对流云结构中许多重要事实,初步得到了关于云中上升气流、下沉气流、液态含水量、降水类型以及雷暴单体发展过程的定量资料。大量观测表明,云内外空气之间存在着强烈的质量、热量、动量和水分的交换，对流运动远非绝热。例如，按照经典气块法的概念，实际上所有热带积云都应能穿入平流层，云顶高度可到气压高度100百帕处或以上。但是在实际大气中，只有少数尺度很大的积云才能真正穿透对流层顶。大多数热带积云顶高处于3～6千米高。按照气块法，这些高度正是气块的浮升力最大的地方。云怎么会不再向上发展呢？肯定有某种类型的拖曳力与浮升力相平衡。而且，云中的液态含水量往往比绝热含水量小，云中温度垂直递减率也往往比湿绝热递减率大。在这些观测事实基础上提出了对流云发展的夹卷理论(Stommel，1947)。1946—1947年美国进行的雷暴研究计划是现代第一个大规模有设计的积云对流综合考察研究计划。根据这个计划的观测结果，概括出气团雷暴发展史的三阶段模式(Byers和Braham，1949)。20世纪50年代初根据质量连续性原理提出了动力夹卷的理论(Houghton，1951)。同时根据室内实验的结果，指出夹卷率与对流单体的半径成反比。在湍流夹卷和动力夹卷假说基础上，提出了对流云发展的气块模式和气柱模式。20世纪60年代根据大量的观测事实，认识到云内外气流的相互作用十分复杂，建立在夹卷理论基础上的积云对流动力学模式对于云内外空气的相互作用处理得过于简单，而且也没有恰当地处理云动力学过程与微物理过程的相互作用。在研究积云对流动力过程时，应该将云体及其环境作为一个流体力学场加以研究。

20世纪60年代对于往往引起局地灾害性天气的强风暴研究有了很大的进展。例如在美国，龙卷和冰雹是美国中西部地区夏季的主要灾害性天气，光是冰雹造成的农作物和其他财产损失每年总计达八亿二千多万美元，引起了广泛的注意。20世纪60年代初期根据多部气象雷达的协同观测，揭示了强风暴的基本回波特征，指出环境风的垂直切变有助于建立和维持稳定持久的强风暴系统，并提出了强风暴的三维结构模式。强风暴中有时云底可观测到冷的上升气流，从而推论出深对流模拟中考虑压力扰动的必要性。

铅直指向脉冲多普勒雷达利用云雾粒子垂直运动的多普勒效应测量空气垂直运动，已为人们探索云中垂直气流场开辟了一种遥感途径。20世纪70年代初双多普勒雷达共轭扫描观测云中气流场的理论方案是云物理研究方法的一个突破。它使人们第一次有可能观测到对流云体内部的气流分布和演变，从而对一些风暴系统中对流的产生、动力结构的演变及其与环境大气的相互作用进行定量的考察。这期间的观测研究特别注重气流场和水汽凝结体分布场，除了用双多普勒雷达观测云中流场以外，还用双波长雷达判定雹区，T-28型装甲飞机穿云直接观测云微物理及动力学参量，并配合投掷式探空仪器测定云中垂直气流等等。为了获得风暴系统的云物理——动力结构及其时空演变的资料，研究强风暴形成、发展的机制和气象效应，这期间研究方法上的一个明显的特点是组织大规模的有设计的综合考察研究计划。例如美国20世纪70年代初为了验证建立在累积带理论基础上的“竞争场”防雹原理而开展的“国家冰雹研究试验”计划(NHRE)，还有“科罗拉多联合冰雹计划”，美国与加拿大合作的“阿尔伯塔冰雹研究计划”，和近几年为研究龙卷、冰雹等灾害性天气而在美国中部开展的“强风暴和中尺度试验”计划(SESAME)，为研究对流风暴的降水效率及其与环境的相互作用，以及动力过程与微物理过程的相互作用而在美国西北部进行的“对流云降水协作试验”(CCOPE)等。采用装备有各种现代云物理仪器的专用考察飞机，有时还配备可以穿越冰雹云的T-28型装甲飞机，可测云内运动场的多普勒雷达，气象卫星特别是地球同步气象卫星和各种遥感仪器。如可测冰雹的双波长雷达，能辨认出非球形固体降水质点的偏振光雷达，监测龙卷等强风暴源的灵敏微压计以及激光、微波探测计、声波探测器等。地面设置加密的地面和高空探测网，有的高空站间距仅 20～30 千米，每隔一个半小时施放一次探空气球。从这些考察研究中得到了许多强风暴的结构资料，促进了强对流问题的研究。许多重要的观测事实和研究成果几乎都是来自这些有设计的综合研究计划。

自 20 世纪 60 年代以来，理论研究有了重大进展，特别是云的数值模拟研究发展很快。这一方面是由于现代化计算机速度和容量不断增大，使人们处理复杂问题的运算能力大大增强，另一方面也由于人们日益认识到，积云对流问题的复杂性是如此之大，以致不求助于计算机演算就很少有希望能正确处理对流系统的理论。无怪乎有人称近 20 年为云物理的"计算机年代"。这不仅对数值模拟研究如此，而且还广泛应用于快速记录并处理大量飞机探测或各种遥感仪器收集到的云物理资料。进展主要表现在：

(1) 在云的动力学模式中引进了参数化的微物理过程，从而开始将云的动力过程与微物理过程结合在一个统一的数值模式中。起初是用连续碰并方程参数化，后来发展到应用随机碰并方程的参数化。它允许采用不同的初始云滴谱，因而可以在模式中对大陆性和海洋性云的微结构差异加以适当处理。在一些模式中且采用各种微物理过程的非参数化方案，较真实地模拟了云中降水形成的物理过程。

(2) 发展了考虑夹卷效应的浮升积云塔的一维积云动力学模式，并据以预报积云动力催化的"可播性"，首次将数值模拟研究应用于人工影响天气试验。

(3) 发展了各种一维时变模式，模式中包括较详细的微物理过程，其中包括冰相过程，并在一定程度上从动力学角度考虑环境空气夹卷等因素对云发展的效应，提出所谓"一维半模式"。

(4) 发展了各种各样的二维轴对称单体积云模式，能较好地模拟出云的某些实测特征。

(5) 在二维面对称模式中考虑水平风速的垂直切变，模拟出积云的生成及生命史。在一些模式中还考虑了山脉加热和过山气流效应。

(6) 通过多普勒雷达及投掷式探空仪观测，发现对流云的垂直结构比现在所认识或假定的要复杂得多，需要更复杂的模式加以模拟。20 世纪 70 年代后期对风速垂直切变环境中的对流云动力学的研究有所进展，指出对上升气流有效的浮力必须适当地与大尺度风垂直切变相匹配。若大气低层存在几千米的暖湿空气楔，而暖湿空气楔与中层对流层的急流相遇，强风暴通常就发生在这种有相当大的位势不稳定且具有较强风垂直切变的环境中。近年来提出了各种考虑风垂直转向切变影响的三维积云对流动力学模式，主要模拟超级单体的发展和结构，有的还着重研究环境风的切变强度及其垂直分布对右行和左行强风暴发展的影响，还讨论了多重分裂风暴的模拟及其动力机制。不过，多数三维模式还处于探索性研究阶段。

三、云动力学研究的关键问题

云的数值模拟研究能在一定程度上模拟出雷暴内部的气流结构和环境大气条件对积云对流的作用。现在有些数值模式以实际资料为初始场，能模拟出与实况大体相似的风暴结构特征和演变过程，使人们对积云对流系统的发生发展机制、结构特征及其与环绕大气的相互作用有较深入的认识。

越来越明显的事实是，中尺度动力学的核心问题是云的动力学问题，特别是积云对流动力学问题。但是，从理论上研究积云对流运动毕竟是一个十分复杂的课题。积云对流动力过程一方面受到各种尺度的大气涡旋运动的制约，另一方面又受到云微物理过程的强烈反馈作用。因此，积云对流过程中既包括了尺度范围很广的各种类型的云和降水质点之间的相互作用，也包括了尺度范围甚至更广的各种尺度大气涡旋之间的相互作用。在降水性对流云中，云质点大小从直径为 10^{-2} 微米的气溶胶粒子到 5×10^{3} 微米的降水质点，对于含雹积雨云甚至可大至 10^{5} 微米，大小相差 7 个量级。另外，与积云生消有关的大气涡旋尺度可以从 10^{-3} 米的湍流耗散尺度，通过对流尺度（$10^{3}\sim10^{4}$ 米）直至 $10^{5}\sim10^{6}$ 米的天气尺度运动。如果积云是强的温带或热带气旋扰动的组成部分，则数值达 9 个量级不同尺度的大气涡旋之间的强烈非线性相互作用，可以在很短的时间尺度中发生。再加上云的动力过程和微物理过程之间极其复杂的热力的和动力的相互作用，使问题变得更为复杂。

所以，积云对流的模拟问题包括了范围极其广泛的湿对流系统中的物理过程和动力过程领域。研究对流风暴降水的主要困难在于许许多多气象现象以多种方式的相互作用。降水的形成显然受云体的尺度和持续时间的制约，从而也取决于边界层性质、大气层结结构和风场等因子，而且还常常与早先由阵雨带来的下沉气流有关。由于积云对流运动一般是非静力、非常定的湍流性很强的运动，需要完全的三维运动方程才能描述它们。目前多数计算机的速度和容量，对于模拟如此复杂的运动过程还不能胜任。因此，积云对流的模拟研究多数都涉及参数化问题，不仅包括基本的云微物理过程的参数化，而且还包括模式所不能分辨的湍流涡旋能量和涡旋通量的参数化。在微物理过程方面，至今仍有许多关键性问题有待解决。需知人们关于云微物理学的许多知识，是建立在实验室的实验基础上的。因此，当把这种实验室可控条件下得到的结果外推到极其复杂的自然条件下的群滴演变过程时，会产生很大的不确定性。例如，在深厚过冷积云中，冰质点的产生和云体冰晶化的不确定性强烈地限制了人们定量地模拟过冷积云的物理过程和动力过程的能力。而云中冰晶起源问题之所以模糊不清，不只是因为成核作用的模式和可能的冰晶繁生机制等基本方面有许多不确定性，还在于不知道对流云里任一给定部分是由来自何处的气块混合组成的。由于积云对流是高度湍流性的，目前人们对于云与周围环境如何混合的物理过程以及云中的湍流通量实际上尚未真正了解。因此，在动力学中如何恰当地表述夹卷效应和湍流应力就成为积云对流模拟问题中一个症结所在。也许这是积云模拟研究中最为困难的问题之一。

但是，基本的困难还是缺乏充分的观测资料。因而，模式中的参数就没有坚实的事实依据，模拟的结果也难以验征，难于用真实的云来检验它，特别是对流云内外的速度场，对于含冰相粒子的深厚对流云更是如此。即使像温度廓线和液态含水量这些比较简单的属性，也缺乏关于它们在云的生命史中如何演变的足够的观测资料来与模式做有意义的比较。甚至为了与模式对比，实测参数应如何取平均，云的“环境”应如何确切描途也还是未解决的问题。总的来说，观测研究还明显落后于模拟试验。因此，尽管近二十年来积云考察的观测设备和技术已有很大进展，人们关于积云对流的动力结构和属性的知识已远非二十年前可比，但它至今仍是积云动力学研究进一步发展的关键。

因此，高质量的中尺度外场综合研究计划是重要的，这不仅是对模式的验证，而且对于了解基本的云物理-动力过程、降水机制以及强风暴的监测和预报都是重要的。但是，外场考察应有模式研究作指导，仅仅是物理观测而没有指导性的模式就难以避免一定的盲目性，不能充分发挥设备潜力。另外，真实云的模拟研究中，过分大的模式由于是高度参数化的，常常不能同时做到精确地考虑云动力学和微物理学，其价值可能变得有限，它们会偏离真实性，并且往往难于估价。所以，模拟研究需要考虑模式的“恰当性”问题，模式应力求为同时进行的好的观测及有用的资料系列所证实。

至于积云对流与大中尺度天气系统之间的相互作用的研究，目前尚处于初期阶段，这里有许多更为复杂的因素和尚未认识的知识领域。由于积云对流对大尺度天气系统演变的重要性，目前短期和长期数值天气预报以及大气环流研究的进一步完善，主要取决于正确考虑大尺度和中小尺度气象因子之间的相互作用。因此，它已成为大气动力学和数值预报研究中的一个热门的专题领域。不过，这已超出本书的范围。在大气动力学研究中，积云对流对天气尺度系统的作用一般是用积云对流参数化的方法表达的，即用描述天气尺度系统的量来表述积云对流作用的总效应。

5.2 《风暴和云动力学》述评

叶家东　范蓓芬

（南京大学大气科学系）

美国科罗拉多州立大学（CSU）的 W. R. Cotton 教授和国家大气研究中心（NCAR）的 R. A. Anthes 博士所著的《Storm and Cloud Dynamics》（1989）中译本《风暴和云动力学》已由气象出版社出版，全书 949 页共 78 万字。这本专著以中尺度气象学的核心领域——中小尺度降水系统为中心，系统地论述了云和降水性中小尺度系统的动力学和物理学，较深刻地阐释了中小尺度气象数值模式的理论基础，概括了 1980 年代这一专题领域最新研究成果，具有较高的科研参考价值，是研究生和大学本科高年级学生学习云动力学和中小尺度气象学的一本好的参考书。书中附有大量反映最新研究进展的参考文献，累计达 1 533 篇。下面就该书的基本内容和特点做一简要的概括和评述。

一、基本概念和参数化

前 6 章讨论云和云系的基本理论和各种次网格物理效应的参数化。第 2 章根据尺度分析导出一组用张量形式表达的适合于云模式或中尺度模式的湿对流系统基本方程组。其中热力学方程引进了一个在液水-冰水绝热转化过程中守恒的热力学变量 θ_{il}，称为冰-液水位温，相当于干系统中的位温 θ。在水分连续方程中引进包括水汽、云水、雨水、冰晶、霰和冰晶聚合体的混合比，且各态自有相应的连续方程，其中包含各种形态水物质之间的转换率，通过微物理过程的参数化方案来计算（第 4 章）。为了闭合湿对流系统方程组，需有计算压力的预报方程。在积云和雷暴系统中，垂直速度太大，需采用非静力平衡法计算压力场，书中介绍了三维云模式中采用的时间分离方案，即将运动方程分解为含声波和不含声波的两部分，用小时步积分含声波部分解出声波，然后利用全弹性连续方程和状态方程建立压力倾向方程求解；受平流时间尺度控制的部分和各标量方程按长时步积分。在非静力平衡模式中采用 $\sigma-Z$ 坐标系，以方便描绘地形。第 3 章全面论述了各种平均理论，并就雷诺平均过程导出雷诺应力和湍流输送项。作者在评述一阶闭合理论的基础上指出，涡动交换理论的基本弱点是物理属性总是顺梯度扩散的，不能解释行星边界层和深对流研究中发现的逆梯度输送的现象，所以对高阶闭合理论尤其是二阶闭合理论做了较为详细的阐释。该章还讨论了中尺度模式网格体的“部分凝结”及其效应问题。

第 4 章讨论云微物理过程及其参数化，包括暖雨过程和冷雨过程的参数化。首先在引进暖云胶性稳定度概念基础上评述各种滴谱拓宽机制、碰并效率、大水滴的自发破碎和碰撞破碎机制。因此用显式模式模拟上述过程引起的全滴谱演变过程十分困难，在较复杂的云和中尺度模式中常采用较简单的参数化方法，但目前的暖云自动转化参数化方案的模拟能力很小，不同方案计算的转化率可相差几个数量级。在雨滴碰并云滴增长过程的参数化方案中对滴谱函数的人为假定使其模拟能力受到限制。云模式或冰相微物理过程的参数化要更为复杂，由于冰质点的形状、密度、运动特性等特性各异，现有的理论很不完善，相应的参数化公式的不确定性更大。因此，在模式中采用何种详细程度的参数化微物理过程需要斟酌。实际上目前云模式中包含的复杂微物理过程，大多数还不能从云物理观测中得到证实。该章还简单讨论了云微物理过程对云动力学的影响。

第 5 章介绍辐射效应及其参数化。适合云或中尺度模式的辐射参数化方案较少，主要是缺乏能明显影响云中辐射传输的云物理特征的有代表意义的资料。目前主要采用“双流近似”方案。对于短波辐射，书中介绍了 Stephens 等发展的参数化方案。对于长波辐射，通常将光学厚度大的云视为黑体，而把光厚度不大的卷云及层积云等看成是“灰体”，定义一有效放射率，根据云中液态水路径的垂直分布确定长波辐射通量垂直廓线。

第 6 章较全面地论述了积云对流参数化。它涉及两个方面，一个是大尺度强迫对积云对流的控制和调节作用，这与降雨率的确定有关；另一个是积云对流对大尺度环境的反馈效应，包括相变潜热效应及热量、水分和动量的垂直输送效应。该章在详细阐述积云参数化的物理依据、观测证据和联结大尺度参数与积云效应之间的数学框架的基础上，对现有的积云参数化方案做了全面深入的评述，包括湿对流调整方案、郭晓岚型方案、Arakawa-Schubert 方案，以及适用于中尺度模式的 Kreitzberg-Perkey 方案和 Fritsch-Chappell 方案。并指出到目前为止，尚没有一种积云参数化方案能反映不同时空尺度上都适用的积云对流与环境之间相互作用的信息。近年来提出用显式模拟云和降水效应的方法避免现有积云参数化方案的随意性。这类能显式分辨对流尺度与中尺度相互作用细节的细分辨模式，有可能成为研究不同尺度系统之间相互作用的最有前途的途径。

二、云动力学

后 6 章占全书篇幅的 2/3，论述各类云和云系的物理学和动力学，结合理论模拟和观测分析中所获取的最新认识论述这一领域的研究现状。第 7 章讨论动力性最小的雾和层积云，把雾作为边界层层状云的一种来看待是有其特色的。该章的重点是海洋性层积云和含云边界层模式，包括整层平均模式、实体型模式、高阶闭合模式、大涡旋模式，以及夹卷不稳定机制和风切变效应。海洋层积云对局地气候起着重要的影响。

第 8 章讨论积云。在分析晴天积云和单体积云结构的基础上，较全面地论述了积云的组织机制、夹卷效应和下沉气流的发动及其在积云组织中的作用、重力波的作用、降水和冰相过程的作用、积云合并机制及其与大尺度辐合的相互作用。

第 9 章积雨云和强对流风暴，是全书篇幅最大的一章。该章较深入地评述了各种对流风暴的结构特征、概念模式、上升气流和湍流、下沉气流和低层外流及阵风锋、风暴的移动和传播机制，包括对流翻转模式和波动- CISK 模式两种自传播理论、风暴的分离、旋转雷暴中气旋和龙卷、雹暴、对流性暴雨和风暴起电与风暴动力学的相互关系等。

第10章中尺度对流系统(MCS)，阐述热带飑线系统、热带云团、中纬度飑线系统和中尺度对流复合体(MCC)四类MCS的结构特征、概念模式和形成机理，指出MCS的发生及其稳定维持涉及一系列复杂过程，关键的问题是中尺度环流的建立和维持，这与深对流的砧状云相互合并、低层中高压形成所创造的自身斜压性以及辐射不稳定化等过程有关。该章还讨论了热带气旋的中尺度结构及其与水汽、热力不稳定度及边界层辐合等环流特征的相互作用。

第11章为温带气旋的中尺度结构和中、高云。在气旋形成的大尺度扰动不稳定基础上概括出三类输送带的概念模式，并讨论了温带气旋不同部位的6类中尺度雨带的结构和成因。该章对中、高云的结构及其数值模拟也做了概括，并从辐射效应角度论述了大尺度模式的中、高云的参数化。

第12章主要讨论受地形强迫产生或加强的冬季降水性云和云系。除了过山气流的线性和大振幅理论以外，对地形性降水中的播撒-受播过程、地形对温带气旋及其降水的影响，以及地形云中过冷液态水分布和降水效率都做了较为详细的评述，后者对飞机结冰、云的辐射特性和人工降水机会等方面都有重要的意义。

本书最后指出，1990年代美国正发展新的观测系统，包括新型NEXRAD多普勒雷达系统，风、温、湿垂直廓线仪系统，新的卫星探测系统，以及自动遥测地面观测网，可分辨出中-β尺度系统的特征，并可以构划出水分及其不同相态之间的四维收支图像。这将使强天气研究和短时预报取得实质性进展，并将促进全球大气化学、大气环流和气候变化等方面的研究。

《风暴和云动力学》一书将云物理学、云动力学与中尺度气象学有机地结合起来，是该书最大的特点。全书始终强调观测研究与模拟研究相结合的重要性。书中对各专题领域的论述兼顾各家观点并做有针对性的评论，并给读者留有独立思考的余地，做到针对性与客观性相结合。当然，该书取材主要源自美、欧、澳地区，对亚洲和中国的云和风暴研究状况几乎没有论及。在中译本前言中作者Cotton表示："我希望本书有助于了解这些风暴在我们各自国家之间的相似性和相异性。我盼望更多地了解这些风暴在中国的特征。"我们也希望这书的中译本将有助于国内云物理和中尺度气象领域的读者了解风暴和云动力学的基本理论和国外近年来的研究进展。

5.3　雷暴中尺度对流动力学研究进展

叶家东

（南京大学大气科学系）

一、概述

对流风暴是重要的降水性云系。它是热带主要的降水系统，也是中纬度夏季重要的降水系统，并且是许多局地灾害性天气，如局地突发性暴雨、冰雹、龙卷、下击暴流、雷电和局地雪暴等强烈天气的制造者。观测表明，深厚对流性降水的强度、持续时间和区域范围基本上是受大尺度环境流控制的。它们在促使对流层低层变暖变湿并造成有利于发展深厚对流的静力不稳定条件中起了重要的作用。行星尺度和天气尺度环流还基本确定了大气流场，特别是水平风的垂直切变特征。众所周知，环境风的垂直切变是产生强对流风暴的重要条件，特别是低层风的垂直切变，对于长生命期风暴的维持十分重要，因为它们大大增强了降水蒸发对雷暴出流空气的冷却效应以触发新的对流单体再生。但是，观测证据也表明，并不能单单靠行星尺度和天气尺度环境所建立的环境流场来确定对流的强度和区域范围。例如，宽广的热带海洋地区出现条件性不稳定时，它能维持深厚对流，然而深厚对流系统却是分散地形成于较小的区域。所以，毕竟还是尺度较小的中尺度过程控制深厚降水性对流系统的发生时间、空间分布及强度。中纬度气旋风暴中的中尺度雨带结构和副热带东亚梅雨锋上的中尺度暴雨云团结构特征也表明，天气尺度的锋面和气旋中的降水结构基本上也是由中尺度过程所决定的。而且，深厚积云对流及其降水过程对大尺度天气系统有重要的反馈作用。热带大面积的对流性降水和潜热加热是哈特来环流及季风环流维持和加强的重要原因。对流性降水的潜热加热效应，在热带扰动通过 CISK 型自激不稳定增长过程转化为热带气旋风暴以及维持这种风暴的过程中起了决定性的作用。尽管中纬度气旋风暴主要不是从对流性不稳定，而是从斜压不稳定中得到大部分动能，要确定对流性降水对气旋发展的反作用较为困难。但是，从细网格模式所得到的证据表明，潜热加热能够显著地改变某些温带气旋的发展。在预报具有明显对流成分的中纬度气旋的降水量时，由于未能将气旋风暴中的中尺度涡旋和中尺度雨带结构加以区分，其误差是人尽皆知的。此外，春夏季节中纬度中尺度对流系统十分活跃，与潜热加热、中尺度补偿性下沉运动以及辐射加热有关的各种大气物理过程，能够显著地影响气团性质并从而影响中纬度气旋的状态。

由此可见，对流风暴的研究涉及不同尺度的天气系统之间的相互作用，包括边界层的影响，环境大气流场和热力学特征对风暴形成和结构的效应，云微物理过程的作用，风暴尺度流场特征在降水发展、类型和强度中的作用，环境干空气夹卷的性质和效应，风暴的再生和传播机制，中尺度对流系统的形成和发展，以及风暴系统通过热量、水汽、动量等物理量的输送和改变辐射收支等对环境大气的反馈效应。所有这些，都是当前中小尺度对流动力学研究的重要课题。

近 20 年来，由于计算机技术迅猛发展，计算速度和容量大为增强，已出现一批每秒亿次乃至 10 亿次运算速度和大容量的超级计算机系统，为云模式和中尺度模式的发展提供了良好的计算条件。各类三维云模式和中尺度模式的研究在这期间取得了长足的进展。中小尺度数值模式研究的发展进程表明，云微物理学、辐射和大气边界层等大气物理过程的参数化研究是中小尺度数值模式研究中的核心研究项目之一。

另外，近年来各种现代探测技术，诸如多普勒雷达、气象卫星、不同类型的飞机观测平台，以及各种地基遥感探测技术，包括双通道微波辐射仪、偏振雷达、偏振激光雷达等新技术的发展和应用，使人们认识和研究对流风暴的能力大为增强。鉴于降水性对流风暴常常组织成中尺度系统，为了实时研究这类系统，需要针对具体的研究目标组织专门的中小尺度综合考察研究计划，进行外场试验。从 20 世纪 70 年代的 GATE、NHRE、MONEX、SESAME，到 80 年代的 CCOPE、AIMCS、PRE - STORM、EMEX、TAMEX，以及正在实施的美国 STORM 计划等一系列中小尺度外场试验，积累了大量中小尺度对流风暴系统的结构特征和发展演变规律及其与大尺环境场相互作用的实时观测资料，大大促进了中尺度降水系统和强天气的研究。现代中小尺度气象学中许多重要的观测事实和研究成果几乎都是来自这些有设计的、目的明确的、设备精良的、各方面协调的大型综合外场研究计划，这需要雄厚的财政和技术装备方面的支持。我国华南前讯期暴雨实验和华东中尺度天气试验对我国暴雨和强对流天气的中尺度研究起了重要的推动作用，“七五”计划期间筹建的几个中尺度试验基地为在我国开展进一步的中尺度外场试验奠定了良好的基础。技术的进展还表现在各种遥感探测和飞机观测资料的实时计算机处理和分析技术，以及不同类型资料的计算机同化技术也取得了重大突破。这对资料的实时应用、结构分析和概念模式的提炼，以及云模式和降水预报模式的初始化等无疑是一重大的贡献。下面对近年来在中小尺度对流风暴和风暴系统研究中取得的一些进展做一简要的归纳。

二、对流风暴动力学和微物理学

1. 对流的起源和边界层效应

在条件性不稳定层结大气中积云对流的发动需要一定的启动能源触发对流环流的产生。提供这种启动能源的强迫机制是多种多样的，包括锋面抬升、地形扰动、局地加热和边界层中的热力波动、边界层辐合、夜间边界层大振幅重力波，前期雷暴下沉气流引起的阵风锋，地面有效温度分布不均匀等各种局地扰动因素，以及低空急流、干线等其他次天气尺度扰动。近年来有关积云对流起源和边界层效应的研究取得了一些引人注目的进展。

(1) 锋面结构特征及其触发对流的效应

锋面抬升是积云对流和强风暴的重要触发机制。关于锋面的结构特征，认识上有一个发展过程。根据地面观测记录的分析认为地面锋是一个不连续面。后来高空观测发展，认为高空锋是一条有限宽度的过渡区——锋区，其水平尺度为 100 km 左右，垂直尺度为 1 km量级，锋面是一倾斜过渡区。这种倾斜度大约为 1/100 的锋面系统，如何能在短时期内将气块抬升到自由对流高度，是不无疑问的。近年来的边界层观测发现，有时冷锋系统的水平尺度只有 1 km 量级，锋面上强的垂直运动可以高达 4～20 m/s，边界层中锋面且近乎垂直。锋面结构特征又回到了不连续面的概念。例如 1983 年 9 月 19～20 日一次冷锋的美国科罗拉多州中部过境时，地面中尺度网和边界层铁塔观测表明，锋面的温度梯度集中在很窄的区域内，这在大尺度观测网上是分辨不出来的。锋面过境时，Boulder 大气观象台(BAO)的300 m 铁塔上 50 m 高度处，在锋面经过的极短时间内，温度下降 8 ℃，1 min时间内出现 7 m/s 的风速涌进和风向气旋性转变。在 250 m 高度上，风速涌进和温度下降同样迅速，只是降温方式略有不同。若按锋面移速 15 m/s 推算，锋面在 1 min 时间内过境表明其宽度大约只有 900 m。锋面过境时的位温分析表明，锋的前缘是一个抬高的锋面头，其后方有一个湍流尾，表现出地面冷锋的密度流特性。用 BAO 铁塔上的声风速仪推算的垂直速度，发现 300 m 高度锋面头上的上升速度达 4.5 m/s。这样强的垂直运动常常足以把气块抬升到自由对流高度，并引起高耸的积雨云对流，产生一条相应的窄冷锋雨带，宽度仅约 5 km。问题是，这样强的垂直运动如何能在近乎中性稳定度的环境中存在呢？初步的研究表明，这种深厚的强垂直运动和窄冷锋降水带之所以能发生，是由于与暖区中垂直切变相关联的水平涡度和与锋的引导线上的浮力梯度相关联的水平涡度之间的相互作用引起的。不过，锋面在边界层的结构特征及其成因的研究在理论上还是一个有待进一步深入的研究课题。

实际上，多年来外场观测业已发现许多种类型的、不同于挪威极锋气旋模式的锋面结构。为了深入揭示各种锋面系统的三维结构特征及其演变规律，改善锋面系统中的中尺度降水特征和强天气事件的预报，美国中部地区将于1995年开展一个重要的多尺度的STORM 1冬/春季外场试验研究计划。对各种锋面系统和其中的降水结构特征及其演变规律和控制这些特征及规律的动力学机制和物理学机制进行天气尺度、中尺度和小尺度的综合考察研究，作为整个20世纪90年代的国家STORM计划的一个组成部分。这个计划的预演性外场试验计划STORM－FEST（风暴尺度业务和研究气象学-锋试验系统检验）已于1992年2月1日～3月15日于美国中部地区实施。预计这一系列研究计划的开展，将会大大推动锋面系统及其中的对流风暴的研究取得实质性进展。

（2）边界层涡动与重力波的相互作用及积云对流的起源

关于边界层涡动与其上自由大气的重力波及云之间的相互作用的研究近年来受到重视。Clark等从数值模拟、线性模式和观测分析等各个角度来研究边界层扰动与自由大气重力波的相互作用。线性模式研究表明，在线性强迫正模与重力波活动之间存在很好的相关，并指出这类波动/涡动结构是与热力强迫正模响应相接近的，它具有负的相速度。三维模拟研究指出，从晴天对流问题中可以得到强的重力波响应，这种效应在边界层与自由大气界面上有强的风切变时特别明显，而边界层中水平热力梯度可以是这种风切变的一种可能的源。

热力强迫重力波促使深对流发动的研究表明，深厚积雨云可以通过加热的湿边界与其上的自由大气相互作用而发展。边界层涡旋，它通常作为积云的根，与自由大气中重力波之间的相速度关系在发动深对流云中起了重要的作用。在晴天积云阶段，云顶以上两种波动的相对速度大约是4 m/s，云块的典型水平间距约为12 km，最有利的动力对流增长间隔为50 min，这时重力波的上升气流恰好与积云耦合。这种撞击频率效应在数值模拟中业已发现。数值模拟研究还指出，在低层主要有风速垂直切变的环境中有利于发展分散的胞状型对流，而具有风向垂直切变的环境下，热力强迫重力波与边界层云轴之间耦合产生的上升气流流型具有线性特征，有利于产生带状多单体对流云系。

其他关于地形扰动、干线、下击暴流、大尺度辐合、边界层辐合、夜间边界层大振幅重力波以及下垫面有效温度变化等扰动机制在对流发动中的作用都继续有人在开展研究，关于夜间边界层大振幅重力波的产生机制，鉴于这类重力波常常来自早先存在飑线的地区，推论它是由于飑线中蒸发冷却空气的下沉气流撞击夜间边界层顶引起的，所以是雷暴下沉气流或下击暴流引起的一种间接效应。

2. 夹卷效应

夹卷效应在积云对流动力学中是一个老问题，但出现了一些新论题。

20世纪70年代关于侧向夹卷理论和建立在侧向夹卷概念基础上的一维定常积云模式有过一场争论。Warner等人指责这类模式缺乏内部协调机能，不能同时正确地预报云的含水量和云的发展高度，并为穿透性下沉夹卷理论的发展提供了一些观测证据。这场争论也促使了云模式中降水形成的微物理过程的研究。不过争论各方都认为，夹卷能抑制积云对流的发展。近年来Clark等人用细网格云模式（分辨率≤50 m）研究表明，夹卷确能显著地稀释上升气流的动量和水分含量并抑制对流云的发展。对一个在环境大气高度不稳定的条件下发展起来的 CCOPE 个例(1981 年 7 月 19 日)进行的二维数值模拟结果表明，夹卷使云顶从11 km 高降至 6 km。

鉴于积云对流参数化，特别是中尺度模式的积云对流参数化的成功与否，在相当大程度上取决于对积云中上升气流和下沉气流和夹卷-消卷机制的认识程度，所以夹卷过程和云内外混合效应的研究至今仍是一个活跃的论题。1979 年 Paluch 采用总水混合比和假相当位温进行示踪分析，表明中纬度积云中云气主要源自云下空气与云顶环境空气的混合。但是，近年来有人论证说，当飞机测量仪器的传感器被沾湿时，由上述采用总水分混合比和假相当位温分析所导出的结论就会产生误差。而且，这种方案当有雨滴形成并落出气块时便不适用，所以不能用来检测云发展后期的混合过程。Blyth 等(1988)的分析则认为云顶夹卷与侧向夹卷都是存在的。他们强调，环境空气进入云顶，环绕的上升热力对流泡的核心下沉，最后进入热力泡尾流部分的四周。这种模型将“云顶夹卷”与“侧向夹卷”机制恰当地结合起来。1991 年在佛罗里达地区开展了一个“对流和降水/起电试验”(CaPE)研究计划，主要是研究对流云中流场、水含量场、热力学场以及电场发展之间的相互作用，其中一项重要任务是监测存在于高不稳定度、低切变环境中的积云和积雨云复合体中的夹卷过程及其效应。

有关夹卷的另一个论题是夹卷对云微物理结构和云滴生长过程的效应问题。Paluch 等人致力于研究对流云中夹卷和混合过程与云滴谱发展的关系。现在人们普遍认为夹卷和混合过程会影响云滴，所以对云滴谱的研究反过来开辟了一条认识夹卷和混合的途径。小尺度观测表明，在积云塔发展的旺盛阶段，夹卷混合过程不是均匀的，它包括体积夹卷(Bulk entrainment)和小尺度混合两种过程。前者产生一种粗混合体，其中包含一些数米至数十米的晴空，后一过程才开始改变云的微物理属性。两种过程是同时进行的，但当云发展的活跃阶段许多夹卷空气仍保持微物理结构未混合的状态。因此，云模式中的夹卷参数化方案需重新修正，因为原先这种参数化方案意味着在整个模式网格空间都有瞬时小尺度混合。这种假定倾向于高估混合时的蒸发速率及相应的浮力变化，可以导致穿透性下沉夹卷的超量发展。它也将高估云和夹卷空气中的化学反应速率和清洗效率，并且会影响播云物质在云内的有效散布。

曾经有一种观点认为，夹卷混合能增强暖雨形成速率。据信云滴在稀释的上升云体中能凝结增长得比绝热云体中更大的尺度，理由是稀释云体中因峰值过饱和较低，活化的云核较少而导致云滴浓度较低，从而存活的云滴能分享继续上升释放出来的过饱和水汽而长得较大，并通过进一步的碰并较快地形成暖雨。但是，观测并没有表明稀释云体中有较大的云滴，云滴众数直径近乎定常。这意味着，至少在大陆性云中稀释很少能加速云滴的增长。观测指出，在小尺度混合作用能够大大稀释局地水滴浓度以前，晴空气块体积夹卷先已将混合云区撕裂成小的碎片区域，它们各自保持原先的属性大体不变，在这种状态下，云继续上升相当一段距离后，再通过进一步的小尺度混合使属性均匀化。混合过程的这种二阶段特性能解释为什么夹卷混合并不会促进大陆性云中云滴的增长。

至于切变环境中，强风暴系统的中层环境干空气的后向入流夹卷的成因及其在中尺度对流系统组织化中的效应，在中尺度对流系统的研究中也是一个热门的论题。

3. 对流风暴中冰雹的起源和增长研究

(1) 对"馈云单体"的作用重新引起重视。在雹暴发展过程中,有证据表明主体雹暴外围的侧向云塔有可能是雹胚的重要源区,而从主体雹暴中形成的原始霰粒在其下降过程中由于融化和溅散所产生的次生水滴落入上述馈云单体,能产生再循环增长,可以成为重要的雹胚源。如果这种认识是对的,那么防雹作业时引晶播撒应在主体雹暴外围迎风侧的"馈云单体"顶部进行。

(2) 对雹暴和雹区的观测手段有了改善,一些新的探测技术设备诸如多参数雷达、装甲飞机、机载多普勒雷达、双通道微波辐射仪等已研制并投入外场考察研究。

(3) 进行了一系列雹暴的数值模拟研究,描绘雹胚轨迹、冰雹的增长及其演变,以及雹暴形成的中尺度环境流场特征。

(4) 通过改进的数值模式和测量技术,以及实时资料的处理和显示技术,有可能大大改善对冰雹的"即时预报"。

(5) 采用示踪剂伴随播云物质的试验来评价雹胚形成和冰雹增长的理论,为人工防雹提供可信的原理假设基础。

(6) 雹暴中云起电的测量装置有所发展,还发展了预报这种起电效应的模式。

(7) 认识到对流风暴对于大气化学过程的重要性。

(8) 强烈的冰雹常常发生在中尺度对流体内,从中也可以看出在风暴尺度、中尺度与天气尺度系统之间的相互作用的重要性,并认识到中尺度对流系统(MCS)的研究是与冰雹研究密切相关的。

(9) 1989 年夏在美国北达科他州中部开展代号为 Hailswath Ⅱ 的外场冰雹研究计划,进行雹暴的三维流场、冰晶起源、雹胚源以及小尺度馈云单体回波区的结构、降水和冰雹的形成和发展、雹暴和馈云单体起电、砧状云区研究、大气化学以及防雹播云试验研究。

4. 冰晶聚合体在降水过程中的作用

冰晶聚合体俗称雪花,是固态降水元的一种基本形态,它是由冰、雪晶之间碰撞聚并而成的。冰晶聚合体的形状、尺度、体积视密度和数浓度特性及其动力学性状是云和降水系统微物理研究的重要内容。Jiusto(1971)认为冰晶聚合体是美国东北部地区降雪的最普遍的类型。Hobbs 等(1971)指出在美国喀斯喀特山脉的顶峰地区冬季有一半以上的固态降水是以冰晶聚合体形式落至地面的。我国冬季许多地区降雪多以雪花形态降落地面。近年来一系列观测证据表明,在各种降水性云系中,冰晶聚合体是降水总量中很重要的一种成分。这包括美国北科罗拉多河流域的冬季地形云系(Rauber,1987),温带气旋的中尺度雨带(Matejka 等,1980),加里福尼亚塞拉山区的冬季层状云降水(Stewart 等,1984)。即使在夏季对流风暴和热带降水性云系中,冰晶聚合体也是云内过冷区中重要的降水形成,如科罗拉多雹暴(Heymsfield 和 Musil,1982)、中国南海地区冬季季风云团(Houze 和 Churchill,1984)、孟加拉湾地区夏季季风云团(Houze 和 Churechill,1987),以及成熟的飓风云系(Jorgensen,1984)。我们的分析研究(Yeh 等,1986,1987)也表明,中纬度中尺度对流复合体(MCC)的层状区内,冰晶聚合体是过冷层中主要的降水元形态,冰晶聚并过程是层状区降水的主要增长机制。冰晶之间的聚并过程开始于较高较冷的气层,当聚合体下降接近融化层时,聚并效率大为提高,从而在 0 ℃至−5 ℃层形成一冰晶聚合带,这是 MCC 层状区微结构的重要特征，它会影响层状区雷达回波 0 ℃层亮带的位置。冰晶聚并效率的研究还不完善，不确定的因素很多，诸如冰晶形状，下降速度及其飘移性状，冰表面的干湿程度，环境空气温度，过冷云滴的含量，以及聚合体本身的尺度、形状、视密度和落速等都会影响聚并效率。所以，这是一个比水滴碰并效率更为复杂的问题。早期的一些实验结果相当分散。我们根据 MCC 层状区降水质点微结构的垂直分布诊断的聚并效率是温度（或高度）的函数，在 0 ℃层附近可以超过 1。为了确定聚合体在云砧和上升气流中的源，需要确定不同环境条件下的聚并效率，Lew 等计划在 UCLA 风洞中进行这类实验。关于聚合体在过冷雾中的凇附效率已在该风洞中做过实验。初步结果表明，凇附效率比预期的似乎要低一些。不过，这与实验时的环境条件关系很密切。

关于冰晶聚合体在冰雹形成中的作用，NHRE 的研究发现，冰晶聚合体在科罗拉多风暴的冰雹形成中可能起着重要作用。在 1981 年 8 月 1 日的 CCOPE 风暴的砧状区中也发现大量冰晶聚合体。Lew 和 Heymsfield 用一质点轨迹模式研究霰和雹块增长中聚合体的作用。结果表明，由于此 CCOPE 风暴太强，过冷区中形成的大多数冰晶聚合体在它们能凇附大量过冷云滴以前就被上升气流带至云砧区而不能充当雹胚，故其重要性比 NHRE 风暴中小得多。

湖效应风暴的观测和数值模拟试验都表明，冰晶聚并过程是重要的降水质点增长过程。故此人们就有理由怀疑美国 NOAA 等机构关于湖效应风的播云原理假设是否可行。这一假设是：过量播云产生大量冰晶以减缓凇附过程，雪晶能更深入内陆方才降落地面，从而达到降雪再分布以抑制沿岸局地暴雪的目的。如果播云结果虽能减缓凇附过程，但却加速了聚并过程，就不能满足原来播云的基本目的，这种例子进一步阐明，在对某种云假设付诸实施前，需要了解降水形成的自然控制机制。

5. 降水的动力学效应——微下击暴流

观测证据表明，降水的重力拖曳和融化冷却效应会促成下击暴流的产生。1986 年夏，在美国东南部亚拉马州的亨茨维尔地区进行了微下击暴流和强雷暴（MIST）外场试验，结合多参数雷达观测（双极化、双波长和单程衰减）与三重多普勒雷达观测，研究湿环境下的孤立强风暴中的微物理属性和下沉气流的强迫机制。对 1986 年 7 月 20 日——孤立强雷暴进行个例研究结果，这个对流风暴在地面产生暴雨、豌豆大小的冰雹和微下击暴流（径向风速极值达 30 m/s），结果表明，初始雷达回波高度大大低于融化层，双极化值达 3.5 dB（表明雨滴直径达 3～4 mm），这意味着早期降水增长是通过暖雨碰并机制的。随着风暴增强，水滴冻结，双极化值降至接近 0 dB，碰冻（凇附）过程变成主要的增长机制。这时雷达反射率和单程衰减迅速增加，从而证实降水正在迅速发展。微下击暴流恰好发生在降水核心降达地面之时。应用多参数雷达信息，降雨率和雨水含水量估计分别达到 150 mm/h 和 6 g/m^3。这样高的降水含水量，其重力负荷可发动下沉气流，而当降水核心下降通过融化层时，降水的融化冷却效应会进一步增强下沉气流达到微下击暴流的强度。二维面对称数值模拟结果与雷达观测很一致。在环境大气潮湿的情况下，融化降水质点由于其温度较低，在云下的蒸发冷却效应不很显著。

1987 年夏在科罗拉多中部进行对流起源和下击暴流试验（CINDE），进一步研究该地区下击暴流形成的先期条件和强迫机制以及对流云发展的边界层特征。Knight 和 Lew 利用高分辨率和高灵敏度雷达配合时滞摄影研究新生积云塔中初始回波的产生和扩展，发现冰晶首先在云顶形成并向四周往下扩展增强，然后在熟知的环状涡旋环流的带动下在云中心再次上升。初步的检测表明，高分辨率（数百米）高灵敏度（高于－10 dBZ）雷达在降水质点增长轨迹的研究中能发挥重要作用。MIST 试验还表明，多参数雷达和多普勒雷达等现代探测设备，在对流风暴的微物理过程与宏观动力过程的相互作用研究中也能发挥重要作用。同样，在雹暴中探明雹胚源及其增长过程，以及在降水性云系中探索可播区（过冷液水区）研究中都能发挥重要作用。美国的 NOAA 等机构应用多普勒雷达、偏振雷达、偏振激光雷达和微波辐射仪等遥感设备来研究犹他州南部山区的地形风暴，初步结果表明这类地形性风暴中，可播区（“液态水机会”）是有的，但并不是在整个风暴生存期内都存在的，而要及时有效地将播云剂播撒到这种过冷液水区，需要仔细筹划，采用全新的途径方能奏效。

三、中尺度对流系统研究概况

中尺度对流系统是一类降水性深对流系统，其尺度大于单体雷暴，在中高对流层常常有大片层状云砧覆盖，水平尺度广达数百千米。这种对流云系的典型生命期达 6～12 h，有时系统的层状云砧部分能维持数日之久。中尺度对流系统是大部分热带和副热带海洋地区以及美国高原地区夏季的主要降水系统。中尺度对流系统中的雷暴常常是产生突发性洪水、灾害性强风，以及冰雹和龙卷等灾害性天气的重要源地。近 20 年来，通过一系列外场中尺度试验计划，对热带飑线系统、热带云团、中纬度飑线系统和中尺度对流复合体等一系列不同地区、不同季节和不同天气条件下的中尺度对流系统的结构特征做了许多观测和分析研究。虽然结构的细节各不相同，但已揭示出一些普遍的特征。成熟的中尺度对流系统通常由两部分组成：(1) 由深对流单体构成的引导线或云团；(2) 中尺度降水性层状云砧区，它可以从 700 hPa 高度一直伸展到 200 hPa 或更高。在这两区中的空气运动是很不相同的。在对流区，有低层辐合，存在对流尺度上升气流和对流尺度湿下沉气流，其值大于 5 m/s，对流区的顶部外流引起高空辐散。在层状区，观测和数值模拟发现，层状区云底一般位于 0 ℃层附近，在融化层以上有中尺度上升气流，其下有中尺度下沉气流。与层状区有关的中尺度环流特征包括前向入流和后向入流、暖心结构的中层涡旋、中层辐合、高层辐散和低层辐散等。需要指出的是，上述环流特征都是根据有限的观测资料归纳出来的，有些观测事实和概念尚需不断充实和完善。例如关于中尺度上升气流和下沉气流的分界面，有证据表明在有些中尺度对流系统中不在 0 ℃层附近而是在－5～6 ℃层高度。

关于层状区的微结构，在中纬度飑线系统的对流区与尾随的层状区之间有低层雷达回波最小的所谓过渡带（Transition zone）。这种现象在中尺度对流复合体中也存在。我们分析的一个 MCC 个例表明，在活跃的对流区与层状区之间也有雷达回波最弱的区域。飞机探测表明，这里的含水量并不比尾随层状区中小，只是质点浓度大而平均尺度小，所以回波强度相对较低。

至于中尺度对流系统及相应环流特征的成因，什么情况下形成飑线系统，什么情况下形成对流区随机分布的中尺度对流复合体，中尺度对流系统内部对流区与层状区的相互作用，以及中尺度对流系统与环境大尺度天气系统之间的相互作用等一系列问题，都是正在开展并有待深入的研究课题。

1985年5～6月间在美国中西部开展的PRE－STORM计划是一次重要的以中尺度对流系统为研究对象的中尺度外场试验计划，全名是Oklahoina-Kansas Preliminary Regional Experiment for STORM-Central，是STORM-Central计划的一次预演性外场试验，STORM计划是“风暴尺度业务和研究气象学”（Stormscale Operational and Research Meteorology）的简称，是1984年设计的一项长期的国家中尺度研究计划，其最终目的是在精确度上和时效上改进并提高检测和预报中小尺度天气现象的能力，并将其服务于保护公众利益和为国家经济服务及满足防灾需求。20世纪90年代是美国国家STORM计划实施的关键阶段。利用美国国家天气局在20世纪90年代开展的大规模观测现代化计划的条件开展STORM计划，改善短期（0～48 h）天气预报和监测并增进对降水和其他中尺度过程以及它们在水分循环中的作用的认识。在天气观测网现代化方面，美国将投资40亿美元用于天气观测硬件和软件建设，包括先进的卫星（如GOES－Next），在170个站上安装NEXRAD雷达，自动地面观测系统，飞机观测的ACARS风、温、湿资料，风廓线仪和一套先进的通信和信息处理系统以取代过时的观测装备。预计随着STORM计划的实施，中尺度研究将会进入一个全新阶段。

5.4 积云对流中扰动压力效应的诊断分析①

叶家东　史斌强

（南京大学大气科学系）

提　要　本文用简单的结构模式分析诊断了扰动压力对积云对流的效应。结果表明：由浮力项、平流项和拖曳力项触发引起的扰动气压垂直梯度力与各对应源项具有同等量级，但与其源项的作用力方向相反。扰动气压梯度力在云的中上部为负力，它抑制了云的生长发展；在云的下部为正力，它使云向上的加速度增大。扰动气压梯度力对深对流的影响要大于对浅对流的影响。在云下部的扰动低压中心位于云边缘附近。浅对流与深对流的量级分别达－0.1～－0.2 hPa和－0.2～－0.4 hPa。

一、引言

早期大多数积云对流模式中，习惯上都忽略扰动压力效应，既不考虑扰动压力梯度力项，也不考虑扰动压力的浮力效应，密度浮力用热浮力代替，特别在一维模式里，几乎都这样假定。Ogura和Phillipse（1962）[1]通过尺度分析表明，积云对流中扰动气压梯度力与浮力项具有同等重要性，对深对流来说，扰动压力浮力项与热浮力同量级，扰动压力对温度变化从而对饱和水汽压也有重要影响。近二十年来，通过外场观测和数值试验，人们逐步认识到扰动气压对积云对流的重要性。

List和Lozowski（1970）[2]首先指出对于上升气流达到10米/秒以上的对流云，必须考虑扰动压力的效应。Barnes（1970）[3]在活跃的强风暴上升气流中5.5千米（500 hPa）高度上观测到3 hPa的扰动压力，它对浮力加速度的贡献与－10 ℃高度层2 ℃的扰动温度

① 1984年10月9日收到，1985年7月25日收到修改稿。

所产生的热浮力加速度相当。同时在云底附近则观测到－1 hPa 的压力亏空。Ramond (1978)[4]在多单体风暴中对流单体上风方云边缘用飞机观测到振幅为 1.7 hPa 的扰动压力场。一系列外场观测[5,6]表明，强风暴云底附近有时观测到负的热浮力空气具有向上的加速度。这些观测事实促使人们认真分析扰动压力对积云对流特别是强对流单体进一步发展的动力学效应。

一些研究者根据积云数值模拟试验，分析扰动压力在模式云的发展和动力结构中的作用。Holton(1973)[7]根据浅对流模式论证了对于半径大于 1 千米的积云扰动压力能够明显地抑制云的生长速率，也减缓了云顶部云属性的梯度分布，还指出在夹卷和扰动压力共同作用下，积云发展有一最佳的水平尺度。郭晓岚[8]根据准定常一维模式研究了扰动气压与夹卷对积云发展的效应，也指出夹卷对小积云发展的抑制效应是主要的，而扰动压力的抑制作用则随云尺度增大而增加，考虑这两种因素的共同影响，积云发展有一最佳尺度[3]。其他一些研究者[9,10]根据深对流模式数值试验指出，在云的上部扰动压力梯度力与热浮力作用是相反的，Schlesinger 的模拟试验则表明扰动压力能在云底附近使具有负浮力的空气加速上升[11,12]。

Yau(1979)[13]从浅对流压力诊断方程入手，引进有关物理量的强迫函数求解诊断方程，分析扰动压力的动力学效应，表明扰动压力垂直梯度力与浮力项同量级，在一定的云尺度条件下云下部具有负浮力的空气由于扰动压力梯度力的作用可能具有正的加速度。通常认为，深对流中扰动压力效应更为重要，本文在浅对流扰动压力方程求解的基础上，着重对深对流扰动压力效应及二维扰动压力场进行诊断分析。结果表明，处于生长发展期间的对流云，中上部扰动压力梯度力是负的，对云的发展起抑制作用，云下部是正的扰动压力梯度力区，有助于下部云的抬升发展。云下部的扰动低压中心不在云中心轴上，而处于云边缘附近；云上部的高压中心则位于云轴上。

二、扰动压力诊断方程和强迫函数

1. 扰动压力诊断方程

原始运动方程采用 Ogura 和 Phillipse 根据尺度分析得到的适合非黏性流体的小尺度湿对流方程：

$$\frac{\partial \mathbf{V}}{\partial t}=-\mathbf{V}\cdot\nabla\mathbf{V}-c_p\theta_0\nabla\pi'+\left(\frac{\theta_v'}{\theta_{v0}}-q_l\right)g\boldsymbol{k} \tag{1}$$

浅对流取不可压缩连续方程：

$$\nabla\cdot\mathbf{V}=0 \tag{2}$$

深对流取滞弹性连续方程：

$$\nabla\cdot(\rho_0\mathbf{V})=0 \tag{3}$$

其中：ρ_0 和 θ_0 是基本状态密度和位温；θ_{v0} 和 θ_v' 是基本状态虚位温和虚位温扰动量；q_l 是液态水比含水量。在基本状态位温均匀分布的假定下，运动方程取散度并利用式(2)和式(3)可分别得到浅对流和深对流的扰动压力诊断方程：

$$c_p\theta_0\Delta\pi' = -\nabla\cdot(\mathbf{V}\cdot\nabla\mathbf{V}) + g\,\nabla\cdot\frac{\theta_v'}{\theta_{v0}}\boldsymbol{k} - g\,\nabla\cdot q_l\boldsymbol{k} \tag{4}$$

$$c_p\theta_0\,\nabla\cdot(\rho_0\,\nabla\pi') = -\nabla\cdot(\rho_0\mathbf{V}\cdot\nabla\mathbf{V}) + g\,\nabla\cdot\left(\rho_0\frac{\theta_v'}{\theta_{v0}}\right)\boldsymbol{k} - g\,\nabla\cdot(\rho_0 q_l)\boldsymbol{k} \tag{5}$$

式(4)和式(5)右端第一项可展开为动能强迫项和涡度强迫项，分别反映了伯努利效应和旋转离心力效应[13]，其所对应的扰动压力分别表示对动能和流体旋转的响应，故称之为动压力项；第二、三两项分别表示浮力强迫项和拖曳力强迫项；相应的扰动压力分别表示为π_m'、π_θ'和π_q'。方程(4)和(5)是线性方程，满足叠加原理，故有

$$\pi' = \pi_m' + \pi_\theta' + \pi_q' \tag{6}$$

边界条件：采用轴对称和平面对称两种模式。模式云的底部和顶部是刚性的、自由滑动的边界，所以对于$z=0$和$z=H$分别有

$$\begin{cases} w=0 \quad \theta_v'=0 \quad q_l=0 \\ \dfrac{\partial u}{\partial z}=0 \quad \dfrac{\partial\pi'}{\partial z}=0 \end{cases} \tag{7}$$

侧边界条件：

$$\text{轴对称模式：}\left.\frac{\partial\pi'}{\partial r}\right|_{r=0}=0 \quad \pi'\big|_{r\to\infty}=0 \tag{8}$$

$$\text{面对称模式：}\left.\frac{\partial\pi'}{\partial x}\right|_{x=0}=0 \quad \pi'\big|_{x\to\infty}=0 \tag{8'}$$

2. 强迫函数

周晓平等在层结大气热对流发展的数值试验中曾用数值方法求解压力诊断方程[14]。为了定量估计扰动压力效应，分析扰动压力场的结构，我们设法求出扰动压力方程的解析解。为此，考虑边界条件(7)和(8)，将压力方程(4)和(5)右端各强迫项用经验性的强迫函数表示(表1)。浅对流的强迫函数形式基本上与Yau所取的一致，各物理量的垂直分布都取简单的正、余函数，液态含水量的水平分布按照Warner的观测资料取所谓“大礼帽”形分布，即云内含水量水平分布均匀，云边缘骤降为零。温度水平分布按McCarthy(1974)取高斯型廓线，而速度分量w和u的水平分布则取略加修正的高斯廓线。深对流的强迫函数形式主要对水平速度的垂直分布做了修正，使其满足深对流连续方程(3)，其中取

$$\rho_0 = \rho_{00}\mathrm{e}^{-kz} \tag{9}$$

ρ_{00}为云底$z=0$处的基态空气密度，计算中k取0.1 km^{-1}。表1所列的强迫函数适合处于生长发展阶段的非降水性对流云单体，其中各强迫函数的振幅值和云的尺度参数取积云对流的典型值(表2)。

表 1　各物理量的强迫函数(a 表示云半径,单位:千米)

物理量		轴对称	面对称
浅对流	θ_v'	$\hat{\theta}\mathrm{e}^{\left(\frac{r}{a}\right)^2}\sin(nz)$	$\hat{\theta}\mathrm{e}^{-\left(\frac{r}{a}\right)^2}\sin(nz)$
	q_l	$\begin{cases}\hat{q}_l\sin(nz) & 0<r\leqslant a\\ 0 & a<r\end{cases}$	$\begin{cases}\hat{q}_l\sin(nz) & 0<x\leqslant a\\ 0 & a<x\end{cases}$
	w	$\hat{w}[1-(r/a)^2]\mathrm{e}^{-\left(\frac{r}{a}\right)^2}\sin(nz)$	$\hat{w}\left[1-\left(\frac{x}{a}\right)^2\right]\mathrm{e}^{-\frac{1}{2}\left(\frac{x}{a}\right)^2}\sin(nz)$
	u	$-n\hat{w}\left(\frac{r}{2}\right)\mathrm{e}^{-\left(\frac{r}{a}\right)^2}\cos(nz)$	$-n\hat{w}x\mathrm{e}^{-\frac{1}{2}\left(\frac{x}{a}\right)^2}\cos(nz)$
深对流	θ_v'	$\hat{\theta}\mathrm{e}^{-\left(\frac{r}{a}\right)^2}\sin(nz)$	$\hat{\theta}\mathrm{e}^{-\left(\frac{x}{a}\right)^2}\sin(nz)$
	q_l	$\begin{cases}\hat{q}_l\sin(nz) & 0<r\leqslant a\\ 0 & a<r\end{cases}$	$\begin{cases}\hat{q}_l\sin(nz) & 0<x\leqslant a\\ 0 & a<x\end{cases}$
	w	$\hat{w}\left[1-\left(\frac{r}{a}\right)^2\right]\mathrm{e}^{-\left(\frac{r}{a}\right)^2}\sin(nz)$	$\hat{w}\left[1-\left(\frac{x}{a}\right)^2\right]\mathrm{e}^{-\frac{1}{2}\left(\frac{x}{a}\right)^2}\sin(nz)$
	u	$-\hat{w}\left(\frac{r}{2}\right)\mathrm{e}^{-\left(\frac{r}{2}\right)^2}[n\cos(nz)-K\sin(nz)]$	$-\hat{w}x\mathrm{e}^{-\frac{1}{2}\left(\frac{x}{a}\right)^2}[n\cos(nz)-K\sin(nz)]$

表 2　各强迫函数振幅值和云的尺度参数

	n	H/千米	$\hat{\theta}$/℃	$\hat{q}_l$/(克/千克)	$\hat{w}$/(米/秒)
浅对流	$\frac{\pi}{H}$	4	1	1	5
深对流	$\frac{\pi}{H}$	8	1.5	2	10

三、扰动压力方程的解

1. 浅对流

浅对流轴对称模式,在边界条件(7)和(8)情形下,方程(4)的形式解可取

$$\pi'=\sum_k \pi_k(r)\cos(knz) \tag{10}$$

代入式(3)左端,得

$$c_p\theta_0\Delta\pi'=c_p\theta_0\sum_k\left[\frac{\mathrm{d}^2\pi_k(r)}{\mathrm{d}r^2}+\frac{1}{r}\frac{\mathrm{d}\pi_k(r)}{\mathrm{d}r}-k^2n^2\pi_k(r)\right]\cos(knz)$$

式(3)右端第一项利用连续方程 $\nabla\cdot\mathbf{V}=0$,有

$$-\nabla\cdot(\mathbf{V}\cdot\nabla\mathbf{V})=-\left(\frac{\partial u}{\partial r}\right)^2-\left(\frac{\partial w}{\partial z}\right)^2-2\left(\frac{\partial u}{\partial z}\right)\left(\frac{\partial w}{\partial r}\right)-\frac{u^2}{r}$$

将表 1 的强迫函数代入上式和式(4)右端其余各项,略加整理,可得

$$\begin{cases} c_p\theta_0\left[\dfrac{\mathrm{d}^2\pi_0(r)}{\mathrm{d}r^2}+\dfrac{1}{r}\dfrac{\mathrm{d}\pi_0(r)}{\mathrm{d}r}\right]=\left(\dfrac{4r^4}{a^4}-\dfrac{7r^2}{a^2}+\dfrac{3}{2}\right)\dfrac{n^2\hat{w}^2}{2}\mathrm{e}^{-2\left(\frac{r}{a}\right)^2} \\ c_p\theta_0\left[\dfrac{\mathrm{d}^2\pi_2(r)}{\mathrm{d}r^2}+\dfrac{1}{r}\dfrac{\mathrm{d}\pi_2(r)}{\mathrm{d}r}-4n^2\pi_2(r)\right]=\left(\dfrac{r^2}{a^2}+\dfrac{3}{2}\right)\dfrac{n^2\hat{w}^2}{2}\mathrm{e}^{-2\left(\frac{r}{a}\right)^2} \\ c_p\theta_0\left[\dfrac{\mathrm{d}^2\pi_{1\theta}(r)}{\mathrm{d}r^2}+\dfrac{1}{r}\dfrac{\mathrm{d}\pi_{1\theta}(r)}{\mathrm{d}r}-n^2\pi_{1\theta}(r)\right]=\dfrac{ng\hat{\theta}}{\theta_{v0}}\mathrm{e}^{-\left(\frac{r}{a}\right)^2} \\ c_p\theta_0\left[\dfrac{\mathrm{d}^2\pi_{1q}(r)}{\mathrm{d}r^2}+\dfrac{1}{r}\dfrac{\mathrm{d}\pi_{1q}(r)}{\mathrm{d}r}-n^2\pi_{1q}(r)\right]=-ng\hat{q}_l\delta \quad \delta=\begin{cases}1 & 0<r\leqslant a\\ 0 & r>a\end{cases} \\ c_p\theta_0\left[\dfrac{\mathrm{d}^2\pi_k(r)}{\mathrm{d}r^2}+\dfrac{1}{r}\dfrac{\mathrm{d}\pi_k(r)}{\mathrm{d}r}-k^2n^2\pi_k(r)\right]=0 \quad (k\geqslant 3) \end{cases} \tag{11}$$

$$\begin{cases}\pi'_m=\pi_0+\pi_2\cos(2nz) \quad \pi'_\theta=\pi_{1\theta}\cos(nz) \quad \pi'_q=\pi_{1q}\cos(nz) \\ \pi'_k=0 \quad (k\geqslant 3)\end{cases} \tag{12}$$

所以主要是求如下形式的定解问题:

$$\begin{cases}\dfrac{\mathrm{d}^2\pi_k(r)}{\mathrm{d}r^2}+\dfrac{1}{r}\dfrac{\mathrm{d}\pi_k(r)}{\mathrm{d}r}-k^2n^2\pi_k(r)=g(r) \\ \left.\dfrac{\mathrm{d}\pi_k(r)}{\mathrm{d}r}\right|_{r=0}=0 \quad \pi_k(r)\big|_{r\to\infty}=0\end{cases} \tag{13}$$

本文不直接用格林函数法求解,而是采用参数变易法求得其解为

$$\pi_k(r)=K_0(knr)\int_0^r\frac{I_0(knr)g(r)}{w(r)}\mathrm{d}r+I_0(knr)\int_0^\infty\frac{K_0(knr)g(r)}{w(r)}\mathrm{d}r \tag{14}$$

其中:$I_0(knr)$和 $K_0(knr)$分别为零阶虚宗量贝塞耳函数和零阶虚宗量诺埃曼函数;$w(r)$为 $I_0(knr)$和 $K_0(knr)$的朗斯基行列式:

$$w(r)=\begin{vmatrix}I_0(knr) & K_0(knr)\\ I'_0(knr) & K'_0(knr)\end{vmatrix}$$

根据刘维尔公式有

$$w(r)=w(r_0)\mathrm{e}^{-\int_{r_0}^r\frac{1}{r}\mathrm{d}r}=\frac{r_0}{r}w(r_0)$$

r_0 在定义域内任取。于是式(14)又可写为

$$\pi_k(r)=k_0(knr)\int_0^r\frac{I_0(knr)rg(r)}{w(r_0)r_0}\mathrm{d}r+I_0(knr)\int_r^\infty\frac{K_0(knr)rg(r)}{w(r_0)r_0}\mathrm{d}r \tag{14'}$$

借助柱函数的一些递推关系式,对式(14′)中两个积分做适当处理,最终可得浅对流轴对称模式的解:

$$\begin{cases}\pi_\theta'=\dfrac{1}{c_p\theta_0}\dfrac{\hat{\theta}}{\theta_{v0}}gnG(r,a^2,1)\cos(nz)\\ \pi_m'=\dfrac{n^2\hat{w}^2}{c_p\theta_0}\left(\dfrac{a^2}{32}-\dfrac{r^2}{8}\right)\mathrm{e}^{-2\left(\frac{r}{a}\right)^2}-\dfrac{n^2\hat{w}^2}{c_p\theta_0}\left[\left(1+\dfrac{n^2a^2}{8}\right)G\left(r,\dfrac{a^2}{2},2\right)+\right.\\ \qquad\left.\dfrac{a^2}{32}\mathrm{e}^{-2\left(\frac{r}{a}\right)^2}\right]\cos(2nz)\end{cases}$$

$$\pi_q'=\begin{cases}\dfrac{1}{c_p\theta_0}\dfrac{g\hat{q}_l}{n}\left[1+\dfrac{naK_1(na)I_0(nr)}{r_0w(r_0)}\right]\cos(nz) & 0<r\leqslant a\\ -\dfrac{1}{c_p\theta_0}\dfrac{g\hat{q}_l}{n}\dfrac{naI_1(na)K_0(nr)}{r_0w(r_0)}\cos(nz) & r>a\end{cases} \tag{15}$$

其中

$$G(r,a^2,k)=K_0(knr)\int_0^r\frac{rI_0(knr)\mathrm{e}^{-\left(\frac{r}{a}\right)^2}}{r_0w(r_0)}\mathrm{d}r+I_0(knr)\int_r^\infty\frac{rK_0(knr)\mathrm{e}^{-\left(\frac{r}{a}\right)^2}}{r_0w(r_0)}\mathrm{d}r$$

相仿，浅对流面对称模式以形式解

$$\pi'=\sum_k\pi_k(x)\cos(knz) \tag{16}$$

代入式(4)，所分离出的模 $\pi_k(x)$ 满足下述形式的定解问题：

$$\begin{cases}\dfrac{\mathrm{d}^2\pi_k(x)}{\mathrm{d}x^2}-k^2n^2\pi_k(x)=f(x)\\ \left.\dfrac{\mathrm{d}\pi_k(x)}{\mathrm{d}x}\right|_{x=0}=0\\ \left.\pi_k(x)\right|_{x\to\infty}=0\end{cases} \tag{17}$$

其解为

$$\pi_k(x)=\int_0^xG(x,\xi)f(\xi)\mathrm{d}\xi+\int_x^\infty G(x,\xi)f(\xi)\mathrm{d}\xi \tag{18}$$

其中

$$G(x,\xi)=\begin{cases}-\dfrac{1}{2kn}(\mathrm{e}^{knx}+\mathrm{e}^{-knx})\mathrm{e}^{-kn\xi} & x\leqslant\xi\\ -\dfrac{1}{2kn}(\mathrm{e}^{kn\xi}+\mathrm{e}^{-kn\xi})\mathrm{e}^{-knx} & x>\xi\end{cases}$$

与轴对称模式类似处理，最终可得面对称模式的解：

$$\left\{\begin{aligned}&\pi_\theta'=-\frac{ag\hat{\theta}\sqrt{\pi}}{4c_p\theta_0\theta_{v0}}\mathrm{e}^{\left(\frac{na}{2}\right)^2}\left\{\left[1+\mathrm{erf}\left(\frac{x}{a}-\frac{na}{2}\right)\right]\mathrm{e}^{-nx}+\right.\\&\left.\left[1-\mathrm{erf}\left(\frac{x}{a}+\frac{na}{2}\right)\right]\mathrm{e}^{nx}\right\}\cos(nz)\\&\pi_m'=-\frac{n^2\hat{w}^2x^2}{2c_p\theta_0}\mathrm{e}^{-\left(\frac{x}{a}\right)^2}-\left\{c\,\mathrm{e}^{-2nx}\left[1+\mathrm{erf}\left(\frac{x}{a}-na\right)\right]+\right.\\&\left.c\,\mathrm{e}^{2nx}\left[1-\mathrm{erf}\left(\frac{x}{a}+na\right)\right]+\frac{n^2a^2\hat{w}^2}{4c_p\theta_0}\mathrm{e}^{-\left(\frac{x}{a}\right)^2}\right\}\cos(2nz)\\&\pi_q'=\begin{cases}\dfrac{g\hat{q}_l}{2nc_p\theta_0}\left[2-\mathrm{e}^{-n(a-x)}-\mathrm{e}^{-n(a+x)}\right]\cos(nz) & 0<x\leqslant a\\ \dfrac{g\hat{q}_l}{2nc_p\theta_0}\left[\mathrm{e}^{n(a-x)}-\mathrm{e}^{-n(a+x)}\right]\cos(nz) & x>a\end{cases}\end{aligned}\right. \tag{19}$$

其中 $c=\dfrac{na\hat{w}^2\sqrt{\pi}}{16c_p\theta_0}(3+2n^2a^2)\mathrm{e}^{n^2a^2}$，而 $\mathrm{erf}(x)=\dfrac{2}{\sqrt{\pi}}\displaystyle\int_0^x\mathrm{e}^{-t^2}\mathrm{d}t$ 是误差函数。

2. 深对流

由于式(5)左端

$$\frac{1}{\rho_0}\nabla\cdot(\rho_0\nabla\pi')=\Delta\pi'+\frac{1}{\rho}\frac{\partial\rho_0}{\partial z}\frac{\partial\pi'}{\partial z}$$

因此深对流扰动压力诊断方程是非标准形的，使得求解难以执行。我们引进扰动压力订正项的方法进行处理。以浮力项为例，令

$$\pi_\theta'=\pi_\theta+\pi_{\theta c} \tag{20}$$

其中 π_θ 是在浅对流中得到的解析结果。以式(20)代入深对流浮力扰动压力方程中，则化为求定解问题：

$$\left\{\begin{aligned}&c_p\theta_0\Delta\pi_{\theta c}+\frac{1}{\rho_0}\nabla\rho_0\cdot(\nabla\pi_\theta+\nabla\pi_{\theta c})c_p\theta_0=\frac{g\theta_v'}{\rho_0\theta_{v0}}\boldsymbol{k}\cdot\nabla\rho_0\\&\left.\frac{\partial\pi_{\theta c}}{\partial z}\right|_{z=0}=\left.\frac{\partial\pi_{\theta c}}{\partial z}\right|_{z=H}=0\\&\left.\frac{\partial\pi_{\theta c}}{\partial r}\right|_{r=0}=0\qquad\left.\pi_{\theta c}\right|_{r\to\infty}=0\end{aligned}\right. \tag{21}$$

由于密度的垂直变化较小，订正项 $\nabla\pi_{\theta c}$ 相对于 $\nabla\pi_\theta$ 来说是小项，故求解过程中略去。于是式(21)的压力方程可写为

$$c_p\theta_0\Delta\pi_{\theta c}=c_pK\theta_0\frac{\partial\pi_\theta}{\partial z}-K\frac{g\theta_v'}{\theta_{v0}} \tag{22}$$

π_θ 由式(15)的第一式表达,故

$$\frac{\partial \pi_\theta}{\partial z}=-\frac{1}{c_p\theta_0}\frac{g\hat{\theta}}{\theta_{v0}}n^2G(r,a^2,1)\sin(nz)$$

将 $\sin(nz)$展成傅里叶余弦级数,连同 θ_v'的强迫函数一起代入式(22),可得

$$c_p\theta_0\Delta\pi_{\theta c}=-\sum_{k=0}^{\infty}a_kK\frac{g\hat{\theta}}{\theta_{v0}}\left[n^2G(r,a^2,1)+e^{-\left(\frac{r}{a}\right)^2}\right]\cos(knz)$$

以式(10)类似的形式解代入上式左端,可求得

$$\pi_{\theta c}=\frac{K\hat{\theta}g}{c_p\theta_0\theta_{v0}}\left\{\sum_{\substack{m=1\\k=2m}}^{\infty}\frac{4}{(k^2-1)\pi}\frac{k^2}{k^2-1}[G(r,a^2,k)-G(r,a^2,1)]\cos(knz)-G(r,a^2,1)\sin(nz)\right\}$$

相仿,可求得动压力项和拖曳力项的扰动压力订正项:

$$\pi_{mc}=\sum_{\substack{m=0\\k=2m+1}}^{\infty}\left\{-\frac{8}{(k^2-4)\pi}\frac{Kn\hat{w}^2}{c_p\theta_0}\left[G\left(r,\frac{a^2}{2},k\right)\left(1+\frac{k^2n^2a^2}{16}\right)+\frac{a^2}{16}e^{-2\left(\frac{r}{a}\right)^2}\right]+\frac{8}{(k^2-4)^2\pi}\frac{Kn\hat{w}^2}{c_p\theta_0}\left[G\left(r,\frac{a^2}{2},2\right)\left(2+\frac{n^2a^2}{4}\right)-\left(2+\frac{1}{16}k^2n^2a^2\right)G\left(r,\frac{a^2}{2},k\right)\right]\right\}\cos(knz)$$

$$\pi_{qc}=-\frac{2g\hat{q}_l}{n\pi c_p\theta_0}\left[\frac{K}{n}+\frac{KaK_1(na)I_0(nr)}{w(r_0)r_0}\right]-\frac{4Kg\hat{q}_l}{c_p\theta_0n^2\pi w(r_0)r_0}\cdot\sum_{\substack{m=1\\k=2m}}^{\infty}\left\{\frac{k}{(k^2-1)^2}\left[\frac{na}{k}I_0(nr)K_1(na)+naI_0(knr)K_1(kna)\right]\right\}\cos(knz)$$

$$r\leqslant a$$

$$\pi_{qc}=\frac{2g\hat{q}_l}{c_p\theta_0n\pi w(r_0)r_0}KaI_1(na)K_0(nr)-\frac{4Kg\hat{q}_l}{c_p\theta_0n^2\pi w(r_0)r_0}\cdot\sum_{\substack{m=1\\k=2m}}^{\infty}\left\{\frac{k}{(k^2-1)^2}\left[naI_1(na)\frac{K_0(nr)}{k}+naI_1(kna)K_0(knr)\right]\right\}\cos(knz)$$

$$r>a$$

面对称的情况也做类似处理。

四、数值计算结果分析

1. 不同水平尺度云的扰动气压梯度力

表3列出不同水平尺度的云内各类扰动气压垂直梯度力及其源项的大小变动范围。由表可见,对应于浮力项、平流项和拖曳力项的扰动气压垂直梯度力与各对应源项具有同等量级,浅对流中为源项的0.2～0.9,随着云的水平尺度增大而增加;深对流中可达0.4～1.1

倍。扰动气压垂直梯度力方向一般与其源项的作用力方向相反，所以对处于生长发展中的对流云来说，扰动气压梯度力一般是起抑制作用的。由表 3 还可以看出，面对称模式的扰动气压梯度力普遍比轴对称模式的对应值大。

表 3　云中各强迫项及其扰动压力梯度力变动范围(单位:达因/克)

强迫项			浅对流(H=4 千米)					深对流(H=8 千米)		
		a/千米	1.0	1.5	2.0	2.5	3.0	3.0	4.0	5.0
浮力项	$g\dfrac{\theta_v'}{\theta_{v0}}$		0～3.30					0～4.95		
	$-c_p\theta_0\dfrac{\partial\pi_\theta'}{\partial z}$	轴对称	−1.10～0	−1.59～0	−1.95～0	−2.21～0	−2.41～0	−2.03～0	−2.53～0	−2.92～0
		面对称	−1.80～0	−2.22～0	−2.50～0	−2.63～0	−2.81～0	−3.05～0	−3.44～0	−3.71～0
动压力项	$-\mathbf{V}\cdot\nabla w$		−2.50～2.50					−2.0～2.0		
	$-c_p\theta_0\dfrac{\partial\pi_m'}{\partial z}$	轴对称	0.49～−0.49	0.62～−0.62	0.69～−0.69	0.71～−0.71	0.71～−0.71	0.93～−0.71	0.86～−1.12	0.88～−1.26
		面对称	1.12～−1.12	1.20～−1.20	1.24～−1.24	1.30～−1.30	1.39～−1.39	1.81～−1.92	2.03～−1.93	2.22～−1.81
拖曳力项	$-gq_l'$		−0.98～0							
	$-c_p\theta_0\dfrac{\partial\pi_q'}{\partial z}$	轴对称	0～0.39	0～0.57	0～0.71	0～0.79	0～0.86	0～0.92	0～1.19	0～1.30
		面对称	0～0.62	0～0.76	0～0.85	0～0.90	0～0.93	0～1.36	0～1.55	0～1.68

2. 云中心扰动气压垂直梯度力的垂直分布

计算表明，云的中上部为负的扰动气压梯度力控制，扰动中心位于云轴上部，其值为−1.0～−2.2 达因/克(浅对流)和−1.6～−2.8 达因/克(深对流)(图略)。云内这种负的扰动气压梯度力对云的生长发展起抑制作用，量值随云尺度增大而增加。云下部约三分之一的部位是正的扰动气压梯度力区，中心极值为 0.05～0.3 达因/克(浅对流)和 0.1～0.6 达因/克(深对流)。面对称模式的正扰动气压梯度力极值要比轴对称模式的大 5 倍，这与面对称模式的二维特性有关，云中心受侧边界影响相对讲较小。

3. 扰动气压垂直梯度力的二维分布

图 1 是扰动气压垂直梯度力的二维分布图，可见云中大部分区域为负的梯度力所控制，扰动中心位于云中心上部三分之二的部位，负扰动区水平方向可延伸至云尺度 2 倍以上的范围，云边缘的扰动气压梯度力值约为云中心极值的 40%～70%(图略)。云下部有较小的正扰动气压梯度力区，轴对称模式扰动中心处于云中心底部附近，而面对称模式的扰动气压梯度力中心则偏离云中心，位于下部云的边缘附近。这种正的扰动气压梯度力有助于较冷的云底空气加速上升。

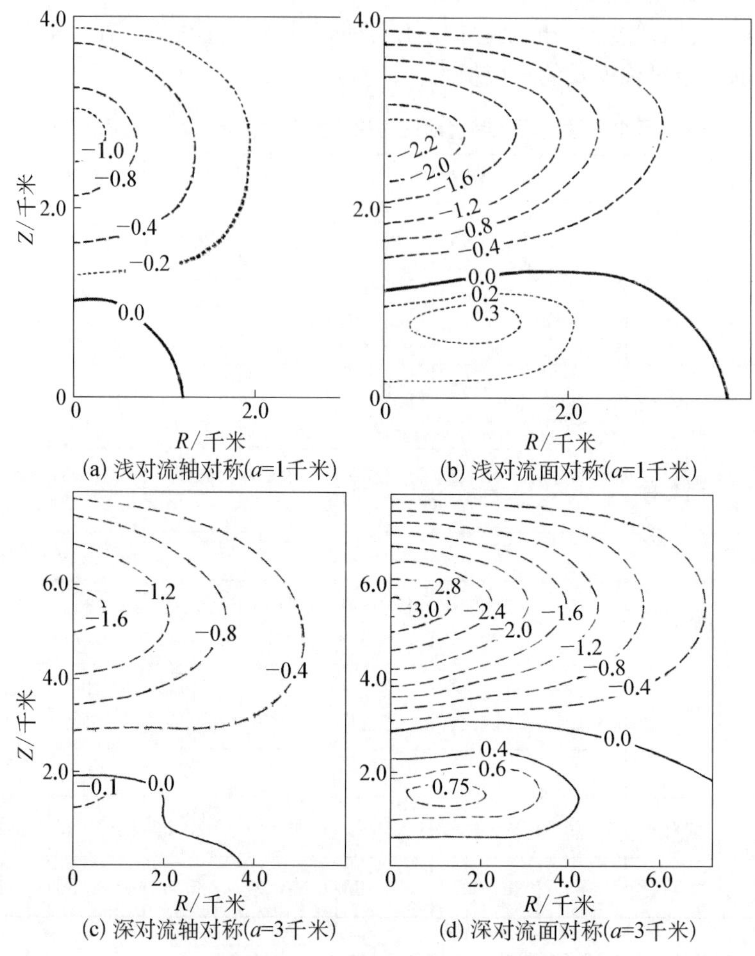

(a) 浅对流轴对称(a=1千米)　(b) 浅对流面对称(a=1千米)

(c) 深对流轴对称(a=3千米)　(d) 深对流面对称(a=3千米)

图 1　扰动压力梯度力分布(单位:达因/克)

4. 扰动压力场二维分布

扰动压力场示于图 2。正扰动压力中心位于云的中心顶部,其值约为 0.2 hPa(浅对流)和 0.3～0.5 hPa(深对流);负中心位于云下部边缘部位,中心极值达－0.1～－0.2 hPa(浅对流)和－0.2～－0.4 hPa(深对流)。由表 1 可知,扰动压力及其梯度力是随云尺度和云物理属性的数值增加而增大的,所以在强对流单体中的扰动压力及其梯度力应比上述数值结果更大。

5. 空气密度垂直变化对扰动气压梯度力的影响

深对流模式中考虑密度订正项的扰动气压垂直梯度力,与不考虑密度垂直变化的数值结果(图略)比较可见,空气密度垂直变化的影响使扰动气压梯度力略有减小,但其量值并不大,一般不超过 0.1 hPa。其相对作用在云下部较大,能使正的扰动气压梯度力减小 15%(面对称)至 50%(轴对称)。

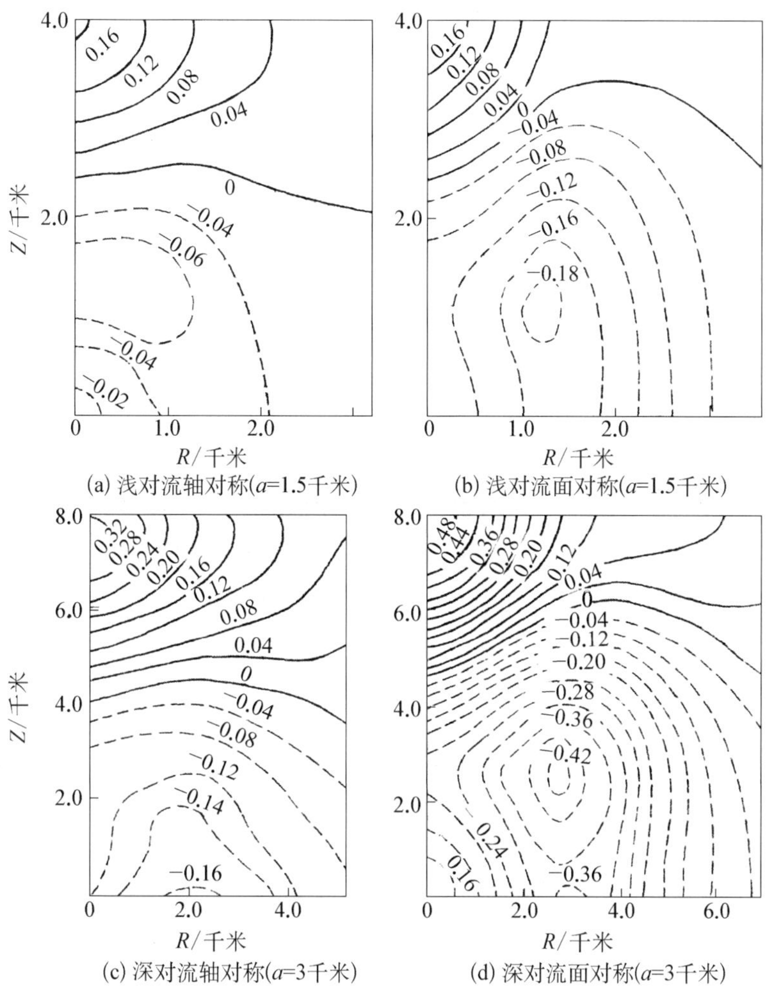

图 2　扰动压力场(单位:hPa)

参考文献

[1] Ogura, Y., and N. A, Phillips, 1962, A scale analysis of deep and shallow convection in the atmosphere, *J. Atmos. Sci.*, **19**, 173—179.

[2] List, R., and E. P. Lozowski, 1970, Pressure preturbations and buoyancy in convective clouds, *J. Atmos. Sci.*, **17**, 169—170.

[3] Barnes, S. L., 1970, Some aspects of a severe right-moving thunderstorm deduced from mesonetwork rawinsonde observation, *J. Atmos. Sci.*, **27**, 634—648.

[4] Ramond, D., 1978, Pressure perturbation in deep convection: an experimental study, *J. Atmos. Set.*, **35**, 1704—1711.

[5] Marwitz, J. D., 1973, Trajectories within the weak echo region of hailstorm, *J. Appl. Meteor.*, **12**, 1174—1182.

[6] Davies-Jones, R, P., and J, H. Henderson, 1975, Updraft properties deduced statistically from rawin soundings, *Pure Appl. Geophysics*, **113**, 787—802.

[7] Holton, J, R., 1973, One dimensional cumulus model including pressure perturbations, *Mon. Wea. Rev.*, **110**, 201—205.

[8] 郭晓岚,1981,大气动力学,江苏科学技术出版社.

[9] Soong, S, T., 1974, Numerical simulation of warm rain development in an axisymmetric cloud model, *J. Atmos. Sci.*, **31**, 1262—1285.

[10] Wilhelmson, R., and Y. Ogura, 1972, The pressure perturbation and numerical modeling of a cloud, *J. Atmos. Sci.*, **29**, 1295—1307.

[11] Schlesinger, R, E. 1973, A numerical model of deep moist convection: Part I, Comparative experiments for variable ambient moisture and wind shear, *J. Atmos. Sci.*, **30**, 835—856.

[12] Schlesinger, R. E. 1978, A three-dimensional numerical model of an isolated thunderstorm: Part I, comparative experiments for variable ambient wind shear, *J. Atmos. Sci.*, **35**, 690—713.

[13] Yau, M.K., 1979, Perturbation pressure and camulus convection, *J. Atmos. Sci.*, **36**, 690—694.

[14] 巢纪平,周晓平,1964,积云动力学,科学出版社.

[15] 王竹溪、郭敦仁,1979,特殊函数概论,科学出版社.

DIAGNOSTIC ANALYSIS OF THE EFFECT OF PERTURBATION PRESSURE ON CUMULUS CONVECTION

Ye Jiadong　Shi Binqiang

(Department of Atmospheric Sciences, Nanjing University)

Abstract The effect of perturbation pressure on cumulus convection is analysed diagnostically with simpler structure models. The results show that the perturbation pressure force induced by buoyancy, velocity advection and dray are of the same order of magnitude as the corresponding resource terms and act against them. The perturbation pressure force is negative in the middle-upper parts of cloud, which suprces the growth of cloud, and is positive in the lower part which promotes the cloud upward acceleration. The effects of perturbation pressure froce are more important in deep convection than that in shallow convection. The low centers of perturbation pressure near the base of cloud are located at the lateral edge of cloud, the magnitudes reach −0. 1——0. 2 hPa for shallow and −0. 2——0. 4 hPa for deep convection, respectively.

5.5 一个缓慢移动的中尺度对流复合体内层状降水区的微结构分析①

叶家东　范蓓芬

(南京大学大气科学系,南京,210093)

W. R. Cotton　M. A. Fortune

(科罗拉多州立大学大气科学系,美国)

提　要　本文对一个中纬中尺度对流复合体层状降水区的微物理结构,结合雷达、卫星和其他飞机观测资料进行了分析。结果表明,MCC 层状区内某些部位盛行冰晶聚合体,它们分布在相当厚的过冷气层内(0.5～−14 ℃或更冷)。冰晶聚并过程是层状区内降水质点增长的主要

① 1990 年 1 月 9 日收到,12 月 18 日收到再改稿。

机制。它起源于较高较冷的气层，在冰晶聚合体下降途中聚并效率渐趋增强，在 0 ℃层附近形成——大的冰晶聚合带。

层状区中云滴液态含水量一般低于 0.3 $g \cdot m^{-3}$。0 ℃层以下降水质点数浓度较低，平均为 0.8 L^{-1}（2D－P 资料）和 2.3 L^{-1}（2D－C 资料），相应的平均体积中值直径分别为 1.0 和 0.6 mm。在 0～－10 ℃气层内，冰质点平均数浓度为 27 L^{-1}（2D－P 资料/和 133 L^{-1}（2D－C），远大于 0 ℃层以下的雨滴数浓度，相应的平均体积中值直径为 0.8 和 0.4 mm。冰质点数浓度随高度向上增加，在飞机垂直探测的顶部（6 600 m 高度）观测到最大数浓度为 52 L^{-1}（2D－P 资料）和 289 L^{-1}（2D－C 资料）。冰质点大小则相反，是随高度下降而增大的。在 0 ℃层附近冰晶聚合体较大较多，冰质点中 15%以上是聚合体，2D－P 探头观测的冰质点平均体积中值直径达 1.8 mm。

滴谱分析表明，负指数律分布能较好地拟合所有观测的降水质点大小谱分布。对水滴，斜率参数 λ 平均为 17（±3.6）cm^{-1}，相对变差不超过 20%，在云模式研究中可以近似地假定 λ 是常数。然而，对冰质点样本，λ 值可相差 3 倍以上，不能当作常数处理。至于截距参数 N_0，不论是水滴还是冰晶样本，都是变量，其值可有 2～3 个数量级之差。但是，N_0 与 λ 之间数值上相关很好，据此可以将降水质点谱简化为单参数分布。

关键词：中尺度对流复合体；层状区；微物理学；冰晶聚合体。

一、引言

自从 1980 年 Maddox[1] 发现并定义了中尺度对流复合体（MCC）以来，这一类中尺度对流系统引起了人们广泛的关注。研究表明，成熟的中尺度对流系统通常由两部分组成，一是由深对流单体组成的引导线或云团，二是中尺度降水性层状云区。在这两区中的空气运动是很不相同的。在对流区，有低层辐合、高层辐散，存在有组织的对流尺度上升气流和湿下沉气流，其值一般可大于 5 m/s。在层状区，云底以上有一中尺度上升气流，融化层以下有一中尺度下沉气流，对应的有中层涡旋和中层辐合等中尺度环流特征。层状区是中尺度对流系统中的重要组成部分，它提供了近一半的降水，在 MCC 中对流活动中止后，整个系统尚能维持数小时之久。所以，研究层状区中降水微物理过程是 MCC 研究中的一个组成部分。

1984 年美国国家海洋大气管理局环境研究室（NOAA/ERL）的天气研究计划（WRP）与科罗拉多州立大学大气合作研究所（CIRA）合作进行的中尺度对流系统飞机研究计划（AIMCS）是第一次对中纬度缓慢移动的夜间中尺度对流系统进行飞机观测研究的外场试验，其目的是研究中尺度对流系统的动力学和微物理学，包括系统内微物理过程与动力学过程的相互作用。本文对其中一个个例的层状降水区微物理结构，结合雷达、卫星及其他飞机观测资料进行了分析，并探讨了层状区降水质点增长机制和降水质点谱及其参数化问题。

二、天气条件及观测概况

这个 MCC 形成于美国大平原衣阿华地区。7 月 14 日 12 GMT，西部有一短波槽及与之对应的地面低压发展，低压伴随着冷锋，850 hPa 面上短波槽前一股南风低空急流带来暖湿空气，极值达到 20 m/s。7 月 15 日 00 GMT 前，冷锋加强并推进至衣阿华中部。这种形势导致 850 hPa 上衣阿华南部有较强的暖湿空气平流及正的涡度平流，有利于在该地区发展中尺度对流系统。夜间，对流活动组织成东西向的对流线，垂直于南风急流。在衣阿华中部(20:30～23:30 GMT)和南部(02:30～04:30 GMT)数次观测到冰雹。对流线的东北侧层状区中飞机观测到较大的降水。

从增强红外卫星云图看出，7 月 14 日 19 GMT 云系已显示出 MCC 的“起始”阶段特征。7 月 15 日 03 GMT 系统达到“最大”阶段，云顶温度低于－52 ℃区域的面积达到 21 万 km^2 [图 1(a)，06:31]。这个 MCC 持续了 15 个小时，于 7 月 15 日 10 GMT 消散。

雷达观测表明，夜间对流云系中对流单体发展，云盖彼此相连并形成层状云区。雷达反射率强度不太大，不超过 50 dBZ。7 月 15 日 04 GMT 前一持续的东西向对流线沿着 41°N 发展，回波顶高达 16 km[图 1(b)，06:35]。此对流线一直维持到 07 GMT。此外，尚有一些孤立的对流单体此起彼伏。层状区迅速东移进入伊里诺斯，而南部移动缓慢，到 05 GMT 左右大部分层状区降水活动已位于 42°N 以南。此后，系统移动十分缓慢，层状区降水活动集中在衣阿华的勃林顿地区。06 GMT 以后东西向对流线已移至 40°N 附近。

(a) NWS雷达回波图　　(b) 1984年7月15日

图例：雷达反射率等值线为 15，41 和 51 dBZ；100/表示回波顶高 10 000 m；回波移动速度矢量单位为 m/s

图 1　MCC 的增强红外卫星云图

观测飞机于 03:10 GMT 从 MCC 的西部边缘进入云系，在 720 hPa 高度上沿对流带北侧层状区中飞行(图 2)，这高度接近层状区云底，观测到连续性中到大雨。飞行途中云系层状区内的温度、湿度和风在同一高度上都有一定程度的起伏，温度介于 10～13 ℃，风向西北向，速度达到 20 m/s。

06:00 至 06:20 GMT，飞机在伯林顿附近爬升至 425 hPa 高度，进行微物理垂直结构观测，探测高度由 2 800 m 至 6 500 m，相应的温度为＋9.6 ℃至－11 ℃。然后在 425 hPa 高度上西飞返航，沿途云系温度为－11～－14 ℃，并观测到＋6 m/s 到－4 m/s 的垂直气流。飞行中飞机未直接穿对流区观测，所以下面的分析限于 MCC 的层状区。

三、层状云区降水微物理特征

用来分析层状降水区微物理结构的资料包括降水质点测量系统(PMS)的 2D－P 和

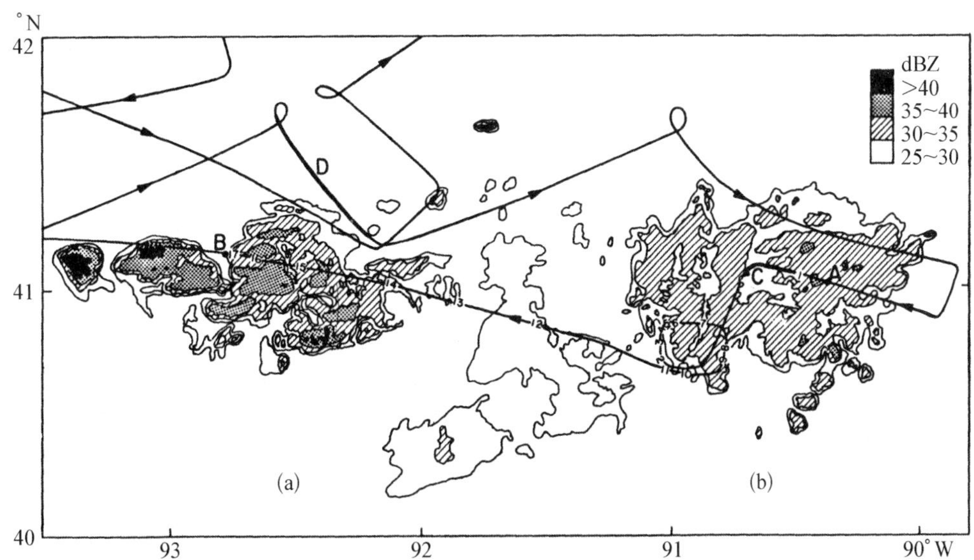

在飞行路径上由A点到B点，数字1～17代表图3中对应序号的微结构样品的取样位置。图中左半部雷达回波(a)是05:05～05:10 GMT飞机位于北侧D处观测的；右半部雷达回波(b)是05:52～05:56 GMT飞机位于C处观测的。B点附近的雷达回波是在飞机飞抵B点附近进行取样观测微结构前1.5小时在D处(2 800 m高度)观测的，取样时该处(6 750 m高度)回波已变为层状性，强度低于25 dBZ。

图2　飞行路径和由机腹雷达观测的回波图(2 800 m高度)

2D－C探头观测的降水质点二维图像。JW热线含水量仪观测的云液态含水量、大气状态参数温度、气压和露点温度，以及NOAA P－3飞机的飞行轨迹和飞行状态参数。此外也应用了机腹平面扫描雷达和机尾垂直扫描雷达资料。根据这些资料，对这个MCC层状降水区的微物理结构做了如下分析：

1. 降水质点的相态结构

2D－P和2D－C探头观测的质点大小范围分别为0.2～6.4 mm和0.05～1.60 mm直径。图3列出一部分二维质点图像的样本及相应的质点数浓度N_T和体积中值直径D_0，还列出对应样本中的冰晶聚合体数浓度AN_T。从图3可以看出，0 ℃层上下质点二维图像的外形特征有着明显的区别。0 ℃层以下高度降水质点主要是水滴，在1.5 ℃层以下高度则全是水滴。0 ℃层以上高度以冰质点为主，在层状区基本上未见到边缘光滑的圆形水滴图像。由二维图像判断融化层厚度约为200 m厚，位于0～1.5 ℃气层之间。

2. 含水量

JW热线含水量仪观测的云滴液态含水量在层状区普遍较低，一般低于0.3 g/m^3，最大值为0.5 g/m^3，是在6 km高度上升气流达5.6 m/s的地区观测到的，估计是在一个小对流泡内取样观测的。由PMS探头观测的二维图像推算的降水质点含水量在融化层以下平均为0.6 g/m^3，而在－3～－12 ℃温度区间平均含水量值为0.8 g/m^3，这里冰粒子的平均视密度以0.1 g/cm^{3}计。图4(a)的曲线A是降水质点含水量的垂直分布廓线，从中可以看出，从0 ℃层向上，降水质点含水量是随高度向上增大的，在5 000 m高度(－5 ℃层)达到极大值1.3 g/m^3，随后向上减小，不过起伏较大，这与垂直探测区是一个衰老的对流云团有关。在融化层，含水量有一极小区，这是由于融化的冰质点落速增大，在这一气层内造成一质量辐散层所致。冷层冰质点含水量(IWC)似乎比预期的层状降水区含水量要

No.	T/℃	H/m	(a)	N_T/L	D_0/mm	AN_T/L
17	−13.9	6 750		0.85	1.24	0.25
16	−13.0	6 728		3.02	1.48	0.70
15	−12.5	6 760		6.10	1.22	1.25
14	−12.0	6 757		26.09	0.82	2.01
13	−12.0	6 779		55.21	0.43	0.35
12	−10.0	6 495		50.87	0.55	0.50
11	−8.0	6 101		31.31	0.63	1.51
10	−6.4	5 844		28.37	0.66	1.46
9	−3.0	5 218		18.64	1.04	2.98
8	−2.8	5 158		19.25	0.95	3.57
7	−2.0	4 653		5.60	1.40	1.46
6	−0.9	4 385		2.18	1.77	0.66
5	−0.3	4 348		2.18	1.77	0.66
4	0.5	4 194		1.08	1.44	0.16
3	1.8	4 005		0.63	0.93	0.01
2	5.4	3 510		0.73	0.78	0.0
1	9.9	2 800		0.38	1.05	0.0

(a) 2D-P图像

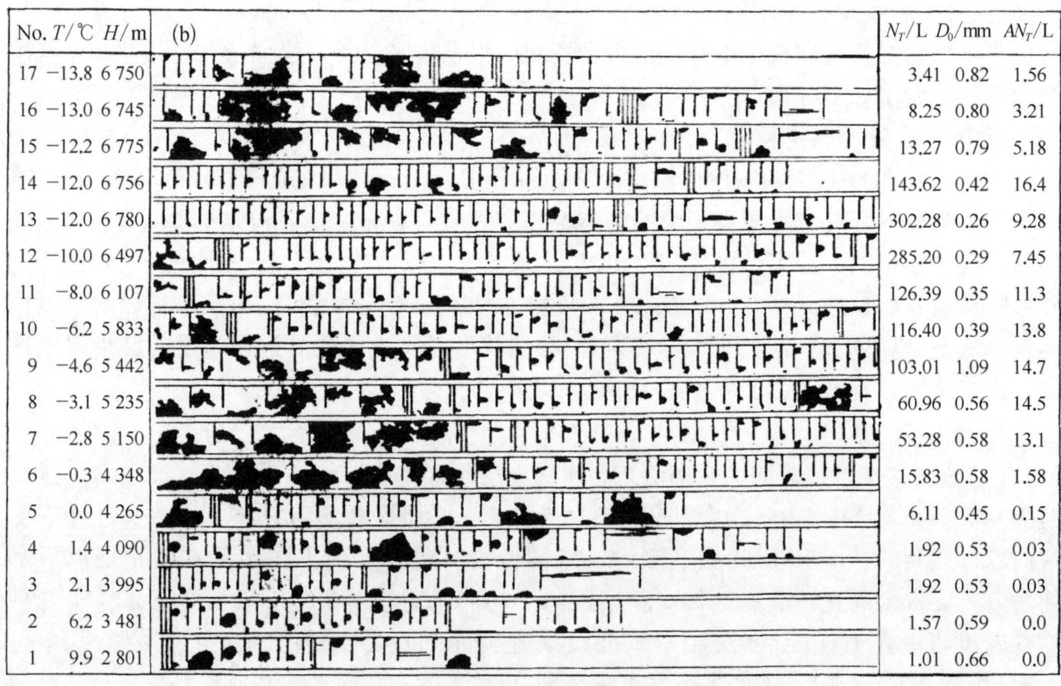

No.	T/℃	H/m	(b)	N_T/L	D_0/mm	AN_T/L
17	−13.8	6 750		3.41	0.82	1.56
16	−13.0	6 745		8.25	0.80	3.21
15	−12.2	6 775		13.27	0.79	5.18
14	−12.0	6 756		143.62	0.42	16.4
13	−12.0	6 780		302.28	0.26	9.28
12	−10.0	6 497		285.20	0.29	7.45
11	−8.0	6 107		126.39	0.35	11.3
10	−6.2	5 833		116.40	0.39	13.8
9	−4.6	5 442		103.01	1.09	14.7
8	−3.1	5 235		60.96	0.56	14.5
7	−2.8	5 150		53.28	0.58	13.1
6	−0.3	4 348		15.83	0.58	1.58
5	0.0	4 265		6.11	0.45	0.15
4	1.4	4 090		1.92	0.53	0.03
3	2.1	3 995		1.92	0.53	0.03
2	6.2	3 481		1.57	0.59	0.0
1	9.9	2 801		1.01	0.66	0.0

(b) 2D-C图像

图3　不同高度(H)和不同温度(T)下的降水质点二维图像样本

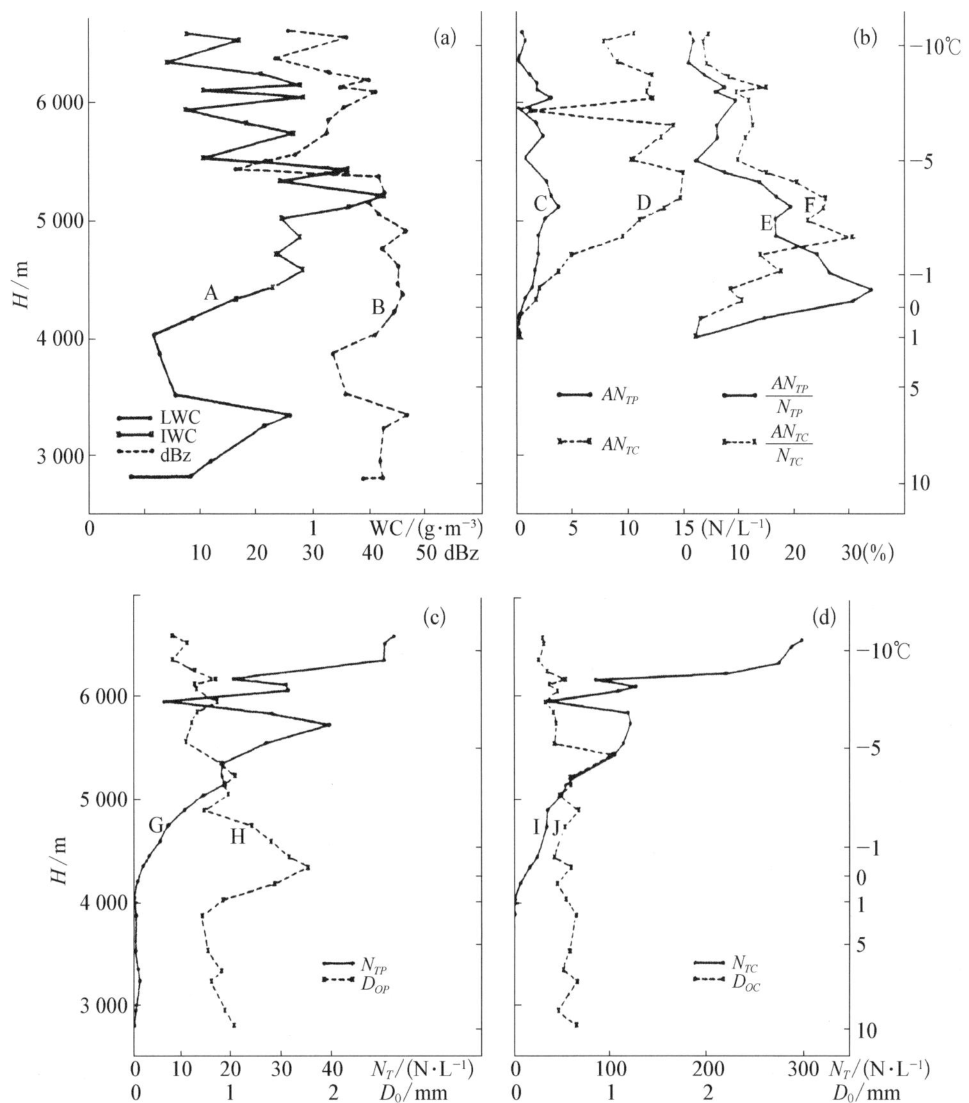

(a) 曲线 A 为液态含水量(LWC)和固态含水量(IWC);曲线 B 为由 PMS 资料推算的雷达反射率因子(dBZ);(b) 曲线 C 和 D 为由 2D－P 和 2D－C 探头观测的冰晶聚合体数浓度(AN_{TP}和AN_{TC});曲线 E 和 F 为其相对浓度$\left(\frac{AN_{TP}}{N_{TP}}和\frac{AN_{TC}}{N_{TC}}\right)$;(c) 曲线 G 和 H 为 2D－P 探头观测的质点总数浓度(N_{TP})和相应的体积中值直径(DOP);(d) 与(c)相同,只是由 2D－C 探头观测的值(N_{TC}和D_{OC})。

图 4　MCC 层状区中微物理量的垂直廓线

大些,这可能与附近尚有对流活动有关。此外,云底较暖(约 10 ℃)、云层较厚也有关系。飞行途中在层状区观测到中至大雨。

3. 降水质点数浓度和尺度

图 4(c)和 4(d)示出由 2D－P 和 2D－C 探头观测的二维质点图像推算的降水质点数浓度 N_T 和体积中值直径 D_0。从中看出,在融化层以下液态水滴数浓度一般都较低,平均为 0.8 L^{-1}(2D－P)和 2.3 L^{-1}(2D－C),相应的平均体积中值直径 D_0 为 1.0 mm(2D－P)和 0.6 mm(2D－C)。

在融化层以上，主要是冰质点，其中包括相当数量的形状不规则的冰晶聚合体。在0～−10 ℃气层内，质点平均数浓度为 27 L^{-1}(2D－P)和 133 L^{-1}(2D－C)，对应的 D_0 分别为 0.8 和 0.4 mm。但是，融化层附近的数浓度却是比较低的，从 4 200 m 到 5 000 m(1～−2 ℃)气层内一般都小于 10 L^{-1}(2D－P)和 35 L^{-1}(2D－C)。质点数浓度大体上是随高度向上增加的，在飞机垂直探测的顶部 6 600 m 高度上观测到最大值 52 L^{-1}(2D－P)和 289 L^{-1}(2D－C)[图 4(c)和(d)中曲线 H 和 J]。

降水质点的大小尺度与数浓度相反，在 0 ℃层以上大致是随高度向上减小的[图 4(c)和(d)中曲线 G 和 I]。在 0 ℃层附近 4 200～4 800 m(0.5～−1.3 ℃)气层内，质点的体积中值直径较大，这里有较多较大的冰晶聚合体。图 4(b)示出冰晶聚合体的数浓度(曲线 C 和 D)及其相对浓度(曲线 E 和 F)，从中可见，在上述气层内冰晶聚合体占总冰质点数的 15％(2D－P)以上。冰晶聚合体对 D_0 值的贡献较大，在 0 ℃层附近 2D－P 探头观测的质点 D_0 达到极值 1.8 mm。5 000 m以上高度，D_0 随高度缓慢减小。由图 4 所列资料结合图 3 的降水质点样本推断，层状降水区 0 ℃层以上的过冷气层内，冰晶聚并过程是降水质点增长的主要机制。随着高度下降接近融化层，聚并过程导致冰质点数浓度减小而质点平均尺度增大的结果。大的冰晶聚合体落至 0 ℃层以上，融化而形成雨滴，这是层状区降水的主要形式。

4. 雷达反射率因子

根据 PMS 2D－P 质点二维图像推算的雷达反射率因子的垂直廓线示于图 4(a)(曲线 B)，在 0 ℃层以上较厚的气层内(4 200～5 300 m)其值都比较大。结合含水量、冰晶聚合体浓度以及总质点体积中值直径 D_0 的垂直廓线(图 4(a～c)中曲线 A，C，H)，可以认为融化层以上雷达反射率因子数值较大主要是由于上述三个因子的联合效应所致。在 0 ℃附近，冰晶聚合体尺度较大引起大的反射率因子。而在高度较高、温度较低的气层，含水量较大和数浓度较大是主要的原因。

在温度略低于 0 ℃的气层内，冰晶大量地聚并形成冰晶聚合体，雷达反射截面增大，导致大的雷达反射率因子，而在温度略暖于 0 ℃的气层内，冰晶聚合体因表面融化而变湿，其表面的介电常数与水滴类似，进一步增强雷达反射率。当冰晶聚合体在下落过程中完全融化时，其体积变小而下落速度变大，从而使其数浓度、平均尺度以及雷达反射率减小。Stewart 等[2]还从质点在融化层以下因融化落速增大导致质点数浓度辐散，以及融化后水滴体积变小、大雨滴破碎等效应来解释雷达反射率的减小。这与本文的降水微结构垂直分布特征是一致的。

5. 冰晶聚合体

冰晶聚合体是固态降水元的一种基本形态。聚合体的形状、尺度、体积密度和数浓度对于云和降水系统的微物理研究具有重要意义。Jiusto[3]认为，冰晶聚合体是美国东北部地区降雪的最普遍类型。Hobbs 等[4]指出在美国喀斯喀特山脉的顶峰地区冬季有一半以上的固态降水是以冰晶聚合体的形式落至地面的。近年来，一系列的观测证据表明，在各种降水性云系中，冰晶聚合体是降水总量中很重要的一种成分，这包括北科罗拉多河流域的冬季地形云系(Rauber)[5]、温带气旋的中尺度雨带(Matejka 等)[6]、加里福尼亚塞拉山脉地区的冬季层状云(Stewart 等)[2]、南中国海地区冬击季风云团中的层状云区(Houzc 等)[7]、科罗拉多的雹暴(Heymsfield 等)[8]，以及成熟期的飓风云系(Jorgensen)[9]。本文前一部分的分析业已表明，冰晶聚合体也是中纬度 MCC 层状区融化层以上的冷层内的主要降水形式，至少在飞机探测高度上限 6 750 m(－14 ℃)以下的层状区中是如此。

从质点二维图像的典型样品图 3 可见，在这个 MCC 层状区探测中，有两个冰晶聚合体集中带，结合图 2 的飞行轨迹推断，这两个不同高度的冰晶聚合体集中带对应于不同的中-β尺度对流雨团。一个在飞机螺旋形爬升时处于 0 ℃至－4.4 ℃区间，对应于第一个雨团(相当于图 2 中 91°W，41°N 的位置)，从垂直加速度仪资料判断，这是一个衰老的对流雨团，观测到的最大垂直气流不超过 6 m/s。前面已经指出，06:00 GMT 时 MCC 内东西向对流线已移至 40°N 附近，飞机探测区域处于对流线北侧的层状区中。在这一地带观测到冰晶聚合体的平均浓度为 1.9 L^{-1}(2D－P)和 8.1 L^{-1}(2D－C)，它们大约占全体降水质点总含水量的 60%以上。冰晶聚合体数浓度最大值为 3.6 L^{-1}(2D－P)和 14.7 L^{-1}(2D－C)，分别位于－3.1 ℃和－4.4 ℃层上[参看图 4(b)中曲线 C 和 D]。

由于冰质点数浓度一般都随高度增加，所以冰晶聚合体的相对浓度(冰晶聚合体浓度对冰晶总浓度之比)在较低的气层达到极大。在 0 ℃至－1 ℃气层内大约有 30%的质点是冰晶聚合体，其平均数浓度为 0.9 L^{-1}，与融化层以下的雨滴数浓度同一数量级。这表明，在 MCC 的层状区，雨滴主要是冰晶聚合体下落通过 0 ℃层高度融化的产物。

由图 4(b)中曲线 C 和 D 还可看出，2D－P 探头观测的较大的冰晶聚合体主要集中在较低的气层，接近 0 ℃层，而 2D－C 探头观测的较小的冰晶聚合体主要集中在较高的高度(－1.4～－3.5 ℃温度区间)。据此推断，在 MCC 的层状区，冰质点的聚并增长过程起源于较高、较冷的气层，在其下降途中聚并效率渐趋增强(这从冰质点数浓度随高度下降和减小、质点平均尺度迅速增大的事实也可得到佐证)，从而在 0 ℃层附近形成一冰晶聚合带，它是层状云中雷达回波亮带的组成成分。在融化层中，冰晶聚合体开始融化表面变湿而体积变化不大，使反射率大增。

当 P－3 飞机离开第一个衰老的对流雨团时，它从 5 500 m 高度向西缓慢爬升至 6 600 m 高度，此其间很少观测到尺度达 3 mm 以上的冰晶聚合体。但是，当飞机抵达云系的西北部(图 2 中 92°W～93°W，41°N 区段)，在云系的边缘地区观测到第二个大的冰晶聚合带，那里原先曾有一对流雨团发展。如图 2 所示的西部雷达回波区，该区是根据 1 小时半以前当飞机在北侧云底平飞时由飞机腹部的平面扫描雷达观测的。06:30 GMT 飞机在 6 700 m左右高度上观测时，该处雷达回波已十分微弱，据此判断这时观测取样的云层是处在消散的对流云团中或其附近的云区。在这一冰晶聚合带，冰晶聚合体的平均数浓度达到 2.0 L^{-1}(2D－P)和14.4 L^{-1} (2D－C)。该处冰晶总浓度较低，不超过 10 L^{-1} (2D－P)和 20 L^{-1}(2D－C)。其中 2D－P 质点有 20%以上是冰晶聚合体。所以，图 3 显示的两个冰晶聚合带实际上并不在同一地区不同高度上观测的，而是在云系的不同部位观测的。

在两个冰晶聚合带，冰晶聚合体主要是由不规则的冰质点和枝状晶体组成的，此外还有看上去有少量淞附的粒子。前面的分析已指出，在这个 MCC 的层状区，云液态含水量不大，一般低于 0.3 g/m^3，这对降水质点的轻度淞附增长是有利的。这种轻度的淞附作用使冰质点表面略为潮湿，有利于冰质点之间的聚并，淞附在冰质点上的过冷云水在冰质点之间的聚合过程中起到一种“粘合剂”的效应。在冰晶聚合带中还观测到极少数像霰粒子那样的质点，表明确有淞附过程在同时起作用，特别是在接近 0 ℃层的较暖的过冷云层中更是如此。

6. *层状区中降水微结构的水平分布*

图 5 示出各微物理要素、垂直气流速度和温度在层状区中的水平分布。这些资料是当

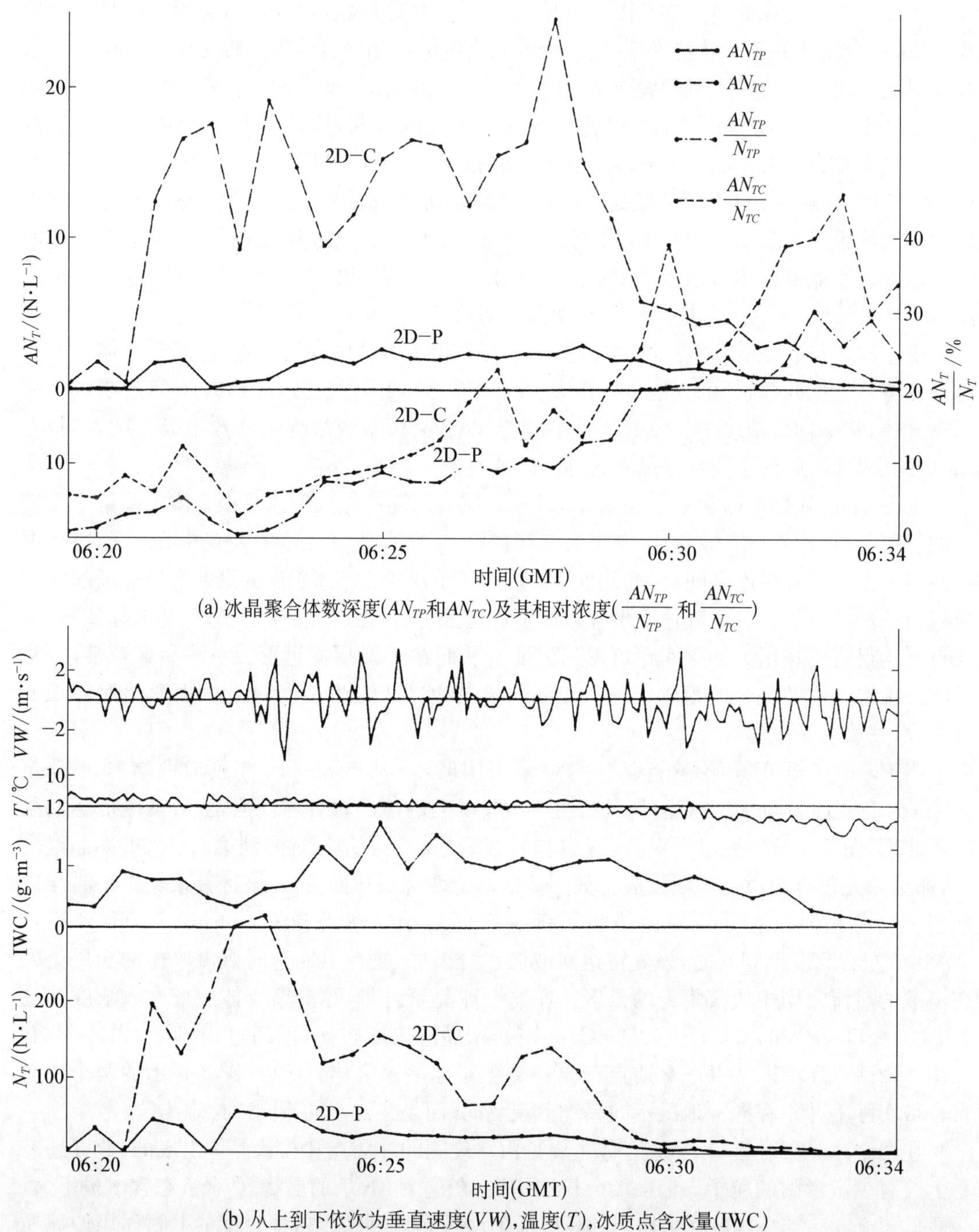

(a) 冰晶聚合体数深度(AN_{TP}和AN_{TC})及其相对浓度($\frac{AN_{TP}}{N_{TP}}$和$\frac{AN_{TC}}{N_{TC}}$)

(b) 从上到下依次为垂直速度(VW),温度(T),冰质点含水量(IWC)和冰质点总浓度N_{TP}和N_{TC}(2D−P和2D−C资料)

图 5　飞机在 MCC 层状区内水平穿云时各微物理量和温度及垂直速度的时间序列分布

飞机在 425～420 hPa(−11.1～−13.9 ℃)高度上在层状区中向西北方向大体上是水平穿云飞行时观测的。图中显示出,各种微物理量包括冰质点含水量(IWC)。冰质点数浓度

(N_T)、冰晶聚合体数浓度(AN_T)及其相对浓度(AN_T/N_T)的水平分布都是不均匀的。由图中可见,在MCC的西北部边缘地区,相当于观测时间06:30～06:34 GMT,观测到大的冰晶聚合体。这里的云是老的并处于消散阶段,垂直速度探测表明这里盛行下沉气流,冰质点数浓度一般都较低,约为10^0个L^{-1}(2D-P)和10^1个L^{-1}(2D-C)量级,其中20%以上是冰晶聚合体,已如上小节所述。这里气温接近-14 ℃,云液态含水量一般都低于JW含水量仪的观测灵敏度(0.1 g/m^3),凇附效应较弱,所以这里冰质点之间的聚并机制与0 ℃层附近的过冷云层中的有所不同。众所周知,-15 ℃附近是枝状冰晶生长的温度区间,而枝状冰晶的分枝叉结构有利于冰晶相互勾联,这种冰晶枝叉之间的机械勾联效应在这一温度区间内的冰晶聚合过程中可能起重要作用[5,10]。

四、降水质点的大小谱分布

本节根据2D-P质点图像资料对直径大于0.2 mm的降水质点大小谱分布特征进行分析。滴谱资料样本容量为78,均在MCC的层状降水区中获取。应用Marshall-Palmer分布型

$$N(D)=N_0 e^{-\lambda D} \tag{1}$$

对资料进行拟合分析。式中:$N(D)$是直径为D的质点大小分布密度函数;N_0和λ是分布参数,分别称为截距参数和斜率参数。拟合分析发现上述负指数律分布对水滴和冰质点大小谱分布拟合都相当好。对于融化层以下的液态水滴,$\log N(D)$对D的线性相关系数达-0.96,对融化层以上的冰质点谱分布,相应的线性相关系数为-0.92,线性拟合都相当好。但是,分布参数N_0和λ都不是常数,而是随数浓度和体积中值直径而变的变量。滴谱分布的斜率参数λ和截距参数N_0对质点的体积中值直径D_0和数浓度N_T的线性拟合关系可分别表示为

$$\lambda=-2.51+\frac{1.65}{D_0}, \tag{2}$$

和

$$N_0=\frac{1.80N_T}{D_0}。 \tag{3}$$

上述最佳拟合线的相关系数都达到0.87(图略)。应用这种经验关系,在降水微物理过程的参数化研究中,降水质点大小谱分布的经验公式可表为

$$N(D)=\frac{1.80N_T}{D_0}\exp\left(2.51-\frac{1.65}{D_0}\right)D。 \tag{4}$$

这一经验关系与Cotton等[11]的参数化公式类似,只是他们假定斜率参数λ(或D_0)是常数。

我们现在将M-P分布中的参数与云中可以测量的降水含水量联系起来。虽然由图6(a)可见,λ与含水量WC之间并无明显的相关关系,但根据Kessler(1969)推导的滴谱参数化公式,我们选用变量λ与组合变量$(N_0/WC)^{1/4}$,其散布图示于图6(b),可见无论对冰质点还是水滴样本,线性拟合都相当好,相关系数分别达到0.98和0.73,拟合公式可表为

$$\lambda = a + b\left(\frac{N_0}{\mathrm{WC}}\right)^{\frac{1}{4}}, \tag{5}$$

对水滴样本，系数 $a=0.96$，$b=1.28$；对冰质点为 $a=-7.1$，$b=1.33$。

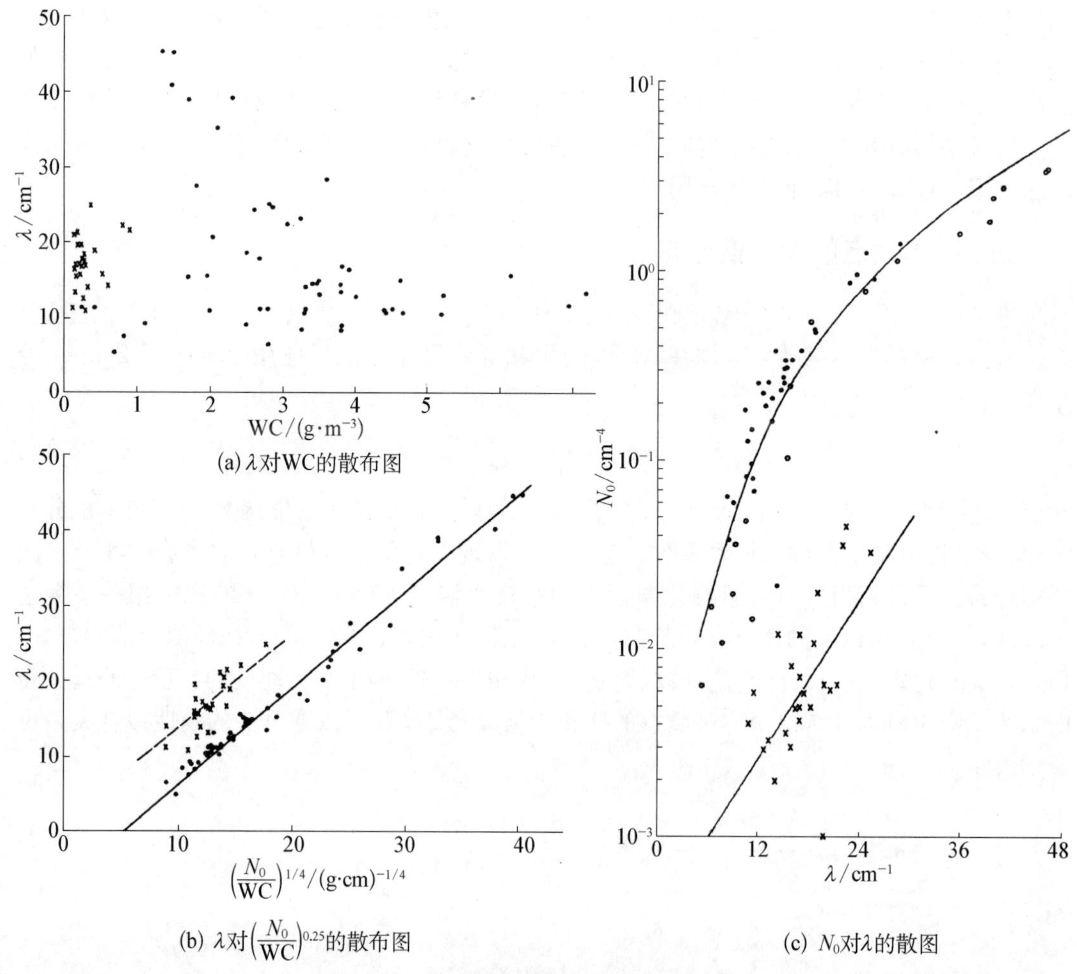

(a) λ 对WC的散布图

(b) λ 对 $\left(\frac{N_0}{\mathrm{WC}}\right)^{0.25}$ 的散布图

(c) N_0 对 λ 的散图

散布点“×”是水滴样本，“·”和“∘”是冰质点样本。

图6　滴谱参数之间的散布图

在云和降水模拟研究中，通常假定降水滴谱分布中的一个参数（λ 或 N_0）为常数，从而将滴谱分布函数简化为单参数分布。图6(c)示出 N_0 对 λ 的散布图，可见对水滴和冰晶质点，N_0 对 λ 的相关关系有明显差异。对水滴，不同样本的斜率参数变化不大，平均值为 $17(\pm 3.6)\,\mathrm{cm}^{-1}$，相对变差不超过20%，所以可以像Cotton等[11]那样，对水滴近似地假定 λ 是常数。然而，对冰质点样本，λ 显然不能当作常数处理，不同样本的 λ 值可相差3倍以上。至于 N_0，则无论是水滴样本还是冰晶样本，都是变量，其值可有2～3个数量级之差。由图6(c)得到 N_0 对 λ 的最佳拟合线（冰质点）为

$$N_0 = 1.1\times 10^{-4}\lambda^{2.8}, \tag{6}$$

拟合线的相关系数达到 0.94。在这种情况下，降水质点谱分布可以只用一个参数 λ 来确定。结合方程(1)、(6)和式(2)，冰质点的大小谱分布可以仅仅根据质点的体积中值直径 D_0 一个参数来推断。

参考文献

[1] Maddox, R. A., 1980, Mesoscale convective complexes, *Bull. Amer. Meteor. Soc.*, **61**, 1374 - 1387.

[2] Stewart, R. E., J. D. Marwitz, J. C. Pace, and R. E. Carbone, 1984, Characteristics through the melting layer of stratiform clouds, *J. Atmos. Sci.*, **41**, 3227 - 3237.

[3] Jiusto, J. E., 1971, Crystal development and glaciation of a supercooled cloud, *J. Rech. Atmos.*, **5**, 69 - 85.

[4] Hobbs, P. V., S. Chang, and J. D. Locatelli, 1974, The dimensions and aggregation of ice crystals in natural clouds, *J. Geoph. Res.*, **79**, 2199 - 2206.

[5] Rauber. R. M., 1987, Characteristics of cloud ice and precipitation during wintertime storms over the mountains of Northern Colorado, *J. Climate and Appl. Meteor.*, **26**, 488 - 524.

[6] Matejka, T. J., R. A. Houze, Jr., and P. V. Hobbs, 1980, Microphysics and dynamics of clouds associated with mesoscale rainbands in extratropical cyclones, *Quart. J. Roy. Met. Soc.*, **106**, 29 - 56.

[7] Houze, R. A., Jr., and D. D. Churchill, 1984, Microphysical structure of winter cloud clusters, *J. Atmos. Sci.*, **41**, 3405 - 3411.

[8] Heymsfield, A. J., and D. J. Musil, 1982, Case study of a hailstorm in Colorado, Part II: Particle growth processes at mid-levels deduced from in-situ measurements, *J. Atmos. Sci.*, **39**, 2847 - 2866.

[9] Jorgensen, D. P., 1984, Mesoscale and convective-scale characteristics of mature hurricanes. Ph. D. dissertation, Colorado State University, Fort Collins, CO 80523, 189 pp.

[10] Ohtake. T., 1970, Factors affecting the size distribution of raindrops and snowflakes, *J. Atmos. Sci.*, 27, 804 - 813.

[11] Cotton. W. R.. G. J. Tripoli, R. M. Rauber, and E. A. Mulvihill, 1986, Numerical simulation of the effects of varying ice crystal nucleation rates and aggregation processes on orographic snowfall, *J. Clim. Appl. Met.*, **25**, 1658 - 1680.

Microphysical Structure in the Stratiform Cloud Region of a Slow Moving MCC

Ye Jiadong　Fan Beifen
(Department of Atmospheric Sciences, Nanjing University, Nanjing　210093)
W. R. Cotton　M. A. Fortune
(Colorado State University Department of Atmospheric Sciences, U. S. A)

Abstract　Microphysical data observed in the stratiform cloud region of an MCC are analyzed together with radar, satellite, and other in situ data collected by a NOAA P-3 research aircraft. The results show, that the aggregates are prevalent in some part of the stratiform cloud region of the MCC; the aggregation process is the predominant precipitation particle growth mechanism in the stratiform cloud region. Aggregation starts in the upper and colder levels but becomes more efficient as the aggregates approach the melting layer. As a result a large aggregation zone is formed near the 0 ℃ level.

Key words: MCC; stratiform cloud region; microphysics; aggregates.

5.6　冷雨过程中的聚并效率问题

叶家东　范蓓芬
(南京大学大气科学系，南京 210093)

冰晶之间的聚并系数是雪花聚并成长过程的主要因子之一，它是碰撞系数与黏合系数的乘积。由于冰粒具有不同的形状、不均衡的垂直运动和水平飘动，它的空气动力学特性及其碰撞系数都远比水滴复杂；冰粒之间的黏合系数，也受众多因子影响，如空气温度、湿度、冰晶形状及表面习性以及是否存在过冷云滴等都会影响冰粒之间的黏合系数，所以通常采用实验研究的方法测定聚并系数。Hallgren 和 Hosler(1960)实验指出在冰面饱和空气中冰粒之间的碰并系数是很低的，一般低于 0.2，且随着气温从－11 ℃降至－26 ℃而降低，而随着气温从－11 ℃升至－6 ℃碰并系数也降低。Latham 和 Saunders(1971)的实验则表明在冰面饱和空气中冰晶的碰并系数近似为常数 0.3，不随气温而变。Passarelli(1978)则从飞机观测资料归纳出，在气温为－12 ℃至－15 ℃范围内，冰晶的平均聚并系数可达1.4±0.6，与上述实验值之间的差异主要是由于自然云中冰晶形状更为复杂、大小尺度谱更宽以及湍流更强所致。而且，实验研究多数是在冰面饱和空气中测试的，自然云中当存在过冷云滴时一般空气湿度对冰面是过饱和的。Hosler 等(1957)业已发现在冰面过饱和环境中测量的聚并系数要高得多。

我们仿照 Passarelli(1978)的方法利用外场云物理观测资料，采用矩守恒方程诊断冰晶聚并效率。对于成熟阶段的降水系统，假定一气层上、下两个高度正处于聚并过程中的降水粒子其总质量通量和相应的雷达回波反射因子通量守恒，据此可以算出该气层内的平均聚并效率。对于成熟阶段处于稳定降水状态具有定常降雪质量通量的系统，所导出的平均聚并系数为

$$\bar{E}=(\lambda^{-1-b}+\lambda_0^{-1-b})\frac{12\Gamma(7+b)\lambda^{a+b}}{(1+b)\pi N_0 hI(b)}。\tag{1}$$

式中：h 是气层的厚度；N_0 和 λ 是气层顶（$Z=Z_0+h$）的降水粒子谱负指数分布的参数；λ_0 是气层底（$Z=Z_0$）的斜率参数；b 是粒子下降速度经验公式 $v(D)=aD^b$ 中的参数；而 $I(b)$ 为积分

$$I(b)=\int_0^\infty\int_0^\infty x^3y^3(x+y)^3\mid x^b-y^b\mid \mathrm{e}^{-(x+y)}\mathrm{d}x\mathrm{d}y。\tag{2}$$

根据 Hobbs 等（1974）的实验，我们选取 $b=0.25$，积分式(2)得 $I(b)=566$。因此，在定常质量通量假定下，只要根据云物理观测资料计算出气层上下界的粒子谱分布参数 N_0 和 λ、λ_0，该气层的平均聚并系数就可以从方程(1)计算出来。

我们根据 1985 年美国 Pre - STORM 计划期间在美国中西部的一个 MCC 成熟期层状云区垂直云物理探测资料计算各高度降水粒子尺度谱分布函数的参数 N_0 和 λ，探测时飞机按照 1～2 米/秒的垂直速度螺旋式下降以追踪冰晶聚并过程，因而可以大体上满足不同高度的雪花具有定常质量通量的假定。图 1 所示是根据不同高度的粒子谱分布资料计算的分布参数散布图和所拟合的回归线 $N_0(Z)$ 和 $\lambda(Z)$，拟合相关系数分别达 0.76 和 0.54，对应的经验关系式为 $N_0(Z)=1.9\times10^{-4}\times10^{0.56Z}$ 和 $\lambda(Z)=-2.30+3.89Z$。式中：Z 的单位为 km，N_0 和 λ 的单位分别为 cm^{-4} 和 cm^{-1}。根据不同高度的 N_0 和 λ 的拟合值由式(1)计算出不同高度气层的平均聚并系数，示于图 1 右侧，对应的聚并系数依温度的经验关系式为

$$\overline{E(T_s)}=10^{0.235+0.0633T_s},\tag{3}$$

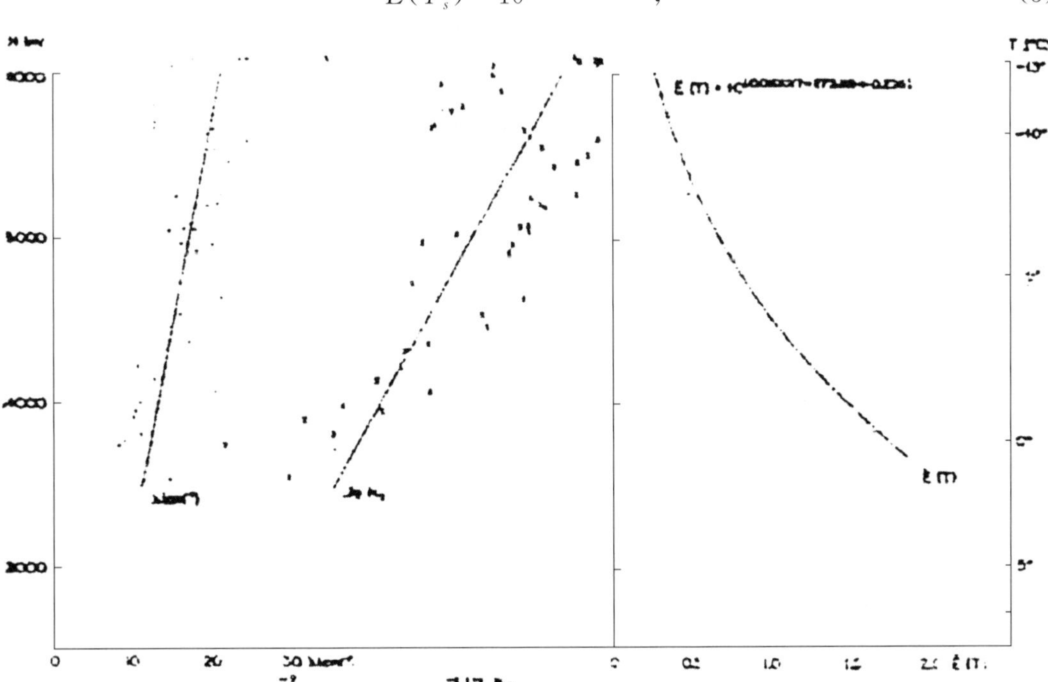

图 1　固态降水粒子该分布的斜率参数 λ 和截距参数 N_0 不同高度的散布图，右侧为诊断的聚并系数 $E(T)$ 分布曲线

拟合线的相关系数达 0.98。式中：$T_s = T - 273.16$，T 是冰粒表面温度，这一经验关系式适用于上述观测资料所在的温度区间 0～－13 ℃。温度更低的气层假定水汽对冰面饱和，仍可按照 Cotton 等(1986)的方法，采用 Hallgren 和 Hosler 的实验结果：

$$\overline{E(T_s)} = \mathrm{Min}[10^{0.035T_s - 0.7}, 0.2]。\tag{4}$$

从图 1 可见，在温度接近 0 ℃层附近(－4～0 ℃)的气层内诊断的聚并系数可以大于 1，这与层状云区 0 ℃层附近常有冰晶聚合带的观测事实相协调。

利用 RAMS 模式对不同聚并系数对冰相微物理过程的效应做数值试验，初值探空资料采用 1985 年 6 月 4 日——MCC 发展前的规测值，图 1 的云物理资料即从该 MCC 的层状区获取。采用水平均匀场进行二维数值模拟试验。控制试验的冰粒聚并系数采用本文拟合的诊断公式(3)和(4)；对比试验聚并系数采用 Cotton 等(1986)所用的两套方案：

对比试验 1：－12～－15 ℃间 $E(T_s)=1.4$；其他用式(4)。

对比试验 2：－12.5～－15.5 ℃区间水面饱和 $E(T_s)=1.4$；其他区间则取 $E(T_s)=0.3$。

控制试验模拟 3 小时期间，冰晶聚合体主要分布在趋于消散的对流单体和云砧层状云区，位于 4～8 千米高度层中，最大比含水量可达 2.5 至 4.0 克/千克(图略)。对比试验 1 云中基本无冰晶聚合体出现，云中只出现少量霰含量，而单体冰晶则满布 7 千米高度以上的冷区，峰值比含水量达 3.0 克/千克(图略)，说明此方案冰晶聚并效率太低；对比试验 2 则云中整个冷区满布冰晶聚合体，峰值比含水量达 4.0 克/千克，模拟 135 分时且达 5 克/千克，而单体冰雪晶和霰含量又过分少(图略)，表现出所用的聚并效率又太高了，这是原 RAMS 模拟中存在的一个缺陷。采用本文诊断的聚并系数模拟的结果则比较合理。

参考文献

[1] Cottor, W. R., et al., J. Clim. Appl. Meteer., 25, 1658 - 1680, 1986

[2] Hallgren, R. E., and C. L. Hosler, Geophys. Monogr. Amer. Geophys. Union. 5:257 - 263, 1960

[3] Hobbs, P. V., et al., J. Geophys. Pes., 79, 2199 - 2206, 1974

[4] Hosler. C. L., et al., J. Met., 14: 415 - 420, 1957

[5] Latham, J. and C. P. R. Saunders, Quart. J. Rey. Meteor. Soc., 96: 257 - 265, 1971

[6] Passarelli, R. E., Jr., J. Atmos. Sci., Passarelli, R. E., Jr., J. Atmos. Sci., 35, 852 - 889, 1978

5.7 中尺度模式的积云对流参数化问题①

叶家东

(南京大学大气科学系，210093，南京)

摘　要　本文对中尺度模式中积云对流参数化的物理依据、观测证据和参数化方案做一概要的评述，以期对中尺度数值模拟研究中正确应用和发展合适的参数化方案有所裨益。

关键词　中尺度模式，积云对流，参数化

分类号　P42

① 本研究得到“八五”攻关项目 85－906 和国家自然科学基金项目(49070244)资助。

0　引言

降水性浓积云中水的相变以及云中强上升气流和下沉气流伴随着大量的能量转换和垂直输送，它们对大中尺度天气系统的动力学和能量学有着重要的影响。即使是非降水性积云和层积云，也能对大尺度天气系统乃至局地气候产生重要影响。所以，在较大尺度的数值模式中就需要定量地考虑这些次网格尺度效应。为此，首先需要确立积云对流效应与较大尺度模式的可分辨参数之间的定量关系，再根据这种关系估计积云对流对可分辨尺度运动的物理效应。这就是积云对流的参数化[1]。它涉及两个重要方面：一个是大尺度强迫对积云对流的调节和控制作用，据此可预报对流降水率；另一个是积云对流对大尺度环境的反馈效应，包括相变潜热效应和热量、水分及动量的垂直输送效应。鉴于目前中尺度数值模式中采用的各种积云参数化方案，有的是从大尺度数值预报模式中引用过来的，有的虽然是针对中尺度模式设计的，但物理依据以及验证这些依据的观测证据都有各自的局限性或缺陷，以致到目前为止，还没有哪一种方案能令人满意地反映不同时空尺度上都适用的积云对流与其环境之间相互作用的信息。因此，认识积云参数化方案的合理性和局限性，对于恰当地应用和改进这些方案是必要的。

1　积云参数化的物理基础和观测证据

为了在数值模式中对积云对流效应进行参数化，首先要求在可分辨尺度运动与次网格尺度积云对流运动之间确实存在某种物理联系，特别是要有能控制或至少是能调节对流过程的大尺度参数，这意味着积云对流是受大尺度动力学和热力学制约的。在热带和温带进行的一系列观测研究表明，积云对流特别是降水性深对流确实是强烈地受环境大气的热力结构和水分结构以及大尺度流场影响的[2-4]。在热带积云生长中，动力因子是比热力因子更为重要的因素。热带太平洋地区积雨云发展的特征性天气条件是弱的层结不稳定度和低层辐合，而晴天则有强的不稳定度和辐散气流[2]。类似的观测证据在GATE期间大西洋东部地区和太平洋西部赤道地区的热带对流系统[3,5]、热带气旋(飓风)云系中都有所发现[6,7]，这构成了郭晓岚参数化方案的基础：总的对流加热是与大尺度水分辐合成比例的[8]。中纬度地区斜压性大，大尺度动力学强迫比热带更强，风的垂直切变也大，空气则更干燥，使得降水蒸发冷却效应比潮湿的热带更强。不过仍然有不少观测证据表明，中纬度地区的积云对流也经常强烈地受大尺度过程控制。在美国中部观测的对流活跃区与低层质量和水汽辐合区吻合[9,10]，传播性飑线对流往往发生在中尺度辐合区中，对流性降水也与地面辐合区相联[11]，类似的证据在中国的梅雨暴雨中也有所发现[12,13]。与热带不同的是，中纬度地区对流降水与层结不稳定度之间存在正相关关系[14,15]，降水效率和对流风暴强度且强烈地受风垂直切变的影响[16,17]。总之，在热带和中纬度地区都有证据表明，天气尺度和中尺度运动对积云对流，特别是降水性深对流有重要的控制或调节作用，其中水汽辐合和上升运动两个大尺度参数与对流性降水有高度的正相关，它们对热带和中纬度地区都是重要的。中纬度湿对流还受大尺度热力不稳定度和风垂直切变的影响。根据这些关系可以由大尺度参数的预报量确定对流降水量。

至于积云对大尺度变量的效应，在利用雷诺平均法导出大尺度(格体面积平均)热力学方程、水汽方程和涡度方程时，从中得出由次网格积云对流效应(包括凝结潜热和垂直通量输送)引起的“视热源 Q_1”“视水分汇 Q_2”和“视涡度源”。通过诊断研究将它们表示为大尺度变量的函数。需要注意的是,这些视源和视汇是由许多大小不一、热力学和动力学属性各异的云共同造成的结果,所以积云参数化方案需考虑云的总体净效应。Yanai 等发展的积云总体模式可以用来估计积云总体的这种累积效应[18]。在这个模式中视热源 Q_1 是辐射冷却、补偿下沉运动引起的绝热增暖以及云水消卷引起的蒸发冷却效应之总和;视水分汇则考虑了补偿下沉运动引起的环境干燥化以及因消卷云中水汽和消卷云水的蒸发引起环境空气的湿润化效应。这构成了 Arakawa-Schubert 积云参数化方案的基础[19]。一系列诊断研究为这提供了经验依据[20-24]。在热带和中纬度地区的诊断研究中,视热源 Q_1 的垂直廓线大体上相类似,多数研究表明 Q_1 在 600~300 hPa 之间有一极大值。中纬度的 Q_1 廓线在对流层低层常常比热带地区的视热源小,这是由于中纬度云底较高、云下蒸发较大引起的。在热带和中纬度的视水分汇 Q_2 的垂直廓线也相类似,最大水分汇出现在 800~600 hPa 高度之间,比 Q_1 廓线中的最大值高度低些。积云的总效应是使大部分对流层变干增暖,这种效应主要是通过云的环境空气的补偿下沉运动引起的绝热增暖造成的,但云水消卷以及热量和水分在湿下沉气流中的输送在低层大气的热量和水分收支中也是重要的因素。积云对流也可以影响大尺度涡度和动量结构。在热带积云对流一般在对流层低层产生一视涡度汇,而在对流层上层产生一视涡度源。中纬度的定量研究较少,但仍然有证据表明积云对流会引起强的动量输送,环境大气对中纬度对流系统的动力响应是大的。中纬度对流系统的热量和水分收支效应与热带中的主要区别包括:(1) 在中纬度水分和温度的存储项(局地变化项)常常与其他项同等大小,而在热带中它们通常是小项,可以忽略;(2) 中纬度云底较高、环境空气较干燥,因而蒸发效应较大;(3) 前面已指出,中纬度系统除了大尺度垂直运动和水分辐合以外,湿对流与静力不稳定度有正相关,这与热带不同。总之,尽管在积云对流、中尺度运动与天气尺度系统之间存在复杂的相互作用,观测研究还是为积云参数化提供了一定的物理基础和经验依据,同时也不难看出,这些基础和依据是有局限性的。

2 粗网格模式积云参数化方案

(1) 湿对流调整方案

假定大气层结不稳定时就产生积云对流,积云对流的反馈效应则使层结调整为中性,对流降水率则由调整前后气柱的水汽含量之差确定。这种方案称为硬对流调整方案[25]。这种方案简单,但未真正涉及调整的物理过程;而且,调整的结果总是使低层空气变得过分干冷,水汽除去太多,而使诊新的降水率太大。故此提出修正的方案,称为软调整方案[25],它假定硬调整只发生在大尺度网格面积的一小部分 a 上,其余$(1-a)$部分面积的热力学状态维持初值不变。a 的数值由调整后的气柱平均相对湿度 f 值采用迭代搜寻法逼近确定,半预报检验发现 $f=82\%$时计算降水率与观测值最一致,但由于最大不稳定度常发生在最大对流之前,所以计算的与观测雨量之间在时间上有位差,只适用于气候模式,这里降雨时间不是最主要的。

近年来 Betts 等[26]发展了一种“新对流调整方案”,假设模式的温度和湿度等热力学结构向某种由观测证据归纳的准平衡态张弛,对深对流和浅对流分别采用不同的准平衡廓线。为了弥补大尺度场变化与对流响应之间的时间滞后效应,引进了张弛时间 τ($\tau\approx1\sim2$ h,取决于不同的模式和不同的降水系统)。这种方案在全球模式、NMC 的 NGM 套网格模式和有限域细网格模式(LFM)中应用,表明对较大的降水预报效果较好[27]。不过有迹象表明这种改善可能主要是由于空间差分方案引起的。

(2) 郭晓岚积云参数化方案

所有建立对流性降水与大尺度水汽辐合之间相关关系基础上的积云参数化方案统称为郭晓岚型方案[8,25,28,29]。郭晓岚方案假定大气条件不稳定和气柱有净水汽辐合时积云对流产生。利用大尺度水汽方程和云水连续方程可建立对流降水率与气柱总水汽水平辐合的函数关系,并假定水汽辐合中有一部分 b 是用来增湿气柱,另一部分$(1-b)$发生凝结并形成降水。气柱的总对流加热率由净凝结潜热释放率和地面感热量确定。为了导出加热率和增湿率的垂直廓线,各种廓型修正方案都假定了相应的归一化垂直分布函数。问题是廓型方案中加热率与增湿率有时并不协调,而且计算的加热率和对流降水率一般偏低。尽管在各种修正方案中对垂直分布函数等做了各种改进而使结果有所改善,但这类方案对云与环境的相互作用机制如夹卷和浮力及下沉气流效应等都没有涉及。而且,在对流系统发展的成熟期,观测的对流降水率有时会超过总水汽辐合[24],意味着 b 值应是负的,这时廓型方案就不适用。

(3) Arakawa-Schubert 参数化方案(简称 AS 方案)[19]

AS 方案是在大气环流模式中发展起来的较先进的积云参数化方案,与郭晓岚方案只考虑大尺度水平辐合不同,AS 方案考虑了积云对流与大尺度环境场之间的相互作用;大尺度强迫函数包括水平平流、垂直平流、辐射以及地面热量和水分通量;积云对流考虑了不同类型积云的分数夹卷率谱,并将其与一个考虑静力能和水汽收支方程的混合层模式联系起来;积云总体对环境场的反馈效应包括云与云之间的补偿下沉运动引起的增暖变干效应和云上部消卷的云水蒸发引起的冷却增湿效应。为了定量表示这些效应,需要确定积云总体质量通量廓线以及云的消卷和消卷云气的温度、水汽和液态水含量。为此提出一个衡量云中累积浮力的大气特性——云底动能产生率,称为云-功函数。大气湿对流不稳定的判据是云-功函数为正值。假定大尺度过程与积云对流效应之间处于准平衡状态,据此可以通过处于准平衡状态的云-功函数由大尺度变量确定云的质量通量。AS 方案对云与环境相互作用的物理过程考虑比较细致,模式的检验结果也比郭型方案好,但云-功函数准平衡假定的依据主要是热带积云的一些观测,另外这个方案比较复杂,计算工作量也较大。

3 中尺度模式的积云参数化

粗网格模式积云参数化方案有两个基本假设:尺度分离假设和准平衡假设。尺度分离假设假定模式可分辨尺度远大于积云尺度,模式网格体中的积云群是作为小尺度"湍流"而考虑其统计效应的。鉴于积云尺度一般为 $10^{-1}\sim10^{1}$ km,模式的分辨率就不能小于 100 km。所以这类参数化方案只适用于中- α 尺度以上的模式。当在网格尺度小于100 km 的中尺度模式中应用这类参数化方案时,上述尺度分离假设就难以自圆其说;准平衡假设假定对流效应与大尺度环境之间是处于某种准平衡状态的,这样才能用大尺度变量描述积云对流效应。对分辨率较小的时空尺度,这种平衡假设有时并不满足。例如中纬度对流系统的降水率有时可能大大超过大尺度水分供应[24],结果云的水分储量会发生大的变化。所以上述粗网格参数化方案用于中尺度模式时,其物理依据是不充分的。考虑大尺度过程与积云对流之间的不平衡状况,近年来根据位势不稳定与在一维积云模式基础上发展了一些中尺度模式的积云参数化方案。下面择要介绍两个:

(1) Kreitzberg-Perkey 方案(简称 KP 方案)[30]

采用一维积云模式模拟对流的反馈效应。当网格中的条件不稳定度超过某一由云厚决定的临界值时发动对流,对流活动一直进行到云底气压与其环境大气压力相同时为止。在有对流维持的网格气柱中,由云模式计算出云的温度、上升气流、液态水和水汽混合比,再计算云与云之间由于补偿下沉运动引起的大尺度量的变化,以及云与环境空气的混合引起的变化,将它们平均分配在大尺度时步(40 min)内而转换成大尺度变量的时变倾向。这一方案在检验时的典型形势下导出的潜热加热能产生较为真实的温度、重力位势和流场扰动。基本效应是使大尺度网格体增热变干,消弱并消除位势不稳定度。

(2) Fritsch-Chappell 方案(简称 FC 方案)[31]

FC 方案也采用一维云模式计算云的属性,与 KP 方案不同的是这里还考虑了湿下沉气流的效应。当有效浮力能为正时产生对流,采用一维积云模式计算湿上升气流和湿下沉气流,根据降水效率与风垂直切变的经验关系确定上升气流凝结体在湿下沉气流中蒸发的部分分数并据以确定降水率。根据云模式计算的各属性和环境属性按照升、降气流和环境的面积加权平均得出大尺度温度、混合比和动量的时变倾向。对流稳定化效应的速率是规定在 0.5～1 h 间隔内消除全部有效浮力能加以确定的。这种方案经初步检验能模拟出对流性中尺度气压系统:参数化的湿下沉气流引起的冷却效应可在低层产生中高压,而补偿下沉增暖效应则可导致中气旋的形成。由于 FC 方案中对流降水率是与有效浮力能密切相关的,所以它只适用于中纬度地区。在热带对流降水与不稳定能量的关系并不密切。FC 方案的主要局限性是深对流引起的全部质量调整必须局限于一个网格体中,结果补偿下沉运动的空间范围受到限制,从而当网格尺度较小时对流效应就会使网格体过分稳定化。所以,这种方案只适用于网格尺度为 20 km 或以上的模式。

由于积云参数化方案都有一些基本假定,所得的动力学和能量学结构有时并不协调。不同方案之间的可比较性也相当差,使得它们的适用性受到限制。例如 MM4 的模拟结果对参数化方案就十分敏感。近年来人们试图用显式模拟云和降水效应的方法避免现有积云参数化方案中的主观随意性。Rosenthal[32] 在一静力平衡飓风模式中发展了一个显式模拟对流加热的方案,较为合理地模拟了飓风的结构特征和 CISK 型增强过程。Orlanski 和 Ross[33] 在用静力平衡模式模拟锋前飑线时采用显式模拟对流加热和蒸发冷却效应的方案,能模拟出飑线的许多大尺度持性。由于模式分辨率较粗,上述显式模拟对流加热的方案可能只适用于由大尺度惯性稳定的环流发动的对流系统。Tripoli[34] 在地形性对流系统的二维细网格(分辨率 1.08 km)RAMS 模式中,采用显式模拟降水物理学和对流加热的方案,并与 FC 方案做了比较。结果证实了中-β 尺度与对流尺度之间并没有明显的尺度间隔,并发现:显式加热的细分辨率模式模拟的中-γ 尺度对流加热振幅较大,但中-β 尺度的加热和输送与参数化方案同数量级。参数化模拟的对流系统常与重力波相耦合,波动的相速度明显偏大。总之,到目前为止,中-β 尺度模式的积云参数化方案还远未完善。看来细分辨率模式采用能分辨对流尺度与中尺度相互作用细节的显式模拟方案,可能是最有前途的途径。

参考文献

[1] Cotton W R and Anthes R A. Storm and Cloud Dynamics. Academic Press. Inc., 1989

[2] Malkus J S and Williams R T. Meteorol Monogr., 1963, 5, 59 - 64

[3] Reed R J, and Recker E E. J Atmos Sci, 1971, 26: 1117 - 1133
[4] HR Cho, Ogura Y, J Aunos Sci, 1974, 31: 2058 - 2065
[5] T Nitta. J Meteorol Soc Japan, 1978, 56: 232 - 242
[6] Ogura Y. Pure Appl Geophys, 1975, 13: 869 - 889
[7] Anfhes R A. Meteorol Monogr, 1982, 41
[8] Kuo H L J. Atmos Sci, 1974, 31: 1232 - 1240
[9] Lewes J M. Mon Wea Rev, 1971, 99: 786 - 795
[10] Hudson H R. J Appl Meteorol, 1971, 10: 755 - 762
[11] Fankhauser J C. J Appl Meteorol, 1974. 12: 1330 - 1353
[12] 叶家东,范蓓芬,宋航,杜京朝. 南京大学学报,1993,56 - 60
[13] 叶家东,范蓓芬,杜京朝,宋航. 南京大学学报,1993,61 - 65
[14] Zawadzki I, Ro C U. J. Appl. Meteorol., 1978, 17: 1327 - 1334
[15] Zawadzki I, Torlaschi E, Sauvageau. R J Atmos Sci, 1981, 38: 1535 - 1540
[16] Fritsch, J M. Pure Appl Geophys, 1975, 13: 851 - 867
[17] Marwitz J D. J Appl Meteorol, 1972, 11: 166 - 201
[18] Yanai M, Esbensen S Cha J. J Atmos Sci, 1973, 30: 611 - 627
[19] Arakawa A, Schubert W H. J Atmos Sci, 1974, 31: 674 - 701
[20] Ogura Y, Cho H R. J Atmos Scl, 1973, 30: 1276 - 1286
[21] Nitta T. J Atmos Sci, 1977, 34: 1163 - 1186
[22] Thompson R M Jr., Payne S W, Recker E E, Reed R J J Atmos Sci, 1978, 36: 53 - 72
[23] Cho H R, Bioxam R M, Cheng L. Atmos: Ocean, 1979, 17: 60 - 76
[24] Lin M S. Ph D Thesis, Colorado State University, 1986
[25] Krishnamurti T N, Ramanathan Y, Pan H-L et al., Mon Wea Rev, 1980, 108: 465 - 472
[26] Betts A K. Quart J R Met Soc, 1985, 112: 677 - 691
[27] Mesinger F, Black T L, Plummer D W, Ward T H. Weather and Forcasting, 1990, 5: 483 - 493
[28] Krishnamurti T N, Nam S L, Pasch R. Mon Wea Rev, 1983, 111: 815 - 823
[29] Arrhes R A. Mon Wea Rev. 1977, 105: 270 - 286
[30] Kreitzberg C W, Perkey D J J Atmos Sci. 1976, 33: 456 - 475
[31] Fritsch J M, Chappell C F. J Atmos Sci, 1980, 37: 1722 - 1733
[32] Rosenthal S L. J Atmos Sci, 1978, 35: 258 - 271
[33] Orlanski I. Ross B B J. Atmos Sci. 1984, 41: 1664 - 1703
[34] Tripoli G Atmos Paper. No. 401, Dept Atmos Sci, CSU, 1986

ON THE PROBLEM OF CUMULUS PARAMETERIZATION IN MESOSCALE NUMERICAL MODELS

Ye Jiadong
(Department of Atmospheric Science, Nanjing University, 210093, Nanjing, China)

Abstract A brief review for the cumulus parameterization schemes on mesoscale numerical model and their physical bases and observed exidences is presented in this paper.

Key words mesoscale numerical model, cumulus convection, parameterization

5.8 云物理模式中数浓度预报方程及其参数化①

叶家东 范蓓芬

(南京大学大气科学系,210093,南京)

摘 要 本文在RAMS云模式的框架中,考虑降水质点之间的自碰并过程和破碎过程等,发展了一套数浓度预报方程及其参数化方案,以改进RAMS模式对云物理过程的模拟能力。

关键词 RAMS云模式,参数化,数浓度预报方程

分类号 P42

0 引言

在云模式的微物理过程参数化研究中,常用负指数分布拟合降水质点谱分布函数,其中有一类云模式假定截距参数 N_0 是常数[1],另一类则假定斜率参数 λ 是常数[2]。我们在中尺度对流系统微物理结构分析研究中发现[3],雨滴谱分布的截距参数 N_0 可改变1个数量级以上,而斜率参数 λ 变化约3倍;至于冰相质点,N_0 和 λ 的变化范围分别达3个数量级和4~7倍(参看文献[3]图1)。所以,一般来讲都不能假定为常数,这样人们就不能根据降水混合比的预报量诊断出降水质点的数浓度。因此有必要在云物理模式中发展一套参数化的降水质点数浓度预报方程。

从理论上讲,由于各种降水质点的自碰并过程和破碎过程只是改变降水质点数浓度而不改变其比含水量;另一些过程如雨滴碰并云滴以及冰质点的淞附增长过程等只改变比含水量而不改其数浓度。所以,降水质点数浓度理论上一般也不能从比含水量加以诊断而需分别加以预报。本文在RAMS模式的框架中,采用与Cotton等[4]关于冰晶数浓度预报相协调的方式,发展了一套参数化的数浓度预报方程,其中降水质点的形态包括冰晶、雨滴、霰粒和冰晶聚合体(雪花),分别用下标i、r、g和a表示,预报方程中各转化项采用RAMS微物理模式的符号,在其前加"N"表示数浓度倾向。模式中影响数浓度的物理过程如平流(ADV)、湍流输送(TURB)、自动转化(CN)、碰并(CL)、溅散(SH)、降落(PR)、凝华(VD)、融化(ML)、冻结(FR)、核化(NU)、次级核化(SP)的参数化基本沿用RAMS中原有的方案[4]并做相应的修正,本文主要导出自碰并(SC_{xx})和破碎(BR)的参数化方案。

1 降水质点的物理属性

1.1 质点大小谱分布

常用的负指数分布

$$N(D)=N_0 e^{-\lambda D} \tag{1}$$

最初由Marshall-Palmer提出,并假定截距 N_0 是常数。在RAMS模式中采用稍为不同的形式

① 本研究得到"八五"攻关项目(85-906)和国家自然基金项目(49070244)资助。

$$N(D)=\frac{N_T}{D_m}\mathrm{e}^{-\frac{D}{D_m}}。\tag{2}$$

式中：N_T 是质点总浓度；D_m 是滴谱的一种“特征”直径并假定其为常数$\left(D_m=\frac{1}{\lambda}\right)$。一些研究从理论上论证了在碰并和破碎效应作用下的雨滴谱是趋向于负指数分布的[5,6]，在凝华和聚并作用下的雪晶谱分布也趋向负指数分布[7]。

1.2 冰晶聚合体密度

冰晶聚合体是一类重要的固态降水形态，特别是层状冷云降火的重要形态。研究表明大的冰晶聚合体（雪花）下降速度与大小基本无关[8,9]，据认为这是由于雪花尺度增大时密度一般趋于减小所致。Magono 和 Nakamura[10] 对各类雪花进行观测以验证尺度与密度的关系。我们对他们的资料做了最小方差拟合，得出下列拟合关系

$$\bar{\rho}_d=0.017D_a^{-1}(cgs),\tag{3}$$

这与 Holroyd[11] 的分析结果一致。对于干雪花和粉状雪花上述拟合曲线的相关系数分别为 0.64 和 0.76。

1.3 降水质点下降速度

与文献[4]相同，不赘述。

2 自碰并过程

Berry 等[12] 将雨滴之间的碰并称为自碰并过程。冰晶以及冰晶聚合体之间也能发生自碰并，通常称为聚并过程。自碰并过程产生较大降水质点而减少小的降水质点，但不改变比含水量。降水质点数浓度因自碰并的减少率可表为

$$\mathrm{NSC}_{xx}=\frac{1}{2\rho_0}\iint N(D_{x1})N(D_{x2})K(D_{x1},D_{x2})\mathrm{d}D_{x1}\mathrm{d}D_{x2}。\tag{4}$$

2.1 雨滴自碰并项

应用 Nickerson 等[13] 的结果，可得雨滴数自碰率

$$\begin{aligned}\mathrm{NSC}_{\mathrm{rr}}=&\frac{1}{2\rho_0}\iint m(D_{\mathrm{r1}})b_s(D_{\mathrm{r1}},D_{\mathrm{r2}})N(D_{\mathrm{r1}})N(D_{\mathrm{r2}})\mathrm{d}D_{\mathrm{r1}}\mathrm{d}D_{\mathrm{r2}}+\\&\frac{1}{2\rho_0}\iint m(D_{\mathrm{r2}})b_{\mathrm{s}}(D_{\mathrm{r1}},D_{\mathrm{r2}})N(D_{\mathrm{r1}})N(D_{\mathrm{r2}})\mathrm{d}D_{\mathrm{r1}}\mathrm{d}D_{\mathrm{r2}}。\end{aligned}\tag{5}$$

式中

$$b_{\mathrm{s}}(D_{\mathrm{r1}},D_{\mathrm{r2}})=\frac{K(D_{\mathrm{r1}},D_{\mathrm{r2}})}{m(D_{\mathrm{r1}})+m(D_{\mathrm{r2}})}$$

$$=\frac{3(D_{r1}+D_{r2})^2}{2\rho_w(D_{r1}^3+D_{r2}^3)}E(D_{r1},D_{r2})[\overline{V(D_{r1})}-\overline{V(D_{r2})}]。\tag{6}$$

以平均体积直径 D_{gr} 与平均直径 D_{fr} 之间的碰并作为自碰并率的一级近似，于是可得

$$\mathrm{NSC}_{rr}=b_s(D_{gr},D_{fr})N_T\bar{r},\tag{7}$$

式中 N_T 和 $\bar{r}$ 是雨滴总浓度和平均比含水量。

2.2 冰晶聚合体自碰并项(聚并项)

Passarelli 和 Srivastava[14] 考虑到相同质量的雪花有下降速度谱，定义了一个“修正”碰并核 K_a。它是通常的碰并核 K 由直径谱分布和落速谱分布加权的加权平均值，表为

$$K_a(D_{a1},D_{a2})=\frac{\pi}{4}\iint(D_{a1}+D_{a2})^2\mid\bar{V}_{a1}-\bar{V}_{a2}\mid E(D_{a1},D_{a2})P(V_{a1}/m_{a1})\cdot P(V_{a2}/m_{a2})\mathrm{d}V_{a1}\mathrm{d}V_{a2}。\tag{8}$$

式中 $P(V_a/m_a)$ 是某一质量为 m_a 的冰晶聚合体的下降速度谱函数。K_a 由两部分组成：

$$K_a(D_{a1},D_{a2})=K_1(D_{a1},D_{a2})+2K_2(D_{a1},D_{a2})。\tag{9}$$

其中

$$K_1(D_{a1},D_{a2})=\frac{\pi}{4}\iint(D_{a1}+D_{a2})^2(V_{a1}-V_{a2})E(D_{a1},D_{a2})\cdot P(V_{a1}/m_{a1})P(V_{a2}/m_{a2})\mathrm{d}V_{a1}\mathrm{d}V_{a2},\tag{10}$$

$$K_2(D_{a1},D_{a2})=\frac{\pi}{4}\iint_0^{V_2}(D_{a1}+D_{a2})^2(V_{a2}-V_{a1})E(D_{a1},D_{a2})\cdot P(V_{a1}/m_{a1})P(V_{a2}/m_{a2})\mathrm{d}V_{a1}\mathrm{d}V_{a2}。\tag{11}$$

与文献[14]不同，我们将冰晶聚合体的平均密度 $\bar{\rho}_a$ 和平均落速 $\bar{V}_a$ 分别取式(3)和 Rogers[15] 的试验结果：

$$\bar{\rho}_a=0.017D_a^{-1},\tag{12}$$

$$\bar{V}_a=C_aD_a^{0.2}。\tag{13}$$

式中 C_a 的值对板状和柱状冰晶组成的聚合体为 0.975，对枝状冰晶聚合体为 0.676。假定冰晶聚合体的落速服从 Γ-分布，即

$$P(V_a/m_a)=\frac{\beta^r}{\Gamma(r)}V_a^{\gamma-1}\exp(-\beta V_a)。\tag{14}$$

于是 $\bar{V}_a=\frac{\gamma}{\beta}$，而均方差 $\sigma_{va}^2=\frac{\gamma}{\beta^2}$。在上述假定下我们导出修正碰并核 $K_a=K_1+2K_2$ 中的 K_1 和 K_2 分别表为

$$K_1(D_{a1},D_{a2})=\frac{\pi}{4C_a\Gamma(\alpha_2)}\sum_{j=0}^{2}C_j\cdot\left[\frac{\Gamma(9-4j+\alpha_1)\Gamma(\alpha_2+4j)}{\beta_1^{9-4j}\beta_2^{4j}}-\frac{\Gamma(8-4j+\alpha_1)\Gamma(4j+1+\alpha_2)}{\beta_1^{8-4j}\beta_2^{4j+1}}\right], \tag{15}$$

$$K_2(D_{a1},D_{a2})=\frac{\Gamma(p)\beta_1^{\alpha1}\beta_2^{\alpha2}}{(\beta_1+\beta_2)^p}\sum_{j=0}^{2}C_j\cdot\left[\frac{F(1,p;\alpha_1+9-4j;k)}{\alpha_1+8-4j}-\frac{F(1,p;\alpha_1+10-4j;k)}{\alpha_1+9-4j}\right]。 \tag{16}$$

式中：$p=9+\alpha_1+\alpha_2$；$k=\dfrac{\beta_1}{\beta_1+\beta_2}$；$C_0=C_2=1$；$C_1=2$。利用这一修正碰并核可导出冰晶聚合体的自碰并项为

$$\begin{aligned}\mathrm{NSC}_{aa}&=\frac{1}{2\rho_0}\int_0^\infty\int_0^\infty N(D_{a1})N(D_{a2})K_1(D_{a1},D_{a2})\,\mathrm{d}D_{a1}\,\mathrm{d}D_{a2}\\&=\frac{N_a^2}{2\rho_0}\overline{K_a(D_{a1},D_{a2})}。\end{aligned} \tag{17}$$

3 破碎过程

根据 Srivastava[16]我们将自发破碎和碰撞破碎方程写成如下形式：

3.1 自发破碎项

只有雨滴会产生自发破碎过程。由此引起的雨滴数浓度改变率可表为

$$\mathrm{NBRS}_{rr}=\frac{1}{\rho_0}\int P(D_r)[Q_0(D_r)-1]N(D_r)\,\mathrm{d}D_r \tag{18}$$

式中：$P(D_r)$是一个直径为 D_r 的水滴自发破碎概率，$Q_0(D_r)$是自发破碎所产生的碎滴平均数，它们可分别表示为[16]

$$P(D_r)=a_1\mathrm{e}^{b_1}D_r, \tag{19}$$

$$Q_0(D_r)=a_2(1-\mathrm{e}^{-b_2})。 \tag{20}$$

式中：$a_1=2.94(10^{-7})$，$b_1=17$，$a_2=62.3$，$b_2=7$，都是常数。于是有：

$$\mathrm{NBRS}_{rr}=a_1N_{0r}[a_2(1-\mathrm{e}^{-b_2})-1]\frac{\mathrm{e}^{(b_1-D_{rm}^{-1})D_a}-1}{(b_1-D_{rm}^{-1})\rho_0} \tag{21}$$

式中：D_a 是最大雨滴直径，D_{rm}和 N_{0r}是负指数分布参数。

3.2 碰撞破碎项

雨滴和冰晶聚合体都可以发生碰撞破碎。对雨滴来说，Gillespie 和 List[6]认为碰撞破

碎是比自发破碎更重要的破碎机制。破撞破碎产生较小的降水质点而使总数浓度增加，增加率可表为

$$\mathrm{NBRC}_{xx}=\frac{1}{2\rho_0}\iint K(D_{x1},D_{x2})[1-E(D_{x1},D_{x2})][s_0(D_{x1},D_{x2})-2]N(D_{x1})N(D_{x2})\mathrm{d}D_{x1}\mathrm{d}D_{x2}。\tag{22}$$

式中：$K(D_{x1},D_{x2})$是碰撞核，$E(D_{x1},D_{x2})$是直径为D_{x1}和D_{x2}的质点之间碰撞后合并的概率，$s_0(D_{x1},D_{x2})$是D_{x1}和D_{x2}质点碰撞破碎后产生的碎滴总数的平均值。

(a) 雨滴的碰撞破碎项

$$\mathrm{NBRC}_{rr}=\frac{1}{2\rho_0}\iint\left(\frac{D_{r1}}{2}+\frac{D_{r2}}{2}\right)^2|\overline{V(D_{r1})}-\overline{V(D_{r2})}|[1-E(D_{r1},D_{r2})]\cdot[s_0(D_{r1},D_{r2})-2]N(D_{r1})N(D_{r2})\mathrm{d}D_{r1}\mathrm{d}D_{r2}。\tag{23}$$

按照 Low 和 List[17]，可得到合并概率$E(D_{r1},D_{r2})$和破碎后碎滴平均数$s_0(D_{r1},D_{r2})$的表达式：

$$E(D_{r1},D_{r2})=a_3\left(1+\frac{D_{r2}}{D_{r1}}\right)^{-2}\exp\left[-\frac{b_3E_t^2(D_{r1},D_{r2})}{(D_{r1}^3+D_{r2}^3)^{2/3}}\right]。\tag{24}$$

式中$a_2=0.778$，$b_3=2.61\times10^6$是常数，而$E_t(D_{r1},D_{r2})$是合并能，可表为

$$E_t(D_{r1},D_{r2})=\frac{\pi\rho_w}{12}\left(\frac{1}{D_{r1}^3}+\frac{1}{D_{r2}^3}\right)^{-1}[V(D_{r1})-V(D_{r2})]^2+\pi\sigma[(D_{r1}^2+D_{r2}^2)-(D_{r1}^3+D_{r2}^3)^{2/3}]。\tag{25}$$

式中σ是水的表面张力，$\sigma=7.28\times10^{-2}$ N/m。关于s_0，我们根据文献[17]的实验资料拟合出一个$s_0(D_{r1},D_{r2})$与$E_t(D_{r1},D_{r2})$之间的经验关系：

$$s_0(D_{r1},D_{r2})=a_4+b_4E_t(D_{r1},D_{r2}),\tag{26}$$

线性拟合的相关系数达0.961。式中$a_4=1.195$，$b_4=0.105$。为计算方便，我们近似地取如下形式的平均值：

$$\overline{E(D_{r1},D_{r2})}=E(D_{fr},D_{gr}),\tag{27}$$

$$\overline{s_0(D_{r1},D_{r2})}=s_0(D_{fr},D_{gr})。\tag{28}$$

式中D_{fr}和D_{gr}是雨滴平均直径和质量众数直径。于是：

$$\mathrm{NBRC}_{rr}=\frac{N_{0r}^2}{4\rho_0}[1-E(D_{fr},D_{gr})][s_0(D_{fr},D_{gr})-2]\cdot|\overline{V_{r1}}-\overline{V_{r2}}|(D_{r1m}^2+D_{r1m}D_{r2m}+D_{r2m}^2)。\tag{29}$$

(b) 冰晶聚合体的碰撞破碎

冰晶聚合体之间以及冰晶聚合体与霰粒之间的碰撞都可导至聚合体破碎。当云中同时存在霰粒和冰晶聚合体时碰撞破碎效应尤为显著[18,19]。我们仿照雨滴碰撞破碎方程导出冰

晶聚合体碰撞破碎方程：

$$\mathrm{NBR_a} = \mathrm{NBR_{aa}} + \mathrm{NBR_{ag}}, \tag{30}$$

这里 $\mathrm{NBR_{aa}}$ 和 $\mathrm{NBR_{ag}}$ 分别表示聚合体之间和聚合体与霰粒之间碰撞破碎引起的聚合体数浓度增加率。与式(23)类似可得

$$\begin{aligned}\mathrm{NBR_{aa}} = &\frac{1}{2\rho_0}\iint K_{\mathrm a}(D_{\mathrm{a1}},D_{\mathrm{a2}})[1-E_{\mathrm{aa}}(D_{\mathrm{a1}},D_{\mathrm{a2}})]\cdot\\ &[s_{0\mathrm a}(D_{\mathrm{a1}},D_{\mathrm{a2}})-2]N(D_{\mathrm{a1}})N(D_{\mathrm{a2}})\mathrm dD_{\mathrm{a1}}\mathrm dD_{\mathrm{a2}},\end{aligned} \tag{31}$$

$$\begin{aligned}\mathrm{NBR_{ag}} = &\frac{1}{2\rho_0}\iint K(D_{\mathrm a},D_{\mathrm g})[1-E_{\mathrm{ag}}(D_{\mathrm a},D_{\mathrm g})]\cdot\\ &[s_{0\mathrm a}(D_{\mathrm a},D_{\mathrm g})-2]N(D_{\mathrm a})N(D_{\mathrm g})\mathrm dD_{\mathrm a}\mathrm dD_{\mathrm g}\\ = &\frac{\pi}{2\rho_0}\iint\left(\frac{D_{\mathrm a}}{2}+\frac{D_{\mathrm g}}{2}\right)^2|V_{\mathrm g}(D_{\mathrm g})-V_{\mathrm a}(D_{\mathrm a})|[1-E_{\mathrm{ag}}(D_{\mathrm a},D_{\mathrm g})]\cdot\\ &[s_{0\mathrm a}(D_{\mathrm a},D_{\mathrm g})-2]N_{\mathrm a}(D_{\mathrm a})N_{\mathrm g}(D_{\mathrm g})\mathrm dD_{\mathrm a}\mathrm dD_{\mathrm g}。\end{aligned} \tag{32}$$

式中 $1-E_{\mathrm{aa}}(D_{\mathrm{a1}},D_{\mathrm{a2}})=q_{\mathrm{aa}}(D_{\mathrm{a1}},D_{\mathrm{a2}})$ 和 $1-E_{\mathrm{ag}}(D_{\mathrm a},D_{\mathrm g})=q_{\mathrm{ag}}(D_{\mathrm a},D_{\mathrm g})$ 分别为聚合体之间和聚合体与霰粒之间的碰撞破碎概率，而 $s_{0\mathrm a}(D_{\mathrm{a1}},D_{\mathrm{a2}})$ 和 $s_{0\mathrm a}(D_{\mathrm a},D_{\mathrm g})$ 是相应的碰撞破碎产生的碎片平均数。可惜，对于冰晶聚合体的破碎概率和碎片平均数及其大小分布都知之甚少。作为一级近似，假定破碎概率函数和每次破碎的碎片平均数与碰撞质点直径的乘积成比例：

$$q_{\mathrm{aa}}(D_{\mathrm{a1}},D_{\mathrm{a2}})=B_{\mathrm a}(T,S)D_{\mathrm{a1}}D_{\mathrm{a2}}, \tag{33}$$

$$q_{0\mathrm a}(D_{\mathrm{a1}},D_{\mathrm{a2}})=B_{\mathrm f}(T,S)D_{\mathrm{a1}}D_{\mathrm{a2}}, \tag{34}$$

$$q_{\mathrm{ag}}(D_{\mathrm a},D_{\mathrm g})=B_{\mathrm{ag}}(T,S)D_{\mathrm a}D_{\mathrm g}, \tag{35}$$

$$S_{0\mathrm a}(D_{\mathrm a},D_{\mathrm g})=B_{\mathrm{fg}}(T,S)D_{\mathrm a}D_{\mathrm g}。 \tag{36}$$

式中：系数 $B_{\mathrm a}$、$B_{\mathrm f}$、B_{ag} 和 B_{fg} 是温度和空气温度的函数，s 是冰面的过饱和度，这些系数的值待定。在上述近似处理下，可以导出冰晶聚合体的碰撞破碎项：

$$\begin{aligned}\mathrm{NBR_{aa}} = &\frac{1}{2\rho_0}\iint K_{\mathrm a}(D_{\mathrm{a1}},D_{\mathrm{a2}})[1-E_{\mathrm{aa}}(D_{\mathrm{a1}},D_{\mathrm{a2}})]\cdot\\ &[s_{0\mathrm a}(D_{\mathrm{a1}},D_{\mathrm{a2}})-2]N(D_{\mathrm{a1}})N(D_{\mathrm{a2}})\mathrm dD_{\mathrm{a1}}\mathrm dD_{\mathrm{a2}}\\ = &\frac{1}{\rho_0}N_{0\mathrm a}^2B_{\mathrm a}(T,s)D_{\mathrm{a}m}^4\overline{K_{\mathrm a}(D_{\mathrm{a1}},D_{\mathrm{a2}})}(2B_{\mathrm f}(T,s)D_{\mathrm{a}m}^2-2),\end{aligned} \tag{37}$$

$$\begin{aligned}\mathrm{NBR_{ag}} = &\frac{\pi N_{0\mathrm a}N_{0\mathrm g}}{8\rho_0}B_{\mathrm{ag}}(T,s)\,|\bar V_{\mathrm g}-\bar V_{\mathrm a}|\,[B_{\mathrm{fg}}(T,s)\Gamma(5)D_{\mathrm{a}m}^3D_{\mathrm{g}m}^3(2D_{\mathrm{a}m}^2+\\ &3D_{\mathrm{a}m}D_{\mathrm{g}m}+2D_{\mathrm{g}m}^2)-2\Gamma(3)D_{\mathrm{a}m}^2D_{\mathrm{g}m}^2(3D_{\mathrm{a}m}^2+4D_{\mathrm{a}m}D_{\mathrm{g}m}+3D_{\mathrm{g}m}^2)]。\end{aligned} \tag{38}$$

现在我们可以利用上述参数化方程和文献[4]中有关的表达式建立不同类型降水质点数浓度的预报方程。

4 数浓度的参数化预报方程

4.1 雨滴数浓度预报方程①

$$\frac{\partial}{\partial t}\left(\frac{N_r}{\rho_0}\right)=\mathrm{ADV}\left(\frac{N_r}{\rho_0}\right)+\mathrm{TURB}\left(\frac{N_r}{\rho_0}\right)+\mathrm{NPR_r}+\mathrm{NCN_{cr}}-\mathrm{NCL_{ng}}-\mathrm{NCL_{rg}}+\mathrm{NML_{gr}}+\mathrm{NSH_{gr}}-\mathrm{NCL_{rcg}}+\mathrm{NML_{ar}}+\mathrm{NSH_{ar}}-\mathrm{NSC_{rr}}+\mathrm{NBR_{rr}}+\mathrm{NML_{ia}}。\tag{39}$$

式中

$$\mathrm{NPR_r}=-\frac{1}{\rho_0}\frac{\partial N_r\bar{V}_r}{\partial z},\tag{40}$$

$$\mathrm{NCN_{cr}}=\left(\frac{N_r}{\rho_0 r_r}\right)CN_{cr},\tag{41}$$

$$\mathrm{CN_{cr}}=\mathrm{CN}3_0^{4/3}T745Ir_c^{7/3}h(R_c-R_{cm}),\tag{42}$$

$$\mathrm{NML_{gr}}+\mathrm{NSH_{gr}}+\mathrm{NML_{ar}}+\mathrm{NSH_{ar}}=\left(\frac{N_r}{\rho_0\overline{r_r}}\right)(\mathrm{ML_{gr}}+\mathrm{SH_{gr}}+\mathrm{ML_{ar}}+\mathrm{SH_{ar}}),\tag{43}$$

$$\mathrm{ML_{gr}}=\left\{\frac{K_{3g}}{L_{i1}}[-\mathrm{KT_s}+\mathrm{DF_v}L_{iv}\rho_0(\overline{r_v}-r_{s1}(T_f))]-\frac{c_wT_s}{L_{i1}}(\mathrm{CL_{cg}}+\mathrm{CL_{rs}})\right\}h(-T_s),\tag{44}$$

$$\mathrm{SH_{gr}}=(\mathrm{CL_{cg}}+\mathrm{CL_{rg}})h(-T_s),\tag{45}$$

$$\begin{aligned}\mathrm{ML_{ar}}&=\left\{\frac{6f_2(Re_a)}{L_{i1}\alpha_1D_{am}}[-\mathrm{KT_s}+L_{iv}\mathrm{DF_v}\rho_0(\overline{r_v}-r_{s1}(T_f))]\overline{r_v}-\left(\frac{c_w}{L_{i1}}\right)T_s(CL_{ca}+CL_{ra})\right\}h(-T_s)\\&=353\{L_{i1}^{-1}D_{am}^{-1}\}f_2(Re_a)[-KT_s+L_{iv}\mathrm{DF_v}\rho_0(\overline{r_v}-r_{i1}(T_f))]\overline{r_v}-\mathrm{CN}35T_s(\mathrm{CL_{cs}}+\mathrm{CL_{ra}})\}h(-T_s),\end{aligned}\tag{46}$$

$$\mathrm{SH_{ar}}=(\mathrm{CL_{ca}}+\mathrm{CL_{ra}})h(-T_s),\tag{47}$$

$$\mathrm{NCL_{ri}}=\frac{\pi N_rN_i}{4\rho_0}\overline{E(r/i)}\mid\bar{V}_r-\bar{V}_i\mid(2D_{rm}^2+2D_{rm}D_i+D_i^2),\tag{48}$$

$$\mathrm{NCL_{rg}}=\frac{\pi N_rN_g}{2\rho_0}\overline{E(r/g)}\mid\bar{V}_r-\bar{V}_g\mid(D_{rm}^2+2D_{rm}D_g+D_{gm}^2),\tag{49}$$

① $\mathrm{NCL_{rg}}$表示雨滴(r)碰撞(CL)霰粒(g)并入霰粒而减小雨滴数浓度的速率；$\mathrm{NCL_{rig}}$表示雨滴(r)与冰晶(i)碰撞(CL)转变成霰粒(g)引起的数浓度变化率。以此类推。

$$\mathrm{NCL_{ra}}=\frac{\pi N_r N_a}{2\rho_0}\overline{E(r/a)}\mid\bar{V}_r-\bar{V}_g\mid(D_{rm}^2+2D_{rm}D_{gm}+D_{gm}^2),\tag{50}$$

$$\mathrm{NBR_{rr}}=\mathrm{NBRS_{rr}}+\mathrm{NBRC_{rr}}。\tag{51}$$

式中：$\mathrm{NSC_{ri}}$是由于雨滴之间自碰并引起的数浓度减少速率，由式(7)确定；$\mathrm{NBRS_{rr}}$和 $\mathrm{NBRC_{rr}}$是由于雨滴自发破碎和碰撞破碎引起的雨滴数浓度增加率，分别由式(21)和式(29)确定。其余各项的含义如引言中所述，相应的表达式由文献[4]中对应的比含水量改写成数浓度表达式而确定。

4.2 冰晶数浓度预报方程

冰晶数浓度预报方程与 Cotton 和 Tripeli[4]在 RAMS 云模式中的浓度方程类似：

$$\begin{aligned}\frac{\partial}{\partial t}\left(\frac{N_r}{\rho_0}\right)=&\mathrm{ADV}\left(\frac{N_r}{\rho_0}\right)+\mathrm{TURB}\left(\frac{N_r}{\rho_0}\right)+NPR_i+\\&\mathrm{NNUA_{vi}}+\mathrm{NNUB_{ci}}+\mathrm{NNUC_{ci}}+\mathrm{NNUD_{ci}}+\mathrm{NSP_{ci}}-\\&\mathrm{NML_{ir}}-\mathrm{NCL_{ig}}-\mathrm{NCN_{ig}}-\mathrm{NCL_{ia}}-\mathrm{NCL_{rg}}-\mathrm{NCL_{rg}}。\end{aligned}\tag{52}$$

式中

$$\mathrm{NPR_i}=-\frac{1}{\rho_0}\frac{\partial(N_i\bar{V}_i)}{\partial z},\tag{53}$$

$$\mathrm{NNUA_{vi}}=m_{i0}^{-1}\mathrm{NUA_{vi}},\tag{54}$$

$$\mathrm{NNUB_{ci}}+\mathrm{NNUC_{ci}}+\mathrm{NNUD_{ci}}=m_{i0}^{-1}(\mathrm{NUB_{ci}}+\mathrm{NUC_{ci}}+\mathrm{NUD_{a}}),\tag{55}$$

$$\mathrm{NSP_{vi}}=m_{i0}^{-1}\mathrm{SP_{vi}},\tag{56}$$

$$\mathrm{NUA_{vi}}=\frac{\mathrm{CN10}}{\rho_0}\max\left[\exp(\beta_2 T_s)w\frac{\partial T_s}{\partial z},0\right]h(T_s),\tag{57}$$

$$\mathrm{NUB_{ci}}+\mathrm{NUC_{ci}}+\mathrm{NUD_{ci}}=m_{i0}\frac{F_1}{\rho_0}\left[\mathrm{DF_{cr}}+F_2\left(f_t-\frac{R_v T}{L_{vi}}\right)\right],\tag{58}$$

$$F_1=\frac{4\pi R_c N_c N_a}{\rho_0},\tag{59}$$

$$\mathrm{DF_{ar}}=\frac{\beta_3 T}{R_{ar}\mu}\left(1+\frac{\beta_4 T}{\mathrm{PR_{ar}}}\right),\tag{60}$$

$$F_2=\frac{K}{p}(T-T_c),\tag{61}$$

$$f_t=\frac{0.4\left[1+1.45K_n+0.4K_n\exp\left(-\frac{1}{K_n}\right)\right](K+2.5K_nK_a)}{(1+3K_n)(2K+5K_aK_n+K_a)},\tag{62}$$

$$\mathrm{NSP_{ci}}=\frac{1}{\rho_0}\left[\beta_5\rho_0+4\times10^{-3}\frac{N_c}{r_c}f_3(M_c)\right][f_1(T_i)\mathrm{CL_{ci}}+f_1(T_g)\mathrm{CL_{cg}}+f_1(T_a)\mathrm{CL_{ca}}], \tag{63}$$

$$\mathrm{CL_{ci}}=\frac{\pi}{4}D_i^2V_i\overline{E(c/i)}N_i\overline{r_c}, \tag{64}$$

$$\mathrm{CL_{cg}}=\frac{\pi}{4}\gamma(3.5)D_{gm}^2V_g(D_{gm})\overline{E(c/g)}N_g\overline{r_c}, \tag{65}$$

$$\mathrm{CL_{ca}}=\frac{\pi}{2}\left(\frac{4g\beta_1}{3C_D}\right)^{\frac{1}{2}}\overline{E(c/a)}D_{am}^2N_a\overline{r_c}, \tag{66}$$

$$\mathrm{CL_{ca}}=2.21\left(\frac{g\rho_0}{C_D\alpha_1}\right)^{\frac{1}{2}}\overline{E(a/c)}\,\overline{r_ar_c}D_{am}^{0.25}, \tag{67}$$

$$\mathrm{NML_{ir}}=N_i(\rho_0\overline{r_i})^{-1}\mathrm{ML_{ir}}, \tag{68}$$

$$\mathrm{ML_{ir}}=\frac{\overline{r_i}h(-T_s)}{\Delta t}, \tag{69}$$

$$\mathrm{NCN_{ig}}=N_i(\rho_0\overline{r_i})^{-1}\mathrm{CN_{ig}}, \tag{70}$$

$$\mathrm{CN_{ig}}=\max\left[\mathrm{CL_{ci}}-\left(\frac{N_i}{\rho_0}\right)C_m,0\right]h(T_s), \tag{71}$$

$$C_m=1\times10^{-6}(\mathrm{g\cdot s^{-1}}), \tag{72}$$

$$\mathrm{NCN_{ia}}=N_i(\rho_0\overline{r_i})^{-1}\mathrm{CN_{ia}}, \tag{73}$$

$$\mathrm{CN_{ia}}=0.523D_i^2\overline{V_i}\,\overline{E(T_i)}N_i\overline{r_i}, \tag{74}$$

$$\mathrm{NCL_{irg}}=\frac{1}{\rho_0}\left(\frac{\mathrm{d}N_i}{\mathrm{d}t}\right)\bigg|_{ir}=\int K_{ir}N_iN(D_r)\mathrm{d}D_r, \tag{75}$$

$$\mathrm{NCL_{ig}}=\frac{1}{\rho_0}\left(\frac{\mathrm{d}N_i}{\mathrm{d}t}\right)\bigg|_{ig}=\int K_{ig}N_iN(D_g)\mathrm{d}D_g, \tag{76}$$

$$\mathrm{NCL_{ia}}=\frac{1}{\rho_0}\left(\frac{\mathrm{d}N_i}{\mathrm{d}t}\right)\bigg|_{ia}=\int K_{ia}N_iN(D_a)\mathrm{d}D_a。 \tag{77}$$

式中

$$K_{ir}=\frac{\pi}{4}(D_r+D_i)^2\mid V(D_r)-V(D_i)\mid E(r/i), \tag{78}$$

$$K_{ig}=\frac{\pi}{4}(D_g+D_i)^2\mid V(D_g)-V(D_i)\mid E(g/i), \tag{79}$$

$$K_{ia}=\frac{\pi}{4}(D_a+D_i)^2\mid V(D_a)-V(D_i)\mid E(a/i)。\tag{80}$$

碰并项的最终形式为

$$\mathrm{NCL}_{irg}=\frac{\pi}{4\rho_0}N_iN_r\mid\bar{V}_r-\bar{V}_i\mid\overline{E(r/i)}(2D_{rm}^2+2D_{rm}D_i+D_i^2),\tag{81}$$

$$\mathrm{NCL}_{ig}=\frac{\pi}{4\rho_0}N_iN_g\mid\bar{V}_g-\bar{V}_i\mid\overline{E(g/i)}(2D_{gm}^2+2D_{gm}D_i+D_i^2),\tag{82}$$

$$\mathrm{NCL}_{ia}=\frac{\pi}{4\rho_0}N_iN_a\mid\bar{V}_a-\bar{V}_i\mid\overline{E(a/i)}(2D_{am}^2+2D_{am}D_i+D_i^2)。\tag{83}$$

4.3 霰粒数浓度预报方程

$$\frac{\partial}{\partial t}\left(\frac{N_g}{\rho_0}\right)=\mathrm{ADV}\left(\frac{N_g}{\rho_0}\right)+\mathrm{TURB}\left(\frac{N_g}{\rho_0}\right)+\mathrm{NPR}_g+\mathrm{NCN}_{ig}+\mathrm{NCN}_{ag}-\mathrm{NML}_{gr}+\mathrm{NCL}_{rig}+\mathrm{NCL}_{rsg}。\tag{84}$$

式中

$$\mathrm{NPR}_g=-\frac{1}{\rho_0}\frac{\partial(N_g\bar{V}_g)}{\partial z},\tag{85}$$

$$\mathrm{NCN}_{ig}=\left(\frac{N_g}{\rho_0\overline{r_g}}\right)\mathrm{CN}_{ig},\tag{86}$$

$$\mathrm{CN}_{ig}=\max\left[\mathrm{CL}_{ci}-\left(\frac{N_i}{\rho_0}\right)c_m,0\right]h(T_s),\tag{87}$$

$$\mathrm{NCN}_{ag}=\left(\frac{N_g}{\rho_0\overline{r_g}}\right)\mathrm{CN}_{ag},\tag{88}$$

$$\mathrm{CN}_{ag}=\max\left[\mathrm{CL}_{ca}-\left(\frac{g\rho_0}{C_{Dg}\mathrm{D}_{gm}}\right)^{1/2}\left(\frac{1}{\rho_w^2\rho_s}\right)^{1/6}\overline{r_cr_a},0\right]h(T_s),\tag{89}$$

$$\mathrm{NCL}_{rig}=\frac{\pi}{4\rho_0}N_rN_i\overline{E(r/i)}\mid\bar{V}_r-\bar{V}_i\mid(2D_{rm}^2+2D_{rm}D_i+D_i^2),\tag{90}$$

$$\mathrm{NCL}_{rag}=\frac{\pi}{4\rho_0}N_rN_a\overline{E(r/a)}\mid\bar{V}_r-\bar{V}_a\mid(D_{rm}^2+D_{rm}D_{am}+D_{am}^2)。\tag{91}$$

4.4 冰晶聚合体数浓度预报方程

$$\frac{\partial}{\partial t}\left(\frac{N_a}{\rho_0}\right)=\mathrm{ADV}\left(\frac{N_a}{\rho_0}\right)+\mathrm{TURB}\left(\frac{N_a}{\rho_0}\right)+\mathrm{NPR}_a+\mathrm{NCN}_w-\mathrm{NCN}_{ag}-\mathrm{NCL}_{ag}-\mathrm{NCL}_{crg}-\mathrm{NML}_{ar}-$$

$$\mathrm{NSC_{aa}} + \mathrm{NBR_{aa}} + \mathrm{NBR_{aga}}。 \tag{92}$$

式中

$$\mathrm{NPR_a} = -\frac{1}{\rho_0} \frac{\partial(N_a \bar{V}_a)}{\partial z}, \tag{93}$$

$$\mathrm{NCL_{arg}} = \frac{\pi}{2\rho_0} \mid \bar{V}_r - \bar{V}_a \mid \overline{E(r/a)} N_a N_r (D_{am}^2 + D_{am} D_{rm} + D_{rm}^2), \tag{94}$$

$$\mathrm{NCL_{ag}} = \frac{\pi}{2\rho_0} \mid \bar{V}_g - \bar{V}_a \mid \overline{E(g/a)} N_a N_g (D_{am}^2 + D_{am} D_{gm} + D_{gm}^2)。 \tag{95}$$

其中自碰并项 $\mathrm{NSC_{aa}}$ 与碰撞破碎项 $\mathrm{NBR_{aa}}$ 和 $\mathrm{NBR_{ag}}$ 分别由式(17)、(37)和(38)确定。

参考文献

[1] Lin Y-H et al. J Climate Appl Meteor. 1983; 22: 1065 - 1092

[2] Cotton W R et al. J Rech Atmos, 1982; 16: 295 - 320

[3] Ye Jiadong. Fan Beifen. Froc 10th Intern Cloud Phys Conf. FRG. 1988; 383 - 385

[4] Cotton W R et al. J Climate appl Meteor. 1986; 25: 1658 - 1680

[5] Srivastava R C. J Atmos Sci. 1971; 28: 410 - 415

[6] Gillesic J R, List R. Proc Int Cloud Phys Conf. Boulder. 1976; 472 - 477

[7] Passarelli R E Jr. J Atmos Sci. 1978; 35: 882 - 889

[8] Livinov I V. Izv Akad Nauk SSSR ser Geaphiz 1956; No. 7: 853

[9] Magono C. Scient Rep Yokohama Univ, 1953; Ser 1, No. 2, 18 - 40

[10] Magono C. Nakamura T. J Meteor Soc Japan. 1965; 43(3): 139 - 147

[11] Holroyd. I E W. Proc Int Conf Wea Modific Canberra, 1971; 135 - 140

[12] Berry E X, Reinhardt R L. J Atomos Sci. 1874; 31: 1814 - 1824

[13] Nickerson E C et al. Mon Wea Rev. 1986; 114. 398 - 414

[14] Passarelli R E. Srivastaa R C. J Atmos Set. 1979; 36: 484 - 493

[15] Rogers D C. Thesis M S. 35PP Dept of Atmos Res. Univ of Wyoming, 1974

[16] Srivastava R C. J Atmos Sci. 1978; 35; 108 - 117

[17] Low T B. List R. Atmos Sci. 1982; 39: 1606 - 1618

[18] Hobbs PK. Farber R J. Rech. J Atmas. 1972; 6: 245 - 258

[19] Vardiman L. J Atmos Sci. 1978; 35: 2168 - 2180

THE CONCENTRATION PROGNOSTIC EQUATIONS AND ITS PARAMETERIZATION OF VARIOUS PRECIPITATION PARTICLES TYPES IN CLOUD MODEL

Ye Jiadong　Fan Beifen

(Department of Atmospheric Sciences, Nanjing University, 210093, Nanjing, China)

Abstract　Considering the self-collection and breadup processes and other processes of various types of precipitation particies, a set of prognostic equations and corresponding parameterizations of varied precipitation particles concentrations are cleveloped in the frame of RAMS cloud model for enhanced its simulation power.

Key words RAMS cloud model, parameterization, concentration prognostic equation

5.9 MICROPHYSICS OF THE STRATIFIED PRECIPITATION REGION OF A MESOSCALE CONVECTIVE SYSTEM

Jia-dong Yeh, Michael A. Fortune, William R. Cotton
Department of Atmospheric Science, Colorado State University, Fort Collins, CO 80523

1. INTRODUCTION

The program of airborne investigations of mesoscale convective systems (AIMCS) is the first research effort directed primarily to obtain aircraft measurements of nocturnal mesoscale convective systems (MCS). As part of the AIMCS - 84 program, the hydrometeor measurements in the nimbostratus region of an MCS were analyzed together with radar, satellite, and other data collected by the NOAA P - 3 research aircraft, to determine the thermodynamic and microphysical structures of the stratified precipitation region. The size spectra of precipitation particles and its parameterization are also discussed.

2. SYNOPTIC CONDITIONS AND OVERVIEW OF DATA COLLECTION

On 1200 GMT 14 July 1984, a short-wave trough and associated surface low was located in Wyoming and Nebraska with a cold frontal system extending through Minnesota and South Dakota into central Wyoming. The southerly jet brought warm moist air at 850 mb east of the front. After 0000 GMT 15 July, and during the night, the cumulonimbi organized into an east-west convective line perpendicular to the southern jet. Hail was observed several times in central Iowa (2030 - 2330 GMT) and southern Iowa (0230 - 0430). The stratiform region northeast of the line produced much precipitation.

The enhanced infrared satellite imagery showed the "initiation" of classical MCC characteristics at 1900 GMT and the "maximum" stage was reach at 0300 GMT with an area of 210 000 km^2 of cloud tops colder than −52 ℃. The MCS had decayed by 1000 GMT and lasted about 15 hours.

The national weather service radar data showed a persistent E-W convective line developed by 0400 GMT along latitude 41N with the highest echo tops at 16 km. After 0600 GMT the convective line had moved near latitude 40N.

At 0333 GMT the aircraft entered the storm system at the western edge of the MCS. Then the aircraft flew at about 720 mb along the northern flank of the convective line.

From 0600 to 0620 GMT, it climbed to 423 mb near Burlington, Iowa for a sounding through the stratiform cloud from height of 2 800 m to 6 750 m, and temperature from +9.6 ℃ to −11.8 ℃. Then the aircraft flew homeward at 420 mb through the stratiform cloud (Fig.1), at temperature between −11 ℃ and −14 ℃, and with vertical motions typically between +1 m/s and −1 m/s, but occasionally much larger.

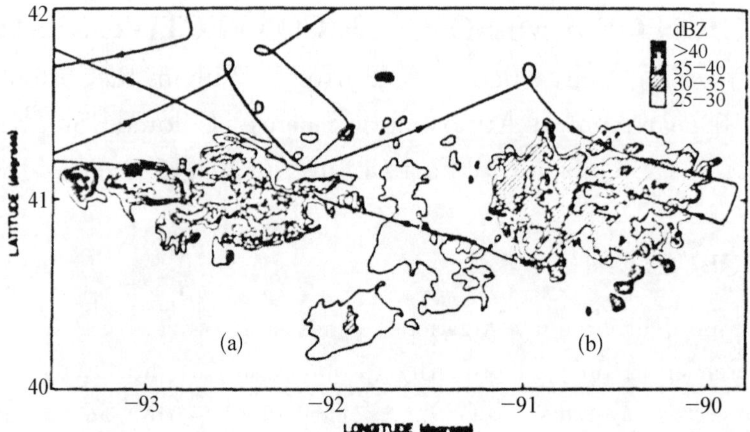

Fig.1 Flight track and radar echo patterns from the lower fuselage radar. From point A to point B on tho flight track, the numbers 1 to 17 give the locations of the samples in Fig.2. The echo pattern in the right half of the figure was observed by the aircraft from location C (heavy line). Tha pattern to the left of longitude 92°W was observed from location D. This pattern was observed one hour before the time that the nearby microphysical samples were taken.

3. MICROPHYSICAL CHARACTERISTICS IN THE STRATIFIED PRECIPITATION REGION

3.1 Water phase

The PMS "2D-C" instrument measures particles in the size range from 0.05 mm to 1.60 mm, and the "2D-P" instrument, from 0.2 mm to 6.4 mm. Some examples of hydrometeor images at various levels are presented in Fig.2. The melting layer as determined by the habits of 2D-images is about 200 m thick with temperature ranging from 0 ℃ to 1.5 ℃.

3.2 Water content

Measurements of cloud droplet liquid water content by the Johnson-William probe are low, generally less than 0.3 g/m^3. The largest value of 0.5 g/m^3 is found in a strong updraft of 5.6 m/s at a height of 6 km. The average water content of precipitation inferred from the PMS probes amounted to only 0.6 g/m^3 beneath the melting layer, and 3.2 g/m^3 above it (Fig.3(a)).

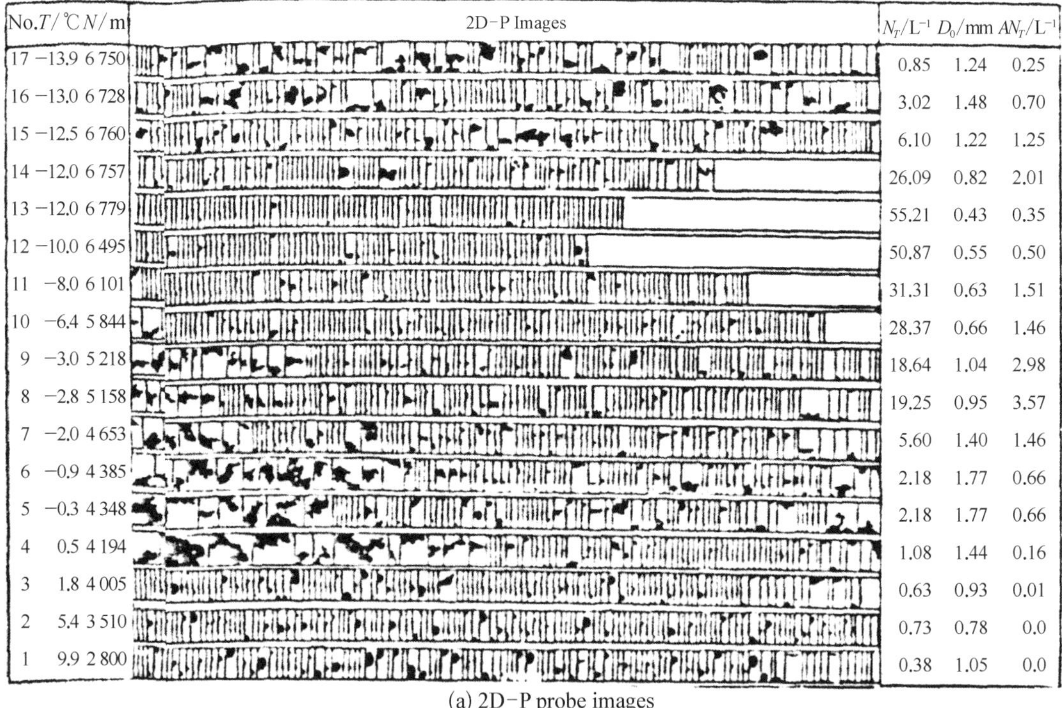

(a) 2D−P probe images

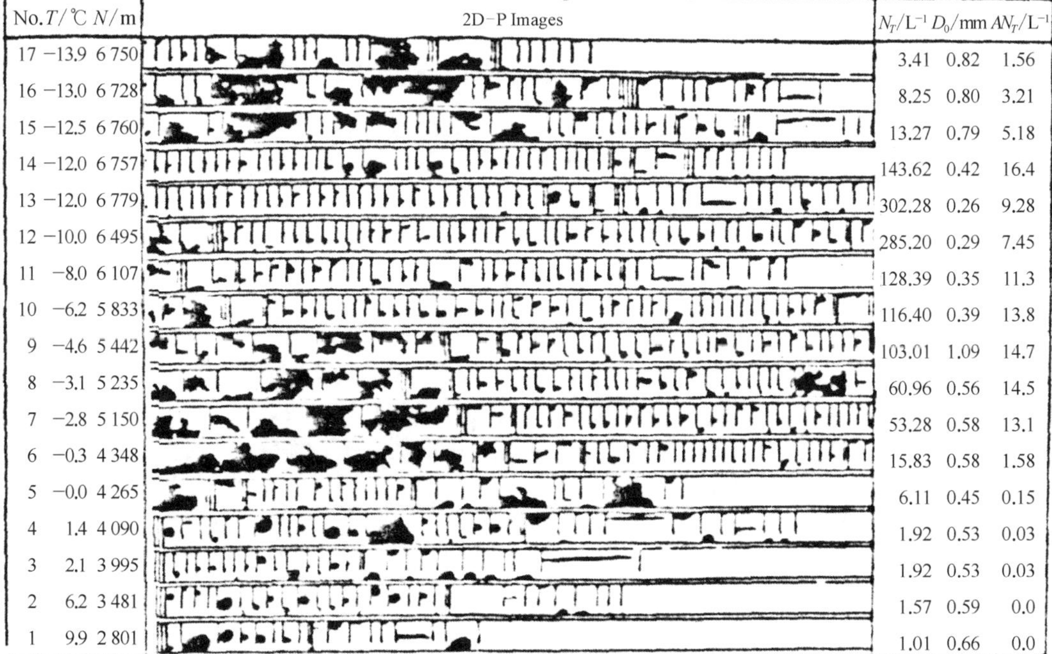

(b) 2D−C probe images

Fig.2 Examples of 2D—images of hydrometeors at various levels and temperatures (D_0 is the median volume diameter; N_T and AN_T are the number concentrations of ice particles and aggregates)

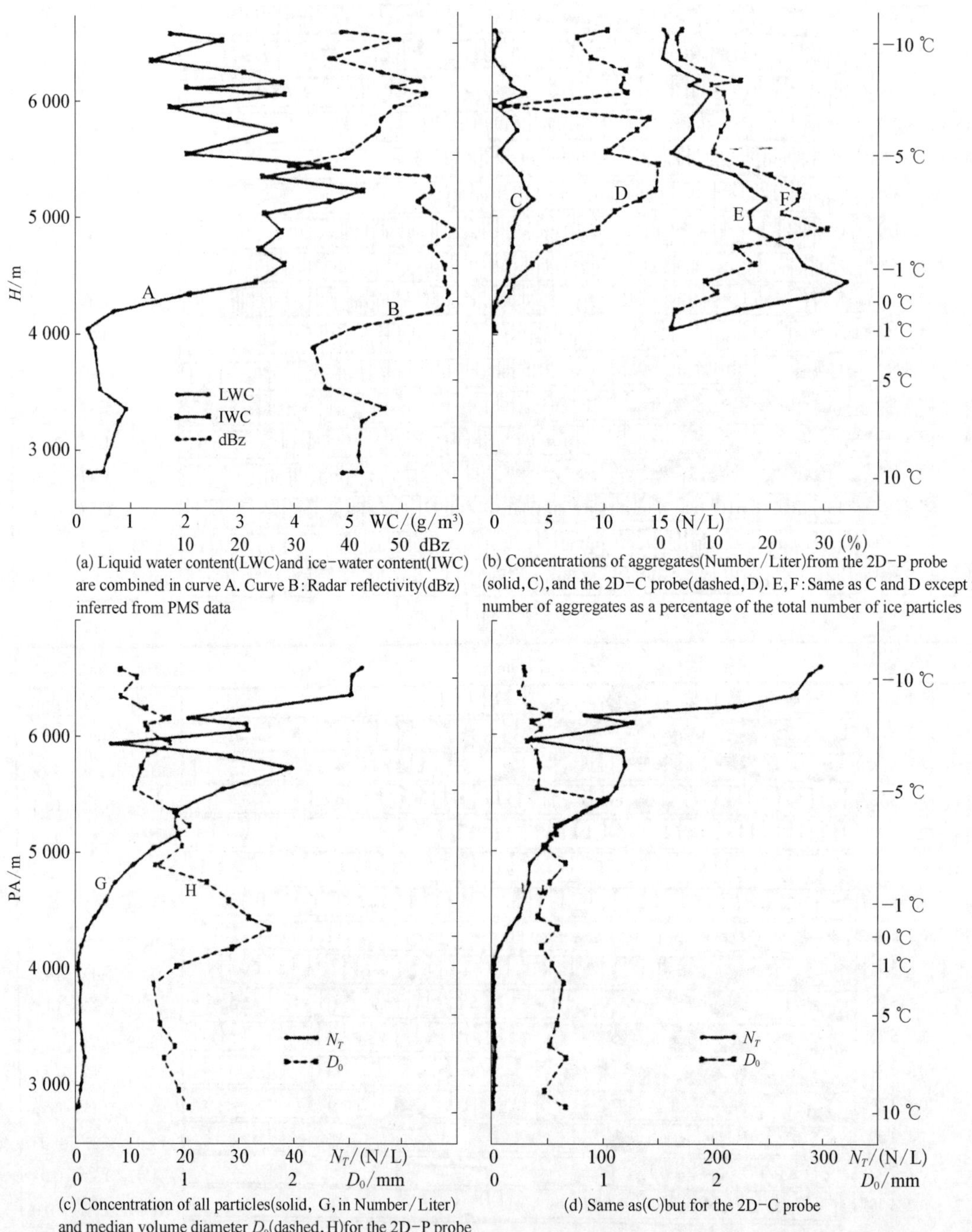

(a) Liquid water content(LWC)and ice-water content(IWC) are combined in curve A. Curve B: Radar reflectivity(dBz) inferred from PMS data

(b) Concentrations of aggregates(Number/Liter)from the 2D-P probe (solid, C), and the 2D-C probe(dashed, D). E, F: Same as C and D except for number of aggregates as a percentage of the total number of ice particles

(c) Concentration of all particles(solid, G, in Number/Liter) and median volume diameter D_0(dashed, H)for the 2D-P probe

(d) Same as(C)but for the 2D-C probe

Fig.3 Vertical profiles of microphysical variables, as a function of height (left axis) and temperature (right axis)

3. 3 Radar reflectivity factor

The radar reflectivity was computed from the PMS images. It attains the greatest relative values in the layer from 0. 5 ℃ to −3. 5 ℃ (4 200—5 380 m) or the so called bright band (Battan, 1973), but especially at heights above the melting level. Comparing the profiles (Fig.3) of water content (Curve A), aggregate number concentration (Curve C), and median volume diameter (Curve H) of 2D－P data suggest that the large radar reflectivity factors are due to the combined effects of these three variables. In the lower part of the bright band, the predominant contributor is the large size of the aggregates, and in the upper part it is the large number concentration of aggregates.

3. 4 Particle concentration and size

Figs. 3(c) and (d) show that the water drop concentrations are low. The average concentrations are 0. 8 L^{-1} (2D－P) and 2. 3 L^{-1} (2D－C), while the median volume diameter D_0 averaged 1. 0 mm (2D－P) and 0. 6 mm (2D－C) beneath the melting layer. Above this layer only ice particles are observed, and there are quite a few aggregates at least up to the highest flight level at −14 ℃. At sub-freezing temperature the concentration of ice particles and their median diameters change markedly with height. The ice particle number concentrations increase with height up to the highest level observed.

In contrast, the median volume diameter D_0 typically decreased with height above the 0 ℃ level. The zone of the largest diameters is located in the layer from 4 200—4 800 m (0. 5 ℃ to −1. 3 ℃) where large aggregates are prevalent(Fig.2). Fig.3(b) shows that the relative concentration of large aggregates in this layer exceeds 15%. The largest diameter of 1. 8 mm is found at the 0 ℃ level. Along with the 2D images in Fig.2, we conclude that the aggregation process above and near the 0 ℃ level is the predominant mechanism of growth of precipitation particles and responsible for the lower concentration of ice particles.

3. 5 Aggregates

In Fig.2, two zones of large aggregates can be seen in the stratiform region of the MCS. One zone lies between 0 ℃ and −4. 4 ℃ where the mean concentrations of aggregates are 1. 9 L^{-1}(2D－P) and 8. 1 L^{-1}(2D－C). They account for over 60% of the total water content of all hydrometeors. The peak concentrations of aggregates are encountered at −3. 1 ℃ (2D－P) and −4. 4 ℃ (2D－C)(Fig.3(b)). The mean aggregate concentration of 0. 9 L^{-1} in the layer from 0 ℃ to −1 ℃ is the same order of magnitude as the concentration of water drops just beneath the melting layer. This substantiates the expected result that the rain drops in the stratiform precipitation of the MCS are mainly melted aggregates falling from the aggregation zone just above the 0 ℃ level.

The large aggregates observed by the 2D－P probe are mainly concentrated in the lower layer near the 0 ℃ level, and smaller aggregates observed by the 2D－C probe are more populous in a higher layer from 4 900 m to 5 350 m (−1.4 ℃ to −3.5 ℃) where the concentration is 12.5 L^{-1} (Fig.3(b)). This implies that aggregation growth began in the colder layer.

The largest aggregates ($D>3$ mm) were rarely seen in the highest levels (5 500 m to 6 600 m) of the vertical sounding. But when the aircraft traversed the northwestern part of the cloud system at the 420 mb level, the second zone of large aggregates was encountered in the temperature ranging from −11 ℃ and −14 ℃. The mean concentrations of aggregates were 2.0 L^{-1} and 14.4 L^{-1} from the two probes. Aggregates accounted for more than 20% of all the ice particles in the edge region of the MCS, where the total ice concentrations falls to less than 10 L^{-1} (2D-P) and 20 L^{-1}(2D-C), after the time of 0630 GMT (Fig.4).

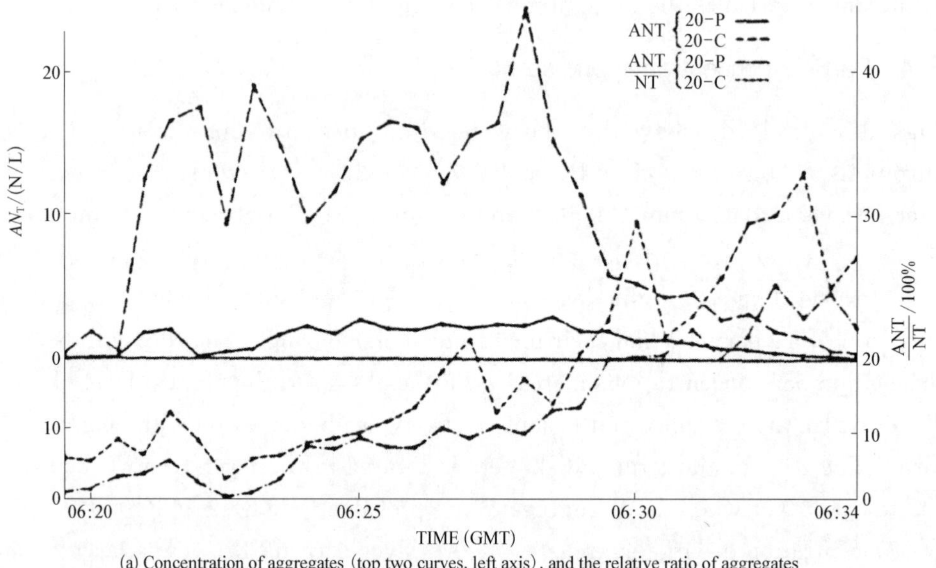

(a) Concentration of aggregates (top two curves, left axis), and the relative ratio of aggregates to all ice particles(bottom two curves, right axis)

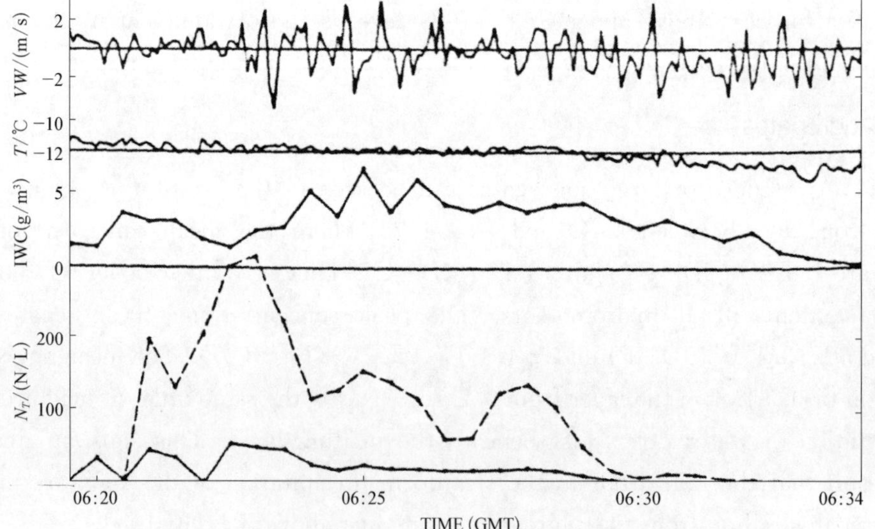

(b) Vertical air velocity(VW), temperature, ice-water content, and ice particle concentrations

Fig.4 Time series of microphysical data obtained during horizontal flight in the stratiform region at the 420 mb level.

In both of the aggregation zones, the predominant aggregates consist of irregular rimed particles and dendrites. The low cloud droplet water contents (typically less than 0.3 g/m^3) favor light riming of particles, which may be more effective for promoting aggregation than unrimed vapor grown crystals. The cloud droplet water may act as an adhesive agent in the aggregation of ice particles.

Fig.4 illustrates the observations obtained during horizontal flight through the edge of the MCS at 420 mb level. The ice particle water content (IWC), and concentrations of ice particles and aggregates are not uniform horizontally. The largest aggregates are encountered in the northwest edge of the MCS during the sounding from 0630 GMT to 0634 GMT, where the cloud is in a decaying stage. Also, vertical velocities shown in Fig.4(b), were generally downward. The aggregation process in this region may be different because here the cloud droplet water contents are generally less than the sensitivity of the JW probe. The riming process is inefficient, and the mechanical interlocking of crystal branches suggested by Oktake (1970) and Rauber (1985) may be the main mechanism for the formation of aggregates.

4. SIZE SPECTRA OF PRECIPITATION PARTICLES

Seventy eight samples of the size spectra of hydrometeors larger than 0.2 mm in diameter were obtained. The Marshall-Palmer distribution,

$$N(D)=N_0 \exp(-\lambda D) \tag{1}$$

holds well in the entire measured region both for the water drops and ice particles, although the slope parameter λ and the intercept parameter N_0 vary with the number concentration N_T and median volume diameter D_0 of the particles. Here $N(D)$ is the size distribution density function of a particle with a diameter of D. Measured spectra fit the M-P distribution very closely over the entire sampled size range beneath the melting level. The average linear correlation coefficient for $\log N(D)$ is -0.96 for water drops and -0.92 for ice particles. The variation of λ can be expressed by

$$\lambda=-2.51+\frac{1.65}{D_0} \tag{2}$$

with the best fit line having a correlation coefficient of 0.87, for $D_0 < 2$ mm, and the variation of N_0 by

$$N_0=\frac{1.80N_T}{D_0} \tag{3}$$

with a correlation coefficient also of 0.87. These relations are similar to the distribution functions used by Cotton et al. (1985) in which

$$N_0=\frac{N_T}{D_m} \text{ and } \lambda=\frac{1}{D_m},$$

where D_m is a characteristic diameter of the particle spectrum.

There does not seem to be an obvious relation among λ, N_0, and the water content (WC). However Kessler (1969) derived a formula for which we fitted an empirical relation of the form

$$\lambda = a + b\left[\frac{N_0}{\text{WC}}\right]^{1/4}, \tag{4}$$

where a, b are different for water and for ice. The relation proved to be fair for water drops; the correlation coefficient was 0.73 when $a = 0.96$ and $b = 1.28$. The correlation was very good for ice particles; the coefficient was 0.98 when $a = -7.1$ and $b = 1.33$. Fig.5 shows the degree of linear correlation.

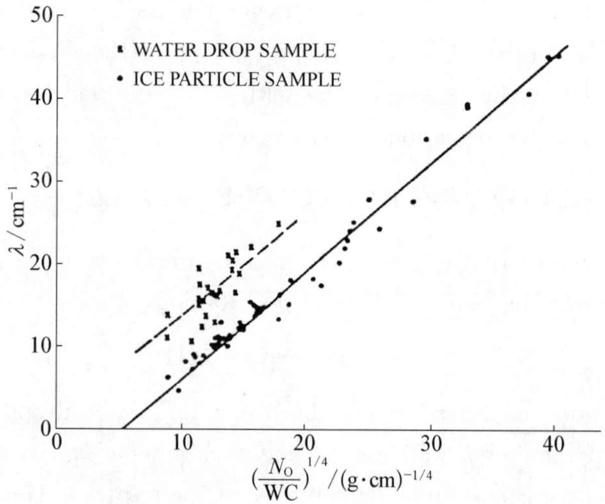

Fig.5 Scatter diagrams of slope parameter λ plotted as a function of combined variable $(N_0/\text{WC})^{1/4}$. Solid line is the best fit for ice particle samples; dashed line is the best fit for water drop samples.

A scatter plot of N_0 as a function of λ (not shown) shows that λ does not vary much for water drops: its mean value was 17 (± 3.6) cm^{-1}, so λ may be considered a constant as suggested by Cotton et al. (1985) beneath the melting level. For the ice particle samples, however the distribution fits

$$N_0 = (1.1 \times 10^{-4})\lambda^{2.8} \tag{5}$$

with a correlation coefficient of 0.94. We can infer the size spectra of ice particles from the median diameter D_0 by combining Eqs. (1), (2), and (5).

5. CONCLUSIONS

(1) The melting layer in the stratiform region of the MCS is about 200 m thick at temperature from 0 ℃ to +1.5 ℃. The bright band is located from +0.5 ℃ to −3.5 ℃

with most of it at height above the 0 ℃ level. It is caused by the large size of aggregates just above the melting level, and the large concentration of aggregates higher up.

(2) The evidence suggests that the aggregation process occurred over a considerable depth (4 000 - 6 700 m) and range of temperatures (at least down to −14 ℃). Two zones of large aggregates were observed. One zone lies between 0 ℃ and −4.4 ℃ and another is located in the edge region of the MCS at temperature from −11 ℃ to −14 ℃. The predominant aggregates consisted of irregular rimed ice particles and dendrites. Low values of cloud liquid water ($<0.3\ g/m^3$) may help to form rime that acts as an adhesive agent. In the colder zone, the mechanical interlocking of crystal branches may be the mechanism responsible for aggregation.

(3) The Marshall-Palmer distribution holds well both for water drops and ice particles. The slope parameter λ is approximately constant for water drops, whereas the intercept parameter N_0 varied over a range of one order of magnitude. For ice particles, λ and N_0 both vary as a function of ice water content.

6. ACKNOWLEDGEMENTS

We thank the Hurricane Research Division of NOAA, especially Robert Black, for providing software for analysis of PMS images; the Weather Research Program (WRP) of NOAA, especially David Jorgensen, for providing software and computer resources for analysis of data from the aircraft flight; and both WRP and the Office of Aircraft Operations for managing a successful AIMCS experiment. Participation in AIMCS was supported by NOAA project #NA81RAH00001 while the analysis was supported by NSF Grant ATM - 8512480.

REFERENCES

Battan, L. J., 1973: Radar Observation of the Atmosphere. 324 pp., University of Chicago Press, Chicago, IL.

Cotton, W.R., and R.A. Anthes, 1985: The dynamics of clouds and mesoscale weather systems. Part I: The dynamics of clouds. To be published by Academic Press.

Kessler, E., 1969: On the distribution and continuity of water substance in atmospheric circulations. Meteorl. Mon.. Amer. Meteorol. Soc., 10, 84pp.

Oktake, T., 1970: Factors affecting the size distribution of raindrops and snowflakes. J. Atmos. Sci., 27. 804 - 813.

Rauber, R. M., 1985: Physical structure of northern Colorado River Basin cloud systems. Ph. D. dissertation, Colorado State University, Dept, of Atmospheric Science, Fort Collins, Colorado 80523.

5.10 COMPARISON OF THE MICROPHYSICS BETWEEN THE TRANSITION REGION AND STRAIJFORM REGION IN A MESOSCALE CONVECTIVE COMPLEX

Jia-Dong Yeh, Bei-Fen Fan, Michael A. Fortune, William R. Cotton
Department of Atmospheric Science, Colorado State University, Fort Collins, CO 80523

1. INTRODUCTION

The mesoscale convective complex (MCC) is an important category of mesoscale convective systems, one that accounts for more than half the amount of summer rainfall in the Great Plains and midwestern United States (Fritsch and Maddox, 1981). Maddox (1980) identified the MCC by the time and space scale of its stratiform cloud shield as viewed by satellite, but did not address internal structural characteristics. The flow structure and radar reflectivity fields of these storm systems have been analyzed by a few investigators (Houze and Rutledge, 1986; Cunning et al., 1986; Smull and Houze, 1985, 1986), but relatively little is known about the microphysical structure of them. In this paper we primarily analyze the microphysical data of an MCC collected from the NOAA P-3 research aircraft on 4 June 1985 during the PRE-STORM. We then interpret the observations with respect to the mesoscale structure of the storm system.

2. RADAR ECHO PATTERN AND AIRCRAFT OBSERVATION

The MCC of 4 June 1985 was a rapidly evolving and fast moving MCC, which started to form by 0630 GMT and decayed at about 1400 GMT. Early in its lifecycle, numerous convective cells developed according to a mode described as "random" by Blanchard and Watson (1986). At 0800 GMT some mature convective echoes merged, and formed a trailing stratiform region in the western part of the MCC. Some convective cells organized into meso-β scale sub-systems (Fortune and McAnelly, 1986). Between the mature convective clusters there was a region in which the radar echo intensity is a minimum, we termed it a "transition region", as it will transform later to a rather uniform stratiform region. This region of weak echo may be similar to the transition zone defined by Smull and Houze (1985, 1986) for a squall line system but somewhat different in shape. The microphysical data of two segments of horizontal flight by the NOAA P-3 aircraft at 5 100 m MSL have been analyzed to illustrate the difference between the transition region and stratiform region. One segment lies through the transition region and the stratiform region from 0844 GMT to 0858 GMT (three sub-segments Ⅰ, Ⅱ and Ⅲ in Fig.1). The other segment lies in the stratiform region in the rear portion of the MCC from 0918 GMT to 0933 GMT. In addition, two vertical soundings were made in the stratiform region from

3 000 m to 6 100 m MSL in the time 1 034 GMT to 1 113 GMT(A in Fig.1) and 1 123 GMT to 1 135 GMT (B in Fig.1), respectively. The vertical microphysical soundings were also made in the rear stratiform region of the MCC, since the storm system has moved to the north of Wichita after 1 030 GMT.

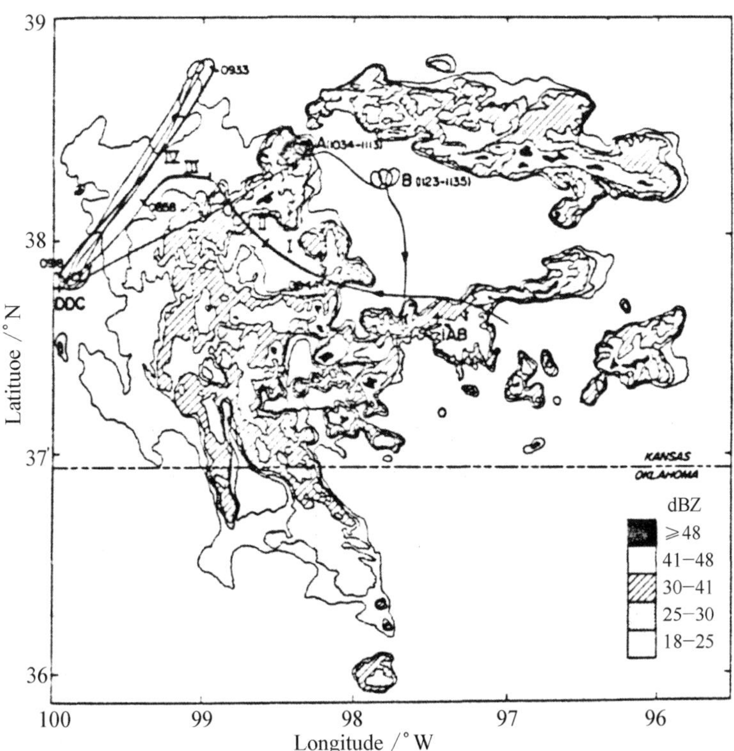

Fig.1 Digitized NWS radar echo pattern of the MCC of 4 June at 0845 GMT and the flight track. Labels Ⅰ, Ⅱ, Ⅲ and Ⅳ are horizontal sampling segments of the aircraft; A and B are vertical sounding locations at 1034—1112 and 1123—1135 GMT, respectively. (At these times the radar echo was stratiform in nature at A abd B)

3. MICROPHYSICS IN THE TRANSITION REGION AND STRATIFORM REGION

The 2D particle measuring system images of precipitation particles and related microstructures are given in Figs.2 and 3. We found the following differences between the transition region and the stratiform region in the MCC:

(1) The concentration of all precipitation particles are about 20 L^{-1}(2D-P probe) and 120 L^{-1}(2D-C probe) in the transition region, and only 5 L^{-1}(2D-P probe) and 20 L^{-1} (2D-C probe) in the stratiform region. On the other hand the sizes of particles are greater in the stratiform region: The average median volume diameter is 0. 73 mm (2D-P probe) and 0. 37 mm (2D-C probe) in the transition region, and 1. 32 mm (2D-P probe) and 0. 63 mm (2D-C probe) in the stratiform region. Most of the precipitation particles larger

No.	Time	T/℃	2D-P Images	N_T/L^{-1}	D_0/mm	AN_T	CN_T
1	0844^{00}	−7.9		0.56	0.88	0.20	0.06
2	0845^{00}	−8.0		0.37	0.73	0.33	0.15
3	0846^{00}	−7.3		4.15	0.75	0.10	0.05
4	0846^{20}	−7.9		12.21	0.76	0.56	0.04
5	0846^{40}	−8.0		14.89	0.78	0.55	0.72
6	0847^{00}	−7.8		19.00	0.74	0.96	0.60
7	0847^{35}	−7.7		18.00	0.50	0.52	0.70
8	0847^{50}	−6.7		12.04	0.63	0.13	0.53
9	0848^{10}	−7.4		20.53	0.61	0.40	0.66
10	0848^{30}	−7.4		11.85	0.80	0.45	0.45
11	0848^{50}	−7.5		11.85	0.81	0.62	0.34
12	0849^{10}	−7.5		27.05	0.72	0.95	0.38
13	0849^{30}	−7.5		23.70	0.77	1.14	0.55
14	0849^{50}	−7.5		26.51	0.68	0.73	0.52
15	0850^{10}	−8.2		23.77	0.67	0.90	0.61
16	0850^{50}	−8.5		20.08	0.79	1.25	0.30
17	0850^{50}	−8.5		17.89	0.93	1.00	0.34

(a) Example of 2D-images of hydrometeors in the transition region of the MCC on 4 June 1985

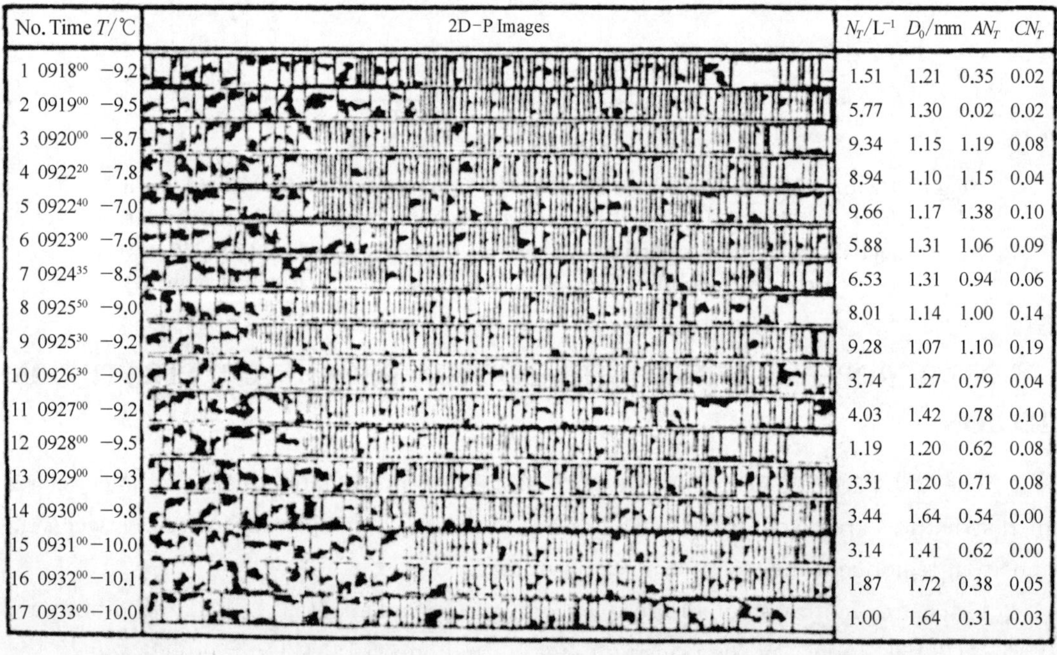

No.	Time	T/℃	2D-P Images	N_T/L^{-1}	D_0/mm	AN_T	CN_T
1	0918^{00}	−9.2		1.51	1.21	0.35	0.02
2	0919^{00}	−9.5		5.77	1.30	0.02	0.02
3	0920^{00}	−8.7		9.34	1.15	1.19	0.08
4	0922^{20}	−7.8		8.94	1.10	1.15	0.04
5	0922^{40}	−7.0		9.66	1.17	1.38	0.10
6	0923^{00}	−7.6		5.88	1.31	1.06	0.09
7	0924^{35}	−8.5		6.53	1.31	0.94	0.06
8	0925^{50}	−9.0		8.01	1.14	1.00	0.14
9	0925^{30}	−9.2		9.28	1.07	1.10	0.19
10	0926^{30}	−9.0		3.74	1.27	0.79	0.04
11	0927^{00}	−9.2		4.03	1.42	0.78	0.10
12	0928^{00}	−9.5		1.19	1.20	0.62	0.08
13	0929^{00}	−9.3		3.31	1.20	0.71	0.08
14	0930^{00}	−9.8		3.44	1.64	0.54	0.00
15	0931^{00}	−10.0		3.14	1.41	0.62	0.00
16	0932^{00}	−10.1		1.87	1.72	0.38	0.05
17	0933^{00}	−10.0		1.00	1.64	0.31	0.03

(b) Example of 2D-images of hydrometeors in the stratiform region of the MCC on 4 June 1985

Fig.2 Examples of 2D-images of hydrometeors in the transition region and stratiform region of the MCC on 4 June 1985 (N_T, AN_T and GN_T are the number concentrations of total precipitation particles, aggregates and graupcls, respectively. D_0 is the median volume diameter of total particles)

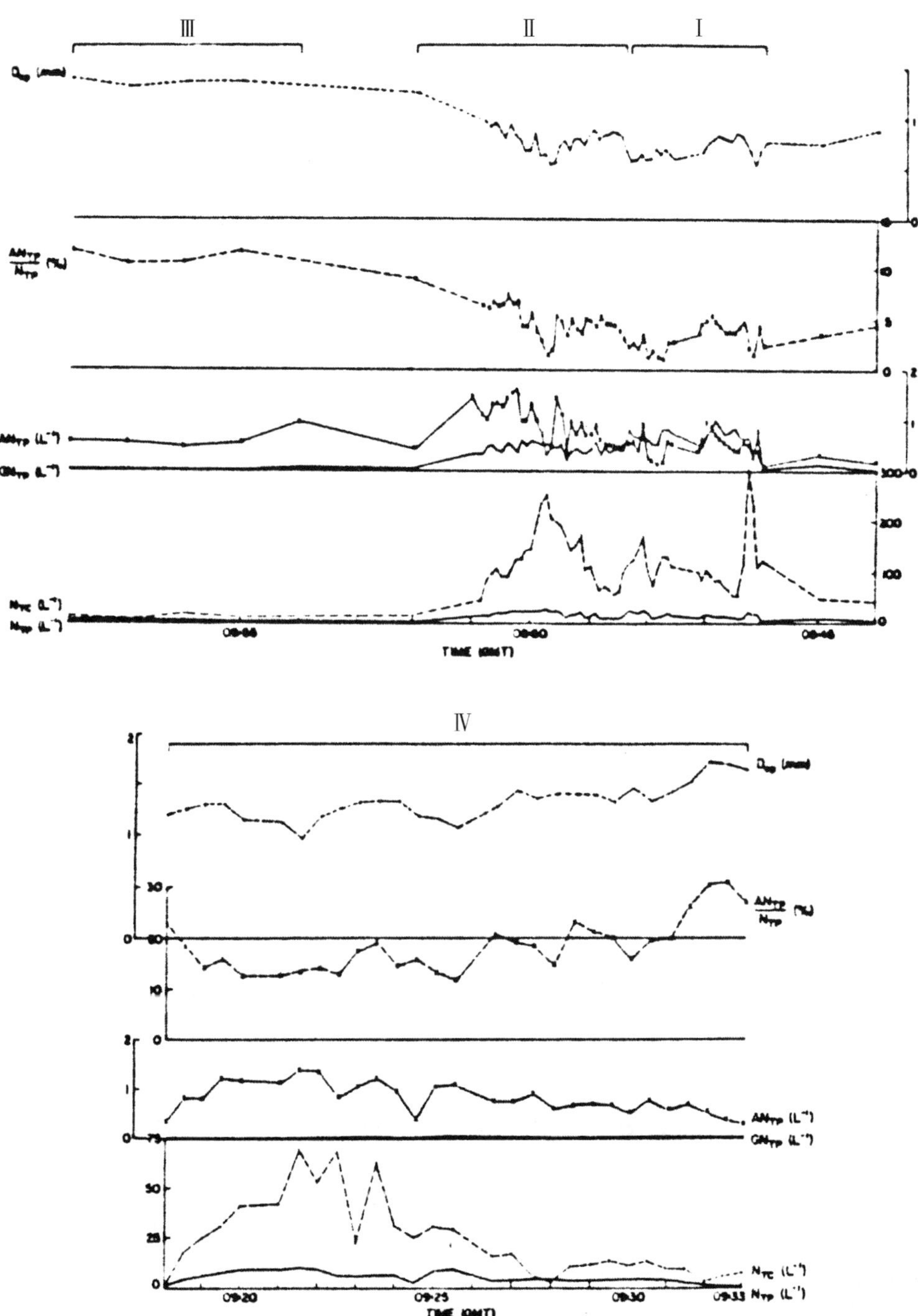

Fig.3 Time series of microphysical data obtained during horizontal flight in the transition region (segment Ⅰ and Ⅱ) and stratiform region (segment Ⅲ and Ⅳ) at the 530 mb level (N_{TP}, AN_{TP} and GN_{TP} are the number concentrations of total particles, aggregates and graupels for 2D-P probe, respectively. D_{0P} is the median volume diameter of the particles)

than 1 mm in diameter are of the shape of aggregates and graupel particles in the transition region, whereas only aggregates predominate in the stratiform region.

(2) In the eastern portion of the transition region where convective clouds are in rapid development nearby (segment Ⅰ in Fig. 1), the concentration of graupel particles (0.66 L^{-1} for 2D-P probe), is slightly greater than that of aggregates (0.51 L^{-1} for 2D-P probe). In the western part of the transition region (segment Ⅱ in Fig. 1), graupel concentrations are lower than those of aggregates (0.46 and 0.92 L^{-1}, respectively). As a whole the concentration of graupel is the same order of magnitude as aggregates in the transition region, whereas in the stratiform region, it exceeds that of graupel by up to an order in magnitude (0.79 L^{-1} for aggregates and only 0.07 L^{-1} for graupel particles). The main precipitation formation mechanisms are riming and aggregation in the transition region, as it appears that graupel particles are detrained from nearby convective cells. The aggregation process predominates in the stratiform cloud.

(3) Aggregates account for about 20% of the total ice particles in the stratiform region, and only about 4% in the transition region. Analysis suggests that the coexistence of dense, rapidly falling graupel or frozen raindrops that originated in the convective cells along with fragile aggregates and ice crystals in the transition region may promote the collisional breakup of aggregates and ice crystals forming in the region. The relatively lower radar reflectivity in the region appears to be a result of smaller precipitation particle sizes. Graupel particles are few in the stratiform region, thus more large aggregates can form by aggregation of ice crystals without being broken up in collision with graupel particles. Collisional breakup of aggregates and ice crystals still affects the microphysical structure to a certain extent, however. Both in the transition and stratiform regions the main components of aggregates are irregular rimed crystals, which may be the fragments of dendrites. Light riming of ice crystals in the presence of small amounts of supercooled liquid water content, on the other hand, may be favorable for the aggregation process (Yeh et al., 1986). The cloud droplets may act as an adhesive agent in the aggregation of ice particles.

4. VERTICAL DISTRIBUTION OF MICROPHYSICS IN THE STRATIFORM REGION

The vertical profiles of microphysical variables (omitted) reveals that the largest concentrations of aggregates lie in the levels from −6 ℃ to −13 ℃ but that the larger sizes are found in lower layers from +1 ℃ to −5 ℃. The number concentration of aggregates relative to the total number concentration of ice particles, is also greater near the melting layer. More than 20% of the particles are aggregates in the layer from 0 ℃ to −2 ℃. We infer that aggregation starts in the upper, colder layer, but that it becomes more efficient near the melting layer in the stratiform region of the MCC.

5. ACKNOWLEDGEMENTS

We thank Robert Black for providing software for analysis of PMS images; David Jorgensen, Jose Meitin and other NOAA/Weather Research Program personnel for providing software and computer resources for analysis of data from the aircraft flight. We also thank Ray McAnelly for his assistence in radar analysis, and Lucy McCall for her drafting work. This research was supported by NSF Grant ATM-8512480.

REFERENCES

Blanchard, D. O., and A. I. Watson, 1986: Modes of mesoscale convection, observed during the PRE-STORM program. Preprints of Joint Sessions, 23rd Conference on Radar Meteorology and Conference on Cloud Physics, AMS, Snowmass, Colorado, J155-158.

Cunning, J. B., D. P. Jorgensen and A. L Watson, 1986: Airborne and ground-based doppler radar observations of a rapidaly moving midlatitude squall line system. Same as Blanchard and Watson, J171-174.

Fortune, M. A., and R. L. McAnelly, 1986: The evolution of two mesoscale convective complexes with different patterns of convective organization. Same as Blanchard and Watson, J175-178.

Fritsch, J. M., and R. A. Maddox, 1981: Convectively driven mesoscale wheather systems aloft. Part I. Observations. J. Appl. Meteorol., 20, 9-19.

Houze, R. A., Jr., and S. A. Rutledge, 1986: A squall line with trailing stratiform precipitation observed during the Oklahoma-Kansas PRESTORM experiment. Same as Blanchard and Watson, J167-170.

Maddox, R. A., 1980: Mesoscale convective complexes. Bull. Amer. Meteor. Soc., 61, 1374-1387.

Smull, B. F., and R. A. Houze, Jr., 1985: A midlatitude squall line with a trailing region of stratiform rain: Radar and satellite observations. Mon. We a. Rev., 113, 117-133.

Smull, B. F., and R. A. Houze, Jr., 1986: Dual-doppler radar analysis of a midlatitude squall line with a trailing region of a stratiform rain. Submitted to J. Atmos. Sci., 44pp.

Yehr J-D., M. A. Fortune, and W. R. Cotton, 1986: Microphysics of the stratified precipitation region of a mesoscale convective system. Same as Blanchard and Watson, J151-154.

5.11 MICROPHYSICS IN A DEEP CONVECTIVE CLOUD SYSTEM ASSOCIATED WITH A MESOSCALE CONVECTIVE COMPLEX-NUMERICAL SIMULATION

Bei-Fen Fan, Jia-Dong Yeh, William R. Cotton

Department of Atmospheric Science, Colorado State University, Fort Collins, CO 80523

G. Tripoli

University of Wisconsin, Department of Meteorology, Madison, WI 53715

1 Introduction

In recent years, increasing observational evidence has revealed that the ice particle aggregation process is one of the predominant mechanisms of precipitation formation in a variety of cloud systems. Airborne cloud microphysical data obtained from midlatitude mesoscale convective systems (MCS) also showed that aggregates are an important component of the precipitation in the stratiform region and transition region of midlatitude mesoscale convective complexes (MCC), especially in the stratiform region (Yeh et al., 1986; 1987). In a simulation of orographic cloud snowfall, Cotton et al. (1986) demonstrated that aggregation plays an important role in controlling the fields of cloud liquid water content, ice crystal concentrations, and surface precipitation amounts. However, when the aggregation model was applied to the simulation of a MCC, the model underpredicted the amount of aggregates over the stratiform region (Chen, 1986). In this paper, a two dimensional version of the Colorado State University (CSU) regional atmospheric modeling system (RAMS) is applied to the simulation of meso-β-scale convective component of a MCC.

2 The Study Case

The case selected for simulation is a meso-β-scale cloud system associated with a PRE-STORM MCC of 4 June 1985. The MCC developed in association with a short wave trough which passed through the mid-west region of the United States. The synoptic situation and sounding data exhibited some important features: (1) The atmosphere was warm and moist in the low levels, and dry in the middle troposphere; (2) the southerly low level jet of 13—15 m/s at 85 kPa level pumped warm moist Gulf air to the area of PRE-STORM network; (3) there was a deep conditionally unstable layer from 90 kPa to 50 kPa level; (4) the wind hodograph showed that wind shear was the strongest beneath the 60 kPa level, above it the southwestly air current prevailed up to the 10 kPa level. The MCC was a rapidly evolving and fast moving, non-linear system, which formed by 0630 GMT and dissipated at about 1400 GMT. Early in its lifecycle, numerous convective cells developed in a random manner. At 0800 GMT the convective echoes started to organize into several meso-β-scale sub-systems (Fortune and McAnelly, 1986). We focus on the simulation of the meso-β-scale convective system. Airborne observations have revealed (Yeh et al., 1987) that the characteristics of ice particles are quite different between the stratiform region and transition region. Most large particles are aggregates and graupel in the transition region, whereas aggregates predominate in the stratiform region. The aggregation process in the stratiform region started at upper and colder levels but become more efficient as the aggregates approached the melting layer.

3 Brief Description of Model Characteristics and Design of Simulation Experiments

A two-dimension version of the CSU RAMS nonhydrostatic cloud model was used to simulate a meso-β-scale convective system. The governing equations and general features of the model system are described in Tripoli and Cotton (1982), Cotton et al. (1982; 1986). Several aspects of the aggregation model developed by Cotton et al. (1986) have been improved in the new version of parameterized microphysics, including the division on the ice crystal class into small, non-precipitating crystals and large snow crystals, the introduction of an empirical formula of collection efficiency of ice particles diagnosed from the observed particle size spectrum (Yeh et al., 1988), and changes in the conversion equation from aggregates to graupel. The numerical experiments are summarized in Table 1.

Table 1 The Numerical Simulation Experiments

Simulation experiment	Microphysics model	Collection efficiency of ice particles
Control experiment	New version	0—−13 ℃: $E(T_s)=10^{0.063T_s+0.235}$ $T<-13$ ℃: $E(T_s)=\text{Min}[10^{0.035T_s-0.7}, 0.2]$
Sensitivity Exp. 1	Version 5	−12—−15 ℃: $E(T_s)=1.4$ otherwise: $E(T_s)=\text{Min}[10^{0.035T_s-0.7}, 0.2]$
Sensitivity Exp. 2	New version	−12.5—−15.5 ℃ and saturation for water: $E(T_s)=1.4$ otherwise: $E(T_s)=0.3$

The 2-D model domain was 90 km long and 19 km high with a grid scale of $\Delta x=1$ km and $\Delta z=0.5$ km. The x-coordinate was selected pointing 60° from north, which is basically consistent with the direction of wind shear from the surface to the 60 kPa level and the observed propagating direction of the MCC. The coordinate frame moved with an average speed of 15 m/s along the x axis. A 10 s timestep was used in the simulation for the first hour, and 7.5 s after. The radiative lateral boundary condition and normal mode radiative top boundary condition were used in the simulation as applied in the orograph cloud simulation by Cotton et al (1986). A cold downdraft initialization method is also used to trigger the cloud circulation. The model was initiated with a sounding data taken from Wichita at 0600 GMT 4 June 1985 just before the MCC developed. A mesoscale updraft beneath 6 km level was added in the simulation with a peak value of 0.40 m/s at the level of 4 km and decreased linearly to 0 m/s at the levels of 0 and 6 km.

4 Dynamical and Thermodynamic Properties of the Simulated Meso-β-scale Convective System

The control experiment reveals the general properties of the convective cloud system. The time-height cross section of vertical motion (Fig. 1) shows that the meso-γ-scale convection is basically unsteady. The main convective cells developed in a time interval of about 30 min. The convective core located at the 7.5—8.5 km levels exhibited peak updraft values of 15.0, 20.5, 19.5 and 19.0 m/s at about 30, 60, 90 and 135 min of simulation time. The convection weakened after 135 min and dissipated at 165 min. The meso-β-scale cloud system exhibited characteristics of a multi-celled convective storm for a period of 3 h (Fig.2). Beneath the updraft, a downdraft developed at about the 8 km level with weaker values of 6—8 m/s. The fact that the main convective activity concentrated at the mid-upper troposphere suggests that the ice-phase process is important to the evolution of the convective system.

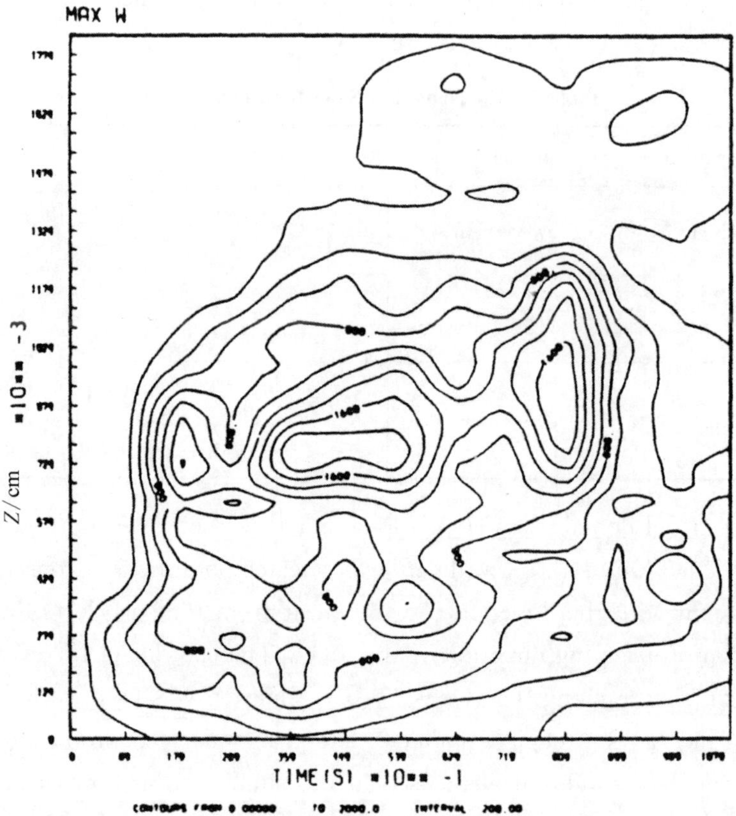

Fig.1 Time-height cross-section of maximum updraft for control experiment (The contour interval is 2 m/s)

The perturbation pressure field exhibits a meso-low of −1.5 mb at low levels at 30 min of simulation time, and reached −7.5 mb at 75 min. The meso-low extended to the upshear side and covered the entire domain beneath the 7.5 km level at the end of the simulation with a peak value of −8.0 mb. A weaker high pressure area developed near cloud top. The meso-high with a peak value of 2 mb at 90 min is located on the upshear side of cloud system at the 8 km to 11 km levels (Fig.3(a)).

The temperature perturbation field shows a warm core in the cloud at the 6 km to 10 km levels. Since the low levels are moist, cooling by evaporation in the downdraft was not evident (Fig.3(b)). Cooling by cloud top evaporation can be seen with peak values of −10 ℃ to −14 ℃.

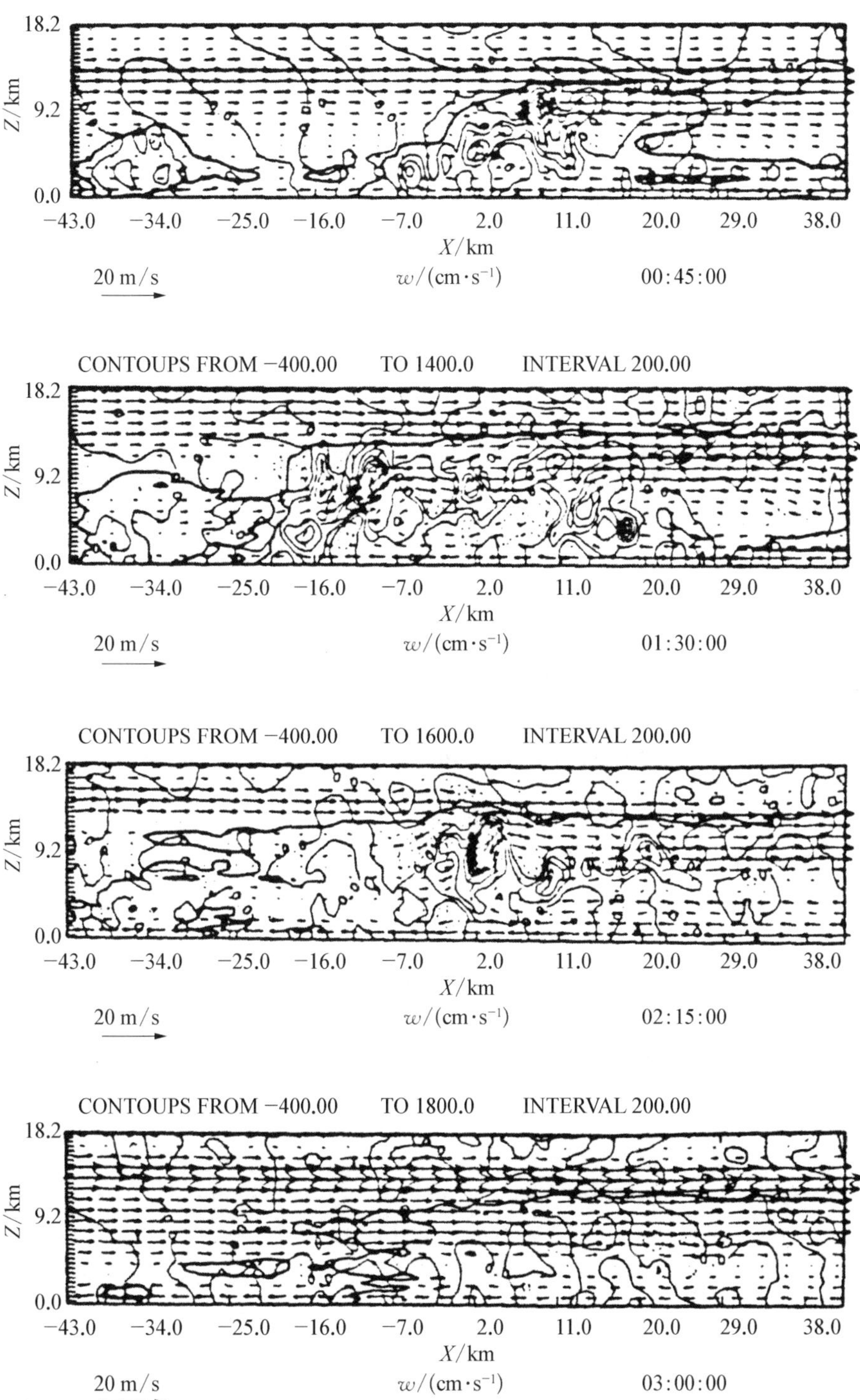

Fig.2 Vertical velocity and wind vector fields for control experiment at 45, 90, 135 and 180 min simulation time (The contour interval is 2 m/s)

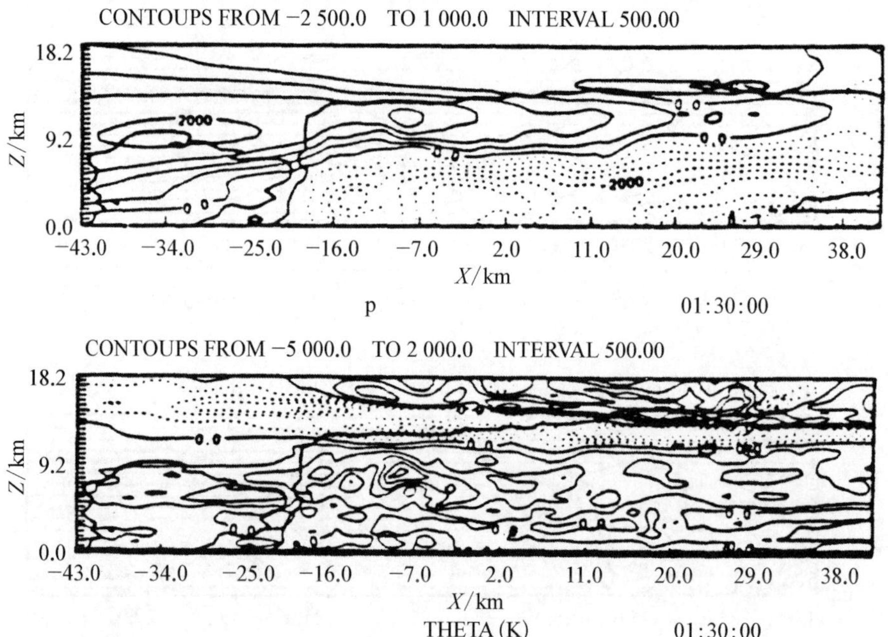

Fig.3 Perturbation pressure (p') and perturbation potential temperature (θ') fields at intervals of 0. 5 mb for p' and 2°K for θ' for control experiment at 90 min simulation time

5 Microphysics of the simulated meso-β-scale convective system

The microphysics show that graupel is primarily concentrated in the strong convective cells which is the main source of convective rainfall after one hour, and aggregates are mainly located in the stratiform region and decaying convective cells which produce the stratiform rainfall. Riming is the predominant precipitation mechanism in convective cells, and aggregation is predominant in the stratiform region. In the transition region, which is near the convective cells and generally composed of decaying convective cells, both riming and aggregation are important. The basic characteristics of precipitation formation is consistent with observation. The results of sensitivity experiments with ice-phase microphysics are summarized as follows:

(1) Fig. 4 shows the contours of the mixing ratio of ice-phase particles at 90 min obtained from the sensitivity Exp. 1 listed in Table 1. It can be seen that the ice crystals occupy the entire cold region above the level of 7 km in the cloud system, and there are no aggregates. Obviously, the aggregation efficiency selected by version 5 of RAMS is too low.

(2) Fig.5 is the same as Fig. 4 except for Exp. 2 in Table 1. Compared with Exp. 1, aggregates dominate the cloud, and pristine crystals and snow crystals are almost totally depleted. The aggregation efficiency is too large.

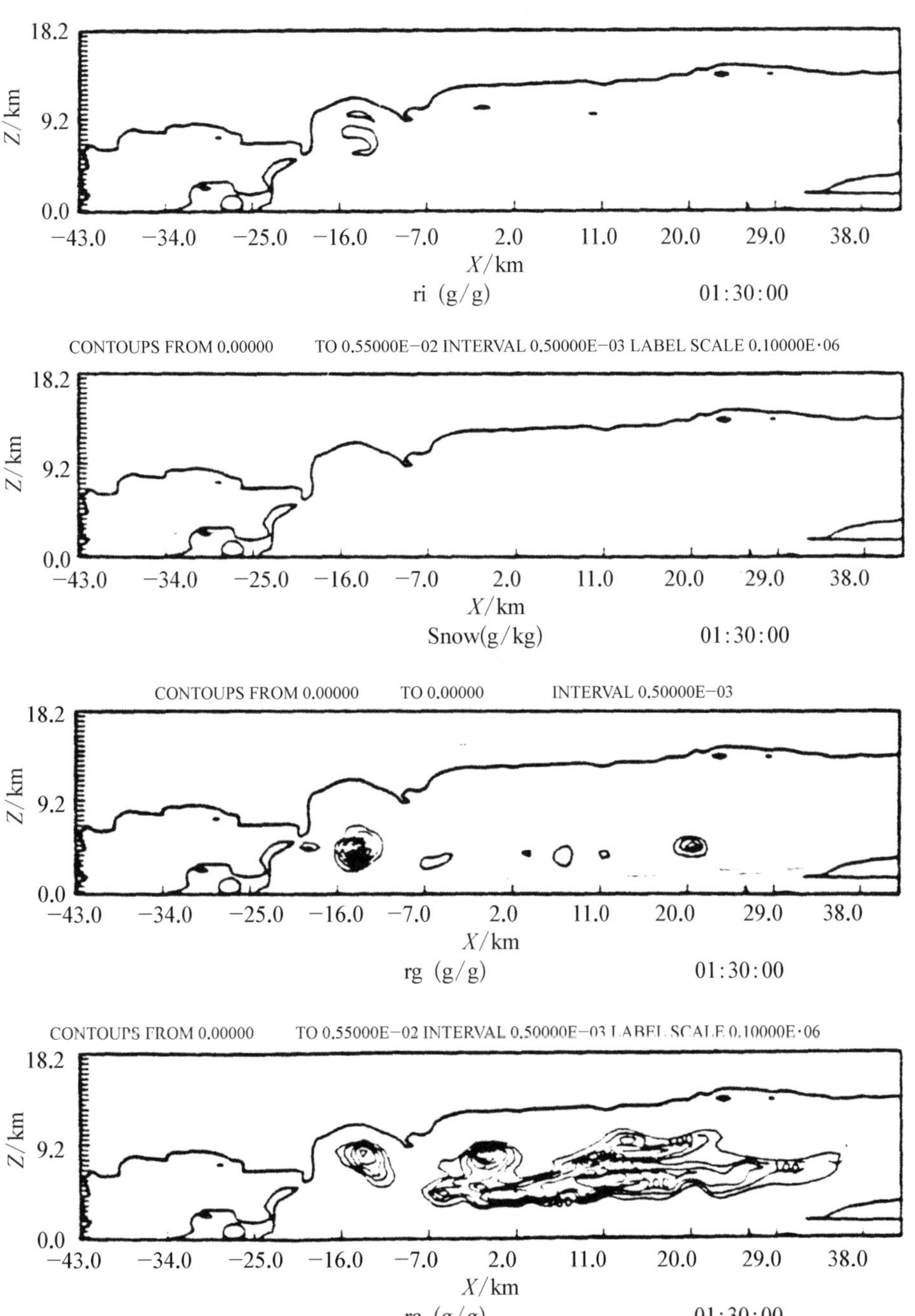

Fig.4 Microphysical structure at 90 min simulation time for Exp.1. (a) contours of ice crystal mixing ratio, (b) graupel mixing ratio and (c) aggregate mixing ratio, all contours at intervals of 0. 5 g/kg.

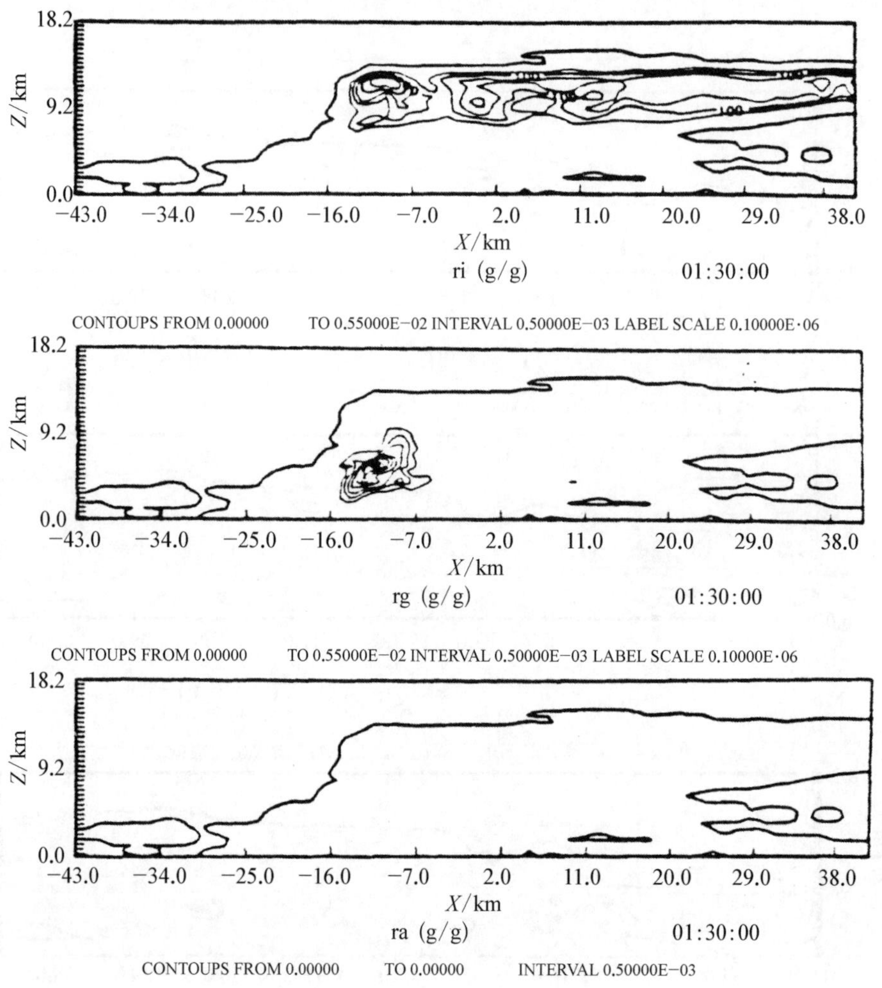

Fig.5 As in Fig.4 except for Exp. 2

(3) The result of the control experiment is given in Fig. 6. Here the microphysical structure using the diagnosed aggregation efficiency is more reasonable. Pristine crystals and snow crystals are mainly above the 7 km level, while graupel are concentrated in the strong convective cells and somewhat in the decaying cells. The aggregates primarily extend in the stratiform region and decaying convective cells in the layer from 3 km to 8 km, which is consistent with the observation that aggregation starts at the upper and colder levels but becomes more efficient near the melting layer.

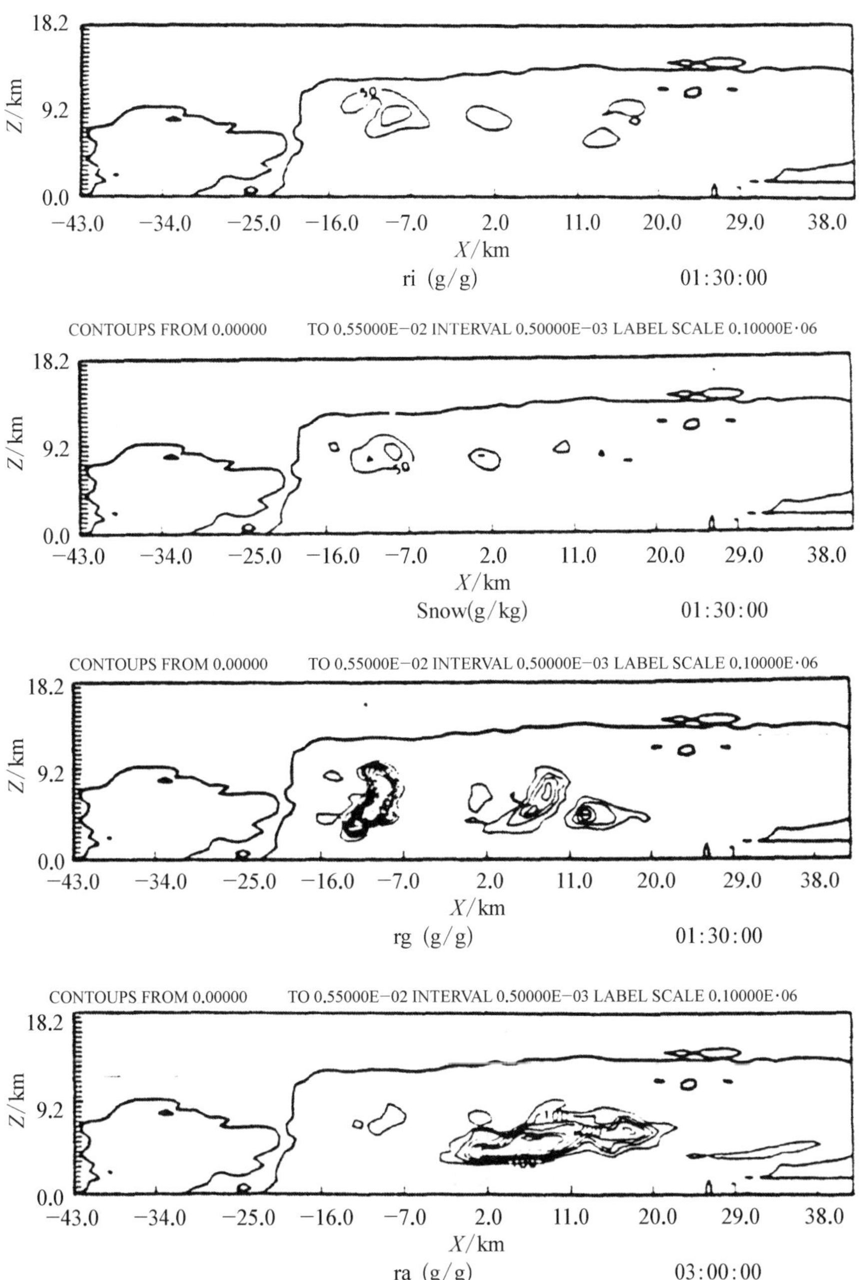

Fig.6 As in Fig.4 except for control experiment

Acknowledgements

This research was supported by National Science Foundation under Grant ♯ATM-8512480 and by the Army Research Office under contract ♯DAAL03-86-K-0175.

References

Chen Sue, 1986: Simulation of the stratiform region of a mesoscale convective system. M. S. Thesis, Colorado State University, Fort Collins, CO 80523.

Cotton, W. R., M. A. Stephens, T. Nekrkoln, and G. J. Tripoli, 1982: The Colorado State University three-dimensional cloud/mesoscale model - 1982. Part II: An ice phase parameterization. J. de Rech. Atmos., 16, 295-320.

Cotton, W. R., G. J. Tripoli, R. M. Rauber, and E. A. Mul-vihill, 1986: Numerical simulation of the effects of varying ice crystal nucleation rates and aggregation processes on orographic snowfall. J. Clim. Appl. Meteor., 25, 1658-1680.

Fortune, M. A., and R. L. McAnnelly, 1986: The evolution of two mesoscale convective complexes with different patterns of convective organization. Preprints of Joint Sessions, 23rd Conference on Radar Meteorology and Conference on Cloud Physics, AMS, Snowmass, Colorado, J175-178.

Tripoli, G. J., and W. A. Cotton, 1982: The Colorado State University three-dimensional cloud/mesosvcale model - 1982. Part I: General theoretical framework and sensitivity experiments. J. de Rech. Atmos., 16, 185-220.

Yeh Jia-Dong, M. A. Fortune, and W. R. Cotton, 1986: Micropgysics of the stratified precipitation region of a mesoscale convective system. Same as Fortune and McAnnelly, J151-154.

Yeh Jia-Dong, Bei-Fen Fan, M. A. Fortune, and W. R. Cotton, 1987: Comparison of the microphysics between the transition region and stratiform region in a mesoscale convective complex. Preprints of Third Conference on Mesoscale Processes, AMS, Vancouver, Canada, 188-189.

Yeh Jia-Dong, Bei-Fen Fan, and W. R. Cotton, 1988: Size distribution of precipitation particles in midlatitude mesoscale convective complex. 10th International Cloud Physics Conference, 15-20 August 1988, Bad Homburg, FRG.

5.12 NUMERICAL SIMULATION OF MICROPHYSICS IN MESO-β-SCALE CONVECTIVE CLOUD SYSTEM ASSOCIATED WITH A MESOSCALE CONVECTIVE COMPLEX

Fan Beifen (范蓓芬), Ye Jiadong(叶家东)

Department of Atmospheric Science, Nanjing University, Nanjing 210093

William R. Cotton, Gregory J. Tripoli[①]

Department of Atmospheric Sciences, Fort Collins, CO 80523 U.S.A.

Received 28 March, 1989

Abstract Numerical simulation of meso-β-scale convective cloud systems associated with a PRE-STORM MCC case has been carried out using a 2-D version of the CSU regional atmospheric modeling system (RAMS) nonhydrostatic model with parameterized microphysics. It is found that the predicted meso-γ-scale convective phenomena are basically unsteady under the situation of strong shear at low-levels, while the meso-β-scale convective system is maintained up to 3 hours or more. The meso-β-scale cloud system exhibits characteristics of a multi-celled convective storm in which the meso-γ-scale convective cells have lifetime of about 30 min.

① Current affiliation: University of Wisconsin, Department of Meteorology, Madison, WI 53715

Pressure perturbation depicts a meso-low after a half hour in the low levels. As the cloud system evolves, the meso-low intensifies and extends to the upshear side and covers the entire domain in the mid-lower levels with the peak values of 5—8 hPa. Temperature perturbation depicts a warm region in the middle levels through the entire simulation period. The meso-γ-scale warm cores with peak values of 4—8 ℃ are associated with strong convective cells. The cloud top evaporation causes a stronger cold layer around the cloud top levels.

Simulation of microphysics exhibits that graupel is primarily concentrated in the strong convective cells forming the main source of convective rainfall after one hour of simulation time. Aggregates are mainly located in the stratiform region and decaying convective cells which produce the stratiform rainfall. Riming of the ice crystals is the predominant precipitation formation mechanism in the convection region, whereas aggregation of ice crystals is the predominant one in the stratiform region, which is consistent with observations. Sensitivity experiments of ice-phase microphysical processes show that the microphysical structures of the convective cloud system can be simulated better with the diagnosed aggregation collection efficiencies.

I. INTRODUCTION

Numerous observational evidences have revealed that the ice particle aggregation process is one of the predominant mechanisms of precipitation formation in a variety of cloud systems (Jiusto, 1971; Hobbs et al., 1974; Matejka et al., 1980; Heymsfield and Musil, 1982; Stewart et al., 1984; Houze and Churchill, 1984; Jorgensen, 1984; and Rauber, 1987). Airborne cloud microphysical data obtained from midlatitude mesoscale convective systems (Yeh et al., 1987) also indicated that aggregates comprise an important component of the precipitation in the stratiform region and transition region of midlatitude mesoscale convective complexes (MCC), especially in the stratiform region.

Cotton et al. (1986) developed an aggregation model in the CSU regional atmospheric modeling system (RAMS) and applied it to the simulation of orographic cloud snowfall. The numerical experiments demonstrated that aggregation plays an important role in controlling the fields of cloud liquid water content, ice crystal concentrations, and surface precipitation amounts. Chen (1986) also applied the model to the simulation of the stratiform region of a midlatitude mesoscale convective complex observed on 14—15 July 1984 during the airborne investigations of mesoscale convective systems (AIMCS) experiment in which the microphysical structures in the stratiform region of it have been analyzed by Yeh et al. (1987). The model predicted the general features of the thermodynamics and dynamics of the cloud system, and demonstrated the effects of the long-wave radiation on the enhancement of the internal cloud circulations in the stratiform region of the MCC. However, the simulation results displayed that the ice-crystal category of water species occupied over the whole stratiform region with the maximum ice mixing ratio of 3.4—3.8 g/kg in the mature stage of the MCC, whereas the aggregate category almost did not be found in the stratiform region of the modeling cloud system, which is inconsistent with the observations. The model underpredicted the amount of aggregates over the stratiform region. This deficiency probably is attributed to some incompleted aspects of aggregation model.

In this paper we applied a two-dimensional version of the CSU RAMS nonhydrostatic cloud model with improved parameterized microphysics to the simulation of meso-β-scale convective cloud system associated with a mesoscale convective complex observed on 4 June 1985 during the PRE-TORM experiment. The microphysical structures in the stratiform region and transition region of the MCC have been analyzed by Yeh et al (1987).

II. DESCRIPTION OF MODEL

The Colorado State University (CSU) regional atmospheric modeling system (RAMS) developed by Cotton's group is front-ended by a preprocessor software package which allows the construction of a Fortran code that is one-dimensional, two-dimensional, or three dimensional and utilizes various options in dynamics, thermodynamics and microphysics. The governing equations and general features of the model system are described in Tripoli and Cotton (1982), Cotton et al. (1982) and Cotton et al (1986). In this case we used a two-dimensional, nonhydrostatic, time-dependent cloud model with parameterized microphysics to the simulation of a meso-β-scale convective system. Several aspects of the aggregation model developed by Cotton et al. (1986) have been improved in the new version of parameterized microphysics. We shall describe here the improvements in the aggregation model.

1. Empirical Estimation of Aggregation Efficiency of Ice Particles

One of the main problems in the simulation of snowflake aggregation is the collection efficiency between ice crystals, which is the product of the collision efficiency and the adhesion efficiency. The collision efficiency is determined by the aerodynamics associated with the flow of air around the collision particles, which is complicated by the nonuniform vertical and horizontal motions of ice particles. The adhesion mechanism is also a complicated process affected by a variety of factors such as the air temperature, humidity, crystal habits, and the presense of supercooled cloud droplets.

Hallgren and Hosler (1960) measured the collection efficiencies of ice particles in the laboratory and showed at ice saturation the collection efficiencies were low, usually less than 0.2, and decreased as the temperature decreasing from −11 ℃ to −26 ℃. Between −11 ℃ and −6 ℃ they decreased with increasing temperature. Latham and Saunders (1971) showed from their experiments, that the collection efficiencies at ice saturation were approximately constant at about 0.3. On the other hand, Passarelli (1978) deduced a mean aggregation efficiency of 1.4±0.6 from aircraft data, over the temperature range from −12 ℃ to −15 ℃. The difference between Passarelli's findings and that obtained from laboratory studies was probably due to the fact that more elaborate crystal shapes and a broader spectrum of crystal sizes plus higher levels of turbulence are found in natural clouds than that in the laboratory. Moreover, most studies were carried out at ice saturation which is probably different with the most situations in the natural clouds. Hosler et al. (1957) noted that considerably higher efficiencies were measured at ice supersaturation such as occurs in a natural cloud containing supercooled cloud droplets.

We therefore applied Passarelli's (1978) scheme to diagnose the collection efficiency in the stratiform region of the PRE-STORM MCC case observed on 4 June 1985 during its mature stage. Passarelli applied the moment conservation equations for the total mass and reflectivity factor fluxes for aggregating snowflakes to calculate the mean collection efficiency. For the case of steady, constant snow mass flux, the resulting explicit expression of the mean collection efficiency $\bar{E}$ in a layer is

$$\bar{E}=(\lambda^{-1-b}-\lambda_0^{-1-b})\frac{12\Gamma(7+b)\lambda^{4+b}}{(1+b)\pi N_0 h I(b)}, \tag{1}$$

where h is the height below a reference level ($z=0$), N_0 and λ are the size distribution parameters of exponential distribution at $z=h$, λ_0 is the slope parameter of the size distribution at the reference level ($z=0$), and b is a parameter in the particle falling speed equation:

$$v(D)=aD^b \tag{2}$$

and

$$I(b)=\int_0^{\infty}\int_0^{\infty}x^3y^3(x+y)^3\mid x^b-y^b\mid \mathrm{e}^{-(x+y)}\mathrm{d}x\mathrm{d}y. \tag{3}$$

Based on the experiments of Hobbs et al (1974), we selected $b=0.25$, then integrating equation (3) obtained $I(b)=566$. Thus, based on the assumption of constant mass flux, the mean collection efficiency in a certain layer can be calculated from Eq. (1) when the particle size distribution parameters N_0 and λ are known at the layer boundaries.

We calculated the values of the size distribution parameters N_0 and λ at various levels based on the data observed in the location A of the stratiform region of the MCC observed on 4 June 1985 as shown in Yeh et al. (1987). We remember that during the time of the vertical sounding carried out at the location A, and cloud system is in its mature stage, and the aircraft was descending slowly with a similar style of termed advecting spiral descent suggested by Lo and Passarelli (1982) through the stratiform region with a descending speed of about 1—2 m/s to trace the growth process of the aggregates, so the assumption of constant mass flux of snowflakes is reasonably satisfied here. The best fit lines for $N_0(z)$ and $\lambda(z)$ exhibit the correlation coefficients of 0.76 and 0.54 as shown in Fig. 1, respectively. The empirical relations are given by

$$N_0(z)=1.9\times10^{-4}10^{0.56z}, \tag{4}$$

$$\lambda(z)=-2.30+3.89z, \tag{5}$$

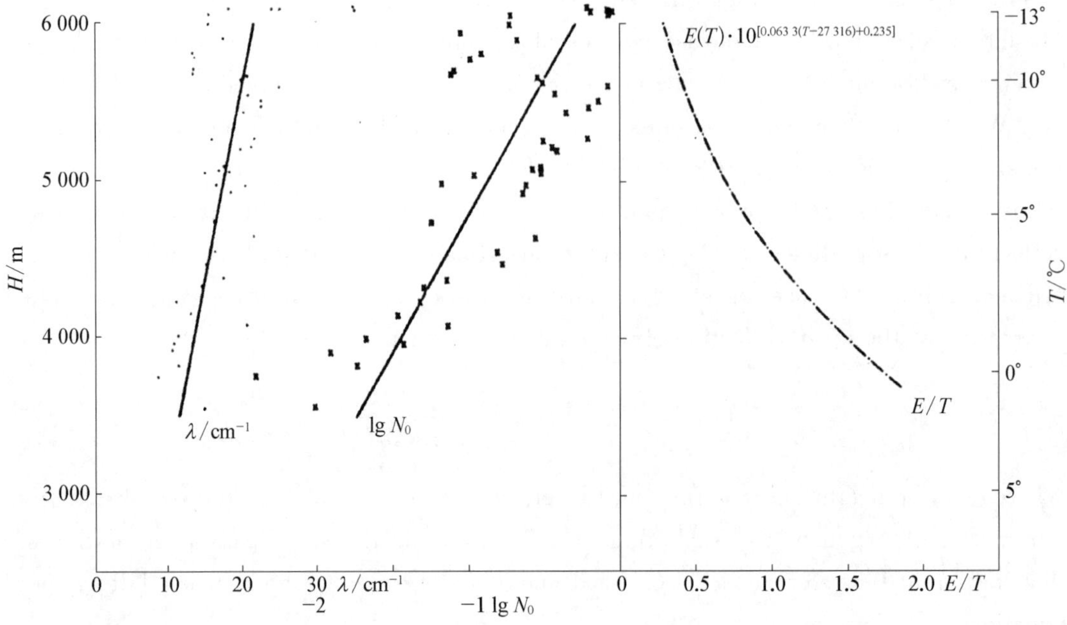

Fig.1 Scatter diagrams of slope parameter λ and intercept parameter N_0 of Marshall-Palmer distribution of hydrometeor size spectra at various levels and corresponding empirical aggregation efficiency as a function of temperature calculated from Passarelli's method (1978)

where z is in km, N_0 and λ are in cm^{-4} and cm^{-1}, respectively. Using the relations the mean collection efficiencies at various layers have been calculated. We obtained an empirical formula as

$$\overline{E(T_s)} = 10^{0.235+0.0633T_s} \tag{6}$$

with a fitting correlation coefficiency of 0.98, or, a simpler linear relation as

$$\overline{E(T_s)} = 1.627 + 0.128T_s \tag{7}$$

with a correlation coefficiency of 0.96, where $T_s = T - 273.16$, and T represents the diagnosed surface temperature of the ice particle. This empirical relation is available in the temperature ranging from 0 ℃ to −13 ℃ in which the microphysical data obtained. The Hallgren and Hosler's results with the temperature dependent coalescence efficiency formula:

$$\overline{E(T_s)} = \min[10^{0.035T_s-0.7}, 0.2] \tag{8}$$

was still used beneath −13 ℃ which is similar with Cotton et al. (1986).

2. The Conversion Term from Aggregates to Graupel

The conversion from aggregates to graupel can happen by aggregate collecting cloud droplets which should result in a concentration tendency for aggregate, graupel and cloud droplet. The calculation is actually made in two parts. The first part is the conversion of cloud droplets to aggregates, i. e., the riming tendency. Then we convert from aggregates to graupel based on that riming tendency. The problem is, however, how much rimed aggregates will be really converted to graupel? This is a difficulty because certainly an aggregate can rime some cloud droplets and still be called an aggregate until some critical point is reached where we start calling it graupel. Since when we see a distribution of aggregates we do not know how much rime they have, we cannot make that decision. One of the approaches to solve the problem approximately is that the total mass of aggregates probably will not become graupel without some significant riming. So, if we require a criterion riming degree not to be exceeded, we can put the excess into conversion. This would only make graupel from aggregates in very vigorous cases. We suggest the rate could be that amount in excess of the deposition and ice crystal collection rates. The physical reasoning being that if riming is a dominant growth process, then the rimed aggregates start looking like graupel. Therefore, we obtain the conversion term from aggregates to graupel as

$$CN_{ag} = [\max(CL_{ca} - (CL_{ia} + CL_{sa} + VD_{va}), 0.) + CL_{ar}]H(T_s), \tag{9}$$

where CN_{ag} is the conversion rate from aggregates to graupel, CL_{ca} is the riming rate of aggregates, CL_{ia}, CL_{sa} and CL_{ar} are the collection rates of pristine crystal, snow crystal and raindrop with aggregates, respectively, VD_{va} is the deposition rate of vapor on the aggregates, and $H(T_s)$ is the Heaviside step function.

III. THE PRE-STORM CASE of 4 JUNE 1985

The case selected for simulation is a meso-β-scale convective cloud system associated with a PRE-STORM MCC of 4 June 1985. The MCC developed in association with a short wave trough system which passed through mid-west region of the United States. An associated cut-off low pressure center was located over southern California which is evident from 700 hPa to 300 hPa levels (Figure omitted). A southwest air current was prevalent to the east of the trough and the low center above 700 hPa level, which brought the warm and drier air to the area of PRE-STORM network. At 850 hPa level, a 13—15 m/s southerly jet pumped warm moist Gulf air northward through central Texas, across the surface front at about the Oklahoma-Kansas border, towards the Panhandle of Texas, northwest Oklahoma, and the western sections of Kansas and Nebraska. The synoptic situation is favorable for the set up of the instability in the mid-low levels of atmosphere and the development of a mesoscale convective system.

Fig.2 shows an analyzed result of the sounding data taken from Wichita at 0600 GMT just before the MCC developing which was used to initialize the model simulations. Some important features are revealed: (1) The atmosphere was warm and moist in the low levels, and dry in the middle troposphere; (2) A southern low level warm moist jet of 13—15 m/s was near 850 hPa level, and a southwestern upper level jet of about 40 m/s near

200 hPa level (Fig.2(b)); (3) A deep conditional instable layer was from 900 hPa level up to 500 hPa level (Fig.2(c)); (4) the wind hodograph displayed that wind veer shear was evident beneath the 600 hPa level, above it the southwest air current prevailing until up to 100 hPa level (Fig.2(d)).

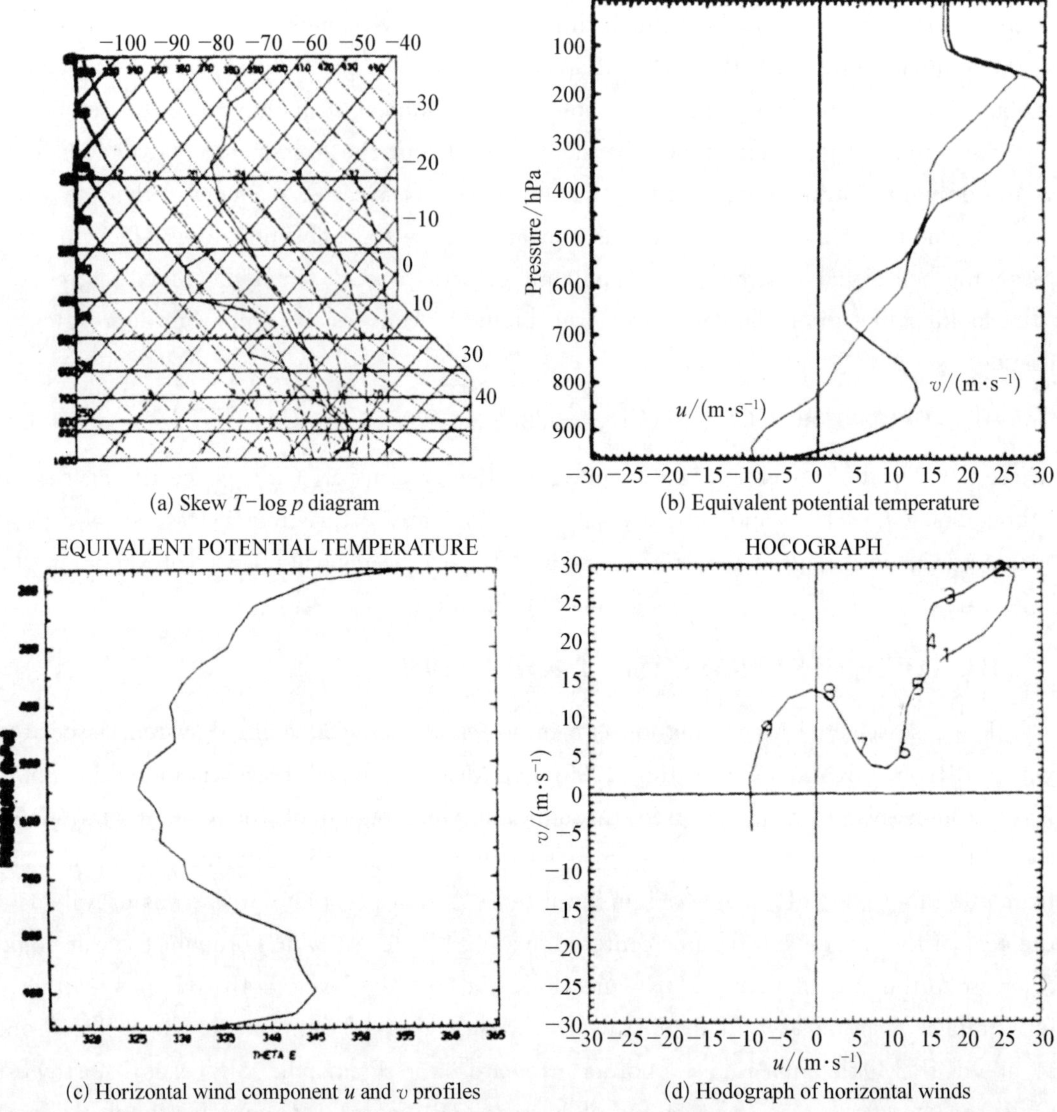

(a) Skew $T-\log p$ diagram (b) Equivalent potential temperature

(c) Horizontal wind component u and v profiles (d) Hodograph of horizontal winds

Fig.2 Vertical profiles of thermodynamics and horizontal wind components u and v for initial sounding from Wichita, Kansas at 06 GMT, 4 June 1985

The MCC was a rapidly evolving and fast moving, non-linear system, which formed by 0630 GMT and dissipated at about 1400 GMT. In the early time of its lifecycle, numerous convective cells developed in a random manner. At 0800 GMT the convective echoes started to organize into several meso-β-scale sub-systems (Fortune and McAnnelly,

1986). The propagation speeds of the MCC were about 15—25 m/s from the direction of about 270°. Under the influence of low level convergence and low level southern jet which located at the western side of the storm system at the 850 hPa level, the meso-β-scale subsystems moved with various direction and speed. In the southwest part, some meso-β-scale convective clusters waved to 60° with speeds of 15—20 m/s, and some northern subsystem moved to east. Some of the meso-β-scale convective cloud systems lasting more than 3 hours although the convective action in it have decayed within its 3 hours lifetime. We focus on the simulations of the meso-β-scale convective cloud system. Airborne observations have revealed as described on Yeh et al. (1987) that the characteristics of ice particle are quite different between the stratiform region and transition region of the MCC. Most of the large particles are aggregates and graupel particles in the transition region, whereas aggregates predominate in the stratiform region. The aggregation process in the stratiform region started at upper and colder levels but become more efficient as the aggregates approach the melting layer.

IV. DESIGN OF NUMERICAL EXPERIMENT

1. Model Domain and Grid Points, Boundary Conditions and Initial Conditions

The two dimension model domain was 90 km long and 19 km high with a grid resolution of $\Delta x = 1$ km and $\Delta z = 0.5$ km. The x-coordinate was selected pointing 60° from north, which is basically parallel to the wind shear vector from surface to 600 hPa level. The coordinate frame moved with an average speed of 15 m/s along the x-axis. A 10 s time step was used in the simulation for the first hour, and 7.5 s after. The radiative lateral boundary condition and normal mode radiative top boundary condition were used in the simulation as applied in the orograph cloud simulation by Cotton et al (1986). A cold downdraft initialization method which attempts to model evaporative cooling from an initial cell is used to trigger the cloud circulation. The model was initialized using the sounding data taken from Wichita at 0600 GMT 4 June 1985 just before the MCC developed. In addition, a mesoscale updraft was applied in the simulation beneath 6 km level with a peak value of 0.40 m/s at the level of 4 km and decreased linearly to 0 m/s at the levels of 0 and 6 km.

2. Numerical Experiments

The numerical experiments are summarized in Table 1. In the control experiment, we applied the new microphysics version developed by Tripoli (1987) with improved aggregation model, in which the diagnosed aggregation efficiency (Eq.6) was used in the temperature ranging of 0—13 ℃ and Hallgren and Hosler's results used under −13 ℃ as described in Section II. Sensitivity experiment 1 used the microphysics model of version 5 of RAMS, same as used as control experiment in Cotton et al (1986). The microphysics model in sensitivity experiment 2 is same with control experiment except for different aggregation efficiency.

Table 1 The numerical simulation experiments

Simulation experiment	Microphysics model	Collection efficiency of ice particles
Control experiment	New version	0—−13 ℃: $E(T_s)=10^{0.063T_s+0.235}$ $T<-13$ ℃: $E(T_s)=\mathrm{Min}[10^{0.035T_s-0.7},0.2]$
Sensitivity Exp. 1	Version 5	−12—−15 ℃: $E(T_s)=1.4$ otherwise: $E(T_s)=\mathrm{Min}[10^{0.035T_s-0.7},0.2]$
Sensitivity Exp. 2	New version	−12.5—−15.5 ℃ and saturation for water: $E(T_s)=1.4$ otherwise: $E(T_s)=0.3$

V. DYNAMICAL AND THERMODYNAMICAL PROPERTIES OF THE SIMULATED MESO-β-SCALE CONVECTIVE CLOUD SYSTEM

The control experiment revealed the general properties of the convective cloud system. The dynamical and thermodynamical characteristics of the simulated convective cloud system in the control experiment are described briefly as follows:

1. Vertical Motion

The time-height cross section of vertical motion (Fig.3) shows that the meso-γ-scale convection is basically unsteady. The main convective cells developed in a time interval of about 30 min. The convective cores located at 7.5—8.5 km levels with peak updraft values of 15.0, 20.5, 19.5 and 19.0 m/s at about 30, 60, 90 and 135 min of simulation time, respectively. The convection weakened after 135 min and dissipated at about 165 min. The meso-β-scale cloud system exhibited characteristics of a multi-celled convective storm for a period of 3 h as shown in Fig.4. In the initiation stage, a small convective cell with a peak updraft of 4.0 m/s formed by 15 min of simulation time at 3 km level, 2 km to the right of domain center. As times advanced, the convective cell rapidly developed to a much stronger one and began to lean more markedly at upper levels. By 30 min, its peak vertical velocity value reached 12.0 m/s at 7 km level. The ice-phase processes at that level have been active, which probably has an important effect to the developing of strong convection as well as vapor condensation. When the convective cell alofted to the upper levels at about 45 min, it moved to about 8 km to the right of center under the influence of upper levels south-western jet current, and the peak value of vertical velocity reached 14 m/s at about 8 km level (Fig.4(a)). At the same time, two new convective cells developed with the peak values of 6.0 m/s at the low levels near the center of domain. These two new cells mainly caused by the convergence of low level wind as well as the first one at the initiated time. They reached their strongest status at about 75 min of simulated real time with a maximum updraft value of 18 m/s. The fourth strong convective cell formed at the left side of the cloud system, 12 km to the left of domain center. It developed much faster than that of the previous cells. In a short range of 15 min it reached its peak updraft value of 16 m/s at about 8 km level, 10 km to the left of the domain center (Fig.4(c)). In the low levels, a new convective cell developed at the location of 18 km to the right of domain center with an updraft value of 10 m/s. After a half hour, the cell reached its strongest updraft value of 12 m/s at about 9 km level. At the same time another convective cell developed at mid-levels near the center of domain and reached its strongest status at 135 min, which peak updraft value was 18 m/s as shown in Fig.4(c). The convective action was beginning to decay after 135 min of simulated real time.

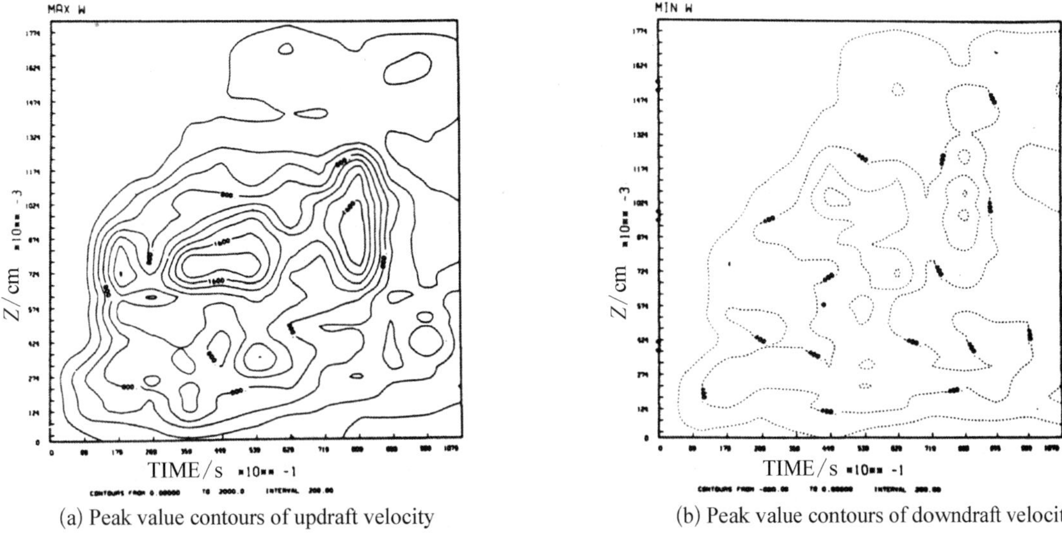

(a) Peak value contours of updraft velocity

(b) Peak value contours of downdraft velocity

Fig.3 Time-height cross-section of peak vertical velocity field. The contour interval is 2 m/s.

Several small and weaker convective cells with peak updrafts not stronger than 6 m/s still developed and maintained to certain range of time at mid-upper levels. The convective action was completely calm down at the simulated real time of 180 min. It is noted from Fig.4 that the interaction of the low level front to rear flow and mid-level rear inflow in the convective region prompted the form of the main convective cells on the left hand of the cloud system. By the main updraft, a downdraft developed at about 8 km level with weaker values of 6—8 m/s. The air flow at the mid-upper levels exhibited a wave characteristics implying that the gravity wave created as a result of meso-β-scale convective overturning turned to affect the air motion structure on the larger meso-β-scale in the troposphere and prompted the formation of the multi-cells storm. The downdraft is weak under the main convective cell since the air is moist in the lower layer, so the effect of evaporation of raindrop may not be evident. The fact that the strong convective activity concentrated at the mid-upper troposphere suggests that the ice-phase process is important to the evolution of the convective system as emphasized by Willoughby et al. (1984) in the simulation of hurricane.

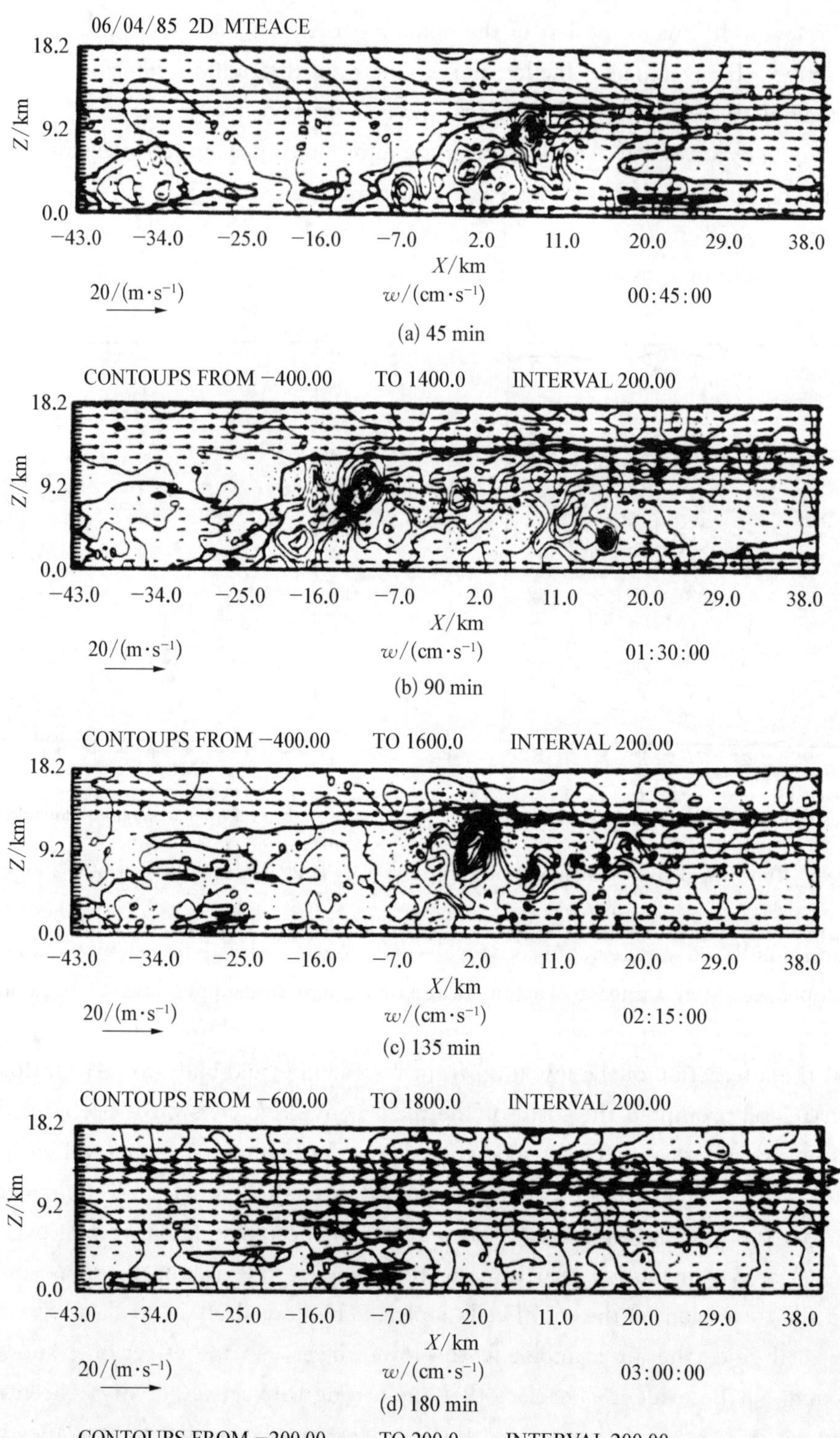

(a) 45 min

(b) 90 min

(c) 135 min

(d) 180 min

Fig.4 Wind vector flow field and contours of vertical velocity at simulated real time of 45 min, 90 min, 135 min, and 180 min. The wind vector is relative to the 15 m/s storm system motion. The solid lines represent the upward motion and the dashed lines represent the downdraft motion. The contour interval is 2 m/s.

2. Horizontal Motion

All the values of horizontal velocity are relative to the moving frame of the simulated cloud system (Figure omitted). In the front side of the mature storm system, inflow mainly concentrated in the low levels, and the outflow is over the mid-upper levels, the mid-level rear inflow evoluted in the mature stage of the storm system. A weaker rear outflow has been found at the upper levels. An upper jet current with peak values larger than 20 m/s is prevalent above the cloud top. By 45 min, an anvil cloud beginning to form and spreaded downwind under the influence of the upper jet current. After 135 min, the upper jet intensified up to 32 m/s or more and merged with the mid-level rear inflow which suppressed the convection and resulted in the dissipating of the storm system.

3. Perturbation Pressure

As the convection developing, the negative perturbation pressure formed and intensified in the lower layer beneath 5 km level after a half hour (Figure omitted). The largest negative perturbation pressure value of 7.5 hPa is found on lower boundary at 75 min. Another lowest perturbation pressure center formed at the end of simulation time. Above 7 km level, the positive perturbation pressure formed by a half hour and suspended to the end of simulation time. Three meso high centers with the peak values of 2.1, 1.6 and 2.6 hPa formed at the levels from 8 km to 11 km at 90, 120 and 150 min, respectively.

The perturbation pressure field (Figure omitted) exhibited that a small meso-high caused by the initial cold downdraft maintained in the first half hour of simulation time near the center of domain with a peak value of 2.0 hPa. A meso-low of −1.5 hPa formed by the first half hour at low levels, 9 km to the center of domain at 30 min of simulation time, and intensified up to −7.5 hPa at 75 min. The meso-low extended to the upshear side and covered the entire domain beneath the 7.5 km level after 90 min of the simulated real time, whereas the intensity of the perturbation low pressure weakened and maintained its intensities of −3.5 hPa to −5.5 hPa for about 75 min. At the end of simulation, the perturbation low pressure reached its peak value of −8.0 hPa. A weaker high pressure area developed near cloud top after 45 min. The meso-high reached its peak value of 2 hPa at 90 min, which located on the upshear side of cloud system at the 8 km to 11 km levels.

4. Perturbation Potential Temperature

In the entire period of storm system developing and evolution, middle-upper troposphere (from 5 km to 10 km) is a warm layer (Figure omitted). Several meso-γ-scale warm cores were basically associated with the action of strong convective cells, especially in first two hours. During the developing of the convective cloud system at the first hour of simulation, a perturbation potential temperature of 4.0 ℃ presented in the first convection cell at about 6 km level, slightly lower than that the strongest convection center. By 45 min, a perturbation temperature of 4.0 ℃ formed out of the cloud in the down-shear side at about 5 km level, which was probably caused by the compensating subsidence out of the cloud. The main warm region still located in the strong convective cell. The warm region extended in the mid-levels as the anvil cloud evolving at 60 min of simulated real time, with a peak anomalous temperature of 8 ℃ at about 8 km level accompanying with the active convective cells. This anomalous warm region lasted to the end of simulation with peak values ranging from 6 ℃ to 8 ℃. The warm region basically located above the 0 ℃ level implying that the ice-phase microphysical processes act an important effect on it.

Since the low levels are moist, cooling by evaporation of raindrops in the low levels was not evident and the warm area almost occupied the entire domain beneath the 9 km level. Cooling by cloud top evaporation can be seen from the simulation with peak anomalous values of −10 ℃ to −14 ℃.

VI. MICROPHYSICS OF THE SIMULATED MESO-β-SCALE CONVECTIVE CLOUD SYSTEM

1. The Control Experiment

Fig.5 illustrated the time series of cloud droplet, rairdrop, graupel, and aggregates mixing ratio. Comparing with Fig.3, the peak values of raindrop and graupel were basically associated with the strong convective cells. Graupel cores were located just over that of raindrop implying the graupel is the main source of convective rainfall. On the other hand, the variation of the peak aggregate value is much smooth with time, implying aggregates was formed in both stratiform region and convective cells. Here we describe the predicted microphysical structures a little bit detailly:

(1) Cloud and rain water fields

The cloud and rain water contents were basically restricted in the warm layer, under the 4.5 km level as expected (see Fig.5(a, b)). The highest values of rain water were seen in the main updraft developed at about 30, 60, 90 and 135 min of simulation time as shown in Fig. 3. The contours of cloud water mixing ratio showed that the liquid cloud water content mainly concentrated in the low levels newly generated convective cells as would be expected due to the condensation occurring there, in which the condensation process in the saturation updraft did not exhaust seriously by collection with precipitation particles since which was lack in the newly generated convective cells.

(2) Pristine crystal and snow crystal fields

The pristine crystals and snow crystals formed by 30 min above the 0 ℃ level. During the developing and mature stages, the peak mixing ratio values of these particles reached 1.0—1.5 g/kg in the stratiform region of the cloud system. The contents of snow crystals were less than 0.5 g/kg after 150 min when the convective action decaying (Figure omitted).

(3) Graupel particle field

The predominant formation mechanism of graupel is riming of freezing raindrop, ice crystal and aggregates. The main growth mechanism of graupel is accretion of raindrop and aggregates by graupel particle itself. Both of them are active in the high water content

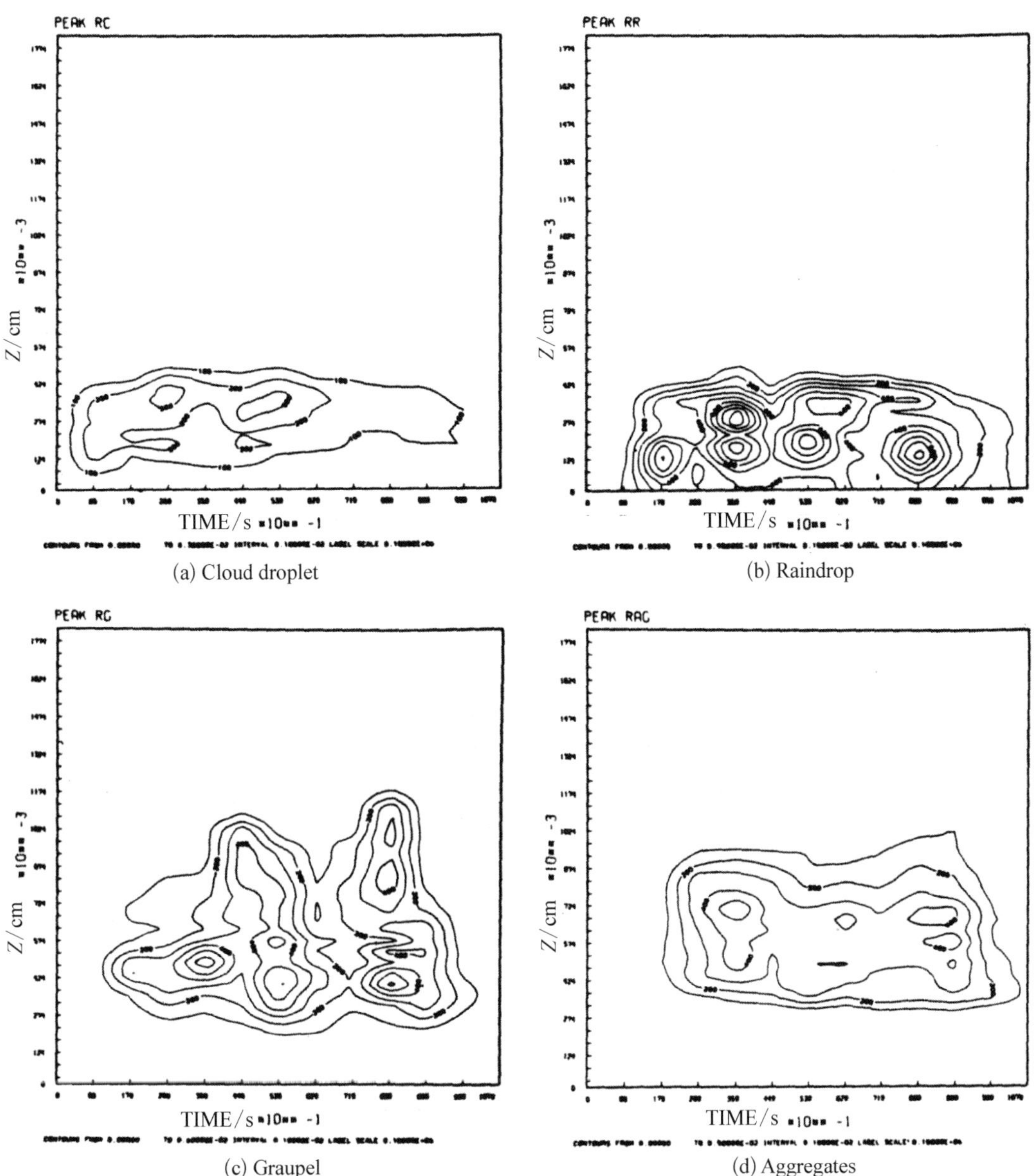

(a) Cloud droplet

(b) Raindrop

(c) Graupel

(d) Aggregates

Fig.5 Time-height cross-section of peak values of various types microphysical variables, cloud droplet, raindrop, graupel, and aggregates. The contour intervals are 1.0 g/kg.

region of cloud droplet and raindrop. So the graupel is primarily concentrated in the strong convective cells which is the main source of the convective rainfall. Contours of graupel mixing ratio are shown in Fig.6. The peak values of 3.5 and 6.0 g/kg of it located at the lower part of main convective cells from 4 km to 6 km levels in the developing stage of cloud system at about 30 and 60 min, and of 5.5 and 4.5 g/kg found from 3 km to 9 km levels in the mature stage at about 90 and 135 min. After 150 min the graupel was beginning to dissipate as convection decaying.

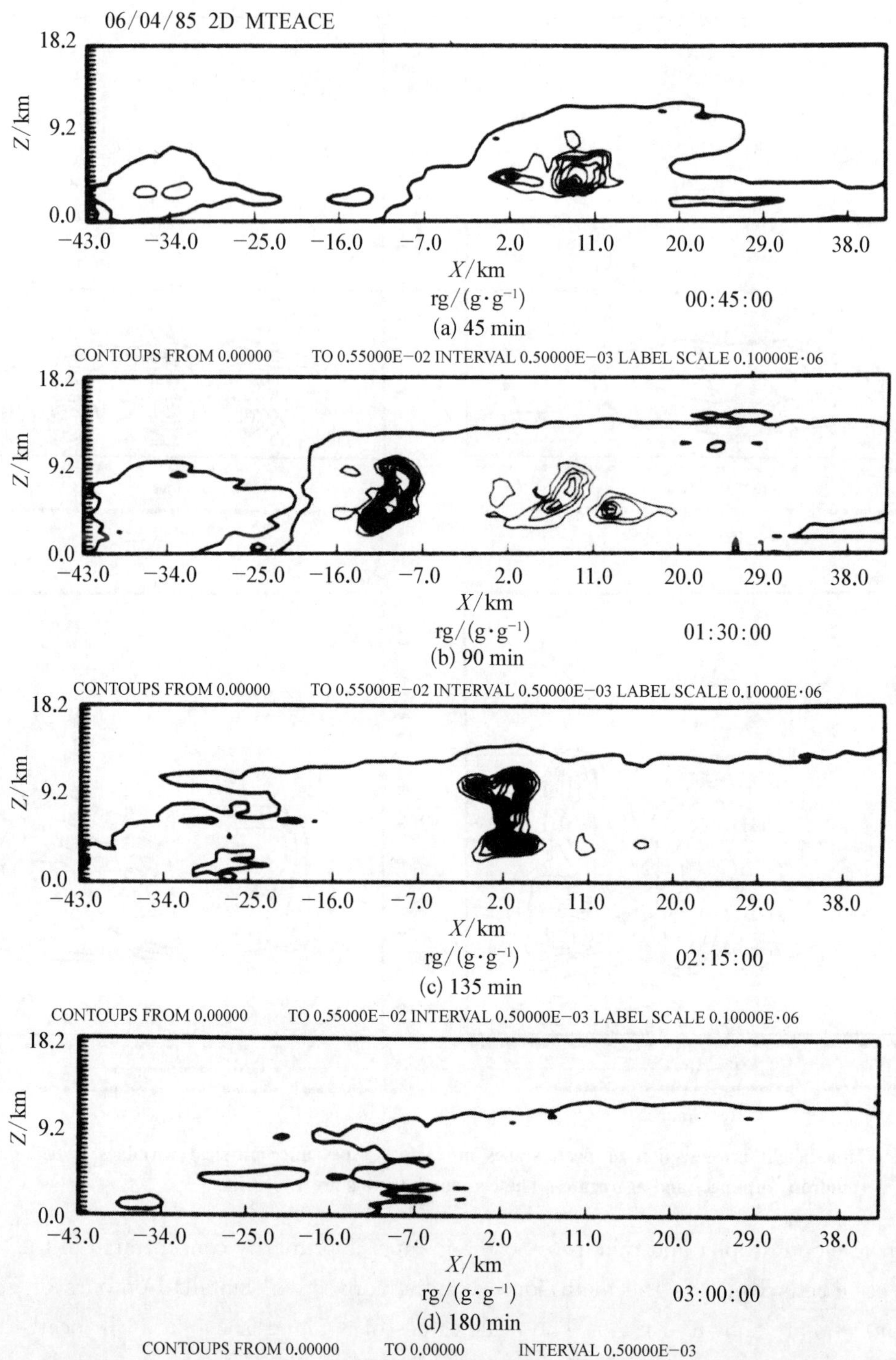

(a) 45 min

(b) 90 min

(c) 135 min

(d) 180 min

Fig.6 Contours of graupel particle mixing ratio for the control experiment at 45 min, 90 min, 135 min, and 180 min. The cloud boundary (condensate in excess of 0. 01 g/kg) is drawn by a heavy dark line. The vertical axis is height in km above mean sea level and the x axis is horizontal distance from the center of domain in km. The contour interval is 0. 5 g/kg.

(4) Aggregates field

Aggregation between ice crystals is a main formation mechanism of aggregates. Observational analysis indicated (Yeh et al., 1987) that, in the MCC, most large precipitation particles are graupel and aggregates in the transition region nearby mature convective cell, whereas aggregates predominated in the stratiform region. Aggregation process occurs over a considerable depth and range of temperature in the stratiform region which is the main source of stratiform rainfall. Fig. 7 showed the simulated contours of

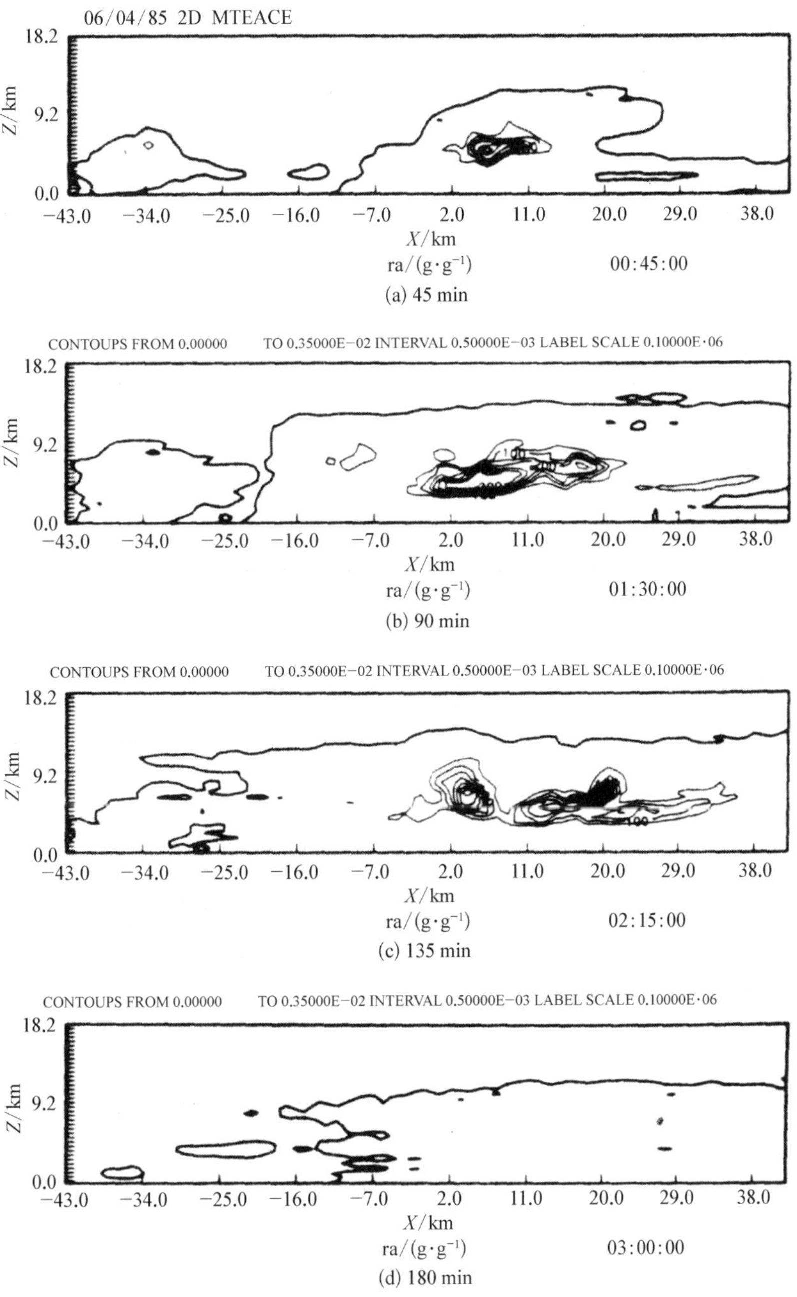

(a) 45 min

(b) 90 min

(c) 135 min

(d) 180 min

Fig.7 Same as Fig.6 except for aggregates

aggregates mixing ratio. It can be seen from Fig.7 that the aggregates spread much larger extent in space than that of graupel. Aggregates primarily located in decaying convective cells and stratiform region of the cloud system at the levels from 4 km to 8 km with peak values ranging from 2. 5 g/kg to 4. 0 g/kg in the developing and mature stage of the storm system.

2. The Sensitivity Experiments

The sensitivity experiments with various aggregation efficiencies are summarized as follows:

In the sensitivity experiment 1 (see Table 1), we assumed the aggregation collection efficiencies were given from that used as a control experiment in Cotton et al. (1986). Fig.8 illustrated the predicted contours of mixing ratio of various ice-phase particles at 90 min obtained from the sensitivity experiment 1. The ice crystals occupy the entire cold region

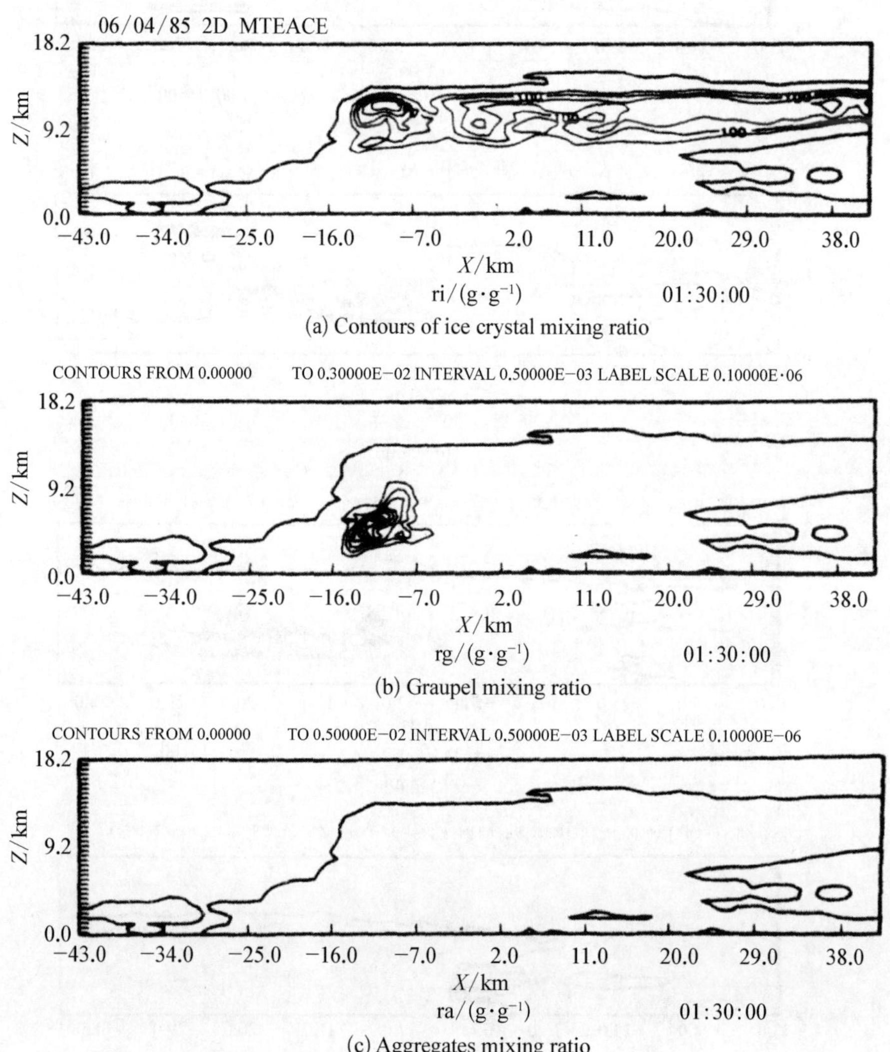

(a) Contours of ice crystal mixing ratio

(b) Graupel mixing ratio

(c) Aggregates mixing ratio

Fig.8 Ice-phase microphysical structure at 90 min simulation time for sensibility Exp. 1 as shown in Table 1. (Contours at intervals of 0. 5 g/kg)

above the 7 km level in the cloud system with a peak value of 3. 0 g/kg located in the main convective cell around 10 km level, whereas there was almost no any aggregates in the entire cloud system. As the cloud system evolution, the peak value of ice crystal water content reached 3. 5 g/kg. Obviously, the aggregation efficiency selected by version 5 of RAMS is too low.

Fig.9 is the same as Fig.8 except for sensitivity experiment 2 in Table 1, in which the

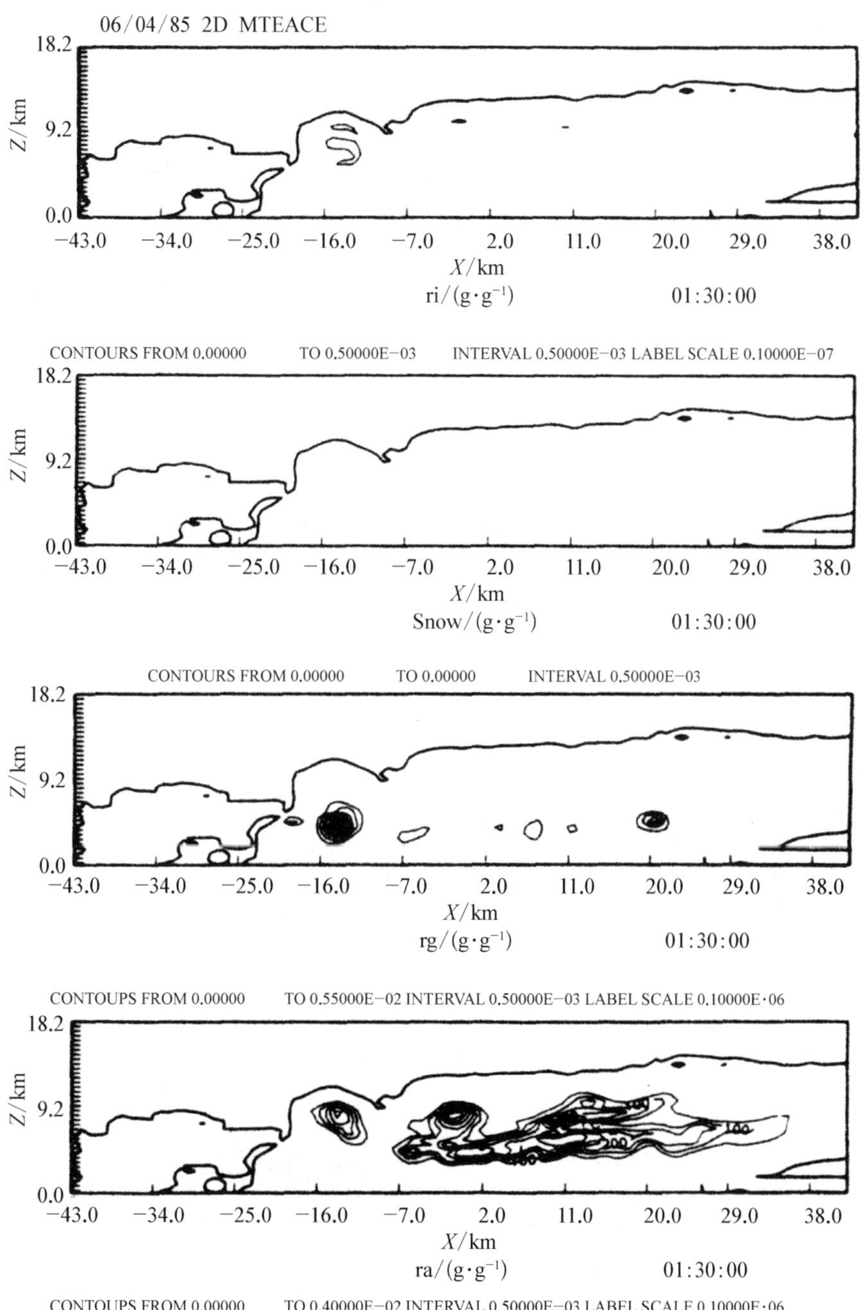

Fig.9 Same as in Fig.8 except for sensibility Exp. 2 as shown in Table 1

Passarelli's diagnosed constant value 1.4 of aggregation collection efficiency was used over the temperature ranging of −12.5—15.5 ℃, and the empirical value 0.3 of Latham and Saunders (1971) was used elsewhere. Compared with Exp.1, aggregates dominated in the whole cold region of cloud system with a peak value of 4.0 g/kg around the 8 km level, and pristine crystals and snow crystals were almost totally depleted. At 135 min, the peak value of aggregates reached 5.0 g/kg in the decaying convective cell near 8 km level. The aggregation efficiencies used in this case is too large.

The results of the control experiment is given in Fig. 10. Here the microphysical

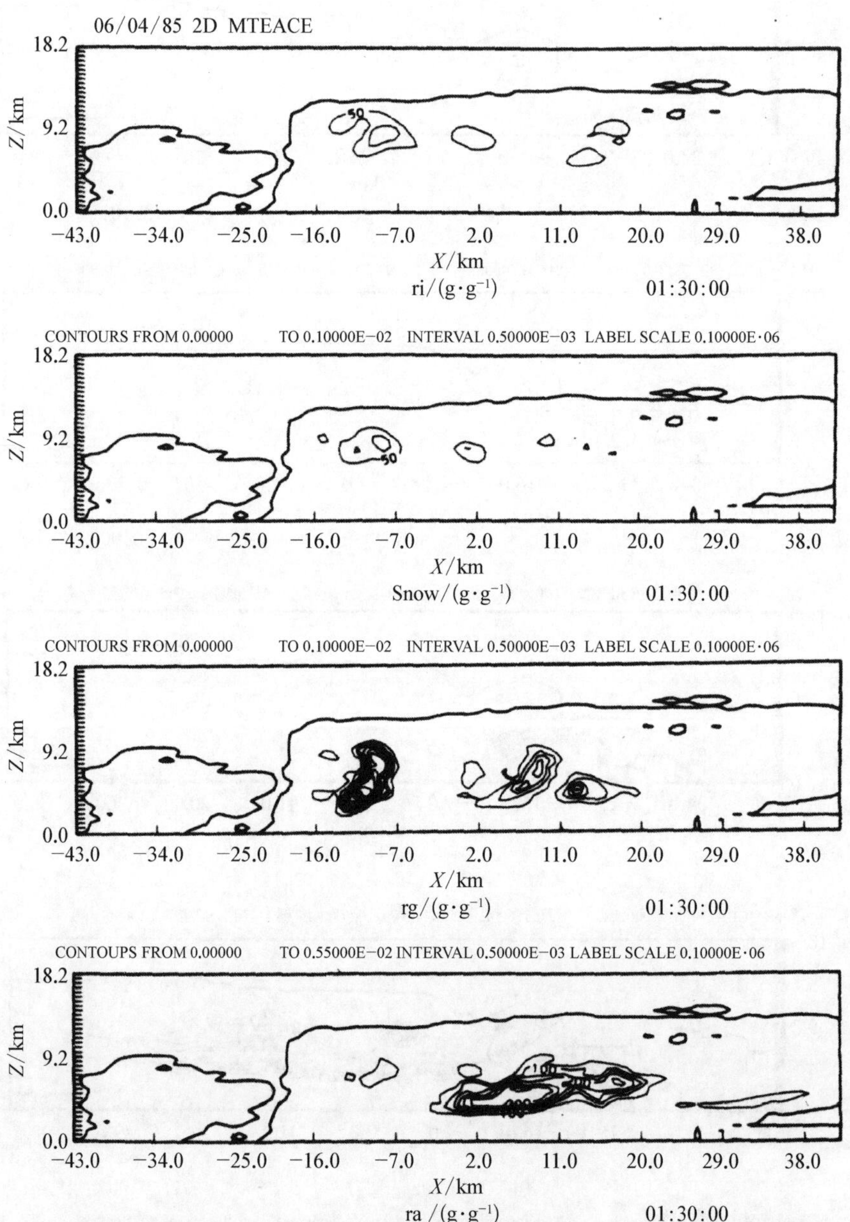

Fig.10 Same as in Fig.8 except for control experiment as shown in Table 1

structure using the diagnosed aggregation efficiency seems to be better. Pristine crystals and snow crystals were found mainly above 7 km level with peak values of 1.0 g/kg. Graupel concentrated in the strong convective cell and extend in a deep layer from 3 km to 9 km with peak value of 5.5 g/kg beneath the strongest convection core which composed the main source of convective rainfall. Some graupel is found in the decaying convective cells. Aggregates primarily extended in the stratiform region and decaying convective cells in the layer from 4 km to 8 km composing the source of stratiform rainfall. The peak aggregates mixing ratio of 3.5 g/kg is located at the lower levels near the melting layer, which is consistent qualitatively with the observations that aggregation starts at the upper and colder levels but becomes more efficient near the melting layer (Yeh, et al., 1987).

VII. CONCLUSIONS

(1) The meso-γ-scale convective phenomena are basically unsteady under the situation of strong shear of wind at low-levels, while the meso-β-scale convective system is maintained up to 3 hours or more. The meso-β-scale cloud system exhibits characteristics of a multi-celled convective storm. The main convective cells (meso-γ-scale) develop in it by a time interval of about 30 min. The fact that the strong convective activity concentrated at mid-upper troposphere suggests that the ice-phase process is important to the development and evolution of the convective system.

(2) Pressure perturbations depict a meso-low after a half hour in the low levels. As the cloud system evolution, the meso-low intensifies and extends to the upshear side and covers the entire domain beneath 7.5 km level with the peak values from 5 hPa to 8 hPa.

(3) Temperature perturbations depict a cold region in the low levels at the first hour, and a much stronger cold layer caused by cloud top evaporation at the upper levels around the cloud top level after one hour. In the middle levels, a warm region is present through the entire simulation period. The warm cores with the peak values from 4.0 ℃ to 8.0 ℃ are basically associated with strong convective cells.

(4) Simulation of microphysics exhibits that graupel is primarily concentrated in the strong convective cells forming the main source of convective rainfall after one hour, and aggregates are mainly located in the stratiform region and decaying convective cells which produce the stratiform rainfall. Riming of ice crystals is the predominant precipitation formation mechanism in the convective cells, whereas aggregation of ice crystals is the predominant one in the stratiform region, which is consistent with observations.

(5) Sensitivity experiments show that the microphysical structures of the convective cloud system can be simulated better by the new version of the parameterized microphysics with the diagnosed collection efficiencies.

This research was funded by the National Science Foundation under Grant ♯ATM-8512480 and by the Army Research Office under contract ♯DAAL03-86-K-0175. We would like to thank Brenda Thompson for her help in processing the manuscript, and Lucy

McCall for her drafting work. All computations were performed on the National Center for Atmospheric Research (NCAR) Cray X - MP48 computer. NCAR is supported by the National Science Foundation of the United States.

REFERENCES

Chen Sue (1986), Simulation of the stratiform region of a mesoscale convective system, M.S. Thesis, Colorado State University, Fort Collins, CO 80523,107pp.

Cotton, W.R., M.A. Stephens, T. Nehrkom, and G.J. Tripoli (1982), The Colorado State University cloud/mesoscale model - 1982, Part II: An ice phase parameterization, *J. Rech. Atmos.*, **16**: 295 - 320.

Cotton, W.R., G.J. Tripoli, R.M. Rauber, and E. A. Mulvihill (1986), Numerical simulation of the effects of varying ice crystal nucleation rates and aggregation processes on orographic snowfall, *J. Clim. Appl. Meteor.*, **25**: 1658 - 1680.

Fortune, M.A., and R.L. McAnnelly (1986), The evolution of two mesoscale convective complexes with different patterns of convective organization, Preprints of Joint Session, *23rd Conference on Radar Meteorology and Conference on Cloud Physics*, AMS, Snowmass, Colorado, J175 - 178.

Hallgren, R.E., and C.L. Hosler (1960), Preliminary results on the aggregation of ice crystals,, *Geophys. Monogr., Am. Geophys. Union*, **5**: 257 - 263.

Heymsfield, A.J., and D.J. Musil(1982), Case study of a hailstorm in Colorado, Part II: Particle growth processes at midlevels deduced from in-situ measurements, *J. Atmos. Sci.*, **39**: 2847 - 2866.

Hobbs, P.V., S. Chang, and J.D. Locatelli (1974), The dimensions and aggregation of ice crystals in natural cloud, *J. Geoph. Res.*, **79**: 2199 - 2206.

Hosler, C.L., D.C. Jensen, and L. Goldshlak (1957), On the aggregation of ice crystals to form snow, *J. Met.*, **14**: 415 - 420.

Houze, R.A., Jr., and D.D. Churchill (1984), Microphysical structure of winter cloud clusters, *J. Atmos. Sci.*, **41**: 3405 - 3411.

Jorgensen, D.P. (1984), Mesoscale and convective-scale characteristics of mature hurricanes, Ph. D. dissertation, Colorado State University, Fort Collins, CO 80523, 189pp.

Jiusto, J.E. (1971), Crystal development and glaciation of a supercooled cloud, *J. Rech, Atmos.*, **5**: 69 - 85.

Latham, J., and C. P. R. Saunders (1971), Experimental measurements of the collection efficiencies of ice crystals in electric fields, *Quart. J. Roy. Meteor. Soc.*, **96**: 257 - 265.

Lo, K.K., and R.E. Passarelli, Jr. (1982), The growth of snow in winter storm: An airborne observational study, *J. Atmos. Sci.*, 39: 697 - 706.

Matejka, T.J., R.A. Houze, Jr., and P. V. Hobbs (1980), Microphysics and dynamics of clouds associated with mesoscale rainbands in extratropical cyclones, *Quart. J. Roy. Met. Soc.*, 106: 29 - 56.

Passarelli, R. E., Jr. (1978), Theoretical and observational study of snow-size spectra and snowflake aggregation efficiencies, *J. Atmos. Sci.*, **35**: 882 - 889.

Rauber, R.M. (1987), Characteristics of cloud ice and precipitation during wintertime storms over the mountains of Northern Colorado, *J. Climate and Appl. Meteor.*, **26**: 488 - 424.

Steward, R.E.. J. D. Marwitz. J.C. Pace, and R.E. Carbone (1984), Characteristics through the melting layer of stratiform clouds, *J. Atmos. Sci.*, **41**: 3227 - 3237.

Tripoli, G.J.. and W.R. Cotton (1982), The Colorado State University three-dimensional cloud/mesoscale model-1982, Part I: General theoretical framework and sensitivity experiments, *J. Reck. Atmos.*, **16**: 185 - 220.

Willoughby. H. E., Han-Liang Jin, S. J. Lord, and J. M. Piotrowicz (1984), Hurricane structure and evolution as simulated by an axisymmetric nonhydrostatic numerical model, *J. Atmos. Sci.* **41**: 1169 - 1186.

Yeh Jiadong. Fan Beifen, M.A. Fortune, and W.R. Cotton (1987), Comparison of the microphysics between the transition region in a mesoscale convective complex, Third Conference on Meso-scale Processes, 21 - 26 August, 1987, Vancouver, D.C., Canada, 188 - 189.

5.13 SIZE DISTRIBUTION OF PRECIPITATION PARTICLES IN MIDLATITUDE MESOSCALE CONVECTIVE COMPLEXES

Jia-Dong Yeh, Bei-Fen Fan, William R. Cotton

Department of Atmospheric Science, Colorado State University, Fort Collins, CO 80523

1 Introduction

Measurements of precipitation particle size distributions above and beneath the melting layer in the transition region and stratiform region of two mid-latitude mesoscale convective complexes during the 1984 airborne investigations of mesoscale convective systems (AIMCS) and the 1985 PRE-STORM experiments were analyzed to understand how well the observed particle size distributions can be approximated by various types of distribution functions, such as exponential and log-normal distributions. The parameterization problem of microphysics in cloud modeling is also discussed.

2 Data

There are 78 samples of the hydrometeor size spectra obtained from the stratiform region of an AIMCS case (15 July 1984) and 149 samples taken from the PRE-STORM case of 4 June 1985. Among them 62 samples were taken from the transition region during the horizontal flight at the 5 150 m level (−6.7—−8.5 ℃), 41 samples were taken from the stratiform region at the same level, and 46 samples were taken during the vertical sounding flight from the 6 100 m (−13.0 ℃) to 3 000 m (+1.2 ℃) levels in the stratiform region.

3 The Exponential Fit to Hydrometeor Size Spectra

The Marshall-Palmer distribution

$$N(D) = N_0 e^{-\lambda D} \tag{1}$$

is applied extensively in the parameterization of microphysics in numerical cloud models. It is determined completely by the slope parameter λ and intercept parameter N_0. Generally, the intercept parameter N_0 is assumed constant, while in some cloud models the slope parameter λ is assumed constant. Observational evidence, however, suggest that both the slope λ and intercept N_0 depend on meteorological conditions, cloud conditions, and time and location of the sample taken in the cloud system (Waldvogel, 1974). The measurements in the mid-latitude mesoscale convective complexes in this study show that the exponential distribution fits the observational precipitation particle size spectra well in the entire measured region. The average linear crrelation coefficients for $\ln N(D)$ and D is -0.96 for water drops and -0.92 for ice particles in the stratiform region of the AIMCS case, and -0.96 and -0.98 for the ice particles in the transition region and stratiform region of the PRE-STORM case, respectively. Fig.1 shows the scatter plots of intercept N_0 as a function of slope λ. It can be seen from Fig.1 that the slope λ does not vary much for water drops in the AIMCS case, its mean value is about 17($\pm$3.6) cm. So λ may be considered approximately a constant as suggested by Cotton et al (1986). The sample size of water drop size spectra from the PRE-STORM case is too small to be meaningful. For ice particles, however, both N_0 and λ are variable. The values of intercept parameter N_0 vary up to 3 orders of magnitude, from 4×10^{-3} cm^{-4} to 3×10^{0} cm^{-4}, and slope parameter λ vary about 7 times on magnitude, from 6 cm^{-1} to 45 cm^{-1}. Therefore, assuming the slope parameter or the intercept parameter as constant in modeling the stratiform region of MCSs is not appropriate. As a consequense, one cannot diagnose the particle concentrations from the prognostic values of mixing ratio. Therefore it is necessary to develop a set of prognostic equations of particle number concentration in a parameterized cloud microphysics model. Here we propose an approximate empirical approach to diagnose the particle concentration from the prognostic value of mixing ratio.

There is some correlation between the intercept N_0 and slope λ as shown in Fig.1, which is reflected by the fact that decreasing values of λ usually correspond to decreasing values of N_0. That is particularly reasonable for the mature cloud system in which the main processes affecting the size spectrum are self collection, aggregation and collisional breakup. The best fit line has a form

$$N_0 = a\lambda^b \tag{2}$$

with a correlation coefficient of 0.94 between $\ln N_0$ and $\ln\lambda$ (AIMCS case when $a = 1.1\times10^{-4}$ and $b = 2.8$), and 0.89 (PRE-STORM case when $a=3.4\times10^{-6}$ and $b = 3.8$).

There does not seems to be an obvious relation between the distribution parameters (N_0 and λ) and the water content (WC). However, assuming an exponential distribution and the definition of water content, we fit an empirical relation of the form

$$\lambda = c + d\left(\frac{N_0}{\mathrm{WC}}\right)^{\frac{1}{4}}, \tag{3}$$

where c, d are different for water and ice particles and depending on the cloud system condition. For AIMCS case, the correlation coefficient is 0.73 for water drops when $c = 0.96$ and $d = 1.28$, and 0.98 for ice particles when $c = -7.1$ and $d = -1.33$. For the PRE-STORM case, the correlation is also pretty good for ice particles, being 0.98 when $c = -2.0$ and $d = -0.86$. Fig.2 shows the degree of linear correlation. Based on Eq.(2) and Eq.(3), the size spectrum parameters can be inferred from the prognostic value of water content WC and, thus, the size distribution and number concentration from the integration of Eq.(1).

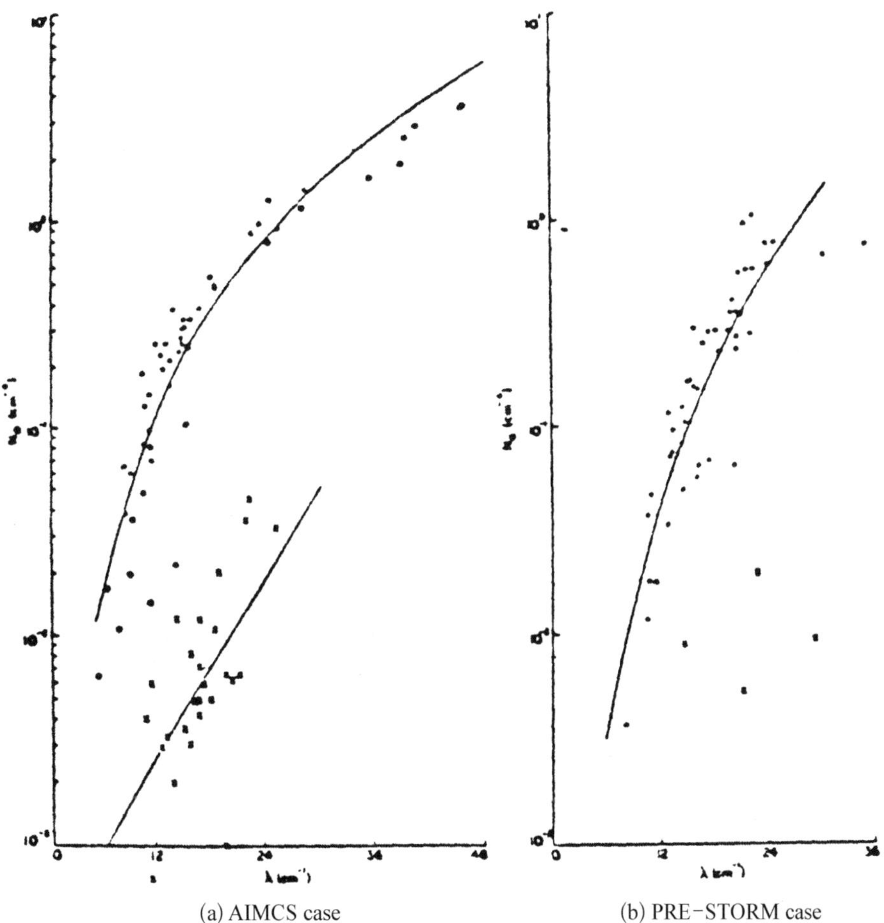

(a) AIMCS case　　(b) PRE-STORM case

Fig.1　Scatter diagrams of intercept parameter N_0 plotted as a function of alope parameter λ of exponential distribution fitting the observed particle size spectra is for AIMCS case and for PRE-STORM case.

4　The Lognormal Fit to Hydrometeor Size Spectra

Although the results of observational and theoretical studies of the evolution of hydrometeor size distributions generally support that the exponential distribution is a good approximation to the size distribution of precipitation particles, important deviation from it have been noted. Feingold and Levin (1986) argued that an exponential distribution cannot

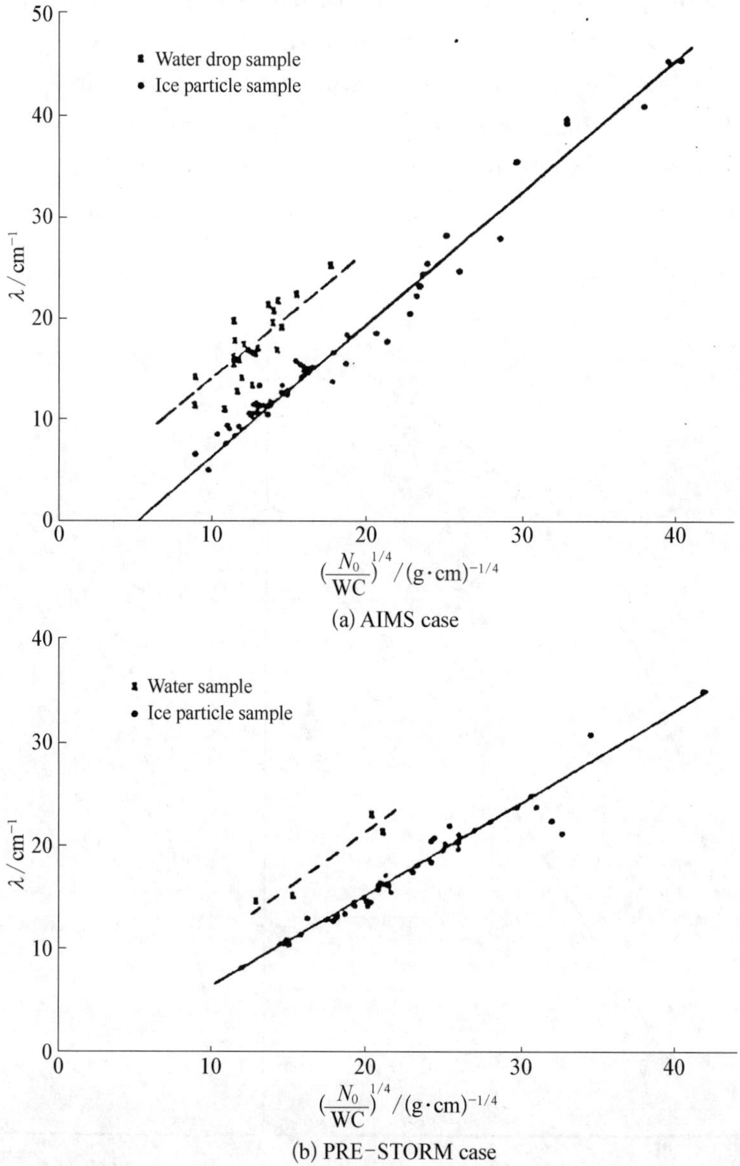

Fig.2 Scatter diagrams of slope parameter λ plotted as a function of combined variable $(N_0/WC)^{1/4}$ for AIMCS case and PRE-STORM case. (Solid line is the best fit for ice particle samples; dashed line is for water drops)

adequately describe the observed raindrop size distributions. They, as well as others in the past, suggested that a lognormal distribution

$$N(D)=\frac{N_T}{\sqrt{2\pi}D\ln\sigma_g}\exp\left[-\frac{\ln^2\left(\frac{D}{D_s}\right)}{2\ln^2\sigma_g}\right] \tag{4}$$

better fits the observational raindrop size distributions. The parameter of the lognormal distribution N_T is total number concentration of particle, D_g is the geometric mean of the particle diameter, and σ_g represents the standard deviation. We are interested in whether the lognormal distribution better fits observed particle spectra than the exponential distribution aloft for the ice particles in the mid-latitude mesoscale convective complexes.

For measuring the goodness of fit of a distribution function to an observational particle-size distribution, we use a relative standard error (coefficient of variability) defined as

$$CV = \frac{1}{N_T}\left\{\frac{1}{n}\sum_{i=1}^{n}[N_i(\text{observe}) - N_i(\text{fit})]^2\right\}^{\frac{1}{2}}, \tag{5}$$

where i represents the ith size category of particles. We use CV to compare the goodness of fit of the lognormal and exponential distribution for the ice particle data taken from horizontal flight in the transition region and stratiform region of the PRE-STORM case. Table 1 presents a comparison of the average relative standard error (CV) for these two distribution functions, the mean values of distribution parameters are also presented in Table 1. The results of the tests indicate that the lognormal distribution is a better fit for ice particles in both the transition region and stratiform region. Fig.3 illustrates the result of applying these two formula to the mean size distributions in the transition and stratiform regions. It can be seen from Fig.3 that the lognormal distribution better approximates the small particle size range, while the exponential model is better in the larger size range.

Table 1 Average values of CV and mean distribution parameters for lognormal and exponential distribution fit in the transition and stratiform regions of PRE-STORM MCC case

	Sample number	Lognormal distribution			Exponential distribution			
		$D_g/(10^{-2}$ cm)	σ_g	$CV_1/10^{-2}$	N_0/cm^{-4}	λ/cm^{-1}	R	$CV_1/10^{-2}$
Transition region	62	3.28	1.90	4.09	0.30	22.5	−0.96	8.63
Stratiform region	41	5.15	2.33	2.56	0.06	13.0	−0.98	3.50

Under the assumption of lognormal distribution, the WC(g/m^3) and radar reflectivity $Z(mm^6/m^3)$ can be represented (Feingold and Levin, 1986) as

$$WC = \frac{\pi}{6}\times 10^{-3} N_T D_g \exp\left(\frac{9}{2}\ln^2\sigma_g\right), \tag{6}$$

$$Z = N_T D_g^6 \exp(18\ln^2\sigma_g). \tag{7}$$

If we can get a certain relation between D_g and σ_g, the number concentration N_T and, thus, the size spectrum of precipitation particle at any time can be inferred from the WC and radar reflectivity Z. Fig.4 exhibit a good correlation between the $\ln\sigma_g$ and $\ln D_g$ (with linear correlation coefficient of 0.91). The best fit line has a form as

$$\sigma_g = 12.65 D_g^{0.56}. \tag{8}$$

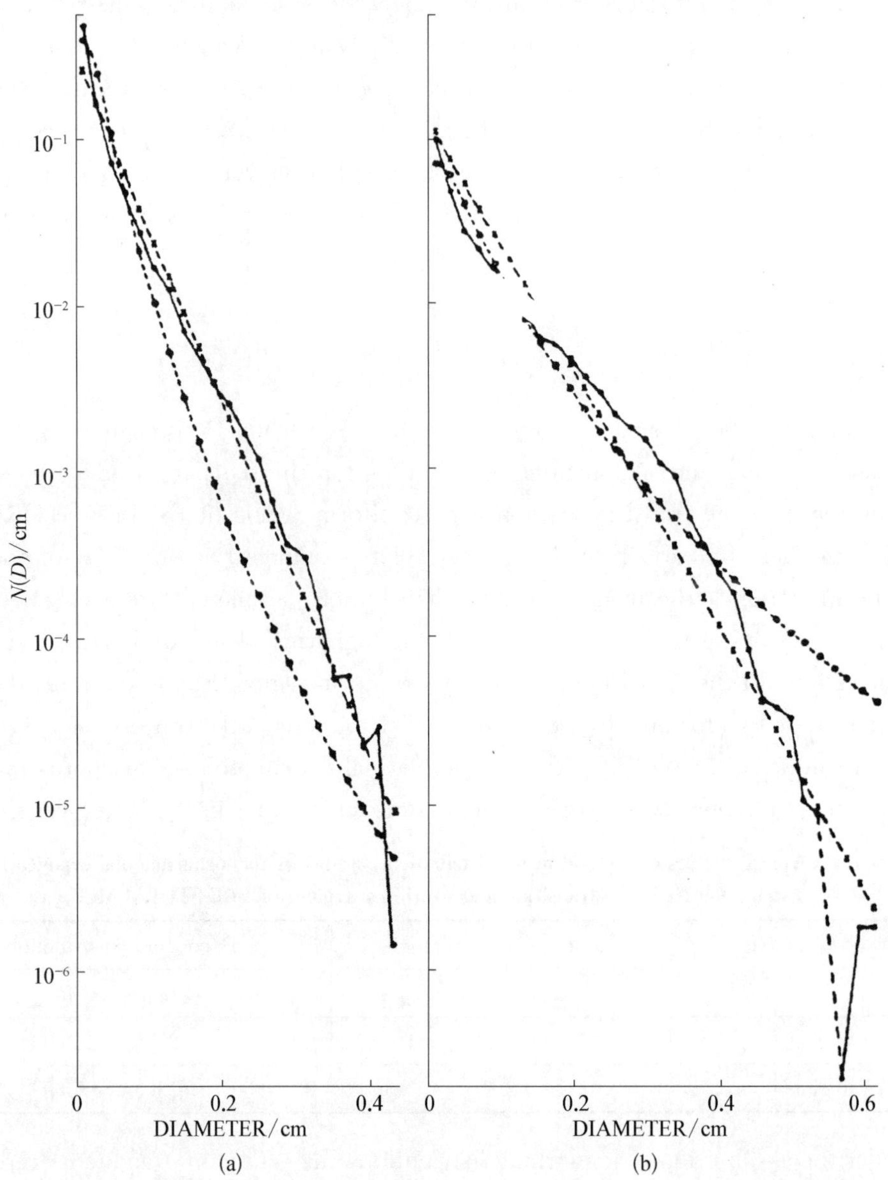

Fig.3 Mean sise distribution in the transition region and stratiform region of PRE-STORM MCC case (solid line), and the corresponding fitting lines of exponential distribution (dashed lines) and lognormal distribution (dotted lines)

Then the particle number concentration can be diagnosed from the WC and radar reflectivity Z *using the equations mentioned above, which should be useful in parameterized cloud microphysics models.*

5 Empirical Aggregation Efficiency

One of the main problems in modeling snowflake aggregation is that of determining the collection efficiency among crystals, which is complicated by the nonuniform vertical

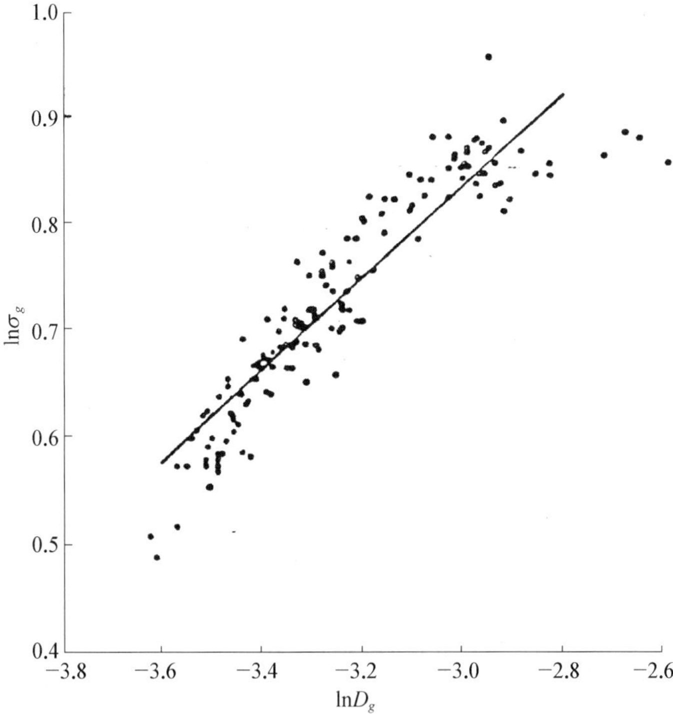

Fig.4 Scatter diagram of lognormal distribution parameters $\ln\sigma_g$ and $\ln D_g$ for PRE-STORM case

and horizontal motions of ice particles, and also affected by a variety of factors, such as the crystal habits, air humidity and temperature, and the presense of supercooled cloud droplets. A few laboratory studies made in restrictive conditions obtained conflicting results. We therefore apply Passarelli's (1978) scheme to estimate the collection efficiency in the stratiform region of the PRE-STORM MCC case during its mature stage. Passarelli used the moment conservation equations for total mass and reflectivity factor fluxes for aggregating snowflakes to diagnose the mean collectıon efficiency. For the case of steady, constant mass flux of snow, the resulting explicit expression of the mean collection efficiency $\bar{E}$ in a layer is

$$\bar{E} = (\lambda^{-1-b_1} - \lambda_0^{-1-b_1}) \frac{12\Gamma(7+b_1)\lambda^{4+b_1}}{(1+b_1)\pi N_0 h I(b_1)}, \tag{9}$$

where h is the height below a reference level ($z = 0$), N_0 and λ the exponential distribution parameters at $z = h$, λ_0 is the value of λ at $z = 0$, b_1 is a parameter in the particle fall speed equation ($v(D)=a_1 D^{b_1}$), and $I(b_1)$ is a function of b_1. In the case of $b_1=0.25$, we obtained $I(b_1)=566$. We calculated the exponential distribution parameters N_0 and λ at various levels during the vertical sounding flight. Using the best fit lines of $N_0(z)$ and $\lambda(z)$ as shown in Fig.5, the mean collection efficiencies at various layers can be calculated. We obtained an empirical formula as

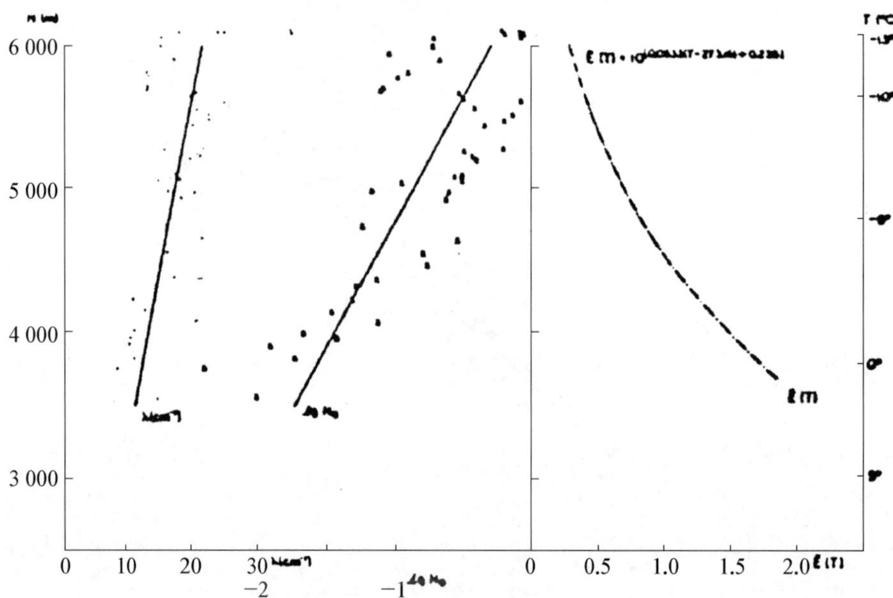

Fig.5 Scatter diagrams of exponential distribution parameters λ and N_0 at various levels in the stratiform region and corresponding regression lines (Right side is the empirical aggregation efficiency calculated from $\lambda(z)$ and $N_0(z)$ as a function of temperature)

$$\overline{E(T_s)} = 10^{0.235+0.0633T_s}. \tag{10}$$

This equation has been used in the simulation of ice-phase microphysics in another paper of this conference (Fan et al., 1988).

Acknowledgements

This research was supported by the National Science Foundation under Grant ＃ATM－8512480 and by the Army Research Office under contract ＃DAAL03－86－K－0175.

REFERENCES

Cotton. W. R., G. J. Tripoli, R.M. Rauber, and E. A. Mulvihill, 1986: Numerical simulation of the effects of varying ice crystal nucleatioin rates and aggregation processes on orographic snowfall. J. Clim. Appl. Meteor., 25, 1658－1680.

Fan, Bei-Fen, Jia-Dong Yeh, W. R. Cotton and G. J. Tripoli, 1988: Micropgysics in deep convective cloud system associated with a mesoscale convective complex-numerical simulation. 10th Internatioal Cloud Physics Conference, 15－20 August 1988, Bad Homburg, FRG.

Feingold, G., and Z. Levin, 1986: The lognormal fit to raindrop spectra from frontal convective clouds in Israel. J. Clim. Appl. Meteor., 25, 1346－1363.

Passarelli, R.E., 1978: Theoretical and observational study of snow-size spectra and snowflake aggregation efficiencies. J. Atmos. Sci., 35, 882－889.

Waldvogel, A., 1974: The N_0 jump of raindrop spectra, J. Atmos. Sci, 31, 1067－1078.

5.14 OBSERVATIONAL STUDY OF MICROPHYSICS IN THE STRATIFORM REGION AND TRANSITION REGION OF A MIDLATITUDE MESOSCALE CONVECTIVE COMPLEX

Ye Jiadong (叶家东), Fan Beifen (范蓓芬)
Department of Atmospheric Science, Nanjing University, Nanjing
William R. Cotton, Michael A. Fortune
Department of Atmospheric Science, Colorado State University, U. S. A.
Received 14 December, 1990

Abstract Cloud microphysical data observed with PMS probes have been combined with radar and other in-situ data collected by a NOAA P-3 aircraft that flew through the stratiform and transition regions of a mesoscale convective complex (MCC). The combined data have been analyzed with respect to the mesoscale structure of the storm systems. The characteristics of ice particles in the transition and stratiform regions were quite different. The ice particle concentrations in the transition region were 4—6 times that found in the stratiform region, and the size of ice particles in the stratiform region was about twice that in the transition region. The relatively lower radar reflectivity in the transition region is a result of smaller particle sizes: The main precipitation particle growth mechanisms are riming and aggregation in the transition region and the aggregation process predominates in the stratiform region referred from the microphysical structures. The aggregation starts in the upper, colder levals but becomes more efficient as the particles approach the melting layer.

Key words: mesoscale convective complex, transition region, stratiform region, microphysical structure, aggregation process

I. INTRODUCTION

Much work has been done in the past decade to investigate mesoscale convective systems (MCSs) in various regions, seasons and weather conditions. Althuogh the detailed structures are different from one system to another, some common characteristics have emerged from these studies. The mature MCS generally consists of (1) a leading line or cluster of deep convective cells and (2) a mesoscale precipitating stratiform-anvil cloud region, which can extend from the 700 hPa level to about 200 hPa or above (Zipser, 1969; 1977; Houze, 1977). The air motion in these two regions is distinctly different. Convective scale updrafts and saturated convective-scale downdrafts generally greater than 5 m/s are present in the convective region associated with low-level convergence. Upper level divergence is also associated with outflow from the tops of the convective cells. In the stratiform region, a mesoscale updraft above the stratiform cloud base and a mesoscale downdraft below the melting layer have been observed and modelled (Leary and Houze, 1979; Brown, 1979; Willoughby et al., 1984; Gamache and Houze, 1982, 1983, 1985;

Rutledge, 1986; Smull and Houze, 1985, 1987; Chen and Cotton, 1988). Mesoscale circulation features develop in association with the stratiform region including front-to-rear and rear-to-front flows, a mid-lever vortex, mid-level convergence, upper level divergence and low-level divergence (Smull and Houze, 1985, 1987; Srivastava et al., 1986; Rutledge et al., 1988).

In tropical monsoon cloud clusters, microphysical observations of ice particles obtained near the − 17 ℃ level in the convective regions showed that ice particle concentrations were of the order of hundreds per liter, and the dominant ice particle growth mechanism appeared to be riming (Houze and Churchill, 1984; Churchill and Houze, 1984). In the stratiform regions of clusters the ice particle concentrations were of the order of 10^0—10^1 per liter and vapor deposition and aggregation growth of ice particles were the dominant particle growth mechanisms, which was also documented in a monsoon depression over the Bay of Bengal in a greater temperature ranging of +5—−25 ℃ (Houze and Churchill, 1984).

Up to now, however, little is known about the cloud microphysical structure of midlatitude mesoscale convective complexes (MCCs) (Maddox, 1980). The 1985 Oklahoma-Kansas PRE-STORM(Cunning et al., 1986) experiment was a program which was designed to study midlatitude mesoscale convective systems. The purpose of the cooperative program was to investigate the development, evolution, and structures of the system through substantially better observations than had previously been available. One of the purposes of the cloud microphysical studies in the program was to discern the interaction of microphysical processes with the dynamical processes within an MCS. The hydrometeor measurements in the stratiform and transition regions of MCSs were made, together with radar and other data collected by the NOAA P - 3 research aircraft, to determine the thermodynamic and microphysical structures in these regions above and beneath the melting level. Direct aircraft penetrations, however, were not made in the convective cores. The flow fields and radar reflectivity fields of several storm systems have been analyzed by a few investigators (Houze and Rutledge, 1986; Cunning et al., 1986; Fortune and McAnelly, 1986), but relatively little is known about their microphysical structures. In this paper we analyze the microphysical data of a system and interpret it with respect to the mesoscale structure, the case was a rapidly evolving MCC composed of discrete cumulonimbus clusters observed on 4 June 1985 in the PRE-STORM network.

II. SYNOPTIC SITUATIONS AND OVERVIEW OF THE MCC

At 1200 UTC 3 June 1985, a stationary front extended from New England southwest through the central Mississippi Valley, acorss the Oklahoma-Kansas border and continuing into east-central New Mexico. A cut-off low pressure center was located over southern California, also evident at both the 500 hPa and 300 hPa levels. A southwest air current was prevalent to the east of the trough and low center. At 850 hPa, a 14 m/s southerly jet flow pumped warm moist air from the Gulf of Mexico northward through central Texas, across the front, toward the Pahandle of Texas, northwest Oklahoma, and the western sections of Kansas and Nebraska. The frontal system moved very little until 1200 UTC 4 June. During the night, a series of MCSs developed and moved across Kansas and Missouri. These large MCSs formed in a region of low-level warm advection over the Texas-Oklahoma Panhandle and western Kansas, east of the 500 hPa low center over the southwestern U. S. After the episode of MCSs, the 850 hPa warm advection appeared to be weakening.

At 1700 UTC 3 June, an MCS developed in south-central Kansas, marking the beginning of an active period. In the early evening another MCS moved into central Kansas from the Texas-Oklahoma Panhandle. By 0630 UTC 4 June, a third MCS formed in the same region and moved into the PRE-STORM network. The MCC was a rapidly evolving and fast moving system without any apparent linear organization. It achieved its maximum areal extent of cloud shield at 1000 UTC and decayed at about 1400 UTC (Figure omitted). The MCC lasted about 7 hours. Early in its lifecycle, numerous convective echoes developed in a random manner. Blanchard and Watson(1986) classified the convective mode as a "randomly organized" convective system. At 0800 UTC some mature radar echoes merged and started to form a trailing stratiform region in the western portion of the MCC. The base of the cloud is at about 3 000 m level. During the aircraft flew through the stratiform region the shield of the MCC has been in the region of −52 ℃ contour referred from the satellite infrared image (Figure omitted), corresponding cloud top is at about 220 hPa level or higher. The convective cells organized into several meso-β-scale clusters, which moved with various speeds and directions, while the meso-α-scale system propagated much faster than any of its components (Fortune and McAnelly, 1986). Between the mature convective clusters there were regions in which the radar reflectivity at low levels was a minimum, less than 18 dBZ (Fig. 1(a)). Minimum reflectivity values less than 25 dBZ were also observed by the aircraft at its flight level of 530 hPa. We refer to it as a "transition region", as it will later transform to a rather uniform stratiform region. This region of weak echo was in the middle of the rapidly expanding, deep middle- and upper-tropospheric cloud, a pattern which may be similar in some respects to the transition zone defined by Smull and Houze (1985, 1987) for a squall line system.

The NOAA P-3 aircraft entered the eastern portion of the MCC at about 0820 UTC at the 530 hPa level. At 0840 UTC the aircraft threaded its way through two evolving meso-β-scale convective clusters, and entered the transition region described above, then flew in the stratiform region on the west side of the MCC. From 0855 to 1025 UTC, horizontal flights were made at the 530 hPa and 460 hPa levels for measuring cloud microphysics in the trailing stratiform region (Fig. 1(a)). The aircraft then tracked the MCC to the northeast and by 1035 UTC started to descend slowly in a spiral downward through the stratiform region from 6 100 m to 3 000 m MSL at a descent rate of about 1 m/s. A second cloud microphysical sounding from 3 000 m to 6 100 m MSL was made from 1123 to

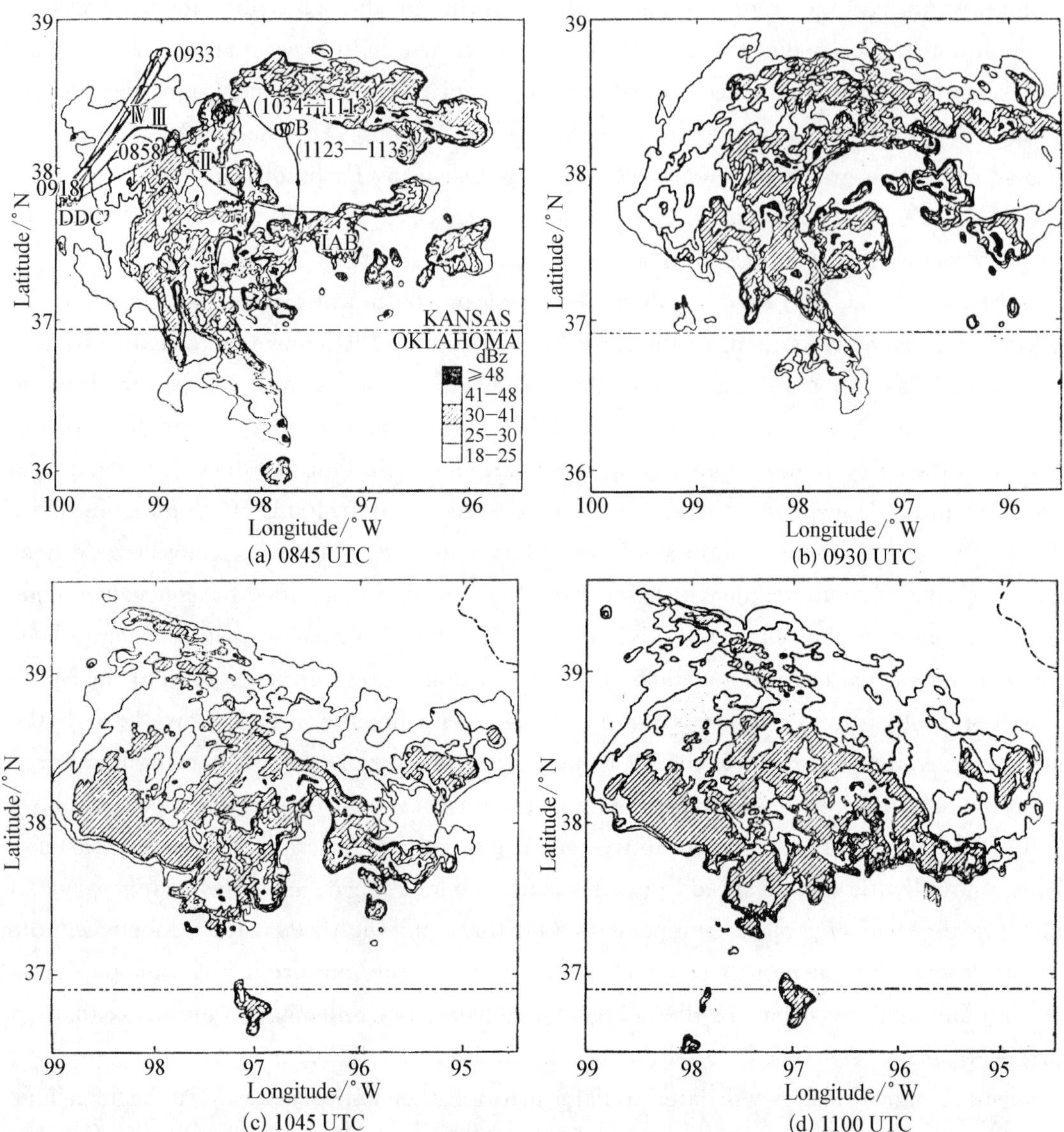

Fig.1 Digitized NWS radar echo pattern of the MCC on 4 June and the flight track of the NOAA P-3 aircraft

1135 UTC in the stratiform region of the dissipating MCC (with much faster ascent speed than the descent speed of the first sounding). The aircraft exited the cloud at about 1139 UTC. A short time later the aircraft encountered a very thin cloud and sampled an abundance of stellar crystals which were significantly different in shape from particles sampled in the stratiform region of the MCC.

III. MICROPHYSICAL CHARACTERISTICS IN THE TRANSITION REGION AND STRATIFORM REGION

The microphysical data observed in four segments of horizotal flight at the 530 hPa level were analysed to illustrate the differences in precipitation particle characteristics between the transition region and stratiform region. The first two segments, Ⅰ and Ⅱ in Fig.1(a), were in the transition region from 0844 to 0852 UTC. The other two segments, Ⅲ and Ⅳ in Fig.1(a) were in the stratiform region from 0854 to 0858 UTC and 0918 to 0933 UTC. The characteristics of precipitation particles observed in these two regions will be examined in this section.

1. Shapes of Precipitation Particle

Some samples of the 2D images of hydrometeors in the transition region and the stratiform region are presented in Fig.2. There was an obvious difference in particle shapes between the two regions. The many small particles in the transition region had irregular, sleet-like, column or needle shapes. Most of the particles larger than 1 mm had the shape of roughly circular graupel particles or of aggregates composed of irregular crystals and dendrites. In the stratiform region, on the other hand, most of the larger particles had the shape of aggregates. Single dendrites were not found in the samples obtained from the stratiform region at the 530 hPa level (from −8 ℃ to −10 ℃), implying that the aggregates contained dendrites which probably fell from higher and colder levels. Rutledge (personal communication) inferred from a vertically-pointing single Doppler radar that a mesoscale updraft with velocities less than 1 m/s was present at the upper levels in the stratiform region. The measured Doppler velocities in the whole layer were negative, which implies that the precipitation particles (whose fall velocities are typically 1.5 m/s to 3.0 m/s) were falling relative to the ground in the stratiform region. Graupel particles were rarely found in the stratiform region, whereas column and needle crystals were present. Irregular particles, however, seemed to be the predominant component in both the transition and stratiform regions of the MCC.

2. Particle Number Concentration, Size and Water Content

The time series of microphysical variables observed during the horizontal flight in these two regions are given in Fig. 3. The microphysical characteristics in the transition region and stratiform region were quite different. Much larger number concentrations and smaller particle sizes were found in the transition region, and lower concentrations and much larger particle sizes were in the stratiform region. The maximum particle concentration was 311 L^{-1} (2D-C probe) and 30 L^{-1} (2D-P probe) in the transition region, and 70 L^{-1} (2D-C probe) and 10 L^{-1} (2D-P probe) in the stratiform region. Whereas the maximum median volume diameter of total particles was 0.93 mm (2D-P probe) and 0.49 mm(2D-C probe) in the transition region, both of them are located near the edge of stratiform region. In the stratiform region, the maximum median volume diameters were 1.71 mm (2D-P probe) and 0.81 mm (2D-C probe), which were almost twice that found in transition region. The average values of particle characteristics—are given in Table 1. It is shown in Table 1 that the average concentration of total precipitation particles was about 18.5 L^{-1} (2D-P probe) and 121.8 L^{-1} (2D-C probe) in the transition region, and only

No.	Time	T/℃	2D-P Images	N_t/L^{-1}	D_0/mm	AN_t	GN_t
1	0844^{00}	−7.9		4.56	0.88	0.20	0.06
2	0845^{00}	−8.0		9.37	0.73	0.33	0.13
3	0846^{00}	−7.3		4.15	0.75	0.10	0.05
4	0846^{20}	−7.9		12.21	0.76	0.56	0.64
5	0846^{40}	−8.0		14.89	0.78	0.55	0.72
6	0847^{00}	−7.0		19.80	0.74	0.96	0.69
7	0847^{35}	−7.7		18.66	0.59	0.52	0.76
8	0847^{50}	−6.7		12.04	0.63	0.13	0.53
9	0848^{10}	−7.4		20.53	0.61	0.40	0.66
10	0848^{30}	−7.6		11.85	0.80	0.45	0.45
11	0848^{50}	−7.5		11.65	0.81	0.62	0.34
12	0849^{10}	−7.5		27.05	0.72	0.95	0.38
13	0849^{30}	−7.5		23.76	0.77	1.14	0.55
14	0849^{50}	−7.5		26.51	0.61	0.73	0.52
15	0850^{10}	−8.2		23.77	0.67	0.99	0.61
16	0850^{30}	−8.5		20.08	0.79	1.25	0.39
17	0850^{50}	−8.5		17.89	0.93	1.09	0.34

(a) 2D−P images of hydrometeors in the transition region

No.	Time	T/℃	2D-P Images	N_t/L^{-1}	D_0/mm	AN_t	GN_t
1	0918^{00}	−9.2		1.51	1.21	0.35	0.02
2	0919^{00}	−9.5		5.77	1.30	0.82	0.02
3	0920^{00}	−8.7		9.34	1.15	1.19	0.08
4	0921^{00}	−7.8		8.96	1.10	1.15	0.04
5	0922^{00}	−7.6		9.66	1.17	1.38	0.10
6	0923^{00}	−7.6		5.88	1.31	1.05	0.09
7	0924^{00}	−8.5		6.53	1.31	0.96	0.06
8	0925^{00}	−9.0		8.01	1.14	1.06	0.14
9	0925^{30}	−9.2		9.28	1.07	1.10	0.19
10	0926^{30}	−9.6		3.74	1.27	0.79	0.08
11	0927^{00}	−9.2		4.01	1.42	0.78	0.10
12	0928^{00}	−9.5		4.19	1.39	0.62	0.08
13	0929^{00}	−9.5		3.31	1.38	0.71	0.06
14	0930^{00}	−9.8		3.44	1.44	0.54	0.06
15	0931^{00}	−10.0		3.14	1.41	0.62	0.09
16	0932^{00}	−10.1		1.87	1.71	0.58	0.05
17	0933^{00}	−10.0		1.09	1.64	0.31	0.05
18							

(b) 2D−P images of hydrometeors in the stratiform region

Fig.2 2D-P images of hydrometeors in the transition region and stratiform region at the 530 hPa level

5.1 L^{-1} (2D-P probe) and 21.8 L^{-1}(2D-C probe) in the stratiform region. On the other hand, the particle sizes were greater in tratiform region; the average median volume diameter was 0.73 mm (2D-P probe) and 0.37 mm (2D-C probe) in the transition region,

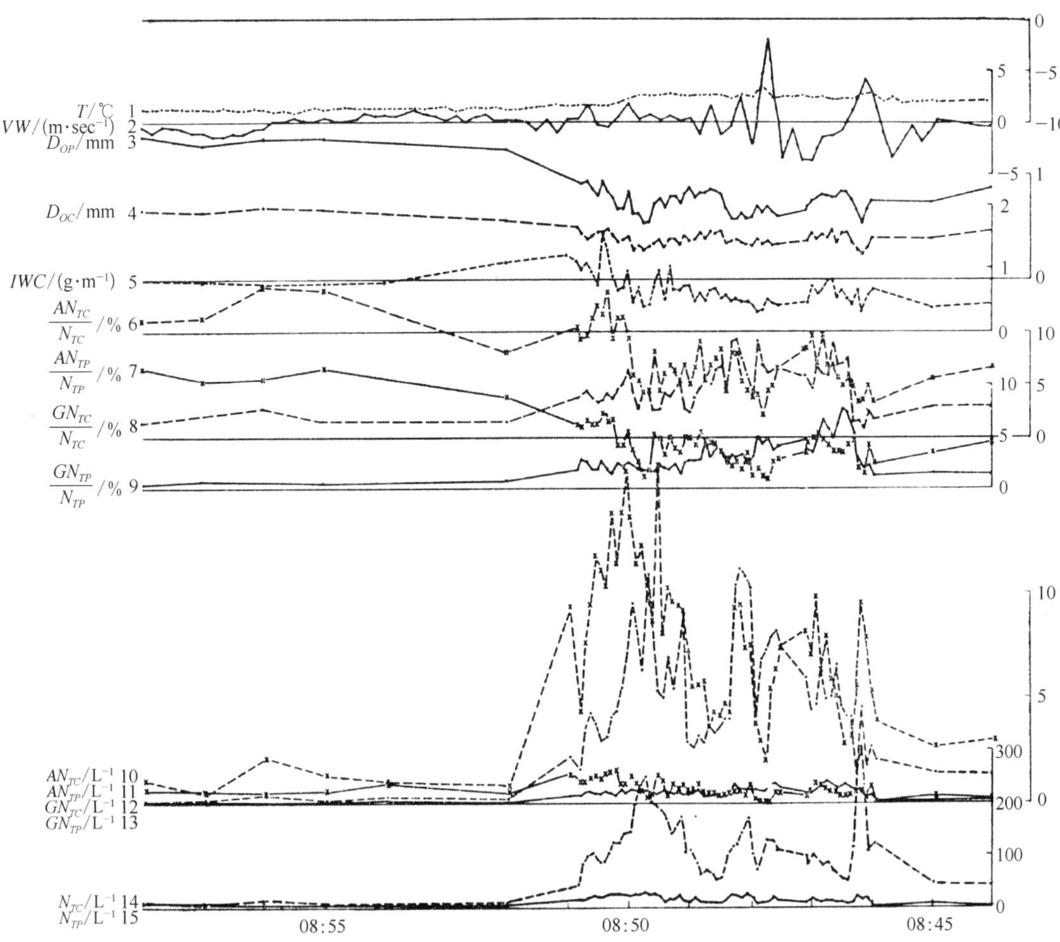

1—Temperature of side-looking radiometer (℃)

2—Vertical air velocity (m • s^{-1})

3—Median volume diameter inferred from the 2D-P probe (mm)

4—Same as 3 but from the 2D-C probe (mm)

5—Ice-water content (g • m^{-3})

6—Ratio of the concentration of aggregate to the total concentration of ice particle for 2D-C probe (%)

7—Same as 6 but for 2D-P probe

8—Ratio of the concentration of graupel to the total concentration of ice particle for 2D-C probe (%)

9—Same as 8 but for 2D-P probe

10—Concentration of aggregate for 2D-C probe (Number/Liter)

11—Same as 10 but for 2D-P probe

12—Concentration of graupel for 2D-C probe (Number/Liter)

13—Same as 12 but for 2D-P probe

14—Total concentration of ice particle for 2D-C probe (Number/Liter)

15—Same as 14 but for 2D-P probe

Fig.3 Time series of microphysical variables obtained during horizontal flight in transition region (segment Ⅰ and Ⅱ) and stratiform region (segment Ⅲ) at the 530 hPa level

and 1.32 mm (2D-P probe) and 0.63 mm (2D-C probe) in the stratiform region. The transition region was characterized by relatively lower reflectivity as shown in Fig.1(a). As a whole, the average water content of precipitation particles was 0.64 g/m^3 in the transition region and 0.54 g/m^3 in the stratiform region, which is the same order of magnitude. The concentrations of precipitation particles in the transition region, however, were 4—6 times that found in the stratiform region. Both the water content and concentration were greater in the transition region. However, we noted above that the average size of particles in the stratiform region was about twice that found in the transition region, thus the relatively lower reflectivity in the transition region was a result of smaller particle sizes.

Table 1 Average values of microphysical variables in the transition region and stratiform region of the MCC on 4 June 1985

Region	Sample number	WC/(g·m^{-3})	N_{TP}/L^{-1}	N_{TC}/L^{-1}	GN_{TP}/L^{-1}	GN_{TC}/L^{-1}	AN_{TP}/L^{-1}	AN_{TC}/L^{-1}	GN_{TP}/N_{TP}/%	GN_{TC}/N_{TC}/%	AN_{TP}/N_{TP}/%	AN_{TC}/N_{TC}/%	D_{OP}/mm	D_{OC}/mm
Ⅰ	25	0.51	16.69	117.03	0.66	5.90	0.51	6.18	4.21	5.67	3.10	5.67	0.69	0.36
Ⅱ	30	0.75	19.98	125.82	0.46	5.08	0.92	9.16	2.43	4.16	4.65	7.82	0.77	0.38
Average	55	0.64	18.49	121.84	0.55	5.45	0.73	7.80	3.24	4.85	3.95	6.84	0.73	0.37
Ⅲ	5	0.87	5.18	10.02	0.04	0.18	0.62	1.17	0.68	1.69	10.46	11.64	1.31	0.64
Ⅳ	29	0.46	5.08	23.84	0.08	0.77	0.83	4.15	1.78	3.75	18.74	19.23	1.32	0.63
Average	34	0.54	5.09	21.81	0.07	0.67	0.79	3.64	1.62	3.45	17.52	18.11	1.32	0.63

For meanings of WC, N_{TP}, N_{TC}, etc., see the caption of Fig.3.

3. Graupel Particles and Aggregates

In the eastern portion of the transition region nearby where convective clouds were evolving rapidly (segment Ⅰ in Fig.1(a)), the average concentration of graupel particles (0.66 L^{-1} for 2D-P probe) was slightly greater than that of aggregates (0.51 L^{-1} for 2D-P probe) as shown in Table 1. In the western part of the transition region (segment Ⅱ in Fig.1(a)), graupel particle concentrations were less than those of aggregates (0.46 and 0.92 L^{-1} on average, respectively). As a whole the concentration of grauple particles was the same order of magnitude as aggregates in the transition region; the average grauple particle concentrations were 0.55 L^{-1} (2D-P probe) and 5.45 L^{-1} (2D-C probe), and the concentrations of aggregates were 0.73 L^{-1} (2D-P probe) and 7.80 L^{-1} (2D-C probe). In the stratiform region, on the other hand, the concentration of aggregates obtained with the 2D-P probe exceeded that of graupel particles by up to an order in magnitude; the 2D-P particle concentration being 0.79 L^{-1} for aggregates and only 0.07 L^{-1} for graupel particles on the average. We inferred from the particle characteristics that the main precipitation formation mechanisms were riming and aggregation in the transition region, with the graupel particles being detrained from nearby convective cells. The aggregation process clearly predominated in the stratiform region.

Aggregates accounted for 4%—6% of the total concentration of ice particles in the transition region, while graupel particles (probably detrained from the vicinity of convective cells) accounted for 3%—5% of the total concentration of particles. The characteristics of much larger concentrations and smaller particle sizes in the transition region may be a result of collision-induced breakup of precipitation particles. Hobbs and Farber (1972) and Vardiman (1978) hypothesized that the coexistence of dense, rapidly falling grauple particles or freezing raindrops along with fragile aggregates and ice crystals promote the collisional breakup of aggreates and ice crystals. Graupel particles are few in number in the stratiform region, thus larger aggregates can form by aggregation of ice crystals without substantial breakup by collision with graupel particles. In Table 1, aggregates comprise about 20% of the total ice particles in the stratiform region. Collisional breakup of aggregates and ice crystals still affects the microphysical structure to a certain extent in the stratiform region, however. Both in the transition and in the stratiform regions the main components of aggregates were irregular, rimed crystals, which may be the fragments of dendrites.

Individual irregular particles, which were most likely fragments, were also prevalent. To illustrate this point, we compared the shapes of particles in and out of the MCC. Fig.4 shows some samples of 2D images of particles observed along the edge of the MCC during 1135 UTC to 1156 UTC. Fig.4(a) was obtained in the stratiform region of the MCC, and Fig.4(b) was obtained in two thin clouds out of the MCC system after 1139 UTC. Along the edge of the MCC, most of the particles were small, irregular, and slightly rimed ice crystals and a few aggregates which were also composed of irregular ice particles. In the thin cloud outside the MCC, however, most of the particles were non-rimed or slightly rimed stellar crystals and a few spatial dendrites. There were also some aggregates composed of two or three interlocked stellar crystals in the thin altostratus cloud. It is implied that the processes of ice particle formation and evolution are much more complex than that in the thin altostratus at the same level. In strafotim region of the MCC, the detrainment of the ice particles from the upper convective region and collisional breakup of ice particles can affect the particle characteristics significantly.

IV. VERTICAL DISTRIBUTION OF MICROPHYSICAL VARIABLES IN THE STRATIFORM REGION

Fig. 5 shows some typical samples of 2D images of precipitation particles at various levels observed at locations A and B in the stratiform region shown in Fig. 1(a). The particle shapes were similar in the layer from 0 ℃ to near −13 ℃. The composition of aggregates was mainly inregular ice crystals throughout the entire layer above the 0 ℃ level. Aggregates larger than 5 mm in diameter were found in the lower levels near the melting layer. Complete single dendrites were not found in the samples obtained during the vertical sounding in the stratiform region even at the top of the sounding near the −13 ℃ level. While studies in the laboratory showed that dendrites should form at the colder temperatures (−13—−17 ℃) in an ice-supersaturated environment, their absence suggests that collisional breakup is modifying the ice particle characteristics substantially.

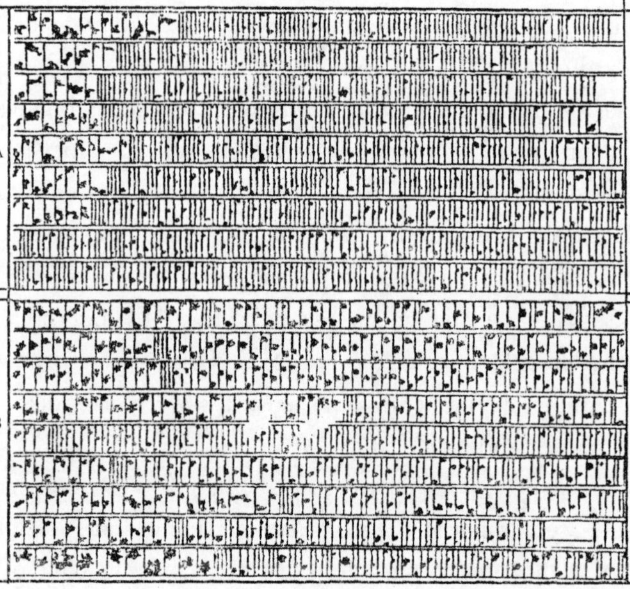

No.	Time	T/℃		2D-P Images	N_t/L^{-1}	D_0/mm	AN_t	GN_t
1	1135^{00}	−12.8			5.92	0.99	1.14	0.0
2	1135^{30}	−12.9			3.68	0.75	0.38	0.0
3	1136^{00}	−13.2			4.50	0.72	0.32	0.0
4	1136^{30}	−13.1			2.51	0.75	0.21	0.0
5	1137^{00}	−13.7	A		2.76	0.72	0.23	0.0
6	1137^{30}	−13.7			1.08	0.78	0.13	0.0
7	1138^{00}	−13.7			1.52	0.75	0.13	0.0
8	1138^{30}	−14.0			1.09	0.61	0.03	0.0
9	1139^{00}	−13.5			0.63	0.59	0.00	0.0
10	1139^{25}	−13.5			1.27	1.35	0.01	0.0
11	1139^{30}	−13.5			0.74	1.29	0.05	0.0
12	1140^{00}	−13.5			0.89	1.22	0.04	0.0
13	1141^{00}	−13.6	B		0.64	1.19	0.01	0.0
14	1152^{00}	−13.3			2.36	0.59	0.04	0.0
15	1153^{00}	−13.3			1.66	0.89	0.13	0.0
16	1154^{00}	−13.3			0.38	1.26	0.10	0.0
17	1155^{00}	−13.8			0.62	0.93	0.01	0.0
18	1156^{00}	−14.6			0.33	1.66	0.04	0.0

(a) 2D images of hydrometeors in the edge portion of the stratiform region of MCC

No.	Time	T/℃		2D-P Images	N_t/L^{-1}	D_0/mm	AN_t	GN_t
1	1135^{00}	−12.8			26.13	0.54	5.31	0.0
2	1135^{30}	−12.9			16.90	0.48	2.94	0.0
3	1136^{00}	−13.2			18.58	0.40	2.84	0.0
4	1136^{30}	−13.1			9.07	0.45	1.43	0.0
5	1137^{00}	−13.7	A		2.72	0.51	0.74	0.0
6	1137^{30}	−13.7			2.14	0.55	0.91	0.0
7	1138^{00}	−13.7			2.20	0.70	0.84	0.0
8	1138^{30}	−14.0			2.68	0.65	0.11	0.0
9	1139^{00}	−13.5			0.81	0.60	0.31	0.0
10	1139^{25}	−13.5			1.61	0.96	0.80	0.0
11	1139^{30}	−13.5			0.59	0.91	0.18	0.0
12	1140^{00}	−13.5			0.92	0.90	0.24	0.0
13	1141^{00}	−13.6	B		0.99	0.89	0.16	0.0
14	1152^{00}	−13.3			4.03	0.55	1.31	0.0
15	1153^{00}	−13.3			0.68	0.74	2.39	0.0
16	1154^{00}	−13.3			1.50	0.69	0.51	0.0
17	1155^{00}	−13.8			1.08	0.75	0.14	0.0
18	1156^{00}	−14.6			0.88	0.59	0.29	0.0

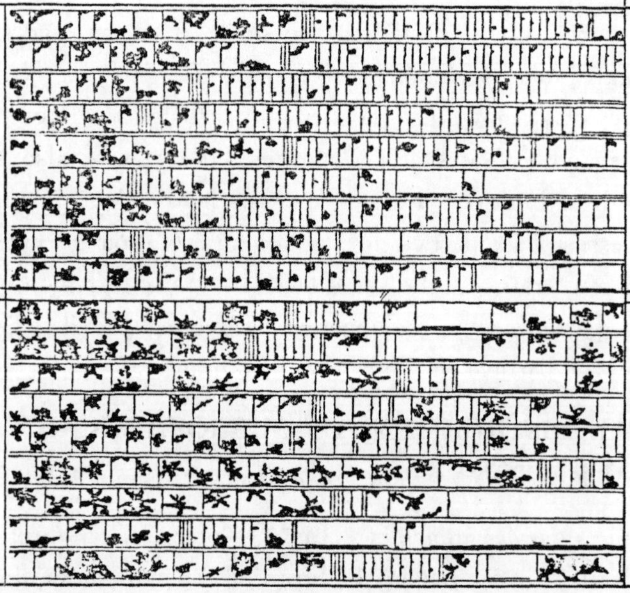

(b) 2D images of hydrometeors in the edge portion in two thin clouds out of the MCC

Fig.4 2D images of hydrometeors in the edge portion of the stratiform region of MCC and in two thin clouds out of the 4 June MCC during the time of 1135 to 1156 UTC

The depth of the melting layer inferred from the 2D images was about 200 m.

The vertical profiles of microphysical variables obtained from location A in Fig.1(a) are shown in Fig.6. The cloud liquid water content measured with a Johnson-Williams hot-wire device is generally less than 0.1 g/m^{-3} in the stratiform region above 0 ℃ level. The water content of ice particles (IWC) inferred from PMS data, in which an ice particle density of 0.1 g/cm^3 has been used, is shown in Fig.6(a). The IWC roughly increased with height; the largest value of 1.42 g/m^3 was located at about 5 000 m (−7 ℃).

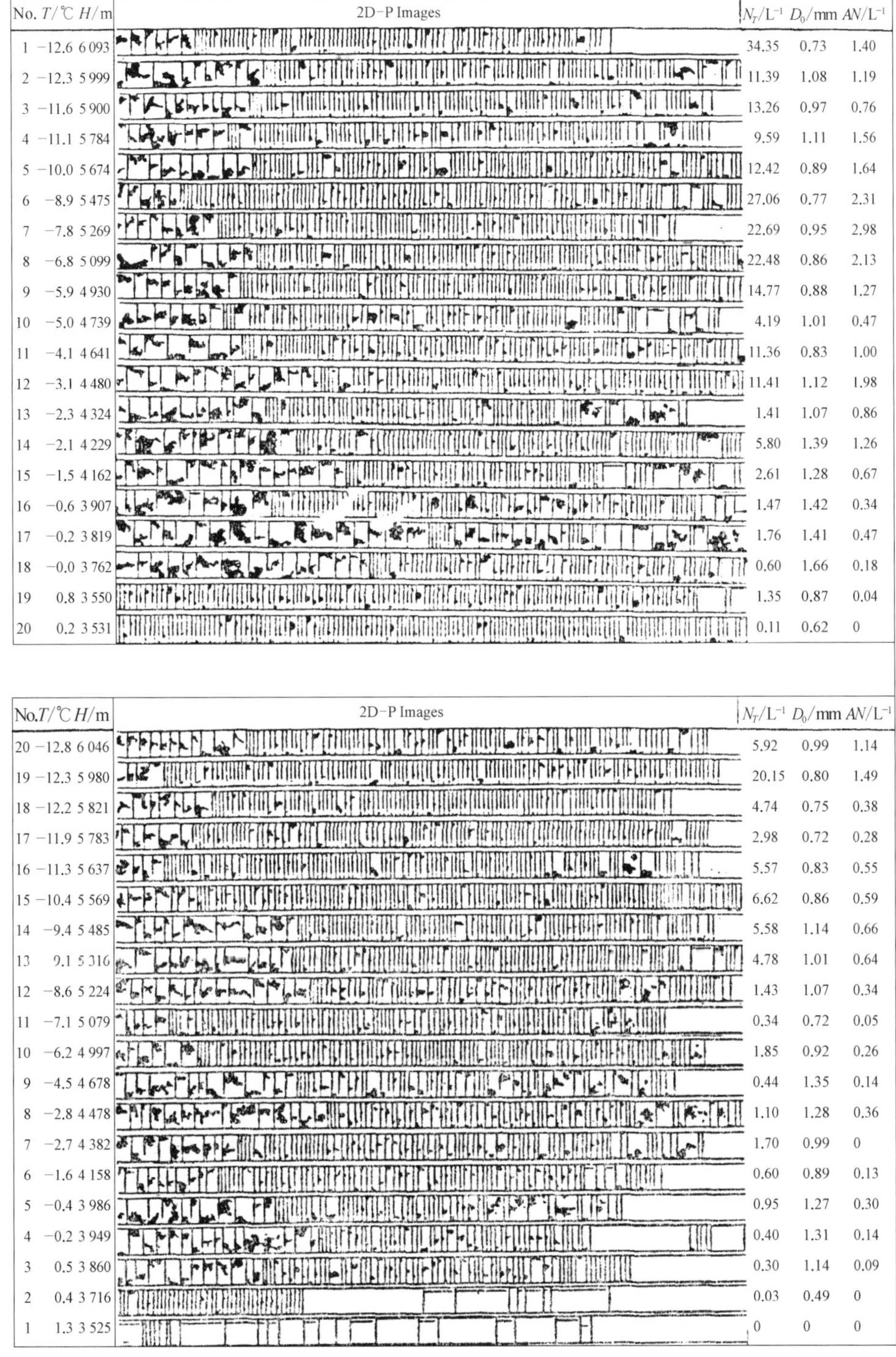

No.	T/℃	H/m	2D-P Images	N_T/L^{-1}	D_0/mm	AN/L^{-1}
1	−12.6	6 093		34.35	0.73	1.40
2	−12.3	5 999		11.39	1.08	1.19
3	−11.6	5 900		13.26	0.97	0.76
4	−11.1	5 784		9.59	1.11	1.56
5	−10.0	5 674		12.42	0.89	1.64
6	−8.9	5 475		27.06	0.77	2.31
7	−7.8	5 269		22.69	0.95	2.98
8	−6.8	5 099		22.48	0.86	2.13
9	−5.9	4 930		14.77	0.88	1.27
10	−5.0	4 739		4.19	1.01	0.47
11	−4.1	4 641		11.36	0.83	1.00
12	−3.1	4 480		11.41	1.12	1.98
13	−2.3	4 324		1.41	1.07	0.86
14	−2.1	4 229		5.80	1.39	1.26
15	−1.5	4 162		2.61	1.28	0.67
16	−0.6	3 907		1.47	1.42	0.34
17	−0.2	3 819		1.76	1.41	0.47
18	−0.0	3 762		0.60	1.66	0.18
19	0.8	3 550		1.35	0.87	0.04
20	0.2	3 531		0.11	0.62	0

No.	T/℃	H/m	2D-P Images	N_T/L^{-1}	D_0/mm	AN/L^{-1}
20	−12.8	6 046		5.92	0.99	1.14
19	−12.3	5 980		20.15	0.80	1.49
18	−12.2	5 821		4.74	0.75	0.38
17	−11.9	5 783		2.98	0.72	0.28
16	−11.3	5 637		5.57	0.83	0.55
15	−10.4	5 569		6.62	0.86	0.59
14	−9.4	5 485		5.58	1.14	0.66
13	9.1	5 316		4.78	1.01	0.64
12	−8.6	5 224		1.43	1.07	0.34
11	−7.1	5 079		0.34	0.72	0.05
10	−6.2	4 997		1.85	0.92	0.26
9	−4.5	4 678		0.44	1.35	0.14
8	−2.8	4 478		1.10	1.28	0.36
7	−2.7	4 382		1.70	0.99	0
6	−1.6	4 158		0.60	0.89	0.13
5	−0.4	3 986		0.95	1.27	0.30
4	−0.2	3 949		0.40	1.31	0.14
3	0.5	3 860		0.30	1.14	0.09
2	0.4	3 716		0.03	0.49	0
1	1.3	3 525		0	0	0

Fig.5 2D images of hydrometeors at various levels and temperatures in the stratiform region at locations A and B (see Fig.1(a)) of the MCC

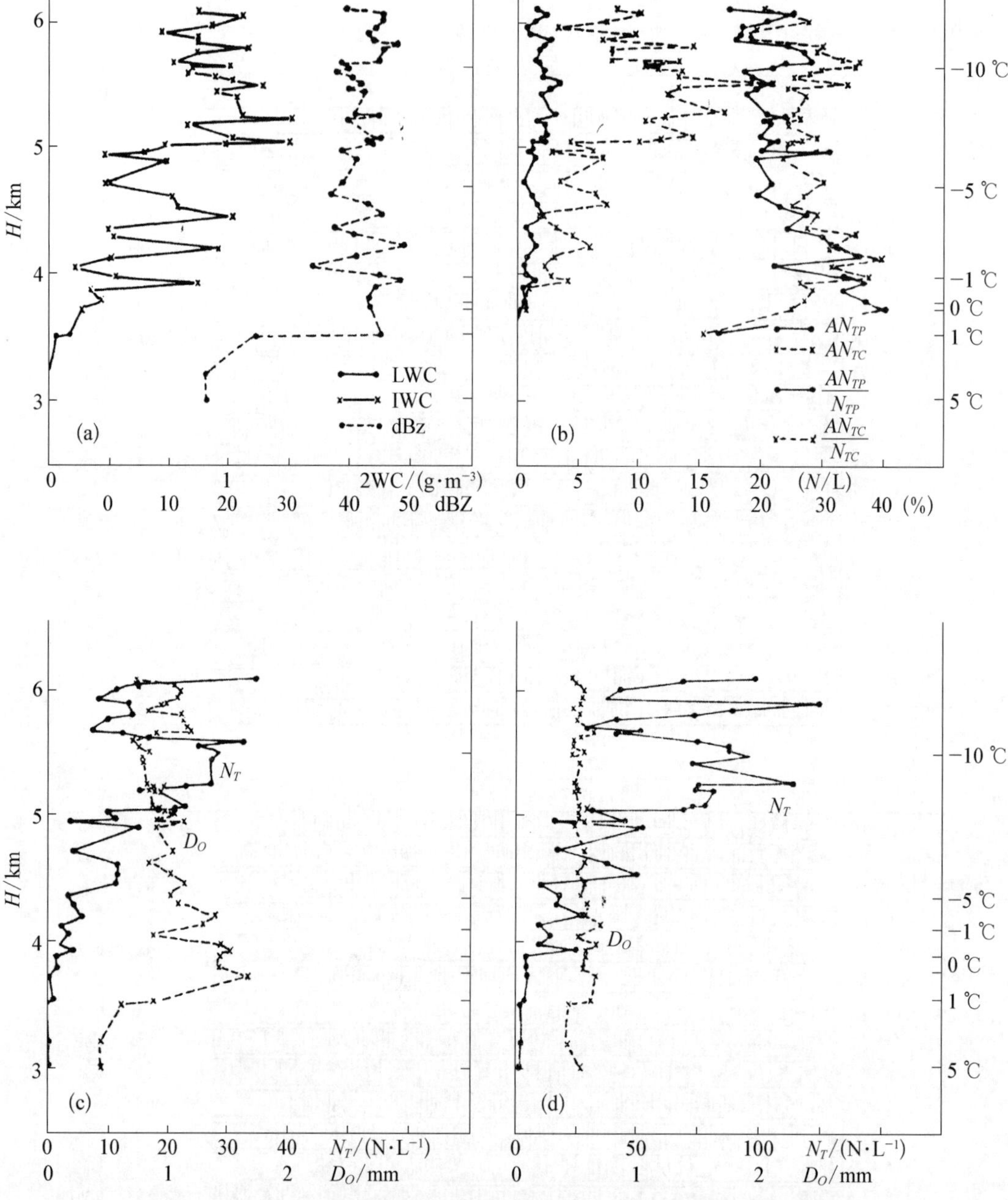

Fig.6 Vertical profiles of microphysical variables as a function of the height and temperature at location A as shown in Fig.1(a)

The average values of IWC were 0.75 g/m^3 and 0.21 g/m^3 in locations A and B. Beneath 0 ℃ level, the average liquid water content of raindrops was 0.13 g/m^3 in location A and less than 0.01 g/m^3 in location B. Light rainfall was observed at the 3 000 m level in the stratiform region during the sounding. The average particle concentrations, as shown in Table 2, were 15.2 L^{-1} (2D-P data) and 50.8 L^{-1} (2D-C data) in location A, and 3.21 L^{-1}

(2D-P data) and 10.3 L^{-1} (2D-C data) in location B. The concentration in location A was about 5 times larger than that in location B, but the average sizes in these two areas were similar. The average median volume diameter was about 1 mm for 2D-P particles and 0.5 mm for 2D-C data. It is noted that the time of vertical sounding at locations A and B is different. Both of A and B located at the western part of the stratiform region, the relative location in the MCC system is similar. The difference of average particle characteristics between these two regions depicted in a certain degree the evolution of microphysical characteristics in the stratiform region of the MCC at different stages. The main difference is in the particle number concentration and water content. The lower ice water content in location B is mainly owing to the concentration decreased, the mean size is similar as mentioned above. In the decaying stage of the MCC, mesoscale downdraft in the middle-lower layer strengthened, small particles evaporated rapidly promoted the particle number concentration decreasing fast. Large particle has much longer life time, so that it can maintain the average particle size decreasing not too fast in the decaying stratiform region.

Table 2 Average and maximum values of particle characteristics observed at the locations A and B in the stratiform region as shown in Fig.6 during the vertical soundings

	Location	Sample number	Average value							Maximum value						
			WC/ $(g\cdot m^{-3})$	N_{TP}/ L^{-1}	N_{TC}/ L^{-1}	AN_{TP}/ L^{-1}	AN_{TC}/ L^{-1}	D_{OP}/ mm	D_{OC}/ mm	WC/ $(g\cdot m^{-3})$	N_{TP}/ L^{-1}	N_{TC}/ L^{-1}	AN_{TP}/ L^{-1}	AN_{TC}/ L^{-1}	D_{OP}/ mm	D_{OC}/ mm
Cold	A	45	0.75	15.2	50.8	1.41	7.73	0.99	0.52	1.42	45.2	124.1	3.20	20.4	1.66	0.68
Layer	B	40	0.21	3.21	10.3	0.44	2.63	0.96	0.55	0.70	20.2	59.7	1.60	10.1	1.45	0.73
Warm	A	4	0.13	0.57	1.34			0.59	0.50	0.33	1.35	2.77			0.87	0.61
Layer	B	3	0.00	0.01	0.03			0.42	0.26	0.00	0.03	0.07			0.49	0.36

Both the total precipitation particles and aggregate concentrations increased with height as shown in Fig. 6(b, c). The largest concentrations resided in the upper levels, from −8 ℃ to −13 ℃, but the largest particle sizes were found in lower levels from +1 ℃ to −2 ℃. The concentration of aggregates relative to the total ice particle concentration was also greater near the melting layer. More than 20% of the particles were aggregates in the layer from 0 ℃ to −2 ℃. They melted as falling through the melting layer and were the main source of the stratiform rainfall. As in many other srtatiform cloud systems, the aggregation process starts in the upper, colder layer, but becomes more efficient as it approaches the melting layer. As a result, an aggregation zone is formed near the 0 ℃ level.

V. SUMMARY AND CONCLUSIONS

We conclude the following:

(1) The melting layer was about 200 m deep over the temperature ranging of 0—+1.5 ℃ in the stratiform region of the MCC. The cloud droplet liquid water content was generally less than 0.1 g/m^3. The water content of precipitation particles was about 0.1 g/m^3 beneath the melting layer. Above the 0 ℃ level, the average water contents of precipitation particles were about 0.8 g/m^3 in location A of the PRE-STORM case, and only 0.2 g/m^3 in the dissipating stratiform region of the PRE-STORM case.

(2) The raindrop concentrations are typically low in the stratiform region beneath the melting layer with mean values of 1.3 L^{-1} (2D-C probe), and corresponding average median volume diameters are 0.5—0.6 mm. Above the height of the 0 ℃ level, the concentration of precipitation particles was much higher, with a mean value of 51 L^{-1} in location A of the PRE-STORM case. The precipitation particle concentration increased with height up to the top of the sounding, where a peak concentration of 124 L^{-1} in the PRE-STORM case.

(3) Precipitation particle characteristics in the transition region and stratiform region are obviously different. The average concentrations of total precipitation particles were about 20 L^{-1} (2D-P probe) and 120 L^{-1} (2D-C probe) at the 530 hPa level (−6.5—−8.5 ℃) in the transition region, and only 5 L^{-1} (2D-P probe) and 20 L^{-1} (2D-C probe) at the same level in the stratiform region. The particle sizes, on the other hand, were greater in the stratiform region: the average median volume diameter was 0.73 mm (2D-P probe) and 0.37 mm (2D-C probe) in the transition region, and 1.32 mm (2D-P probe) and 0.63 mm (2D-C probe) in the stratiform region. Aggregates account for about 20% of the total ice particles in the stratiform region, and only 4% in the transition region. Graupel particles account for about 4% of total ice particles in the transition region and only about 2% in the stratiform region.

(4) Irregular crystals and lightly rimed particles were common shapes of ice particles in both transition region and stratiform region of the MCC. In contrast, a thin cloud outside of the main MCC system contained predominantly stellar crystals and spatial dendrites with a few aggregates composed of pristine crystals. Most of the precipitation particles in the MCC, larger than 1 mm in dimension are aggregates and graupel particles in the transition region, whereas only aggregates predominate in the stratiform region. The main precipitation formation mechanisms are riming and aggregation in the transition region. It appears that graupel particles are detrained from neighboring convective cells. The aggregation process predominates in the stratiform region. Complete single dendrites were rarely observed in the PMS samples, implying that collisional breakup of ice crystals and aggregates affect the particle characteristics in both regions of the MCC, especially in the transition region where the coexistence of graupel particles and aggregates or ice crystals favor collisional breakup. Actually, the concentrations of precipitation particles in the transition region are 4—6 times that found in the stratiform region, and the average sizes of particles in the stratiform region are about twice that found in the transition region.

The relatively lower radar reflectivity in the transition region was a result of smaller precipitation particle sizes, and the smaller sizes and denser number concentrations of precipitation particles are probably caused by the collisional breakup among the graupel particles, aggregates and ice crystals.

(5) The aggregation process occurred over a considerable depth (up to 6 100 m) and range of temperatures (at least as cold as −13 ℃). The largest concentration of aggregates resided in the upper levels from −6 ℃ to the top of the sounding at −13 ℃, but the larger-sized aggregates were found in lower layers from +1 ℃ to −5 ℃. The relative concentration of aggregates was also greater near the melting layer. More than 20% of the particles are aggregates in the layer from 0 ℃ to −2 ℃. We inferred that aggregation starts in the upper, colder layer, but becomes more efficient as the particles approach the melting layer in the stratiform region of MCC. Light riming of ice crystals in the presence of small amounts of supercooled cloud liquid water content ($<0.3\ g/m^3$) may enhance the aggregation process because the collected cloud droplets may act as an adhesive, agent in the aggregation of ice particles. In the colder zones, mechanical interlocking of dendritic crystal branches appeared to be the dominant mechanism responsible for aggregation.

We thank Robert Black for providing software for analysis of PMS data; David Jorgensen, Jose Meitin and other NOAA/Weather Research Program personnel for providing software and resources for analysis of data from the aircraft flight. We also thank Ray McAnelly for his assistance in radar analysis and helpful suggestions in preparing the manuscript, Brenda Thompson for typing assistance, and Lucy McCall for her drafting work. This research was supported by National Science Foundation Grant ATM - 8512480 and Army Research Office Contract # DAAL03 - 86 - K - 0175.

REFERENCES

Blanchard, D.O. and Watson, A.I. (1986), Modes of mesoscale convection observed during the PRE-STORM program, Preprints of Joint Sessions, 23rd Conference on Radar Meteorology and Conference on Cloud Physics, AMS, Snowmass, Colorado, J155 - 158.

Brown, J.M. (1979), Mesoscale unsaturated downdrafts driven by rainfall evaporation: A numerical study, *J. Atmos. Sci.*, **36**: 313 - 338.

Chen, S. and Cotton, W.R. (1988), The sensitivity of a simulated extratropical mesoscale convective system to longwave radiation and ice-phase microphysics, Accepted in *J. Atmos. Sci.*, Dec. 1988.

Churchill, DD. and Houze, R.A. Jr. (1984), Development and structure of winter monsoon cloud clusters on 10 December 1978, *J. Atmos. Sci.*, **41**: 933 - 960.

Cunning, J.B. (1986), The Oklahoma-Kansas Preliminary regional experiment for STORM central, *Bull. Amer. Met. Soc.*, **67**: 1478 - 1486.

Cunning, J.B., Jorgensen, D.P. and Watson, A.I. (1986), Airborne and ground-based Doppler radar observations of a rapidly moving mid-latitude squall line system, Preprints of Joint Sessions, 23rd Conference on Radar Meteorology and Conference on Cloud Physics, AMS, Snowmass, Colorado,

J171－174.

Fortune, M.A. and McAnelly, R.L. (1986), The evolution of two mesoscale convective complexes with different patterns of convective organization, *ibid.*, Preprints of Joint Sessions, J175－178.

Gamacbe, J.F. and Houze, R.A. Jr.(1982), Mescscale air motions associated with a tropical squall line, *Mon, Wea. Rev.*, **110**: 118－135.

Gamache, J.F. and Houze, R.A. Jr., (1983), Water budget of a mesoscale convective system in the tropics, *J. Atmos. Sci.*, **40**: 1835－1850.

Gamache, J.F. and Houze R.A. Jr. (1985), Further analysis of the composite wind and thermodynamic structure of the 12 September GATE squall line, *Mon. Wea. Rev.*, **113**: 1241－1259.

Hobbs, P.K. and Farber, R.J. (1972), Fragmentation of ice particles in clouds, *J. Rech. Atmos.*, **6**: 245－258.

Houze, R.A. Jr. (1977), Structure and dynamics of a tropical squall-line system observed during GATE, *Mon. Wea. Rev.*, **105**: 1540－1569.

Houze, R.A. Jr. and Churchill, D.D. (1984), Microphysical structure of winter cloud clusters, *J. Atmos. Sci.*, **41**: 3405－3411.

Houze, R.A. Jr. and Rutledge, S.A. (1986), A squall line with trailing stratiform precipitation observed during the Oklahoma-Kansas PRE-STORM experiment, Preprints of Joint Sessions, 23rd Conference on Radar Meteorology and Conference on Cloud Physics, AMS, Snowmass, Colorado, J167－170.

Leary, C.A. and Houze, R.A. Jr. (1979), Melting and evaporation of hydrometeors in precipitation from the anvil clouds of deep tropical convection, *J. Atmos. Sci.*, **36**: 669－679.

Maddox, R.A. (1980) Mesoscale convective complexes, *Bull. Amer. Meteor. Soc.*, **61**: 1374－1387.

Rutledge, S.A. (1986), A diagnostic modeling study of the stratiform region associated with a tropical squall line, *J. Atmos. Sci.*, **43**: 1356－1377.

Smull, B.F. and Houze, R.A. Jr. (1985), A midlatitude squall line with a trailing region of stratiform rain: Radar and satellite observations, *Mon. Wea. Rev.*, **113**: 117－133.

Smull, B.F. and Houze, R.A. Jr., (1987), Dual-Doppler radar analysis of a midlatitude squall line with a trailing region of a stratiform rain, *J. Atmos. Sci.*, **44**: 2128－2148.

Srivastava, R.C., Matejka, T.J. and Lorello, T.J. (1936), Doppler radar study of the trailing anvil region associated with a squall line, *J. Atmos. Sci.*, **43**: 356－377.

Vardiman, L. (1978), The generation of secondary ice particles in clouds by crystal-crystal collision, *J. Atmos. Sci.*, **35**: 2168－2180.

Willoughby, H.E., Jin, H.L., Lord, S.J.and Piotrowicz, J.M. (1984), Hurricane structure and evolution as simulated by an axisymmetric, nonhydrostatic numerical model, *J. Atmos. Sci.*, **41**: 1169－1186.

Zipser, E.J, (1969), The role of organized unsaturated convective downdrafts in the structure and rapid decay of an equatorial disturbance, *J. Appl, Meteor.*, **8**: 799－814.

Zipser, E.J. (1977), Mesoscale and convective-scale downdrafts as distinct components of squall-line circulation, *Mon. Wea. Rev.*, **105**: 1568－1589.

5.15 MICROPHYSICAL STRUCTURE OF A SLOW-MOVING MESOSCALE CONVECTIVE COMPLEX IN THE STRATIFORM CLOUD REGION

Ye Jiadong(叶家东), Fan Beifen(范蓓芬)
(Department of Atmospheric Science, Nanjing University, Nanjing 210093)
William R. Cotton, Michael A. Fortune
(Department of Atmospheric Science, Colorado State University, Fort Collins, CO 80523, U.S.A.)

Chinese manuscript received 9 January, 1990; revised 18 December, 1990

Microphysical data observed in the stratiform cloud region of a mesoscale convective complex (MCC) are analyzed together with radar, satellite and other data collected by a NOAA P-3 research aircraft. The results show that the aggregates are prevalent in some part of the stratiform cloud region of the MCC. The aggregation process is the predominant precipitation particle growth mechanism in the stratiform cloud region. Aggregation starts at the upper and colder levels but becomes more efficient as the aggregates approach the melting layer, as a result, a large aggregation zone is formed near the 0 ℃ level.

The cloud water content is usually less than 0.3 g • m^{-3}. The number concentration of precipitation particle beneath the 0 ℃ level is on the average value of 0.8 L^{-1} (2D-P data) and 2.3 L^{-1} (2D-C data); the corresponding mean diameters are 1.0 mm and 0.6 mm, respectively. Above the 0 ℃ level, the concentration of ice particle is much higher, with a mean value of 27 L^{-1} (2D-P data) and 133 L^{-1} (2D-C data) in the temperature ranging from 0 ℃ to −10 ℃; the corresponding mean diameters are 0.8 mm and 0.4 mm, respectively. The ice particle concentration increases with height up to the top of the sounding, where a peak concentration of 52 L^{-1} (2D-P data) and 289 L^{-1} (2D-C data) was observed. Aggregates account for about 15% of the total ice particles near the 0 ℃ level where the mean diameter of the total ice particles is 1.8 mm.

Analyses of the particle size spectra show that the negative exponential distribution fits well in the observed precipitation particle samples. The slope parameter λ is approximately constant for water drop with mean value of 17(±3.6) cm^{-1}, whereas for the ice particle λ varied over a range of about 3 times of magnitude. The intercept parameter N_0 is variable for both of water drop and ice particle samples, and the values varied over a range of 2—3 orders of magnitude. There is a good correlation between the parameter N_0 and λ. Based on this correlation, the particle size spectra can be simplified as a single parameter distribution approximately.

Key words: mesoscale convective complex, stratiform cloud region, microphysics, aggregates.

1. INTRODUCTION

Since Maddox[1] found and defined the mesoscale convective complex (MCC) in 1980, extensive attentions on the investigation of MCC have been attracted in the past decade. Studies have revealed that the mature mesoscale convective system (MCS) generally consists of two parts, one is a leading line or cloud cluster composed of deep convective cells, and the other is a mesoscale precipitating stratiform cloud of several hundred kilometers in horizontal dimension. The air motions in the two regions are distinctly different. In the convective region, organizing convective scale updraft and moisture downdraft generally greater than 5 $m \cdot s^{-1}$ are present associated with the low-level convergence and upper-level divergence. In the stratiform cloud region, a mesoscale updraft above the stratiform cloud base and a mesoscale downdraft beneath it have been observed, corresponding mesoscale circulation characterized by the mid-level mesovortex and mid-level convergence associated with the front-to-rear and rear-to-front relative flows. Stratiform cloud is an important consistent portion in a mesoscale convective system, which contributed about a half rainfall of the MCS. After the active convection decreased, the stratiform cloud portion of the system can survive for several hours or even several days in some occasions. The study of microphysical structure in the stratiform region of MCC is therefore an important aspect in the investigation of mesoscale convective system.

The program of airborne investigations of mesoscale convective system (AIMCS) conducted jointly by the weather research program (WRP) of NOAA/ERL's environmental science group and the cooperative institute for research in the atmosphere (CIRA), Colorado State University, is the first research of aircraft measurements of slow-moving nocturnal mesoscale convective system (MCS). The goals of the program were to determine the capabilities of research aircraft to operate near and within the MCSs and to begin studies of the dynamics of these systems through substantially better observational data than had been available. One of the primary purposes of the cloud physics studies in AIMCS was to discern the interaction of microphysical processes with the dynamical processes within an MCS. As part of the AIMCS-84 program, the hydrometer measurements in the nimbostratus region of an MCS were made and analyzed together with radar, satellite, and other data collected by the NOAA P－3 research air-craft, to determine the thermodynamic and microphysical structures of the stratified precipitation region above and below the melting level. The size spectra of precipitation particles and its parameterization are also discussed in this paper.

2. SYNOPTIC CONDITIONS AND WATCH OF THE MCS FOR 15 JULY, 1984

At 1200 UTC 14 July, 1984, a short-wave trough and associated surface low were located in Wyoming and Nebraska. The main surface low pressure center was near Lake Winnipeg in Canada, with a cold frontal system extending south and southwestward through Minnesota and South Dakota into central Wyoming. A southerly jet brought warm moist air at 850 hPa east of the front and the short-wave trough moved with peak speed of 20 $m \cdot s^{-1}$ near southwestern Kansas. At 500 hPa, an associated short wave was evident through the western Dakotas and eastern Montana. By 0000 UTC 15 July, 1984, the cold front intensified and pushed through central Iowa, and a surface low was centered over northern Kansas. This led to concentrated warm moist air advection and positive vorticity advection flowing over the 850 hPa front in southern Iowa. These conditions are favorable for the development of a mesoscale convective system. At 0000 UTC 15 July, 1984, strong convection in northern Iowa extended into central Iowa; during the night, the convection organized into an east-west line perpendicular to the southern jet. Hail was observed several times in central Iowa (2030—2330 UTC 14 July, 1984) and southern Iowa (0230—0430 UTC 15 July, 1984). The stratiform cloud region northeast of the line produced much strong precipitation.

From the enhanced infrared satellite imagery, we can see that the cloud system of 1900 UTC 14 July, 1984 showed clearly the "initiation" characteristics of classical MCC and the "maximum" stage was reached at 0300 UTC, 15 July 1984 in southeastern Iowa, northeastern Missouri and northwestern Illinois with an area of 210 000 km^2 of cloud tops colder than −52 ℃. The MCC had decayed by 1000 UTC 15 July, 1984 and lasted about 15 hours (Fig.1(a)).

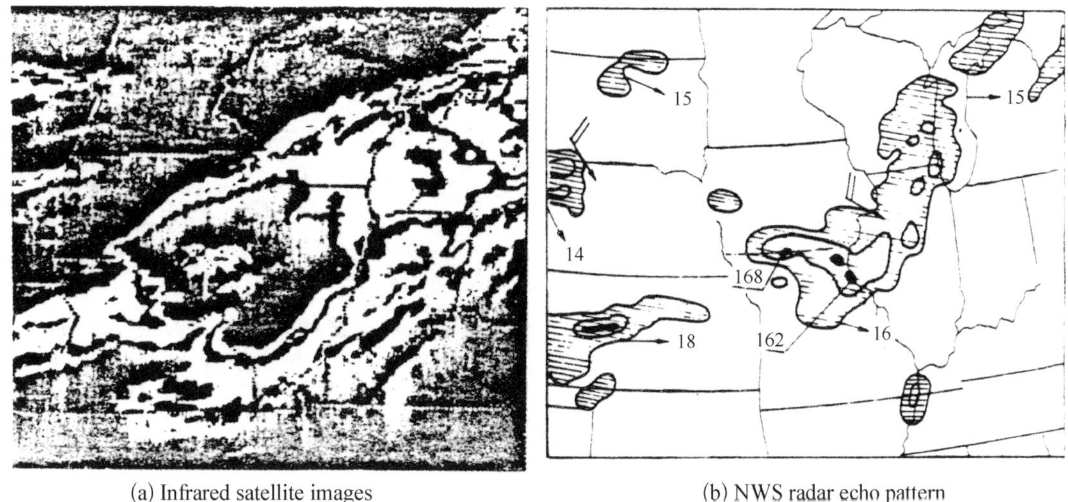

(a) Infrared satellite images

(b) NWS radar echo pattern

Fig.1 Infrared satellite images and NWS radar echo pattern (b) of the MCC on 14—15 July 1984 (Contours of radar reflectivity are for 15, 41 and 51 dBZ; 100/represents the top of radar echo is 10 000 m; unit of vector of radar echo speed is m · s^{-1})

National weather service (NWS) radar data showed, during the evening, the transition from reflectivity pattern characteristics of convective cells to those characteristics of stratiform cloud, as the anvil rain moved eastward rapidly. The reflectivity levels were not too strong, generally at levels 3—4 (40—50 dBZ). At 0400 UTC 15 July, 1984, a persistent E-W convective line developed along latitude 41°N with highest echo top of 16 km(Fig. 1(b)). This line persisted by 0700 UTC 15 July, 1984. Some isolating convective cells remained. The northern portion of the stratiform cloud region moved rapidly eastward into northern Illinois, and the southern portion moved much slowly, so that by 0500 UTC 15 July, 1984 or so, most anvil activity in Iowa occurred south of latitude 42°N. Thereafter, the system moved much slowly and anvil activity concentrated near Burlington, Iowa at 0530 UTC, then moved slowly southeastward. After 0600 UTC, the E-W convective line moved near latitude 40°N. At 0144 UTC, the aircraft took off and entered the storm system at the western edge of the MCC at approximately 0310 UTC just east of Omaha, Nebraska. Then, the aircraft flew at about 720 hPa on the northern flank of the intense east-west convective line which was near the base of the stratiform cloud of the MCC. Four Doppler radar sawtooth legs were flown at this level for anvil experiments (see Fig.2). Beneath the stratiform cloud at 720 hPa level, nearly continuous moderate to heavy rainfall was observed by the aircraft group. The temperature fluctuated between 10 ℃ and 13 ℃, and the wind blew from WNW at speed up to 20 m • s^{-1}.

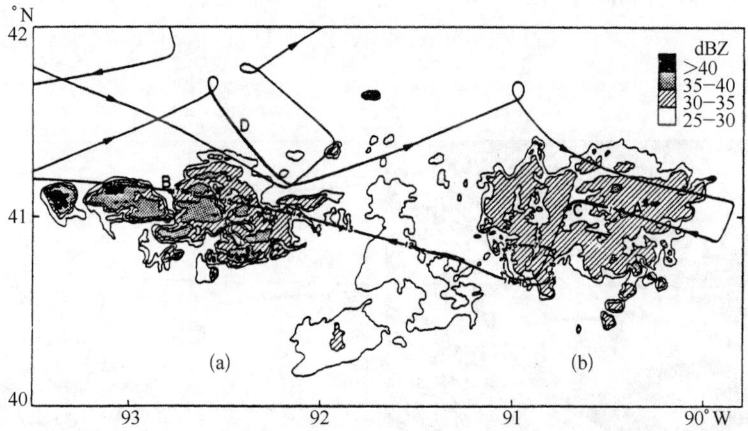

Fig.2 Flight track of aircraft and lower fuselage radar echo patterns at 2 800 m level of the MCC on 14—15 June 1985 (From Points A to B, the numbers 1 to 17 on the flight path are particle sample numbers in Fig. 3. (a) Radar echo observed at the time of 0505—0510 GMT and (b) Radar echo observed at the time of 0552—0556 GMT)

From 0600 to 0620 UTC, the aircraft climbed up to 425 hPa level near Burlington, Iowa for microphysics measurements as well as sounding data through the stratiform cloud heights of 2 800 m to 6 750 m, and temperature ranging from 9. 6 ℃ to −11 ℃. Then the homebound aircraft flew at 425—420 hPa through the stratiform cloud region in which temperature is from −11 ℃ to −14 ℃, and vertical motion is from 6 to −4 m • s^{-1}.

3. MICROPHYSICAL CHARACTERISTICS OF PRECIPITATION IN STRATIFORM CLOUD REGION

The data used to analyze the microphysical structure in the stratiform cloud precipitation region including the images measured by particle measurement system (PMS) 2D-P and 2D-C probes; cloud liquid water contents measured by Johnson-Williams (JW) probe; state parameters such as temperature, pressure and dew point; and navigational variables of NOAA P－3 aircraft. The radar echo data observed with the lower fuselage radar and tail radar are also utilized in this study. The determination methods and sensitivities of these variables can refer to the NOAA Technical Memorandum REL RFC-7 and 8[2,3].

Based on these data, the microphysical structure in the stratiform cloud region of the MCC will be analyzed as follows.

3.1 Water Phase of Hydrometers

The ranges of particle size, measured with 2D-P and 2D-C probes are 0.2—6.4 mm and 0.05—1.60 mm in diameter, respectively. The 2D-P and 2D-C data were objectively edited with the program developed by Robert Black. The 2D-P images were classified into two categories of water drop and ice particle, and 2D-C images were classified into four categories of water drop, ice particle, graupel and column particles according to particle shape. In fact, however, most of the particles classified in the ice particle category seem to be like the large aggregates, and a few particle images which looked like aggregates were classified into other categories. So we distinguished the aggregates by visually inspecting the shadow images of particle. In addition, most of the particle images classified in the water drop category by the program above 0 ℃ level are actually ice particles, so we still classified them in the ice particle category. Some samples of hydrometer images obtained with 2D-P and 2D-C probes at various levels combined with the corresponding particle number concentrations (N_T), median volume diameters and associated aggregates number concentrations in the sample (AN_T) are presented in Fig.3.

There is distinct evolution in particle characteristics above and beneath the melting level. Fig.3 shows clearly that the hydrometers in the warm layer (beneath 0 ℃ level) are predominantly water drops, and are composed almost completely of water drops at temperatures warmer than 1.5 ℃. On the other hand, the ice particles prevail above 0 ℃ level. The melting layer determined by the habits of 2D-images is about 200 m thick with temperature ranging from 0 ℃ to 1.5 ℃.

3.2 Water Content

Cloud water content measured by JW probe is low, generally less than 0.3 g · m^{-3}. The largest liquid water content of 0.5 g · m^{-3} is found in a strong updraft of 5.6 m · s^{-1} at the height of 6 km. The water content of precipitation particles inferred from the PMS probe amounts to only 0.6 g · m^{-3} on average beneath the melting layer, and 0.8 g · m^{-3} is encountered near the levels with temperatures ranging from −3 ℃ to −12 ℃, here the apparent ice particle density of 0.1 g · cm^{-3} is used. The LWC and IWC illustrated in Fig.4(a) represent the liquid water content and ice water content of hydrometers. It is noticed that the water contents of ice particles (IWC) appear to be generally larger than expected in the cold layer of the stratiform cloud region. It is perhaps caused by much thick depth of the stratiform cloud layer of the MCC and active convection nearby. Nearly continuous

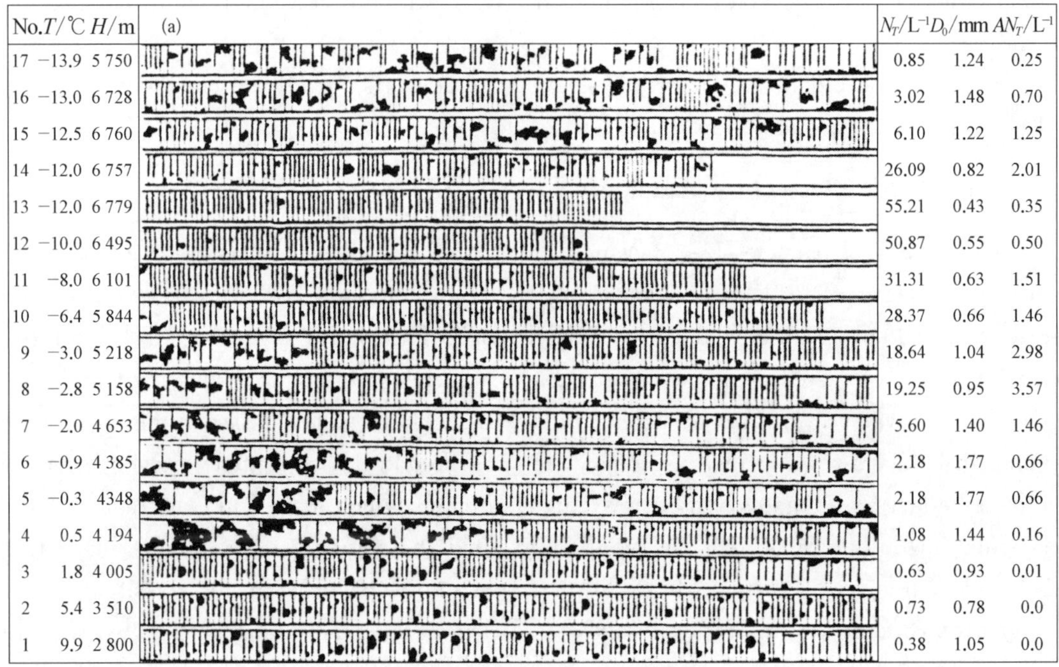

(a) 2D−P images

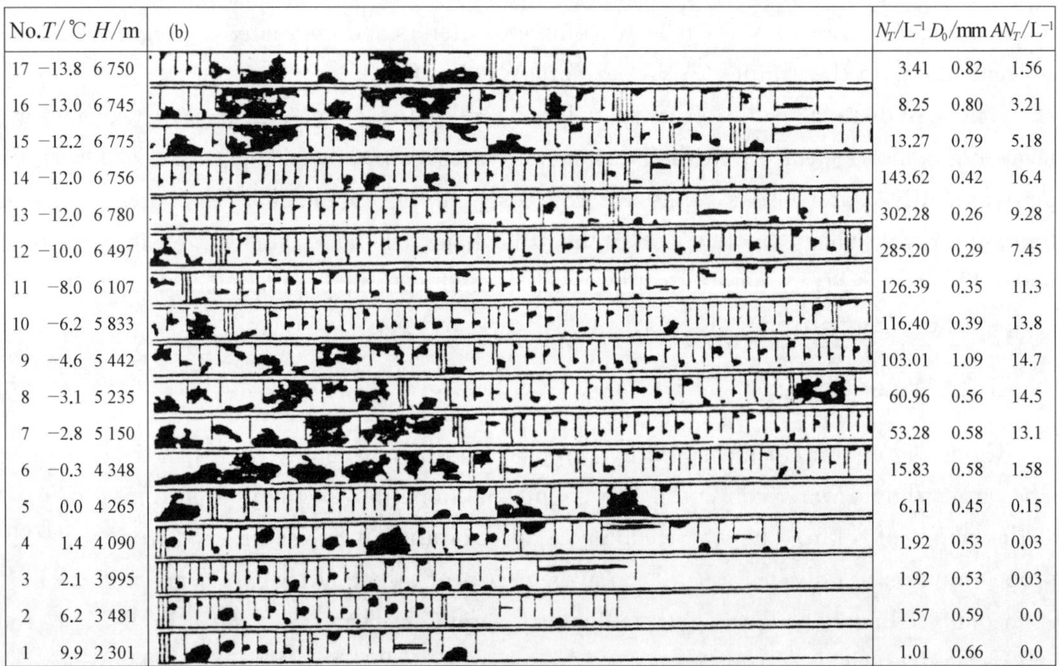

(b) 2D−C images

Fig.3 2D images of hydrometers at various levels and temperatures in the stratiform cloud region of the MCC on 14—15 July 1984

moderate to heavy rainfall was observed beneath the stratiform cloud at 720 hPa as mentioned in Section 2.

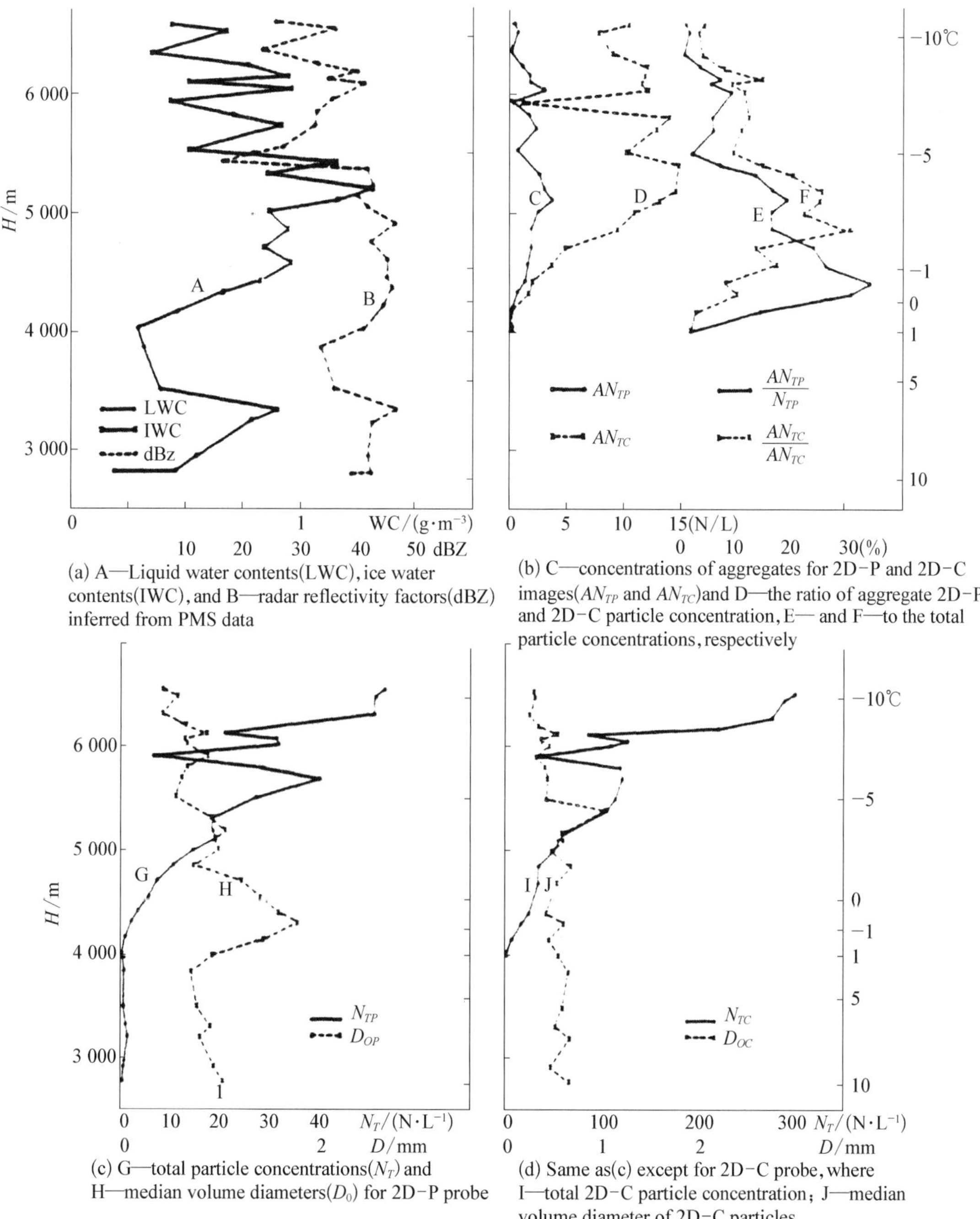

(a) A—Liquid water contents(LWC), ice water contents(IWC), and B—radar reflectivity factors(dBZ) inferred from PMS data

(b) C—concentrations of aggregates for 2D-P and 2D-C images(AN_{TP} and AN_{TC}) and D—the ratio of aggregate 2D-P and 2D-C particle concentration, E— and F—to the total particle concentrations, respectively

(c) G—total particle concentrations(N_T) and H—median volume diameters(D_0) for 2D-P probe

(d) Same as(c) except for 2D-C probe, where I—total 2D-C particle concentration; J—median volume diameter of 2D-C particles

Fig.4 Vertical profiles of microphysical variables as a function of height and temperatures in the stratiform cloud region of the MCC on 14—15 July 1984

3.3 Particle Number Concentration and Size

The particle number concentrations (N_T) and median volume diameter (D_0) inferred from 2D-P and 2D-C images during the vertical sounding as function of height and temperature are shown in Figs. 4 (c, d). It can be seen from these figures that the water drop number concentrations are obviously low, with averaged N_T of 0.8 L^{-1}(2D-P) and 2.3 L^{-1}(2D-C), and averaged D_0 of 1.0 mm (2D-P) and 0.6 mm (2D-C) beneath the melting layer. Above this layer the dominant particles are ice particles with quite a few aggregates at least in the sampling zone, which is restricted to temperature warmer than −14 ℃ (Fig.4). In the temperature ranging from 0 ℃ to −10 ℃, the mean concentration of ice particles are 27 L^{-1} in the D_0 of 0.8 mm (2D-P) and 133 L^{-1} in the D_0 of 0.4 mm (2D-C). The ice particle number concentrations, however, are low, typically less than 10 L^{-1} (2D-P) and 35 L^{-1} (2D-C), between 4 200 m and 5 000 m (1— −2 ℃), and increased with height up to 6 600 m. The maximum concentration is encountered at the top of the vertical sounding at 6 600 m level (−11 ℃) with the peak values of 52 L^{-1}(2D-P) and 289 L^{-1}(2D-C).

In contrast with the number concentration, the median volume diameter D_0 of particle typically decreased with sounding height above the 0 ℃ level. The large D_0 zone is located in the layer of 4 200—4 800 m (0.5— −1.3 ℃) associated with the large aggregates (Fig.4(c)). Fig.4(b) shows the relative concentration of aggregates (ratio of aggregates concentration to total ice particle concentration) which are greater than 15% in this layer. The peak D_0 value of 1.8 mm is found at 0 ℃ level. Above 5 000 m level, the D_0 of 2D-P particles is decreased slowly with height increasing. Combined with the 2D images of particle illustrated in Fig.3, we inferred that the aggregation process above and near the 0 ℃ level is the predominant mechanism of precipitation particle growth which is responsible for the decrease of ice particle number concentration and the increase of the mean particle size. The large aggregates melted into water drop as it fell through the melting layer, and this is the main form of the stratiform cloud precipitation.

3.4 Radar Reflectivity Factor

The radar reflectivity factors computed from the PMS 2D image data together with the water content of hydrometer as functions of height and temperature are illustrated in Fig.4(a). The large radar reflectivity factor is located in the layer of 0.5— −3.5 ℃ (4 200—5 380 m) with main part above the 0 ℃ level. Comparing with the profiles of water content (IWC), aggregate number concentration (AN_{TP}) and median volume diameter (D_0) of the 2D-P data, it suggests that the large radar reflectivity factors are mainly contributed by the combined effect of these variables. In the lower part of the large reflectivity layer, the predominant contributor is the large size of aggregates. At colder temperatures the large reflectivity is mainly caused by both the high water content and large aggregate number concentration.

Previous investigations have suggested that at temperatures slightly colder than 0 ℃ the high radar reflectivity is a result of vigorous aggregation of ice particles, creating large aggregates with large radar cross sections[4]. In addition to the large radar cross sections due to large particle size, at temperatures slightly warmer than 0 ℃, the surface of the large aggregates becomes wetted. The wetted surface of the aggregates yields a dielectric constant similar to water instead of ice, thus further enhancing the radar reflectivity. Complete melting of the aggregates occurs lower in the melting layer, thus losing their large shape and also acquiring a higher terminal velocity and resulting in a decrease in particle concentrations and the radar reflectivity. Steward et al. [5] also explained the decrease of reflectivity beneath the melting layer by the divergence of particle concentration due to the increase of falling speed of particle as they melt, and the decrease on particle size as large aggregates melt and raindrops breakup. This is consistent with the profile of B in Fig.4(a), in which the lowest values of reflectivity and water content, which might be termed the "dark band", are located just beneath the melting layer corresponding to the accelerating zone of the falling speed of melting particle.

3.5 Aggregates

An aggregate is a fundamental type of frozen precipitation element. The investigation of the shape, dimension, bulk density and number concentration of aggregates is important for the study of cloud and precipitation systems. Jiusto[6] commented that aggregation snowfall is the most common type of snowfall in the northeastern United States. Hobbs et al. [7] indicated that more than half the mass of the solid precipitation that reaches the ground near the crest of the Cascade Mountains in the winter is the form of aggregates. In recent years, more and more observational evidences suggest that aggregated ice particles comprise an important component of the total precipitation falling from a variety of cloud systems, including the winter orographic cloud systems over the northern Colorado River Basin[8], the mesoscale rainband of the extratropical cyclones[9], the winter stratiform clouds in the California Sierra Mountains[5], the stratiform cloud regions of the winter monsoon cloud clusters over the South China Sea[10], hailstorms in Colorado[11], and mature hurricanes[12]. It is shown in this study that aggregates are also the dominant form of precipitation above the melting layer to heights corresponding to temperatures as cold as −14 ℃ (the coldest temperature sampled by the NOAA P-3) in the stratiform cloud region of the MCC.

The typical samples of 2D particle images obtained at various heights are shown in Fig.3. It is shown clearly that there are two zones of large aggregates in the stratiform cloud region of the MCC. One layer lies between 0 ℃ and −4.4 ℃ where the mean concentrations of aggregates are 1.9 L^{-1}(2D-P) and 8.1 L^{-1}(2D-C). They account for over 60% of the total water content of all hydrometer particles. The peak values of aggregate number concentration of 3.6 L^{-1}(2D-P) and 14.7 L^{-1}(2D-C) are encountered at −3.1 ℃ and −4.4 ℃ level, respectively (see Fig.4(b)).

Combined with the flight track shown in Fig.2, we inferred that these two aggregate zones are located in different areas associated with two meso-β-scale convective clusters. The first aggregate zone is associated with a decaying convective cluster in which the peak value of vertical velocity does not exceed 6 $m \cdot s^{-1}$.

Because the ice particle number concentrations increase with altitude, the peak value of relative concentrations of aggregates (concentration ratio of aggregates to total ice particles) is reached in the lower part of the layer. About 30% of the particles are aggregate in the layer from 0 ℃ to −1 ℃ for 2D-P data. The mean aggregate concentration of 0.9 L^{-1} in the layer is the same order of magnitude as the water drop concentration just beneath the melting layer. This indicates that the rain drops in the stratiform precipitation region of the MCC are mainly the result of melting of aggregates falling from the aggregation zone just above the 0 ℃ level.

It is also observed from the profiles of aggregate number concentration of 2D-P and 2D-C data in Fig.4(b) that the large aggregates obtained from the 2D-P probe are mainly concentrated in the lower layer nearby 0 ℃ level, and that smaller aggregates obtained from 2D-C probe are more populous in a higher layer of 4 900—5 350 m (−1.4— −3.5 ℃) with a mean concentration of 12.5 L^{-1}. This implies that ice particle aggregation growth originated at a colder and higher level of the stratiform cloud region. Decreasing ice particle concentrations at lower altitudes is also an evidence for the enhanced aggregation in their descent from a colder layer.

The large aggregates ($D>3$ mm) are rarely found in the layer of 5 500—6 600 m in the vertical sounding. But when the aircraft crossed over the region in the northwestern part of the system at 425 hPa level (see Fig.2), the second zone of large aggregates was encountered in the temperature ranging from −11 ℃ to −14 ℃. The mean concentrations of aggregates are 2.0 L^{-1} (2D-P) and 14.4 L^{-1} (2D-C). The relative concentration of aggregates is larger than 20% in the marginal region of the MCC, where the ice particle concentrations are less than 10 L^{-1} (2D-P) and 20 L^{-1} (2D-C), respectively. The region coincided with the western radar echo region is shown in Fig.2 in which the sampling of 2D PMS data was observed by the lower fuselage radar at about 1.5 hour before. When the aircraft crossed over the region at 0630 UTC, the radar echo was quite weak there. So we inferred that the second aggregate zone is observed from a decayed convective cluster region or nearby it.

In the two aggregation zones, the dominant aggregates appear to be composed of irregular crystals, dendritic crystals and lightly rimmed particles. The low supercooled cloud droplet water contents (typically less than 0.3 $g \cdot m^{-3}$) in the stratiform cloud region of the MCCs provide a favorable condition for light rimming of precipitation particles, which may be more effective in promoting aggregation than purely dry, vapor-grown ice particles. The cloud droplet liquid water may act as an adhesive agent in the aggregation process of ice particles. A few graupel-like particles are also observed in the two aggregation zones, and this provides further evidence for the rimming growth process, especially in the layer near 0 ℃ level.

3.6 Horizontal Distribution of Microphysical Structures of Precipitation in the Stratiform Cloud Region of the MCC

Fig.5 illustrates the microphysical structures, vertical velocity and temperature data

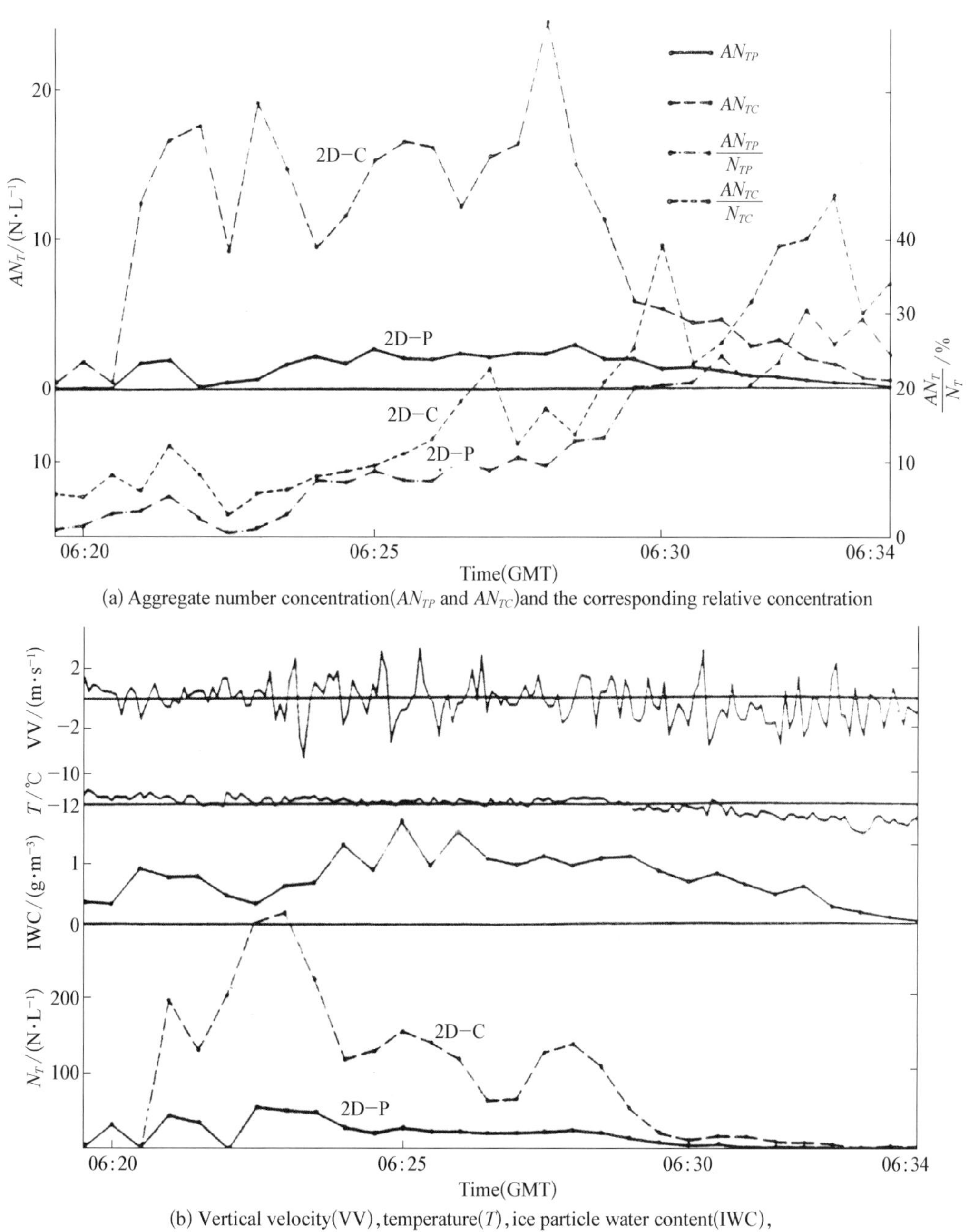

(a) Aggregate number concentration(AN_{TP} and AN_{TC})and the corresponding relative concentration

(b) Vertical velocity(VV), temperature(T), ice particle water content(IWC), and particle number concentration(N_{TP} and N_{TC})

Fig.5 Time series of microphysical variables, vertical speed and temperature obtained during horizontal flight in the stratiform cloud region of the MCC

obtained during the horizontal penetration in the stratiform cloud region of the MCC at 425—420 hPa levels (−11.1—−13.9 ℃). It can be seen from Fig.5 that the microphysical structure including the ice particle water content (IWC), number concentration (N_T) and aggregate concentration (AN_T) is not horizontally uniform. Large aggregates are encountered in the northwest marginal region of the MCC during the sounding period of 0630—0634 UTC, where the cloud is old and in a dissipating stage. The vertical velocity computed from the integrated vertical accelerometers is prevailing downdraft. The ice particle number concentrations are low, more than 20% of ice particles are aggregate as mentioned above. The aggregation process of hydrometer in this region may be different from that occurring in the aggregation zone just above 0 ℃ level because the cloud droplet water contents here are generally below the sensitivity of the JW probe. As a result, the rimming process is generally inefficient, and the mechanical interlocking of crystal branches suggested by Ohtake[13] and Rauber[8] may be the main mechanism for the formation of aggregates. The temperature here is near −14 ℃, as well known that the dendrites shape of ice crystal is prevailing nearby the temperature of −15 ℃ which is favorable for mechanical interlocking process.

4. SIZE SPECTRA OF PRECIPITATION PARTICLES IN THE STRATIFORM CLOUD REGION

There are 78 samples of the size spectra of hydrometers larger than 0.2 mm in diameter obtained from 2D-P probe during insitu measurements in the stratiform cloud region of the MCC. Analyses of these data show that the negative exponential distribution (Marshall-Palmer distribution)

$$N(D)=N_0\exp(-\lambda D) \tag{1}$$

holds well in the entire measured region both for the water drops and ice particles, in spite of the parameters of size distribution such as N_0, the number concentration N_T and median volume diameter D_0 of particles. Measured particle-size spectra fit the negative exponential distribution very closely over the entire sampled size range beneath the melting level. The average linear correlation coefficient of $\log N(D)$ on D is −0.96 for water drops, and −0.92 for ice particles above melting level. The variation of slope parameter λ can be expressed as a function of the median volume diameter D_0 as

$$\lambda=-2.51+1.65/D_0, \tag{2}$$

with the best fit line with a correlation coefficient of 0.87 (Figure omitted), for $D_0<$ 2 mm. The variation of intercept parameter N_0 can be expressed as

$$N_0=1.80N_T/D_0 \tag{3}$$

with a correlation coefficient also of 0.87 (Figure omitted). For the parameterization of microphysics, the empirical formula of size spectra of precipitation particle larger than

0.2 mm in diameter is represented as

$$N(D)=\frac{1.80N_T}{D_0}\exp\left(2.51-\frac{1.65}{D_0}\right)D, \tag{4}$$

which is similar to the distribution function formulated by Cotton et al.[14], but the slope parameter λ (or D_0) for ice particle here is not constant as well as intercept parameter N_0.

It seems to be not an obvious relationship between the size spectrum parameters of M-P distribution and the hydrometer water content (Fig.6(a)). However, Kessler's derived a formula[15] and we derived an empirical relation as follows:

$$\lambda=a+b(N_0/\mathrm{WC})^{1/4}, \tag{5}$$

where coefficients a and b are different for water drops and ice particles. The relation proved to be fair for water drops; the correlation coefficient was 0.73 when $a=0.96$ and

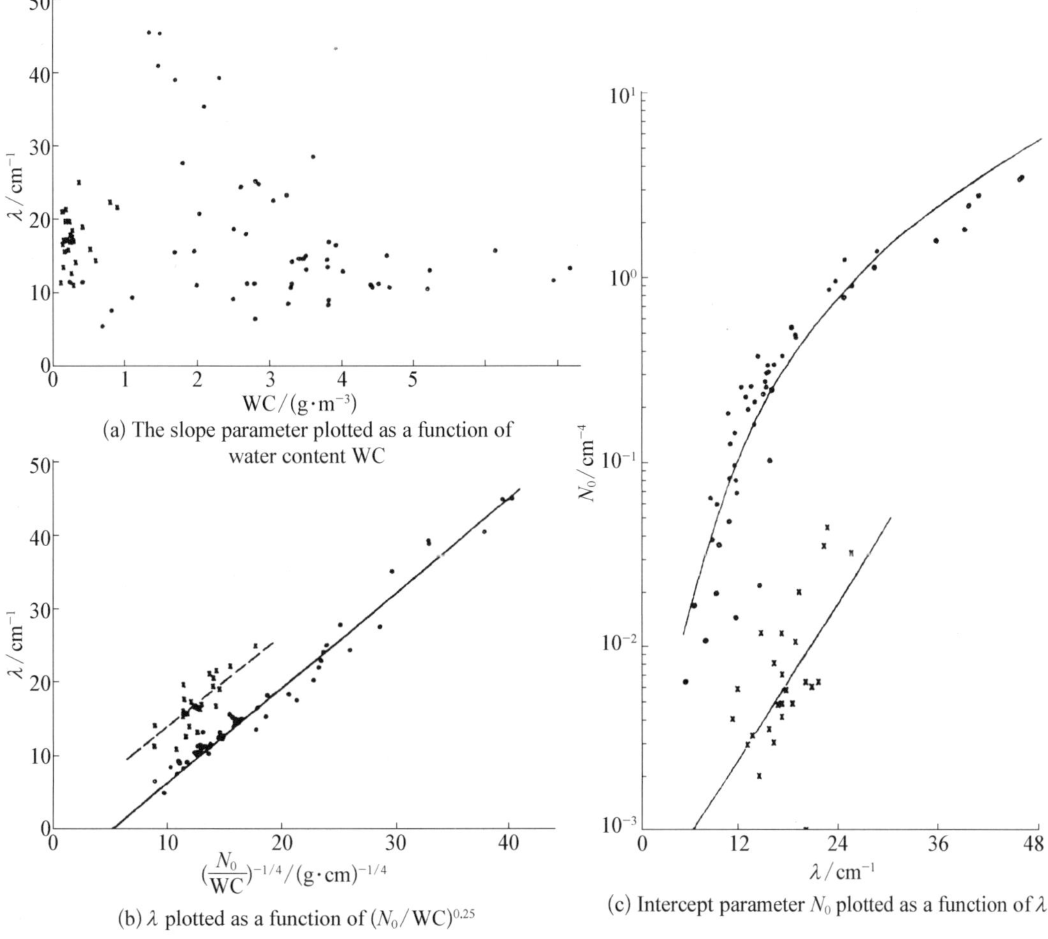

(a) The slope parameter plotted as a function of water content WC

(b) λ plotted as a function of $(N_0/\mathrm{WC})^{0.25}$

(c) Intercept parameter N_0 plotted as a function of λ

Fig.6 Scatter diagrams of particle size-spectra distribution parameters (The scatter point "×" is for the water drop samples, "·" and "o" is for the ice particle samples)

$b=1.28$. The correlation was very good for ice particles; the coefficient was 0.98 when $a=-7.1$ and $b=1.33$. Fig.6(b) shows the scatter diagrams and the corresponding degree of linear correlation.

In the studies of cloud model, it is usually assumed that one of the size spectra distribution parameters (λ or N_0) is constant, so that the size spectrum function becomes a single parameter distribution. Fig. 6 (c) illustrates the scatter diagrams of intercept parameter N_0 plotted as a function of the slope parameter λ of M-P distribution fitting the observed particle size spectra. The scatter diagram of ice particles differs from that of water drops. The slope parameter λ of water drop samples varied not too much, with average value of 17.0 and a standard deviation of 3.6, and the relative deviation of λ is only 20%, so λ for water drops can be assumed approximately to be a constant beneath the melting level of the stratiform cloud precipitation region. For the ice particle samples, however, the slope parameter λ is not obviously constant, which varied about 3 times in amount for different samples. But the intercept parameter N_0, both for the water drop and ice particle samples, varied over a range of 2—3 orders of magnitude. The intercept parameter N_0 for the ice particle samples can be expressed as a function of the slope parameter λ as follows:

$$N_0=1.1\times 10^{-4}\lambda^{2.8}, \tag{6}$$

with the best fit line with a correlation coefficient of 0.94. In this case, the particle size-spectra function can be determined by only one parameter such as λ. Combined Eqs. (1), (6) and (2), the ice particle size-spectra can be inferred from the value of the median volume diameter D_0.

REFERENCES

[1] Maddox, R.A., 1980: Mesoscale convective complexes, *Bull. Amer. Meteor. Soc.*, **61**, 1374 - 1387.

[2] Merceret, F. J. and H. W. Davis, 1982: The determination of navigational and meteorological variables measured by NOAA/RFC WP3D Aircraft, NOAA Technical Memorandum ERL EFC - 7,21pp.

[3] Merceret, F. J., 1982: The sensitivity of variables computed from RFC WP3D flight data to fluctuations in the raw data inputs, NOAA Technical Memorandum ERL EFC - 8, 43pp.

[4] Houze, R.A. Jr., 1981: Structures of atmospheric precipitation systems: A global survey. *Radio Science*, **6**, 671 - 689.

[5] Steward, R. E., J. D. Marwitz, J. C. Pace and R. E. Carbone, 1984: Characteristics through the melting layer of stratiform clouds, *J. Atmos. Sci.*, **41**, 3227 - 3237.

[6] Jiusto, J.E., 1971: Crystal development and glaciation of a supercooled cloud, *J. Rech. Atmos.*, **5**, 69 - 85.

[7] Hobbs, P.V., S. Chang, and J.D. Locatelli, 1974: The dimensions and aggregation of ice crystals in natural clouds, *J. Geoph. Res.*, **79**, 2199 - 2206.

[8] Rauber, R.M., 1987: Characteristics of cloud ice and precipitation during wintertime storms over the

mountains of Northern Colorado, *J. Climate and Appl. Meteor.*, **26**, 488 - 524.

[9] Matejka, T.J., R.A. Houze, Jr. and P.V. Hobbs, 1980: Microphysics and dynamics of clouds associated with mesoscale rainbands in extratropical cyclones, *Quart. J. Roy. Met. Soc.*, **100**, 29 - 56.

[10] Houze, R.A. Jr. and D.D. Churchill, 1984: Microphysical structure of winter cloud clusters, *J. Atmos. Sci.*, **41**, 3405 - 3411.

[11] Heymsfield, A.J. and D.J. Musil, 1982: Case study of a hailstorm in Colorado, Part II: Particle growth processes at mid-levels deduced from in-situ measurements, *J. Atmos. Sci.*, **30**, 2847 - 2866.

[12] Jorgensen, D.P., 1984: Mesoscale and convective-scale characteristics of mature hurricanes, Ph.D. dissertation, Colorado State University, Fort Collins, CO 80523, 189pp.

[13] Ohtake, T., 1970: Factors affecting the size distribution of raindrops and snowflakes, *J. Atmos. Sci.*, **27**, 804 - 813.

[14] Cotton, W.R., G.J. Tripoli, R.M. Rauber and E.A.Mulvihill, 1986: Numerical simulation of the effects of varying ice crystal nucleation rates and aggregation processes on orographic snowfall, *J. Clim. Appl. Met.*, **25**, 1658 - 1680.

[15] Kessler, E., 1969: On the distribution and continuity of water substance in atmospheric circulations, *Meteor. Mon., Amer. Meteor. Soc.*, **10**, 84pp.

5.16 ANALYSIS OF A MESOSCALE CONVECTIVE SYSTEM PRODUCING HAILS AND HEAVY RAINS OVER THE JIANG-HUAI RIVER BASIN IN CHINA

Yeh Jiadong[1], Fan Beifen[1], Tang Xunchang[2], Du Jingchao[2], Song Hang[2]

1 Department of Atmospheric Science, Nanjing University, Nanjing;

2 Jiangsu Meteorological Bureau, Nanjing

1. Introduction

From the evening of 3 May 1988 to the early morning next day, a severe convective storm system developed in region of Jiang-Huai river basin in East China. The storm system formed in northern portion of Jiangsu and Anhui provinces. During the ensuing 8—10 hours of its lifetime as the storm system propagated southward, a series of severe weather events such as heavy rainfall, hails and severe wind gusts produced. The peak values of daily rainfall amount come up to 90—97 mm, and the heavy rainfall area in which the daily amount is more than 50 mm located at the region of Jiang-Huai river basin in a long and narrow zone of about 600 km × 100 km. The heavy rainfall is abrupt with a duration of 1—3 h. Maximum rainfall density in the center region of heavy rains is up to 35—46 mm/h. At the same time hails have been observed in several sites of Moncheng, Fengyang, Xuyi and Xinghua. Strong surface wind gusts were also found in more sites with a peak value of 24 m/s. This paper concentrates on the analyses of large scale weather conditions favorable for generating the storm system, the surface mesoscale flow field, and radar echo structure of the convective storms. Then a brief discussion of the storm system generation and organization mechanisms, and its propagation characteristics is given.

2. Large Scale Circulation Situation

Early in the May 1988, the northern portion of Asia is controlled by polar vortex, the polar frontal zone progress southerly to 45°N—60°N. Circulation in midlatitude of the East Asia is prevailing west flow, and a series of short wave troughs propagate eastward constantly. At 0800 LST 3 May, a trough in westerlies is propagating eastward rapidly with a speed of about 30 km/h through Mongolia plateau, and spreading southward to the regions of Northeast China and North China. At 2000 LST when the severe convective storm initiating, the northern portion of Jiang-Huai river basin is affected by the tail of the trough (Fig.1). The air flow in 500 hPa level at this time has transformed into WNW direction. Cold air mass begin to intrude into Jiang-Huai river basin at midlevels of troposphere. However, southwest air flow is still prevailing in lower level characterized by a strong low-level advection of heat and moist air providing an abundant supply of moist static energy.

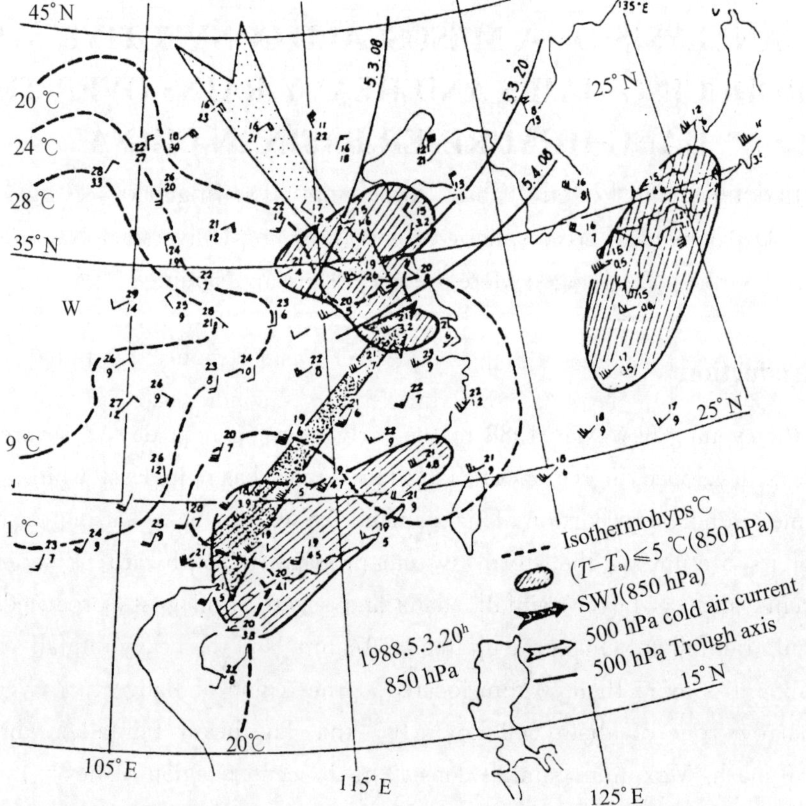

Fig.1 Large scale circulation situation in the 850 hPa and 500 hPa levels

Coordinating with the propagating eastward of 500 hPa trough and spreading southerly, the surface cold air mass in Mongolia region is also strengthened in its moving course eastward, resulting in a cyclone formed from the low-pressure disturbance. Pressure in the cyclone center descents to 990 hPa at 0800 LST 3 May. The cyclone spreaded southerly in East Asia affects the regions of Northeast China and North China. The tail of cold front sweeps easterly across the northern portion of Jiang-Huai river basin, and interacts with a warm shear line associated with the West China inverted trough, providing a large scale weather situation favorable for the development of severe convective storm system.

Low-level SW jet stream: Early in May the West Pacific subtropical high ridge on 500 hPa level locates at 19°N, the ridge top represented by 588 isobar pushes westward to 98°E. At the lower levels over the South China and Jiang-Huai rive basin, the West Pacific subtropical high establishes itself far enough westward to drive warm-moist air northeastward over these areas. This circulation especially favors the development of low-level SW jet laden with warm moist air. At 2000 LST, the SW jet moves northward and strenthens with a peak value of 18 m/s. The SW LLJ axis has veered to a more southwesterly direction, the southern component of jet stream is enhanced evidently and exhibits the flow becomes increasing convergent along the jet. The Jiang-Huai river basin region is in the left-exit zones which is favored zone for upward motion and severe weather occurrence. Accompanying the strenthened of jet and convergence of moist air, a rich water vapor region is formed in Jiang-Huai river basin and Shandong peninsula at 2000 LST (Fig.1). The dew-point deficits at Jinan, Xuzhou and Nanjing are 1. 6, 2. 6 and 3. 2 ℃, respectively. On the surface, the effect of SW jet is also evident. Under the influence of the lasting SW warm-moist air flow, a sustained temperature increasing is found from 29 April to 3 May. The temperatures in several sites approach to the highest levels in the historical records. As an example, the peak value of temperature in 3 May in Nanjing is up to 32. 2 ℃. The sounding data over Jiang-Huai river basin region exhibit that a deep layer from surface to 350 hPa level (0—8. 5 km MSL) is conditionally instable, the CAPE reaches up to about 4. 0 kJ/kg in Fuyang at 2000 LST just before the initiation of the convective storm system. Low levels beneath 700 hPa is controlled by warm-moist SW air flow, the dew-point deficits are 3—8 ℃. The air is drier on the 500 hPa level under the influence of northern cold flow, the dew-point deficits reach up to 10—19 ℃. Once a suitable trigger factor occur under the instabilized stratification condition, strong convective overturning will immediately break out.

3. Mesoscale Surface Flow Field

The initial forcing of the severe convection system is attributed to the movement southerly of dry-cold air in the middle troposphere over northward-moving moist warm low-level air. This creates a region of conditional unstable air as described above. Low-level thermal advection, in addition to creating a narrow tongue of warm-moist air as shown in Fig.1 contributes to surface convergence. Thus, a narrow zone of conditionally unstable air with a mesoscale surface convergence triggers deep convection. The mesoscale surface convergence is obvious in the surface flow field (Figure omitted). Eastern wind prevails in most area portion of Jiangsu province under the influence of the West-China inverted trough, south-west of it in the regions of southern Anhui province and flanks of Changjiang River (Yangtze River) is prevailing southwestern wind. In addition to the influence of favorable topography of Dabieshan Mountain, a southern flow pushes into Caohu Lake Plain east of Dabieshan Mountain and arrives to a hilly land south of Benbu and Fengyang. At 2000 LST the surface southern winds in Lujiang and Dinyuan are SSW 6.3 m/s and SSE 4.3 m/s, respectively. Surface mesoscale convergence of wind speed and direction is evident in the region of south flank of Huaihe river accompanying a warm-moist tongue with peak values of 31.4 ℃ and 32.7 hPa for temperature and vapor pressure in the warm-moist tongue, respectively. In there the southern air flow encounters the eastern air flow came from Jiangsu province to form a mesoscale convergence zone, in addition to the favorable topographic effect of hilly land, especially favors to trigger severe convective weather occurrence. Resuits in this region become an important convective storm originating source.

The low-level mesoscale convergence is not only a trigger factor, but also is an important organization mechanism of severe convective system. Analyses of hourly surface rainfall associated with the severe storm system and of surface air flow field reveal that the convective precipitation region is closely related with the surface mesoscale convergence zone. At 2000 LST the surface mesoscale convergence center is in the region north of Huaihe river with a peak value of divergence of $-6.1\times10^{-5}\ s^{-1}$. An isolated severe thunderstorm develops violently in Moncheng with an hourly rainfall amount of 46.2 mm and produces hails. After 2000 LST, several isolated abrupt convective precipitations occur in Huaibei, Xuzhou and Guoyang with hourly rainfall amounts of 20.6, 17.7 and 27.9 mm, respectively. Since 0000 LST 4 May, cold air flow strengthens and pushes southerly to a line along Benbu, Baoyin and Sheyang. The convective precipitation region is organized in a mesoscale band with peak hourly rainfall values of 41.3, 25.8 and 20.8 mm that is aligned by a mesoscale surface convergence zone in which the peak divergence values are of $-8.4\times10^{-5}\ s^{-1}$ and $-8.7\times10^{-5}\ s^{-1}$ in Sihong and Guannan as shown in Fig. 2a. After that time, the convective precipitation band moves southerly following the strengthened surface mesoscale convergence zone (Figs. 2 (b, c)), and produces hails in Xinghua. At 0400 LST 4 May, the surface mesoscale convergence zone has arrived to the south flank of Changjiang River, the peak values of divergence in convergence zone reach -9×10^{-5}, -14.4×10^{-5} and -12.1×10^{-5} s (Fig. 2(c)). The corresponding hourly rainfall band is following the convergence zone to a belt along Changjiang River from Nanjing, Zhenjiang, to Hai-an, as shown in Fig.2(c). Thus it can be seen the severe

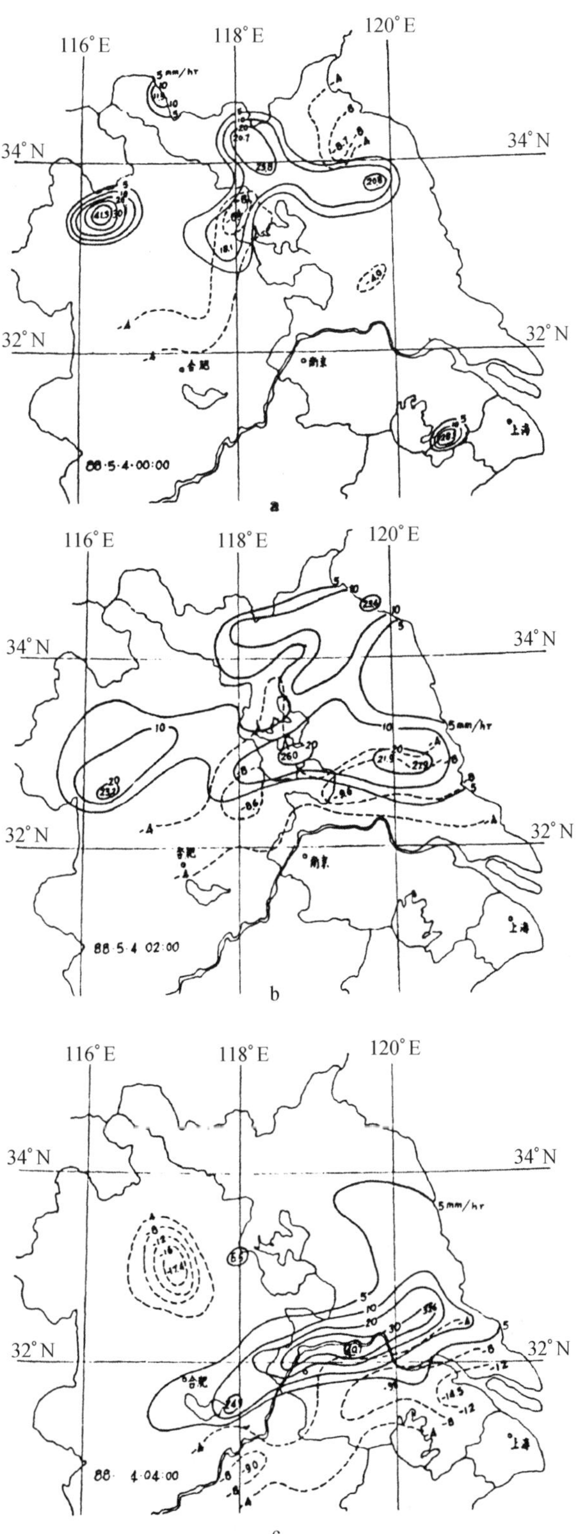

Fig.2 Surface mesoscale convergence zone and convective precipitation band with the hourly rainfall amount larger than 5 mm (The solid line denotes the contour of hourly rainfall amount in mm/h, and the dashed line is the contour of surface divergence value (Unit is 10^{-5} s^{-1}) in the convergence zone)

convective storm producing heavy rains and hails is triggered by a mesoscale convergent air flow in low-level under the favorable large scale circulation situation, heat and moisture environment, and local topographic conditions. It is then organized to a mesoscale convective system along the surface convergence zone and propagates southerly under its forcing. So the propagation mechanism of the convective system seems to be a forced propagation, remember the evaporation effect of raindrop in lower layer is probably weak since the air in the low level is moist as found from the sonnding data.

4. Radar Echo Structure Characteristics

Several isolated strong convective echoes developed in the region of Huaihe river basin from 1900 LST to 2200 LST 3 May and produced local abrupt convective rainfall as mentioned above. After 2200 LST, since the cold air flow intrude into this region and the surface mesoscale convergence zone intensify, a series of convective echoes develop and organize to form a mesoscale convective system following the surface mesoscale convergence zone. Fig. 3 shows the CAPPI radar echoes at varied levels represented as a portion of the convective system. It can be seen from Fig. 3 that the anvil cloud regions produced by the severe convective storms B and C, and several smaller convective cells including the decaying storm cells A developed by 1930 LST 3 May have merged at higher levels as shown in Figs.3(c, d) to form a mesoscale convective cluster with an area of about 300 km×200 km. The severe convective storms B and C are in mature stage by 2200 LST 3 May. The echo top of strong echo region with reflectivity of 45 and 65 dBZ in these two storms reach up to heights upper than 18 km and 15 km levels, respectively (Figs.3(c,d)). The RHI echo structures of these two severe storms are shown in Figs.4(b,c,d) and Fig.5. Storm B initiated at about 2000 LST. The radar echo top of it at the mature stage reaches up to 19 km (MSL), and the width of strong echo region with reflectivity of 65 dBZ is about 14 km which top is up to 16 km height. Hail falling from the severe storm B have been observed in Xuyi. When the storm approaches to the Hongzehu Lake the storm is intensified under the influence of warm-moist air over lake. The severe storm B is consisted by 3—4 cells having various intensity and life-cycle stages, which looks like a multicell severe storm. Sounding data taken at 2000 LST 3 May in Xuzhou and Fuyang showed that the vertical shears of horizontal wind are about 2.7×10^{-3} s^{-1} and 2.2×10^{-3} s^{-1} from surface to 14 km height, which is favorable for the development of multicell severe storm. Storm C located at the place of 120 km west of storm B, is another lasting vigorous severe convective storm. The sustained time period of its echo top exceeded 17 km is longer than 2.5 hours. The reflectivity of strong echo center is also reaches 65 dBZ and produced hails in the region of Fengyang. Storm C is also consisted by 2—4 convective cells and has evident structure characteristics of multicell severe storm in aspects of echo intensity, lasting range of life-cycle, and hail shooting. After 0030 LST 4 May the reflectivity in strongest echo center is still maintained at the level of 65 dBZ (Figs.5(c,d)).

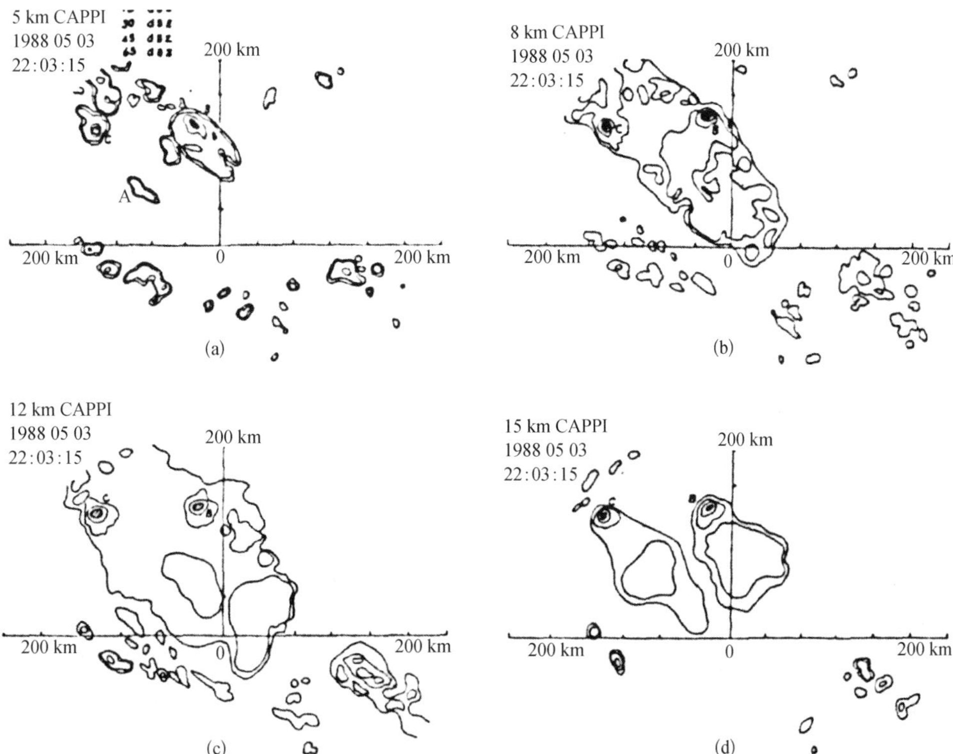

Fig.3 CAPPI radar echoes of the severe convective storms B and C at varied levels (The contours denotes the reflectivity of 15, 30, 45, and 65 dBZ, respectively. The origin is in Nanjing)

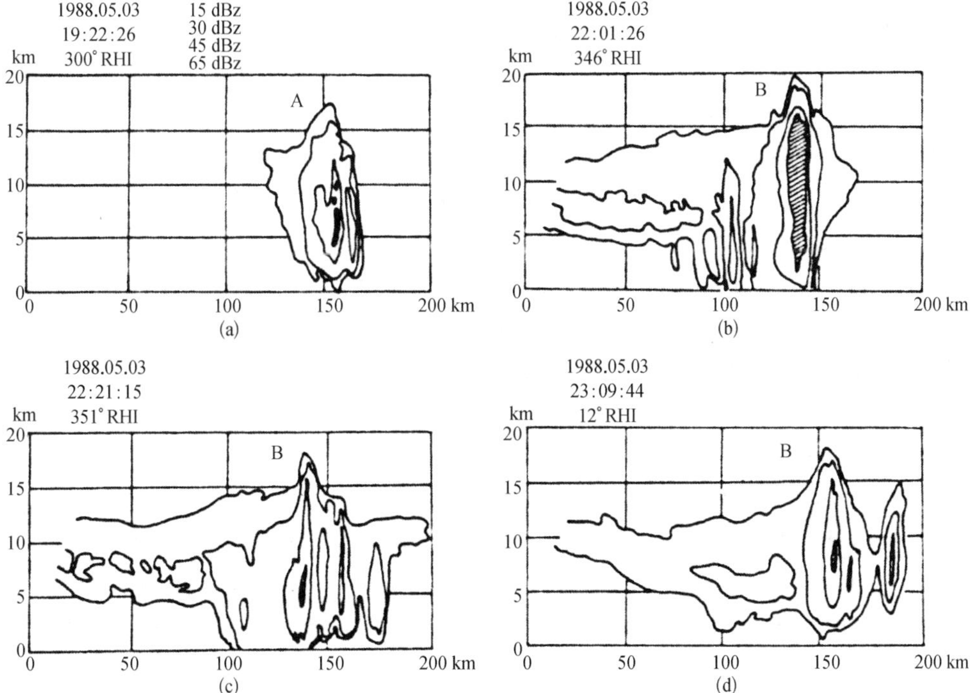

Fig.4 RHI radar echoes of the severe convective storm B and an earler developed storm A (The contours are same as in Fig.3)

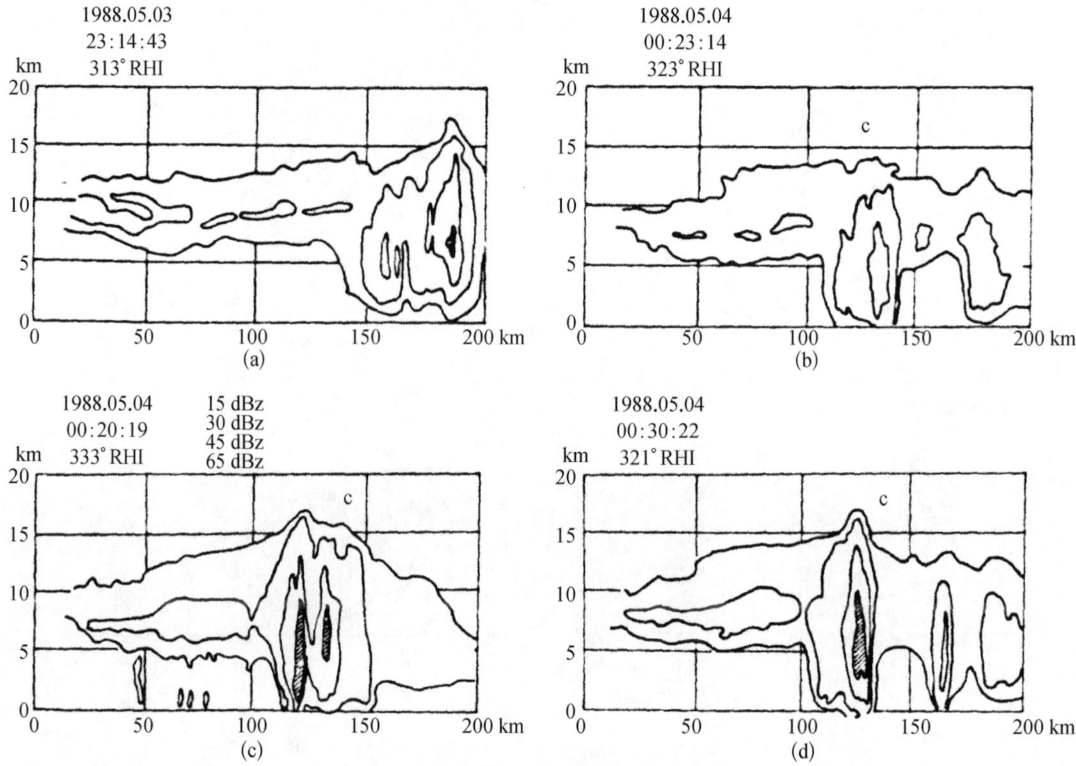

Fig.5 Same as Fig.4 except for the storm C

The stratiform anvil cloud region of mature storm is located at a layer with the cloud base from 4 km to 6 km height, and the top of it from 10 km to 14 km. The depth of stratiform anvil cloud is 5—10 km. The main part of it is in the levels above 0 ℃ level. In the lower portion of the anvil cloud region from 5 km to 8 km height, a radar echo belt with reflectivity exceeding 30 dBZ has been found (Figs.4 and 5). It appears to be caused mainly by the aggregates as found in the stratiform region of MCC (Yeh et al., 1991). A weaker echo region is also presented between the stratiform anvil cloud region and convective region similar to that found in MCS in the United States (Smull and Houze, 1985, 1987; Yeh et al. 1987). Forthermore, the sounding data in the region of Jiang-Huai river bails at 2000 LST exhibit that the environmental atmosphere is conditionally instable under 9 km height MSL and is moister with dew-point deficits of 4—10 ℃ in the layer of 7—10 km MSL. Thus the upper divergent flow caused by convection should be in more higher levels as 10—15 km MSL to form the anvil cloud, which is hard to explain the fact that the stratiform anvil cloud mainly concentrates at the layer of 5—10 km MSL. Observations of a squall line system have shown that a deep layer of convergence exists between 6.0 and 10.0 km height in tailing stratiform region. The inferred vertical motion field from the divergence profile showed that, above 6.5 km, a deep layer of ascent is inferred that is similar to that found in the stratiform region of tropical and midlatitude MCSs (Srivastava et al., 1986). Numerical simulation of a deep convective system has shown that convective latent effects will be able to cause a mesoscale warm center which can promote the formation of mesoscale convergence and ascent motion in the middle layer of troposphere (Fan et al., 1990). Thus, the resultant stratiform anvil cloud caused by the mesoscale ascent motion is active in dynamics and can usually last a longer range of time after the convection decaying.

This research is supported by the Natural Science Foundation of China in the number of 49070244.

REFERENCES

Smull, B.F., and R.A.Jr.Houze, 1985: A midlatitude squall line with a trailing region of stratiform rains: Radar and satellite observations, Mon. wea. Rev., 113: 117 - 133.

Smull, B.F., and R.A.Jr. Houze, 1987: Dual-Doppler radar analysis of s midlatitude squall line with a trailing region of a stratiform rain, J. Atmos. Sci., 44: 2128 - 2148.

Srivastava, R.C., T.J. Matejka, and T.J. Lore-llo, 1986: Doppler radar study of the trailing anvil region associated with a squall line. J. Atmos. Sci., 43: 356 - 377.

Fan Bei-Fen, Yeh Jia-Dong, W.R.Cotton, and G.J.Tripoli, 1990: Numerical simulation of microphysics in meso-scale convective cloud system associated vith an MCC, Adv. Atmos. Sci., 7: 154 - 170.

Yeh Jia-Dong, Fan Bei-Fen, M.A. Fortune, and W.R. Cotton, 1987: Comparison of the mice rophysics between the transition region and stratiform region in a mesoscal convective complex, Preprints to the Third Conference on Mesoscale Processes, Vancouver, B.C., Canada, 21 - 26 August, 1987, 188 - 189.

Yeh Jia-Dong, Fan Bei-Pen, W.R.Cotton, and M.A.Fortune, 1991: Observational study of microphysics in the straiform region and transition region of a midlatitude mesoscale convective complex, ACTA METEOROLOGICA SINICA, 5: 527 - 540.

5.17 三峡洪水和登陆台风极值的统计推断*

叶家东　范蓓芬

（南京大学大气科学系，210093，南京）

摘　要　应用Gumbel极限分布理论对长江三峡最大洪水和在台湾地区登陆的台风（热带气旋）的最大风速进行统计推断。结果表明，长江三峡大坝工程设计要在1%～1‰的风险度下抵御未来可能发生的特大洪水袭击，设计洪水的洪峰流量标准应在$10.5\times10^4\ m^3/s$以上。这一结论为古洪水考察检测的资料所证实。对我国沿海登陆的热带气旋分析表明我国热带气旋登陆频数最大的区域是海南岛和雷州半岛地区，而台风登陆频数最大则在中国台湾岛。对台湾地区登陆台风极值风速统计推断表明，沿海百年寿命的大型建筑工程的设计风力（风速）达130 m/s以上时，工程建筑所承受的风险度是1%。

关键词　洪水，台风，极值重现期，统计推断

分类号　P426.16

* 本研究得到“八五”攻关项目（85 - 906）资助。

本文台风登陆部分资料由张裕华提供，插图由金仪璐绘制。

收稿日期：1995 - 06 - 04

第一作者简介：叶家东，男，1936年10月生，教授，大气物理专业，发表有“方差不相等的双样本回归分析”等论文。

0 引言

台风袭击引起的主要致灾因素是大风、风暴潮和洪涝；暴雨引起的主要致灾因素是洪涝。从防灾的角度着眼，在建设沿海或沿江城市建筑、工矿企业及居民生活设施时，特别是在大型水利工程设计中，需要对各种可能的致灾因素及其破坏力进行恰当的估计，使得工程建筑既有足够的抗灾能力又经济实惠。例如海堤的强度和高度，能抵御可能出现的最大风暴潮的袭击；水库高坝建设中则需要估计可能出现的最大洪水，从而设计适当的水库防洪库容和泄洪道泄量，并使坝基强度和坝高能顶得住建库后长时期内可能出现的最大洪峰流量的冲击。统计表明有30%的垮坝是洪水漫顶所致，设计不周的险坝一旦出事，垮坝洪水冲击波将给下游人民和社会带来毁灭性灾害。河南"75·8"暴雨引起的严重灾害就是鲜明的例子。大型工程设计中需要估计的上述种种可能出现的最大洪水、最强风力以及最高风暴潮等等，在水利工程建设中通常称为设计洪水、设计风力和设计潮位等。从统计学角度看，台风引起的最大风力、最强风暴潮及与之密切相关的用来表征台风强度特征之一的中心最低气压，以及台风和其他暴雨的最大日雨量、流域最大过程总雨量以及所引起的最大洪峰流量等，都是气象要素或水文要素的极值或极值的组合函数问题。而上述工程设计中需要估计的未来长时期内可能出现的最大洪水、最强风力、最高潮位及台风中心最低气压等，就是极值的统计推断问题。为了进行这类统计推断，首先需要分析气象、水文要素极值的概率分布，然后根据统计理论进行合理的统计推断，并导出这种推断的准确率或失误率，后者就是工程设计中的风险度。

1 极值的概率分布和重现期

通常在水文、气象上将洪水事件或台风袭击事件看成是离散型事件，且满足二项分布[1]或泊松分布[2]。例如，T 年一遇的洪水 Q 事件的概率是 $P=\frac{1}{T}$，其中的 T 称为洪水 Q 的重现期。相应地，Q 不出现的概率为 $q=1-P=1-\frac{1}{T}$。如果工程有效服务年为 n 年，n 年内发生 T 年一遇洪水的概率就是工程设计所承受的风险度，也称失事概率 U。由概率论中独立试验序列概率理论[3]知

$$U=1-q^n=1-\left(1-\frac{1}{T}\right)^n。\tag{1}$$

据此推断，如工程寿命 n 为100年，则当工程设计洪水标准的重现期 T 为100，200，500和1 000年时失事概率分别为63.4%，39.4%，18.1%和9.5%。这表明百年寿命的工程在抵御千年一遇的洪水中所承受的风险度接近10%，并不太小。一般，洪水事件是一种稀有事件，出现的概率很小。由概率论知[3]，若当 $n\to\infty$ 而 $P\to 0$ 时有 $nP\to\lambda$（常数）时，则 n 年中洪水 Q 出现 k 次的概率满足泊松分布。于是，n 年中至少发生一次洪水 Q 的概率为

$$P_n=1-\mathrm{e}^{-\lambda},\tag{2}$$

P_n 就是失事概率 U。就上例而言，$P=\frac{1}{T}$，于是 $\lambda=nP=\frac{n}{T}$。对于工程寿命 $n=100$ 年，设计洪水标准的重现期 T 为 100，200，500，1 000 和 2 000 年的失事概率分别为 63.2%，39.3%，18.1%，9.5%和 4.9%，与二项分布概型推断的结论基本一致。

问题是，表征洪水事件 Q 的水文气象变量实际上是一些连续型统计变量，如洪水流量、流域过程总雨量等；相仿，表征台风袭击事件的气象变量是台风中心最低气压、台风中心附近最大风速、台风引起的风暴潮位以及暴雨强度等等。因此，极值统计推断问题的恰当提法应该是：在一定的重现期内，最大洪峰流量超过某一限值的概率，或台风中心最低气压低于某一限值的概率，余此类推。这类连续型变量的极值分布实际上就是研究右侧概率或左侧概率的分布问题[4]。在 x 的分布函数为 $F(x)$ 的情况下，由概率分布知最大值和最小值的重现期 $T(x)$ 可分别表示为

$$T(x)=\frac{1}{1-F(x)} \tag{3}$$

和

$$T(x)=\frac{1}{F(x)}, \tag{4}$$

即重现期是右侧概率或左侧概率的倒数。重现期内的极大值 x_n 和极小值 x_1 则可分别由下二式解出

$$F(x_n)=1-\frac{1}{T} \tag{5}$$

和

$$F(x_1)=\frac{1}{T}。 \tag{6}$$

可见，要计算极值的重现期或重现期值，必须对 x 的极值概率分布 $F(x)$ 或密度函数 $f(x)$ 进行统计推断。

2　极值的概率分布

根据次序统计量分布理论知[4]，原始分布函数为 $F(x)$ 的极大值 x_n' 的分布函数为

$$F_n(x)=[F(x)]^n。 \tag{7}$$

显然，$F_n(x)$ 取决于原始分布函数 $F(x)$ 和样本容量 n。由数理统计极值分布理论知[5]，当 $n\to\infty$ 时，极值的渐近分布并不过多地依赖于原始分布函数。对于许多服从指数型分布（如正态分布、对数正态分布、负指数分布以及 Γ-分布等）的气象要素来说，极值的一种近似极限分布就是所谓耿贝尔分布(Gumbel)[5]，其分布函数和概率密度函数分别为

$$F(x)=\exp(-\mathrm{e}^{-y}), \tag{8}$$

$$f(x)=a\exp(-y-\mathrm{e}^{-y})。 \tag{9}$$

其中 $y=a(x-b)(-\infty<x<\infty)$，$a$ 是尺度参数，b 是概率密度分布的众数，可分别表示为

$$a=\frac{\sigma_y}{\sigma_x},\quad b=m_x-m_y\frac{\sigma_x}{\sigma_y}。\tag{10}$$

其中 m_x、m_y 和 σ_x、σ_y 分别为随机变量 x 和 y 的数学期望和均方差。由次序统计量分布理论知，将 N 个样本极大值按大小顺序排列为 $x'_1\leqslant x'_2\leqslant\cdots\leqslant x'_m\cdots\leqslant x'_N$，居于第 m 位的次序统计量 x'_m 的分布满足

$$F(x'_m)=\exp[-\mathrm{e}^{-n(x'_m-b)}]=\frac{m}{N+1}。\tag{11}$$

由于 $y_m=a(x'_m-b)$，所以由式(11)可得

$$y_m=-\ln\left(-\ln\frac{m}{N+1}\right),\quad m=1,2,\cdots,N。\tag{12}$$

由这些 y_m 值可以计算平均值 $\bar{y}$ 和均方差 s_y 作为 m_y 和 σ_y 的估计值，由上可知，$\bar{y}$ 和 s_y 只与 N 有关，其值可查表求得[5]。例如，当 $N=32$ 和 83 时，$\bar{y}$ 和 s_y 值分别为 0.538 0 和 0.557 4 以及 1.119 3 和 1.196 0。

分布函数确定后，就可以根据式(5)和式(8)求出对应于特定重现期 T 的极值

$$x_T=b-\frac{1}{a}\ln\left[-\ln\left(1-\frac{1}{T}\right)\right],\tag{13}$$

相应于一定重现期值 x_T 的重现期 T 则可表为

$$T=\frac{1}{1-\exp[-\mathrm{e}^{-a(x_T-b)}]}。\tag{14}$$

极小值的概率分布：通过变换可将极小值问题转化为极大值问题，因为

$$x_1'=\min(x_1,x_2,\cdots,x_n)=-\max(-x_1,-x_2,\cdots,-x_n)=-x_n',\tag{15}$$

于是有 $F_1(x)=1-F_n(-x)$。对于耿贝尔分布

$$F_n(-x)=\exp[-\mathrm{e}^{a(x+b)}]\quad(-\infty<x<\infty),\tag{16}$$

相仿，T 年一遇的极小值可由下式求出

$$x_T=\frac{1}{a}\ln\left[-\ln\left(1-\frac{1}{T}\right)\right]-b。\tag{17}$$

3 长江三峡洪峰日均流量极值及其重现期的统计推断

通常，将长江上游控制站宜昌观测的日均流量超过 $4.5\times10^4\ \mathrm{m^3/s}$ 的洪峰定为三峡洪水标准。表 1 是 1957—1983 年间满足这一标准的观测样本，容量为 32。

表 1　宜昌站 1957—1983 年期间日均流量≥4.50×10⁴(m³/s)的洪水资料[6]

Table 1　Flooding data for the daily average flow≥4.5×10⁴(m³/s) on Yichang station in period of 1957—1983

序号	流量/(10^4 m^3·s^{-1})	序号	流量/(10^4 m^3·s^{-1})	序号	流量/(10^4 m^3·s^{-1})	序号	流量/(10^4 m^3·s^{-1})
1	5.35	9	5.56	17	5.76	25	45.5
2	4.71	10	4.88	18	4.72	26	5.46
3	5.95	11	5.11	19	4.53	27	6.95
4	5.35	12	4.97	20	5.15	28	4.51
5	5.18	13	4.97	21	5.19	29	5.04
6	5.32	14	4.55	22	6.10	30	5.90
7	5.09	15	4.84	23	4.84	31	5.04
8	5.23	16	5.96	24	4.93	32	5.26

由表 1 的资料得到样本平均值和标准差分别为 $\bar{x}=5.217$ 和 $s_x=0.5365$。现在以样本统计量 $\bar{x}$ 和 s_x，以及前面讨论的当 $n=32$ 时由表查得的随机变量 y 的样本统计量 $\bar{y}=0.5380$ 和 $s_y=1.1193$ 分别取代相应的数学期望 m_x，m_y 和标准差 σ_x，σ_y，并代入式(10)即可求得 $\hat{a}=2.086$ 和 $\hat{b}=4.959$。为了计算对应于不同重现期 T 值的洪水极值，需要对式(13)做一订正。因为表 1 资料表明，1957—1983 年共 27 年中出现 32 次洪水，即满足日均流量≥4.50×10⁴ m³/s 标准的洪水其年均频数 $f=1.185$，相应的重现期 $t=0.844$ 年，所以前面导出的式(3)，(4)，(5)，(6)和(13)，(14)，(17)中的重现期 $T(x)$ 均需以 $T'(x)=tT(x)$ 取代。这样由式(13)计算的对应于不同重现期 $T(x)$ 值的洪水极值 x_T 列于表 2。在这里，T 年一遇的洪水 x_T 的统计含义是洪峰流量达到或超过 x_T 的洪水出现的概率是 $\frac{1}{T}$，若设 $T=1\,000$ 年，意即平均在 1 000 年中含有一次机会出现这类特大洪水，是统计平均的意思。这并不意味着在 100 年或 500 年内不会发生洪峰流量≥x_T 的洪水。于是，现实的问题是，在工程有效服务期(工程寿命)内，出现达到或超过 x_T 值洪水的概率有多大？这就是工程设计的风险度。下面对风险度(失事概率)进行统计推断。

表 2　长江三峡 *T* 年一遇的洪峰流量极值的统计推断值

Table 2　Statistical inference values x_T of the extreme flooding flow with a recurrence intervals of *T* on the Three-Gorges of Yangtze River

$T(x_T)$/年	50	100	200	500	1 000	1 500	2 000	3 000
x_t/(10^4 m^3·s^{-1})	6.747	7.082	7.416	7.856	8.189	8.383	8.521	8.716

将洪峰流量 x 大于等于某一洪水标准值 x_T 这一随机事件是否出现当作离散型随机事件，并设它服从泊松分布，于是，对于不同工程寿命(n)的三峡大坝工程设计，在不同的风险度(P_n)下所能抵御的洪峰极值流量 x_T 可以由式(2)和式(13)联立求出，计算结果列于表 3。由表可见，工程寿命为 1 000 年的三峡大坝在抵御 10.5(10⁴ m³/s)流量的特大洪水事件中所

承受的风险度(失事概率)大约是百分之一。若要求在千分之一风险度下设计百年寿命和千年寿命的三峡大项,则设汁洪水标准分别达 10.5(10^4 m^3/s)和 11.6(10^4 m^3/s)。表 4 所列资料是长江三峡古洪水考察研究中利用古洪水沉积物"陈江泥"样品,由国家地震局地壳应力研究所根据放射性碳同位素(C_{14})测年法估算的近 3 000 年历史上出现过的几次特大洪峰流量资料[6]。由表 4 可见,在近 3 000 年的历史中确实出现过洪峰流量为 10.50(10^4 m^3/s)的特大洪水事件,这与表 3 计算的设计洪水极值的推断值相符。只有将大坝的设计洪水标准定在 10.5(10^4 m^3/s)以上,工程才有较高的安全度(较小的风险度)。对于千年寿命的大坝工程,设计洪水标准需达 11.5(10^4 m^3/s)以上。

表 3　不同寿命 n(年)的三峡大坝工程在不同风险度 P_n(%)下所能抵御的洪峰极值(10^4 m^3/s)

Table 3　Design flooding flow extreme values (10^4 m^3/s) with various risky probability P_n(%) and for defferent available service range n (years)

n	P_n			
	10	5	1	0.1
	h			
100	8.245	8.590	9.372	10.478
200	8.578	8.923	9.704	10.810
500	9.017	9.362	10.143	11.249
1 000	9.349	9.6974	10.476	11.582

表 4　长江三峡古洪水考察特大洪水资料

Table 4　The extreme flooding flow data in the history inferred from the investigation of ancient flood on the Three-Georges of Yangtze River

样本号(地点)	估算年代	洪峰流量/(10^4 $m^3 \cdot s^{-1}$)
黄陵庙	1870 年	10.50
清水湖	距今 2850±220 年	9.08
西湾村	距今 1937±407 年	8.72
高家溪	距今 2230±263 年	9.85
高家溪	距今 2420±295 年	10.20

4　登陆台风最大风速的统计特征及其重现期的统计推断

按惯例将致灾热带气旋分为 3 类:台风(中心附近最大风速>32.6 m/s),热带风暴(最大风速介于 17.2～32.5 m/s)和热带低压(最大风速介于 10.8～17.1 m/s)。为了反映我国沿海不同地区热带气旋登陆的频繁程度,我们按海岸线大约 150 km 的区间划分登陆区段,并标以区段序号,如图 1 所示。将 1949—1991 年期间 43 年中热带气旋在各区段登陆的频数分布直方图示于图 2,其中实线表示登陆台风的频数,虚线是登陆热带气旋总频数。由图可见,我国热带气旋登陆频数最大的区域是海南岛和雷州半岛东侧地区,而台风登陆频数最大的则在台湾岛东海岸。现在我们着重分析在台湾地区登陆的热带气旋最大风速的统计特征，其频数分布列于表 5。下面对最大风速的极值重现期进行统计推断。由表 5 的资料可计算登陆热带气旋（台风）中心附近最大风速的样本统计量 $\bar{x}=36.9$ m/s 和 $s_x=13.3$ m/s，样本容量 $n=83$。由前面讨论知，对样本容量 $n=83$，相应的 y 变量的样本统计量 $\bar{y}$ 和 s_y 可以查 Gumbel 分布表求得，分别为 $\bar{y}=0.557$ 和 $s_y=1.196$。以上述样本统计量近似取代相应变量的数学期望 m_x、m_y 和均方差 σ_x、σ_y，从而可由式(10)算出 a 和 b 的估计值 $\hat{a}=0.089$ 和 $\hat{b}=29.180$。另外，由表 5 知，1949—1991 年 43 年期间共发生 83 次热带气旋登陆事件，年均发生频率 $f=1.93$ 次/年，相应的重现期 $t=0.518$ 年。与前相类似，以 $T'(x)=tT(x)$ 取代式(13)中的 $T(x)$，即可计算对应于不同重现期 $T(x)$ 值的台风极值风速 x_T，计算结果列于表 6。由表可见，百年一遇的台风极值风速为 73 m/s 以上。所以 1959 年台东登陆台风最大风速 80 m/s 属于百年一遇特大台风。为了估计沿海大型建筑工程需要设计成能抵御多大的强台风袭击方为安全，我们不能完全仿照前面设计洪水标准的统计推断方案，直接利用表 5 的全部资料由式(2)和(13)计算不同寿命(n)的工程建筑在不同风险度下能抵御的台风极值风速 x_T。因为对某一沿海地点，并不是所有登陆台风都对它有影响的。我们设登陆台风的影响区域为 150 km，为使统计推断有较多样本容量，选择图 1 中标号为 B 的区段进行统计，这包括表 5 中在花莲和新港登陆的热带气旋，样本容量为 28。由表 5 中相应资料计算在 B 区登陆的热带气旋中心附近最大风速样本平均值和均方差分别为 $\bar{x}=40.32$ m/s 和 $s_x=11.66$ m/s，相应的 Y 变量的样本统计量由 Gumbel 分布表求得为 $\bar{y}=0.543$，$s_y=1.105$。由式(10)算出 a 和 b 的估计值 $\hat{a}=0.0948$，$\hat{b}=34.681$。由表 5 知，在 B 区登陆的热带气旋 43 年中有 28 次，年均发生频数 $f=0.651$，相应的重现期 $t=1.536$ 年。于是以 $T'(x)=tT(x)$ 取代式(13)中的 $T(x)$ 计算对应不同 $T(x)$ 值的台风极值风速 x_T，计算结果列于表 7。显然，台湾东海岸中段(图 1 中 B 段)登陆的 T 年一遇的台风极值风速要比整个中国台湾岛登陆的平均极值风速强。由式(2)和(13)计算东海岸 B 区不同寿命(n)的工程建筑在不同风险度(P_n)下能抵御的台风极值风速 x_T，结果列于表 8。可见，百年寿命的大型建筑工程的设计风力(风速)需达 130 m/s 以上，方能有 99% 的安全系数(即这类建筑所承受的风险度为 1%)。

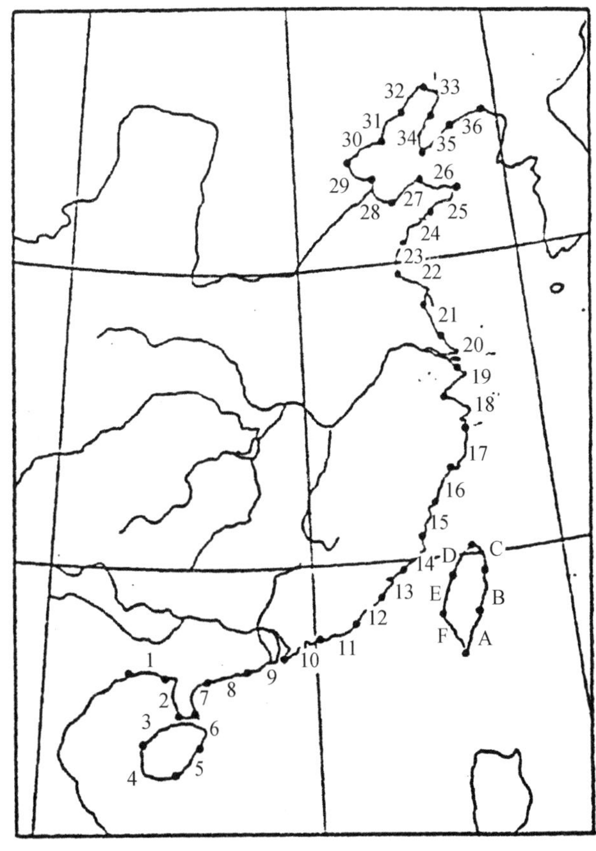

图 1 我国沿海按 150 km 间距划分的台风登陆区段及其序号

Fig.1 The typhoon strike areas and its rank numbers, 150 km in length along the coastline of the China

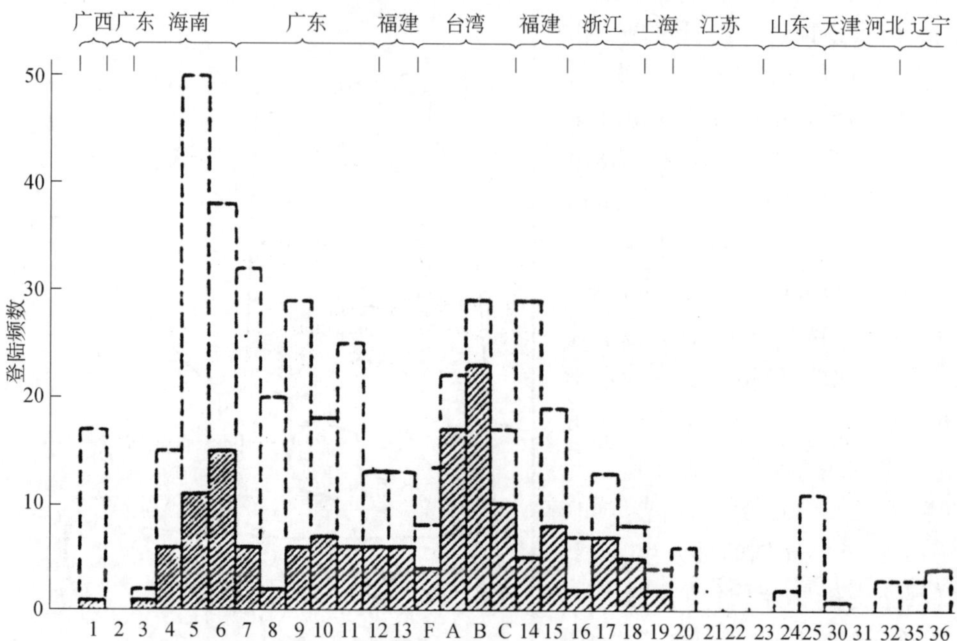

图 2　我国沿海按 150 km 间距划分的区段中热带气旋(虚线)和台风(实线)登陆的频数分布直方图(1949—1991 年)

Fig.2　Histograms of the landfall frequence numbers of tropical cyclones (clashed lines) and typhoon (solid lines) in the strike areas, 150 km in length along the coastlines of China

表 5　1949—1991 年期间在台湾地区登陆的热带气旋最大风速频数分布表

Table 5　Frequence numbers of the extreme winds in the landing tropical cyclones on Taiwan area during the period of 1949—1991

最大风速/($m \cdot s^{-1}$)	11～20	21～30	31～40	41～50	51～60	61～70	71～80
出现频数	12	17	25	20	7	1	1

表 6　台湾登陆的台风 T 年一遇极值风速 x_T 的统计推断值

Table 6　Statistical inference values (x_T) of the extrem winds of landing typhoon on the Taiwan area with a recurrence interval of T (year)

$T(x_T)$/年	50	100	200	500	1 000
x_T/($m \cdot s^{-1}$)	65.4	73.3	81.0	91.2	99.0

表 7　台湾东海岸 B 段登陆台风 T 年一遇极值风速 x_T 的统计推断值

Table 7　Statistical inference values (x_T) of the extrem winds of landing typhoon on the east coastline Taiwan area with a recurrence interval of T (year)

$T(x_T)$/年	50	100	200	500	1 000
x_T/($m \cdot s^{-1}$)	80.4	87.8	95.1	104.8	112.1

表 8 不同寿命 n(年)的台湾东海岸 B 段大型建筑工程在不同风险度 P_n(%)下所能抵御的洪峰极值风速 x_T(m/s)

Table 8 Design extreme wind velocity (x_T m/s) with various risky probability P_n(%) and for defferent available service range n(years) typhoon-resistant building on the east coastline of Taiwan area

n	P_n			
	10	5	1	0.1
	x_T			
100	107.0	114.6	131.8	156.1
200	114.3	121.9	139.1	163.4
500	124.0	131.6	148.8	173.1
1000	131.3	138.9	156.1	180.4

参考文献

[1] 詹道江,邹进上. 可能最大暴雨与洪水. 水利电力出版社. 1983

[2] Simpson R H Riehl H. The Hurricane and Its Impact. Louisiana State University Press, USA, 1981

[3] Meyer P L. 概率引论及统计应用(中译本). 高等教育出版社. 1986

[4] 马开玉,丁裕国,屠其璞,么枕生. 气候统计原理与方法. 气象出版社. 1993

[5] Gumbel E J. Statistics of Exztremes. Columbia University Press, New York. 1958

[6] 章淹,黄忠恕,范钟秀,宋肇英编. 长江三峡致洪暴雨与洪水的中长期预报. 气象出版社. 1993

STATISTICAL INFERENCES FOR THE MAXIMAL FLOOD IN THE THREE-GORGES OF YANGTZE RIVER AND THE EXTREME WIND CAUSING BY LANDING TYPHOON

Ye Jiadong　Fan Beifen

(Department of Atmospheric Science, Nanjing University, 210093 Nanjing)

Abstract Statistical inferences for the maximal flood in the Three-Gorges of Yangtze River and the extreme winds causing by landing typhoon are performed in this paper using the Gumbel theory for extreme value distribution. It is revealed that for the project of Three-Gorges great dam of Yangtze River, with a risky probability of 1.0%—0.1% as a flood control installation, the criterion of design flooding flow should be more than that of 10.5×10^4 m^4/s, which has been verified qualitatively by the investigation of ancient flood on Yangtze River. Analyses for the statistical features of landing tropical cyclones on the coastal areas in China reveals that the maximal frequent landfall tropical cyclone is occurred on the east coastline of Hainan Island and Leizhou Peninsula, and east coastline of Taiwan Island of China for typhoon. Statistical inference for the extreme winds causing by landing typhoon shows that, on the east coastline of Taiwan area, the typhoon-resistant building standard to protect against extreme winds more than that of 130 m/s has a risky probability of 1.0% in an available building service range of 100 years.

Keywords flood, typhoon, extreme value recurrence intervals, statistical inference

5.18 对流性暴雨过程及其中-α尺度数值模拟试验*

叶家东 范蓓芬

(南京大学大气科学系,210093,南京)

宋 航 杜京朝

(江苏省气象台,210008,南京)

摘 要 对1987年6月7日发生在东南沿海的一次对流性暴雨过程的天气背景、降水和雷达回波特征进行分析,指出低空西南急流为暴雨区提供了充足的水汽和不稳定能量;伴随着冷锋出海,在浙闽沿海地区锋面出现中尺度波动扰动,触发形成这场对流性暴雨。应用RAMS静力平衡模式对这次暴雨过程进行三维中-α尺度数值模拟试验。结果表明,模式能较真实地模拟低空辐合线、锋后冷高压500 hPa槽线和地面冷锋的发展和移动;模拟的中低空流场特征与实况相当吻合;模式预报的上升运动区和对流降水带与沿海对流雨带定性相符。模拟的对流降水带移动较快,发展时间也偏早,这与模式采用的积云参数化方案及下垫面条件处理较简单有关,且未能恰当考虑热力不稳定效应所致。

关键词 对流性暴雨,中-α尺度数值模拟,RAMS模式

分类号 P426.62

0 引言

1987年6月7日下午至8日凌晨,我国东南沿海和台湾地区,伴随冷锋出海有一次强降水过程,在冷锋过境时的浙南-闽北沿海地区和地处锋前暖区的台湾北部地区分别产生最大日雨量超过100 mm的暴雨。日雨量大于50 mm的暴雨区集中在浙东仙居、黄岩至闽北福州地区南北长约400 km、东西宽约100 km的沿海地带(图1)。强降水持续时间7 h左右,集中在7日傍晚至午夜(图略)。台湾北部地区的暴雨则发生在7日下午14时以后,属于地形性影响下的锋前暖区对流性降水(图1)。

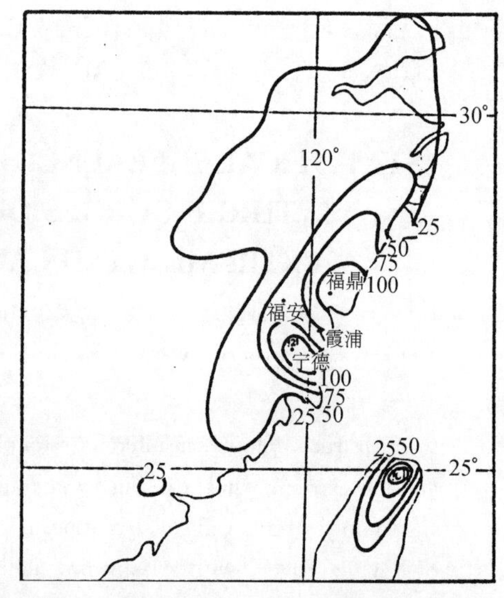

图1 1987年6月7日~8日24小时雨量图(mm)

Fig.1 Daily rainfall amount (mm)

本文利用美国科罗拉多州立大学的区域大气模拟系统(RAMS)静力平衡模式对这次对流性暴雨过程进行三维中-α尺度数值模拟试验,检验这套模式系统模拟东亚副热带地区中尺度降水系统的能

* 本研究得到"八·五"攻关项目(85-906)和国家自然科学基金项目49070244资助;本文雷达回波和探空资料由黄世明提供,本文插图由金仪璐绘制。

收稿日期:1995-06-21

第一作者简介:叶家东,男,1936年10月生,教授,大气物理专业,发表有"Observational study of microphysics in the stratiform region and transition region of a midlatitude mesoscale convective complex"等论文。

力，并对这次对流性暴雨的形成机理进行初步探讨。

1 对流性暴雨的天气背景和回波特征

6月7日8时，中空欧亚大陆为二槽一脊形势。高压脊稳定在乌拉尔山地区，西太平洋副热带高压则向东南方后撤，东亚大槽缓慢东移并明显加深，从我国东北向西南偏南方向越过黄海、长江下游直至云贵高原(图2(b))，在700 hPa和850 hPa上低槽加深尤为明显(图略)。暴雨区处于槽前西南气流中，东南沿海和台湾海峡有12 m/s的西南低空急流，为暴雨区提供充足的水汽和不稳定能量。θ_m 分析表明，从华南有一明显的暖湿舌伸向长江下游，中心最大 θ_m 值达350°K(850 hPa)和351°K(700 hPa)，未来的暴雨区恰在其下风方向地带。同时，槽后华北地区有一股强冷空气南下，中空是西北气流，而在700 hPa和850 hPa面上由于低槽加深并伸向西南，冷空气流呈北和北偏东方向南下，与来自副热带高压西北侧的西南低空急流在闽赣地区交汇(图略)。与此相对应，地面图上冷高压位于河套地区，稳定少动而强度略趋减弱；地面冷锋则位于东海至闽粤沿海一线，北段出海移速加快(达26 m/s)，南段受浙闽粤沿海山地丘陵阻滞，移动较缓慢(约14 m/s)。由于7日低空西南急流位于台湾海峡地区，其左侧有明显的气旋性切变，再加上台湾海峡下塾面(洋面)摩擦阻力明显比沿海丘陵山地小，使暴雨区在7日20时前产生一中尺度锋面波动扰动(图2(a))。地面冷峰在台湾海峡和福建沿海地区的这一波动扰动很可能是触发该地区这次对流性暴雨过程的中尺度辐合环流机制。暴雨中心区的强降水恰好在7日20时前后开始。

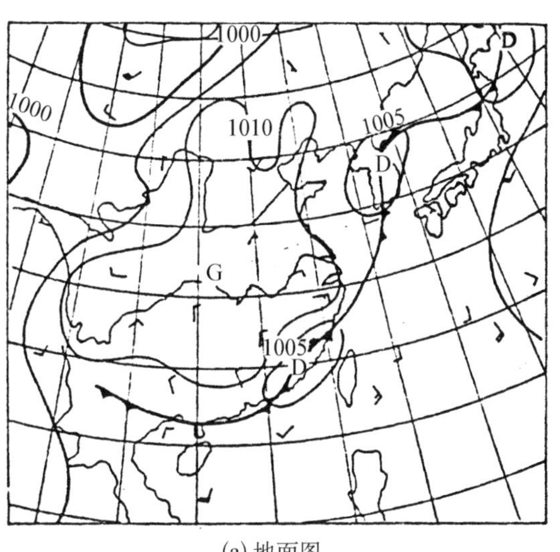

(a) 地面图

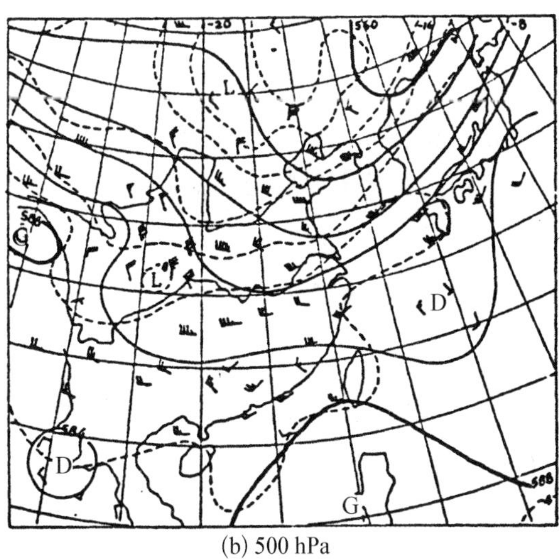

(b) 500 hPa

图2 1987年6月7日天气形势图(20时)

Fig.2 Synoptic situation at 20.00 LST of 7 June, 1987

关于低空急流在台湾地区的梅雨暴雨中的作用已有人做过分析[1]，在6月7日这场与冷锋出海相关联的对流性暴雨的发动过程中，槽前西南低空暖湿急流不断向暴雨区输送感热、水汽和潜热能加上中高空槽后偏此冷空气的侵入，结果使得该地区大气层结变得十分不稳定。R_i 数分析表明，暴雨区低空处于 $R_i<0$ 的不稳定区，最大值达−46.2。福州地区探空分析表明，从6日20时至7日20时，地面增温4 ℃，大气层结从基本上是中性层结转变为几乎整个对流层大气为条件性不稳定层结(图略)。200 hPa

高度以下的大气层对流有效位能 CAPE=1.48 千焦耳/千克，抬升凝结高度仅为 950 hPa 高度。所以，一旦低空的锋面波动扰动发展，所形成的中尺度辅合十分容易在该地区激发深对流活动。上面的分析已经指出，低空急流在上述锋面的中尺度波动扰动的发展中起着重要作用。如果不考虑扰动气压强迫、水负荷和夹卷等效应，对流有效位能全部转化为对流上升运动的动能，则平衡高度上的上升气流值可达

$$W_{CAPE}=[2(CAPE)]^{0.5}\approx 54\ m/s。$$

按照 Jorgensen 和 LeMone 的分析[2]，由于水负荷等因素的制约，对热带海洋性对流云系(包括飓风云系，GATE 和 TAMEX 对流云系)观测的对流上升气流大约只有 W_{CAPE} 的十分之一。所以，上述福州地区的对流上升速度值为 5～6 m/s 数量级。

福州地区 713 雷达回波显示(图略)，6 月 7 日 13 时开始在西北象限 200 km 范围内有强度在 30 dBZ 以下、回波顶高 7～9 km 的对流云发展并有阵雨。此时最强回波出现在台湾北部地区的山区，强回波达 47 dBZ，是该地区下午产生的突发性暴雨的主体云系。15 时福州西北的对流回波进一步发展并东移，强回波达 38 dBZ，回波顶高 10～12 km。同时在福建和浙南沿海山区均有对流回波发展。19～20 时对流回波范围扩展，回波强度达 38～45 dBZ，回波顶高仍为 11～12 km(图略)。6 月 8 日当冷锋逼近台湾岛时由设在岛上的多普勒雷达观测的回波结构可以看出[3]，回波顶高也是 9～13 km，强回波达 45 dBZ 以上，不过降水云系中出现明显的层状结构，在 5 km 高度附近有回波亮带，强回波主要集中在 0 ℃层以下的暖区。7 日傍晚开始的沿海地区的暴雨主要是对流性的，回波结构与热带海洋性对流云团及飓风中对流雨带结构类似[2]。

2　模式特点和模拟试验设计

2.1　RAMS 模式特点

本文应用的 RAMS 模式系统是根据三个模式基础上改造、重建而成的新模式系统，其中一个是原 Cotton 的非静力平衡云模式，一个是 Pielke 的静力平衡中尺度模式和另一个由 Tremback 等人发展的静力平衡中尺度模式。它保留了原模式的大部分功能并发展了一系列新的结构特征，成为一套多功能的、机动性较大的模式系统，能模拟从边界层大涡模拟到中尺度对流系统等尺度范围很宽的一系列中小尺度气象问题，并能进行多重双向作用套网格方式运行。

2.2　模拟试验设计

本试验采用 RAMS 静力平衡模式[4]对 6 月 7 日的对流性暴雨过程进行三维中-α 尺度数值模拟试验。模式区域水平方向 2 600 km ×2 600 km，水平格距 $\Delta x=0.5°$，$\Delta y=0.45°$(接近 50 km)；垂直方向 15 km，格距可变，分 27 层(1 000 m 以下格距依次为 25，50，75，100，250 和 250 和 250 m；1 000～5 500 m 之间的格距为 500 m；5 500～7 000 m 间格距为 750 m；7 000～15 000 m 之间的格距为 1 000 m)；水平方向格点数为 54(x 方向)和 53(y 方向)。初始网格点的坐标为 15°E，102.5°E。模式采用过饱和凝结调节法考虑水汽凝结过程但不考虑微物理过程；采用郭晓岚积云对流参数化方案计算对流降水率和对流潜热效应；采

用 Mahrer 和 Pielke 方案考虑晴空和云天的辐射效应;边界层参数化根据地面能量平衡方程导出地面温度,从而诊断近地面层的热通量和水汽通量。能量平衡方程考虑长波辐射和短波辐射通量、潜热和感热通量,以及地下热传导通量,故而又计算了一个 11 层土壤温度预报模式。模式的侧边界条件采用 Klemp 和 Lilly 的辐射边界条件、顶边界刚性。模式初始化利用 1987 年 6 月 7 日 0000UTC(北京时 0800)NMC 资料在等熵面上进行客观分析并内插到网格点上获得初始场。模拟计算时步取 30 s;每 5 min 更新一次辐射倾向;每半小时更换一次积云参数化值。

3 模拟试验结果分析

(1) 从 10LST 开始,东南和华南沿海近地层出现一锋前辐合带,向东南方向移动;北侧移动较快,大约在 12LST 就开始影响台湾北部地区(图 3(a))。6 日下午该地区在此辐合带的影响下加上台湾岛有利的地形和午后热力效应的共同作用下产生了一场突发性局地对流性暴雨。图 6 所示的模式预报的对流降水带北端的运行状况大体与此锋前暖区辐合带的局地暴雨相符。

(2) 5 km 高度上模式预报的流场结构(图 3(b))与观测的 500 hPa 形势图(图 2(b))相当吻合。槽线位于中国东部,随时间缓慢东移并略有加深。槽后盛行西北气流,槽前是明显的西南气流。模式预报的 5.16 km 高度扰动气压场结构(图略)与此也相吻合。

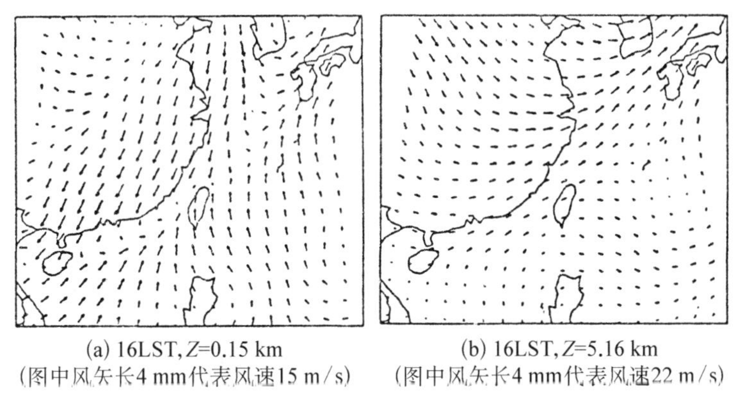

(a) 16LST, Z=0.15 km
(图中风矢长4 mm代表风速15 m/s)

(b) 16LST, Z=5.16 km
(图中风矢长4 mm代表风速22 m/s)

图 3 模拟 4 小时和 8 小时的风矢图

Fig.3 Numerical simulated wind vectors

(3) 由温度梯度(图 4)和湿度梯度(图略)密集带所反映的近地面锋带位于东南沿海和华南地区,并缓慢向东南方向移动,其中北段移速较快,这与图 2(a)所示的地面图实况相当接近;特别在东南沿海和台湾海峡地区温度场反映出地面锋有明显的波动扰动,与观测的锋面波动扰动相当吻合。前面的分析已指出该锋面波动扰动引起的中尺度辐合是这次对流性暴雨的重要触发机制。

(4) 对比模拟的近地面扰动气压场(图 5)与地面天气图(图 2(a))可以看出,地面高压中心位于中国大陆,并缓慢南压,强度有所削弱。这在模拟的扰动气压场中清楚地反映出来;在东南和华南沿海地区,模拟的扰动气压场出现明显的扰动低压槽,这与观测的地面冷锋低压带相对应。上述分析表明,应用 RAMS 静力平衡模式能较真实地模拟出大气流场、高低压系统、低槽和辐合带、地面锋及其中尺度波动扰动的结构特征及其发展和演变的过程。

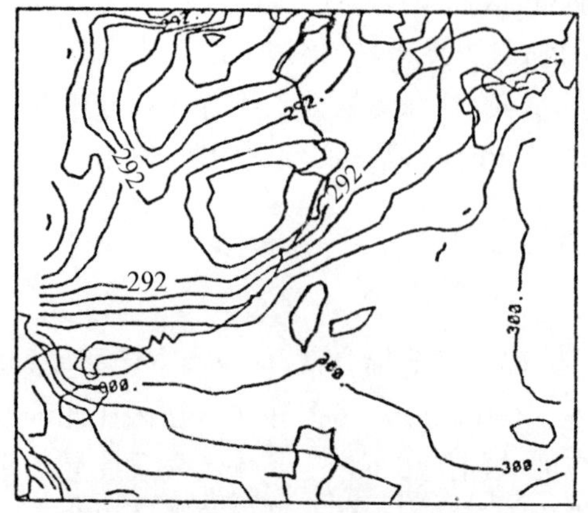

图 4　模拟的近地面温度场

（16LST，$Z=0.15$ km）

Fig.4　Numerical simulated fields of surface temperature

（16LST，$Z=0.15$ km）

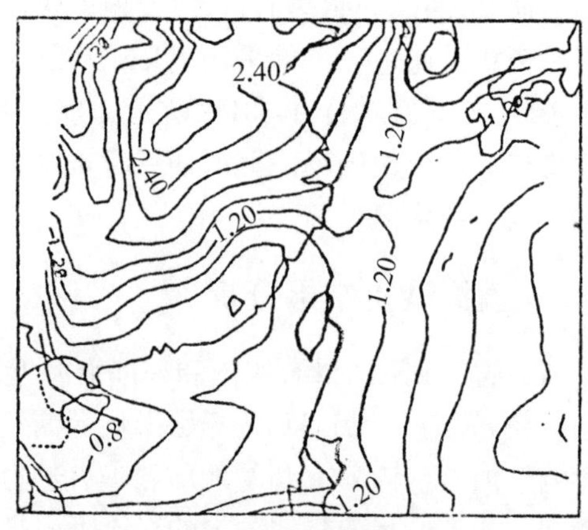

图 5　模拟的近地面层扰动气压场

（16LST，$Z=0.15$ km）

Fig.5　Numerical simulated fields of perturbation pressure

（16LST，$Z=0.15$ km）

（5）与中低层辐合带（图 3(a，b)）相对应的是垂直速度上升区（图略）和对流降水带（图 6）。这一对降水带与观测的沿海对流雨带定性相符（图 1）。模式预报的对流降水带是根据郭晓岚积云参数化方案由水汽辐合诊断确定的，它基本上是锋前辐合带的反映。由图 6 与图 1 的比较分析可以看出，模式预报的对流雨带移速较快，发展时间较实况早。Nitta[5] 在对 GATE 资料的分析中曾指出，最强降水出现的时间要比最大行星边界层辐合出现的时间落后大约 6 小时。如果考虑这一时滞效应，则模拟的对流降水带出现的时间就与实况比较接近。

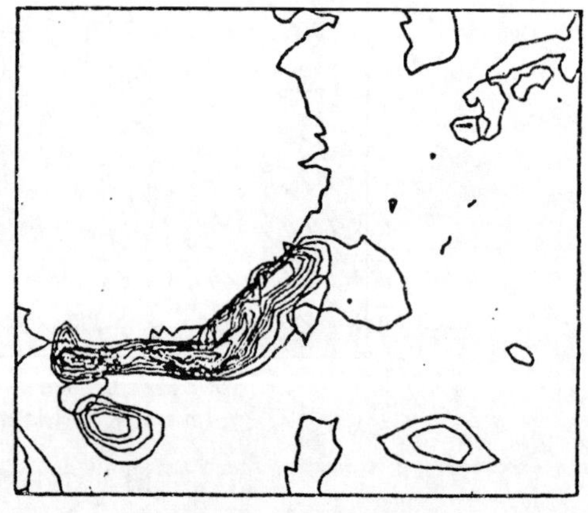

图 6　模式预报的对流降水带（12LST）

Fig.6　Numerical simulated convective precipitation zone (12LST)

需要指出，郭晓岚积云对流参数化方案主要考虑水汽辐合效应，它的观测依据主要源自热带地区，未能恰当地考虑热力不稳定度的效应。前面的分析业已指出，6 月 7 日东南沿海的暴雨区在西南低空暖湿急流的持续影响下，大气层结不稳定十分明显，为这次对流性暴雨提供了充足的水汽和不稳定能量，加上锋面波动引起的中尺度辐合和闽浙山区有利的地形影响，在 7 日傍晚激发这场对流性暴雨，其中低空急流左侧气旋性切变引起的锋面波动是触发这场暴雨的重要因素，这在温度场的模拟中已经反映出来（图 4），加上水汽辐合与对流降水之间的时滞效应订正，对暴雨的落区和发生时间都具有一定的预报能力。不过，郭晓岚积云参数化方案不能恰当地反映静力不稳定和地形等效应。为了较真实地定量模拟对流降水过程，在暴雨区需采用细网格、非静力平衡、能显式模拟对流云降水物理过程的中尺度嵌套模拟方案。目前的中-α尺度静力平衡模式在模拟天气形势和某些中尺度扰动场方面已显示出较强的模拟和预报能力，如要在此框架中改善对流降水的预报，需适当改进对流参数化方案。

参考文献

[1] Chen G T-J, Yu C C. Mon Wea Rev, 1988, 116, 884～891
[2] Jorgensen D P, LeMone M A. J Atmos Sci, 1989, 46: 621～640
[3] Trier S B, Parsons D B, Matejka T J. Mon Wea Rev, 1990, 2449～2470
[4] RAMS Model Formulation. Dept of Atmos Sci, Colorado State University, 1988
[5] Nitta T J. Meteorol Soc Japan, 1978, 56: 232～242
[6] Tremback C, Tripoli G, Arritt R, Cotton W R, Pielke R A. The regional atmospheric modeling system. Proc Intern Conf Develop and Appl of Computer Tech to Environ Studies, 1986, 601～607
[7] Cotton W R, Anthes R A. Storm and cloud dynamics. Academic Press, Inc., 1989

A CONVECTIVE HEAVY RAINFALL EVENT AND ITS MESO-α SCALE NUMERICAL SIMULATION EXPERIMENT

Ye Jiadong　Fan Beifen
(Department of Atmospheric Science, Nanjing University, 210093, Naijing),
Song Hang　Du Jingchao
(Jiangsu Meterological Observatory, 210008, Nanjing)

Abstract　A convective heavy rainfall event occurred on South-East Chinese coast area in June 7, 1987 is analysed. It is revealed that the low level SW jet transport enormous moisture and instability energy over the heavy rainfall area; accompanying surface cold front put out to sea, a mesoscale wavelike frontal perturbation develops along the Zhe-Min coastline and triggered the heavy rainfall event. The Colorado State University regional atmospheric modeling system (RAMS) static equilibrium model is adapted to the mainland China East region for a meso - α scale numerical simulation experiment of the heavy rainfall case. Simulated results show that the model can simulate reasonably the development and moving of low-level convergent line, 500 hPa trough line, surface high, and cold front. The meso-scale wavelike frontal perturbation is also simulated truly. The model predicted updraft band and convective precipitation pand is consistent with observed coastal convective rainfall band qualitatively, but the timing of rainfall band is earlier and its moving is faster than that observed, it seems to be caused by the cumulus parameterization scheme used in model, in which the effects of instability of stratification and retardation of cold front over South-East Chinese coastal mountains couldn't be considered suitably.

Key words　convective heavy rainfall, meso - α scale numerical simulation, RAMS

5.19 江淮流域一次梅雨锋波动暴雨过程中尺度分析*

叶家东 范蓓芬
（南京大学大气科学系）
宋 航 杜京朝
（江苏省气象台）

摘 要 1991年7月3日的暴雨是在梅雨锋波动影响下由一系列中-β尺度对流云团产生的。利用地面加密中尺度资料，雷达、卫星等资料所做的分析表明，中尺度对流雨团的上风方有地面中尺度辐合区相对应，对流潜热效应对波动低压发展和暴雨传播有重要作用；暴雨前期梅雨云系为典型的积层混合型结构，冰晶聚合体淞附过冷水对0 ℃层附近的回波亮带有贡献；3日下午的暴雨是对流性暴雨，暖雨过程在梅雨暴雨降水形成过程中有重要的作用。

关键词 暴雨，中尺度对流雨团，潜热效应，暖雨过程

1991年7月3日江淮流域梅雨锋波动引起了一场特大暴雨。日雨量大于50 mm的暴雨区位于武汉以东江淮流域，东西长达700 km，南北宽200～300 km的狭长地带（图略）。日雨量大于150 mm的特大暴雨区位于天长、高邮、兴化、东台、大丰等地40 km×200 km的地区，其中高邮最大达209 mm，集中在3日下午，每次降水过程持续时间短，伴有雷暴、大风，其中兴化、大丰的大风分别达14 $m \cdot s^{-1}$和24 $m \cdot s^{-1}$，天长的最大雨强达57 $mm \cdot h^{-1}$（16～17时）。

1 大尺度环流形势和层结结构特征

7月2日20时500 hPa低槽东移南下至四川盆地，由于西太平洋副高稳定维持，槽前西南气流明显加强。3日08时850 hPa芷江、武汉、安庆出现22 $m \cdot s^{-1}$的西南急流（图略）。相应地700和500 hPa高度上长沙、武汉、安庆、南京一线也分别有20～24和20～28 $m \cdot s^{-1}$的西南气流（图略）。中低层深厚的西南气流加强促使低层切变线北抬并产生波动，在武汉附近形成一低涡（图略）。20时该低涡中心移至淮河下游，形成大的风向风速辐合，为该地区的暴雨提供有利的动力辐合条件。同时地面华北冷锋南下与梅雨锋汇合，加速静止锋波动发展，3日14时中心气压999 hPa，位于淮河下游地区（图略）与850 hPa低涡相对应。3日下500 hPa槽线在江苏过境，中层冷空气侵入，形成不稳定的层结条件，促使江淮之间的对流性暴雨猛烈发展。

暴雨开始前后，温度层结变化不大，而湿度层结结构变化却十分显著。表1是南京7月初两次暴雨过程前后中低空的温湿度层结特征。暴雨前2日20时，温度层结为弱不稳定，最明显的特点是中低空700～400 hPa气层十分干燥，受西北气流影响，温度露点差达10～20 ℃，6日暴雨过程前的层结也有类似特征。暴雨过程开始后，3日08时的温度层结仍为

---

* （1）本研究得到国家自然科学基金项目49070244和“八五”攻关项目85-906资助；本文插图由金仪璐和张雪林绘制。
（2）本文雷达资料由唐询昌提供。

弱不稳定，中低空盛行西南气流，700～400 hPa 气层风速达 17～24 m·s^{-1}，其结果主要不是输送感热，而是水汽和潜热，从而使中层大气明显变潮湿，温度露点差减小到 0.0～2.8 ℃。在整层大气潮湿的情况下，云系受环境的夹卷影响减小，降水效率和潜热效应大增，这对暴雨和低压系统的发展十分有利。

表 1 南京 7 月份两次梅雨暴雨过程前后中低空层结特征

日时	850 hPa			700 hPa			600 hPa			500 hPa			400 hPa		
	T/℃	T_d/℃	ΔT_d	T/℃	T_d/℃	ΔT_d	T/℃	T_d/℃	ΔT_d	T/℃	T_d/℃	ΔT_d	T/℃	T_d/℃	ΔT_d
2.20	19.6	16.2	3.4	12.2	−8.0	20.2	3.8	−11.0	14.8	−5.0	−19.0	14.0	−12.0	−22.0	10.0
3.08	17.0	16.2	0.8	9.8	7.0	2.8	0.0	−0.8	0.8	−4.7	−7.0	2.3	−13.8	−13.8	0.0
3.20	19.0	18.7	0.3	12.0	10.0	2.0	4.5	2.8	1.7	−3.0	−4.0	1.0	−13.2	−14.0	0.8
5.20	19.2	16.8	2.4	13.0	−1.0	14.0	4.5	−9.0	13.5	−2.0	−18.0	16.0	−12.0	−33.0	21.0
6.08	20.0	19.0	1.0	11.0	9.3	1.7	4.3	2.4	1.9	−4.0	−5.5	1.5	−11.0	−12.8	1.8
6.20	19.2	17.5	1.7	12.0	10.5	1.5	6.0	4.2	1.8	−2.0	−4.5	2.5	−10.0	−23.0	13.0

2 对流云团的中尺度结构和地面中尺度辐合

图 1 中点影区为红外卫星云图所示的−32 ℃和−52 ℃卷云盖的图像，与 850 hPa 图比较可见，在 850 hPa 风速大于 20 m·s^{-1}的急流轴前方，对应于地面静止锋波动中心的辐合区中，大别山区及苏皖两省江淮之间，有一中-α 尺度对流云系。另外在长江以南的大片暖输送带中有 4 条云带，它们只在南岭山区和浙闽丘陵地区产生日雨量不超过 25 mm 的对流性降水，主要的暴雨带集中在沿梅雨锋的对流云带中（图略）。由卫星云图（略）可见，在梅雨锋波动辐合区中，自 3 日 08 时开始，在大别山上空及以东地区先后有 4 个中-β 尺度对流云团发展，水平尺度介于 50～200 km 不等，该日暴雨区的主要降雨量是由这些中尺度对流云团提供的，各个雨团的降水持续时间在 2～3 小时之内。由图 2 可见，中尺度对流雨团与地面中尺度辐合区相匹配，如图 2 所示以高邮和泗洪为中心的两个对流雨团的西南侧分别有中心值为 7.3×10^{-5}和 6.3×10^{-5}秒$^{-1}$的中尺度辐合区相对应，它们对 3 日苏北的暴雨起重要的触发作用。位于蚌埠附近的辐合区与 850 hPa 低涡中心相对应，主要是动力性低压辐合造成的，至于南京附近的辐合区，可能与该地区午前的雷暴雨有关，南京和仪征分别在 8～11 时和 9～12 时降雨 44.6 和 49.6 毫米，由对流潜热释放加热的结果使该地区地面气压比周围低 1～2 hPa，从而引起中尺度辐合。这是一种 CISK 型机制造成的反馈效应，它在梅雨锋波动低压和暴雨的进一步发展和传播中起着重要作用[1]。

3 梅雨云系的雷达回波结构特征

7 月 3 日上午的回波主要是在大片层状回波中嵌入几条对流回波短带（图略）。回波顶高 7～8 km，长 50～100 km，宽仅 10～20 km，强回波中心反射率 50～55 dBZ，限于 5 km 以下的暖层[图 3(a,b)]。这种结构特征在梅雨锋云系中较为普通，表明梅雨锋暴雨过程中暖雨过程的重要性。层状云的回波强度为 25～30 dBZ。在 0 ℃层附近存在明显的回波亮带，强度达 30～35 dBZ，厚度从 500 米至 800 米不等，主要部分位于 5 km 高度以下。3 日 08

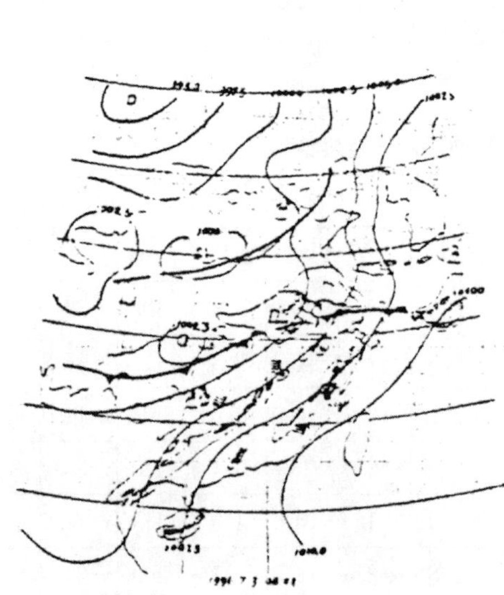

图1　1991年7月3日地面天气图,点影区为增强红外卫星云图(淡影区为−32 ℃卷云盖,深景区为−52 ℃卷云盖)

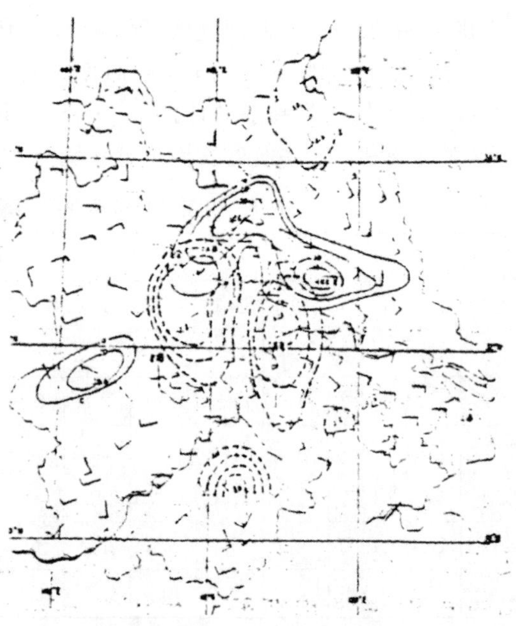

图2　7月3日14时地面中尺度辐合(虚线,单位:$10^{-5}\ s^{-1}$)和对应时间前后2小时雨量图(实线,单位:mm)

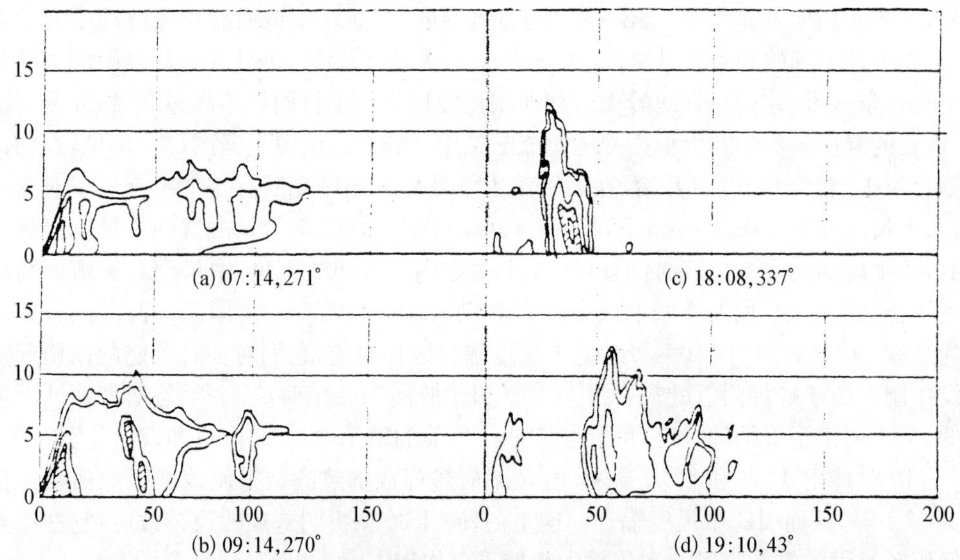

图3　7月3日梅雨云系雷达RHI回波图像(等值线分别代表反射率15,30,45,55和65 dBZ)

时0 ℃层高度约4.5 km,所以回波亮带中有一部分位于0 ℃层高度以上,这不能用通常关于回波亮带是由于冰晶下降到0 ℃层高度开始融化使表面潮湿增大反射率的所谓融化带的概念加以解释。实际上,在梅雨云系云底暖,云中含水量大的情况下,0 ℃层以上较暖的过冷层中有较丰富的过冷水,它们与层状云区0～−5 ℃层中形成的大冰晶聚合体淞附,呈湿增长状态,冰面上附一层水膜,其折射指数与水滴相似,从而能对0 ℃层高度以上的回波亮带的形成做出贡献[2]。

混合型回波结构自10时开始发生变化。13时开始，南京北方和东北方的对流回波带发展，强中心达55 dBZ，但回波顶高仍位于7.5 km。18时以后，南京西北滁县地区强对流回波发展，回波顶高达12千米以上，45 dBZ的强回波区宽达20 km，中心强度达65 dBZ(图3(c))。但值得注意的是，45 dBZ的强回波区顶高仅约6 km，主要位于5 km以下的暖层。这与通常的强对流风暴结构有所不同，那里强回波区可伸展到15 km以上，冷雨过程起着重要作用。一系列对流回波短带东移发展，在苏北先后产生短时大暴雨，暴雨中心雨强达35～57 mm/h，高邮和东台10分钟最大雨强达20 mm以上。

4 结论

(1) 1991年7月3日的暴雨过程主要是由高空槽东移加深、冷空气南下、中低空西南暖湿气流加强促使地面梅雨锋产生波动引起的。暴雨过程中，大气层结弱不稳定、暴雨区整层潮湿、降水效率高，对流凝结潜热释放在梅雨锋带弱斜压区对波动低压的发展和暴雨的传播起着重要作用。

(2) 3日的暴雨主要由相继发展的中-β尺度对流云团产生，雨强大，持续时间短，具有明显的中尺度对流系统降水特征。大别山区是重要的对流云团源地，它们在向东传播过程中地面有明显的中尺度辐合区相对应。中尺变辐合区一般位于同期对流雨团的上游。

(3) 3日上午梅雨云系的结构为典型的积层混合型结构，回波顶高介于6～8 km不等，在大片层状云中嵌入数条对流回波短带。在层状区0 ℃层附近存在明显的回波亮带，冰晶聚合体的淞附作用对0 ℃层高度以上的回波亮带的形成有重要贡献。3日下午主要是对流回波结构，回波顶高可达12 km以上，强回波中心反射率可达65 dBZ，不过强回波区一般均位于5 km高度以下的暖层，表明暖雨过程在梅雨锋暴雨形成过程中起着重要作用，不过这是一个需要从云物理角度进一步研究的问题。

参考文献

[1] Ninomiya, K., T. Akiyama and M. Ikawa, 1988: Evolution and fine stucture of a long-lived meso-α-cale convective system in Baiu frontal zone, Part I: Evolution and meso-β-scale characteristics, J. Met, Soc. Japan, 66, 331 - 350.

[2] Yeh Jia-Dong, Fan Bei-Fen, W. R. Cotton and M. A. Fortune, 1991: Observational study of microphysics in the stratiform region and transition region of a midlatitude mesoscale convective complex. Acta Meteorologica Sinica, Vol. 5, No. 5, 527 - 540.

Mesoscale Analysis of a Heavy Rain in Jiang-Huai River Basin Caused by Wave Motion of Mei-Yu Front

Yeh Jiadong, Fan Beifen
(Department of Atmospheric Science, Nanjing University)
Song Hang, Du Jingchao
(Jiangsu Meteorological Observatory)

Abstract The heavy rain on July 3, 1991 was produced by a series of meso-β scale convective dusters developed under the influence of wave motion in Mei-Yu front. The analysis by using the data from dense surface mesoscale network, radar and satellite shows that the mesoscale rain duster producing heavy rains was accompanied by a surface mesoscale convergence area. The effect of convective latent heat was important in the development of wave depression and propagation of heavy rain area. A typical complex structure of cumulo-stratus was found in the early stage of the heavy rains. The riming of ice aggregates contributes to the echo bright in the stratiform region above the 0 ℃ level. Convective heavy rain was prevailing in later time of July 3, in which the warm-rain process played an important role in precipitation formation.

Key words: heavy rains, mesoscale convective clusters, effect of latent heat, warm-rain process

6　附录

《落基吟》诗词集

弘扬中华传统文化，放眼人生春秋冬夏

目 录

前言

本集收集了近四十年来我的一些诗词习作。我自然不是诗人，诗词习作写得难免比较“业余”，但多少反映了这一时期我的某种心路历程。与其说是写给别人看的，不如说主要是写给自己看的，也是一种为了忘却的记忆。古人云：“苔花米粒小，也学牡丹开”，大体就这么一种情结。仔细想想，也不尽然。生命无论大小，是否美丽，都有自身的生命历程。开花结果，繁衍生息，是天赋的权利和义务，无须学。我比较欣赏赵本山早年扮演的小品《小草》中的那位可笑却也可爱的老太太。人上了岁数，智力体能，包括才艺记忆力，都会衰退，自然规律使然。但她那种乐观向上、永不言老、奋发自强的昂扬生机，着实令人神往，可敬可爱。俗话说，人是要有点精神的。老年人更需要有点精神。“城中桃李愁风雨，春在溪头荠菜花”。小草的情怀，平和、宽厚而倔强。以此自勉吧！也望与老年朋友互勉。

在时代的潮流里，人生恰似一叶小舟，飘飘荡荡，颇难自持，我愿搭载在南京大学这艘巨轮上，劈波斩浪一齐奋进，遮风避雨以图安宁。所以，当我和老伴蓓芬的一位旅居美国的老同学兼老同事(半个世纪前在庐山脚下星子城旁的鄱阳湖里，我曾救过他一命，他一直难以忘怀)，在马里兰远东饭店宴请我们时，忍不住问道：当年回国，是否后悔了？我回答：不！当年回国时没有后悔，如今依然没有后悔。诚然，在蓓芬出国前，以及我们回国后，确实遭受到了种种阻拦和障碍，使你有举步维艰之感。但终究没有能够阻拦住蓓芬出国访问。那个时期，大气科学系确实流失了一批核心骨干师资，仅56届留校任教的11人中，净流出5人，占45%。不可谓不是一大损失。但就我们而论，当年回国是对了，不离开南京大学也对了，无须后悔。“千跃百转山溪笑，迆逦归海自在流”。

近二十余年来，南京大学教职工的住房条件大为改善。但前期我们多少有点保守，一直留恋入住多年的老住房，似有点抱残守缺之嫌。所以，当我们在国外探亲得知南京大学在仙林开辟新校区，并开发一教职工住宅区——南大和园，便立马回国参加住房申购。由于我们位于北京西路二号新村的原住房面积较小，故而在申请住房买房排队时，我们排在退休教职工队伍的第九号，颇感庆幸。住房也点得较为称心。入住以后更感到远离尘嚣而又紧贴南大校园较为舒心。享受到南大赋予的教职工福利，身为南大人的荣幸感油然而生。每年暑假前后在南大校园里总会出现一副醒目的、富含哲理的、动情而充满期待催人奋进的标语，叫作“今日我以南大为荣，明日南大以我为荣”。我在南大学习、工作和生活已经整整六十年了，在我心目中，“昨日我以南大为荣，今日我以南大为幸”。这种情怀，萦绕心头，久久挥之不去。我接触过一些海外校友，他们大多有类似情结。用旅居美国的校友彭沛涛的话来说，就是一种难以割舍的归属感。所以南京大学大气科学系(学院)，旅居美国大华府地区的校友每年一度在马里兰黑山公园举行野餐聚会，凝聚友谊，共叙南大乡情，温馨祥和、其乐融融。“万里幸会皆是缘，南大老乡新气象，不谈海外多浪漫，且说仙林好风光，相约明年重聚首，繁花硕果庆安康。”南大乡情高山阻不断，大洋隔不开，愿南大乡情遍布寰球，绵绵无绝期。

最后，我要感谢我的老伴范蓓芬副教授。我眼神不济，错漏谬误在所难免，得亏她帮我细心审校诗词文稿。我们是大学同班同学，相识相知六十余载，相依相伴半个多世纪。风风雨雨，坎坎坷坷，一路走来，总是相伴相随，无怨无悔。老伴相夫持家，说她“吃苦在先，享受在后”，毫不为过。女儿叶黎也深知父母心，她在国外拼搏奋斗三十年，不畏艰辛，勇往直前，很不容易。在她稍微有点经济实力时，就为我们买高端瑞士表，还出资让我们购置较为宽敞的养老住房，还多次安排或陪伴我们美欧各地旅游，让父母的退休生涯，过得较为充实愉悦。我们老两口金婚时节，女儿带着小外孙 Wesley·王永正，陪我们到卡斯科 CasCo 精心挑选了一枚精致的钻戒，小外孙还认真地给婆婆戴在手指上，光彩夺目，甚是好看，让从来没有时尚享受的老妈妈也风光了一下。现在我们已年届八旬，栖居仙林，颐养天年。当下我们老人间流传一句“四老”口头禅：养老养老，有一个老伴，有一个老窝，有一点老本，有几个老友足矣！我们还有一个时时惦记着我们的女儿和小外孙。虽然远隔重洋，不时通过视频电话沟通。况且我们生活在南大校园氛围内，并不怎么感到寂寞。前不久，暑假刚开始，女儿就和女婿王立明带着小外孙一家三口回国陪我们乘量子号邮轮赴日本周游一圈，让我们两个八旬老人得以放心地领略一番那似曾相识的东方异域风情和邮轮上祥和的休闲生涯。来也匆匆，去也匆匆，没有惆怅，只是感到欣慰。女儿不是闲人，一片心意，父母很知足了，复有何求？“且喜夕阳无限好，无暇品味黄昏近”。

2016 年 9 月 21 日

（农历八月二十一日，八十周岁生日）

于南京仙林南大和园

诗词集

汤山出院别诸病友 1975.9

1975年初夏赴北京大学云雾物理和人工降水训练班讲学，归来后发现血沉抗0指数均高。遂至汤山医院（时称汤山煤矿医院）用温泉疗养，渡过了一段闲逸散淡的休养生涯。因拟于10月份赴西安参加专业会议，乃于9月底出院返校。

九月二十八，出院离汤山。病友来相送，相顾话珍重。望望不见亭，行匆复流连。斜照堤上草，耳迴高原雪。笑踩将军路，却寻朱砂洞。太极八卦掌，温泉浴身健。夜来望北斗，阳台议春秋。台上笑语频，楼前舞流萤，闲逸度日速，麦黄稻复熟。倚天制风云，抽刀断水流。老九不学儒，浮云载去就。驱病思壮游，落叶寄离愁，街头万人丛，何日喜相逢。

汤山将军楼 1975.9

9月某日偕病友结伴游汤山将军楼，遇警卫阻拦，扫兴而还。

朱砂洞旁白玉楼，碧山从中黄花茂。
秦皇位极思东海，将军功成何所忧？

致齐康 1975.9

出院前夕，南京工学院病友齐康即时画了一副苏州园林钢笔画送我。画虽速成，却也惟妙惟肖，足见其人才华横溢，惜在功名场上辛苦拼搏，午方四十六却已头白似雪。

竹林傲游世称贤，白眼高歌望青天。
廿载碌碌心酸泪，头白未品鲈鱼脍。

赠吴金陵 1975.9

一竿风雨一杯酒，汤泉湖畔钓直钩。
秋临佳节斜阳暮，相送无忘新街口。

延安吟 1975.10

云飞天涯去，北征万里尘。霜柿红似花，秋重访圣城。
崎岖高原路，黄陵柏森森。延水忆当年，宝塔思深沉。

宏图王家坪，巨手转乾坤。星聚杨家岭，愚公移山坚。
枣园幸福渠，军民一家亲，南泥湾精神，总理纺纱勤。
实事求是好，艰苦朴素真，煌煌革命业，先辈苦经营。
后人何所虑，饮水莫忘本，先烈血旗红，传统要继承。
深知创业艰，莫学二世秦。夕照宝塔静，暮霭罩山城。
悠悠中华源，忧天向天陈。

湘西行 1976.6

忆昔端阳行湘西，芳风丽日随人意。
喜临洞庭觅桃源，惊越铁山过泸溪。
武陵翠橘扬清芬，绿叶素荣参天地。
岭南青梅湘西杏，杏梅难辨山人喜。
山人喜，阮陵落魄不皱眉。
凝望雏燕八字路，喃喃细语却依依。
苍悟云，潇湘水，烟波洞庭长江汇。
吾跨长虹望江水，斑斑皆是离人泪。
重华南巡无阻拦，吴楚天阔山万里。
山万里，皇英乘风下翠薇。

桂林山水 1978

（一）漓江

明镜青山人如画，轻舟碧波戏彩霞。
信是蓬莱仙山好，不识漓江冠天下。

（二）阳朔

漫天朝霞满江花，群峰起舞披轻纱。
樵歌一曲廻天乐，碧莲峰下好人家。

悼念徐尔灏先生 1978

徐尔灏教授平反昭雪追悼大会有感：

嗟乎徐先师，惨惨从何语。本系寒窗士，负笈西海渡。
赖有红日照，雄鸡唱天晓。万民得其所，国威震寰宇。
勤恳事业计，革心树宏图。百步光明顶，耕云惊神女。
“小小足球队”，环保驰先驱。气象物理化，登高指前途。
育人仰师表，严谨莫含糊。风云变难测，骇浪卷萍浮。
松柏本耿直，桃李无所措。明哲不保身，世道忒坎坷。
天地路漫漫，离骚投汨罗。

死生去就轻，贾生缘人妒。忽念归去来，桃源无觅处。
魂飞东南隅，夫子冤千古。千古有遗恨，失足无人扶。
掩涕暗叹息，炎凉一何殊。浮尘蔽日昏，鬼蜮掀妖雾。
贼林四人帮，祸国殃民蠹。除害谢天地，百花得甘露。
冤屈申明日，痛定泪难阻。四化忆良才，星陨余几许。
春雨催春笋，夜阑闻杜芋。

注：南秀村一带树高林茂，春夏之交夜阑人静，时闻杜鹃鸟啼鸣林间，声声委婉悠扬，扣人心扉，似在告诫世人："哥哥何苦，世路坎坷"。闻之怆然。

答朱李　1979. 4. 16

朱国江、李如祥两位老师鼓动我去申报首批高级职称。我自感年资尚浅，婉谢有感：

群芳烂漫竞相妍，独托幽岩展素心。
感尔春风送雨意，有心浇花待来年。

失眠　1980. 5. 17

无端春风逗叶笑，粗狂车轮碾地跑。
点点疏星照无寐，声声杜鹃催人老。

黄山始信峰　1980. 9

天都莲花五彩桥，罗汉十八竞踊跃。
仙人欲指攀天路，始信峰云自逍遥。

黄山北海　1980. 9

梦笔生花华容改，石猴观海海浮沉。
莲花映日年年盛，天都光明处处春。

成都全国气象年会　1982. 11

露含黄花映秋山，笑迎群英聚锦官。
劫波有涯余勇在，摘星无畏蜀道难。

古田石塔山　1983. 5

(一) 卜算子

绿水青山绕，曲折登山路。欲凌绝顶观沧海，却惹云雾妒。

惊定莫回首，休顾松涛怒。幽谷风生残云去，莺歌留春住。

(二) 如梦令

常忆石塔天柱，耸入云幕深处。但闻燕语频，不见伊住何处。风舞，风舞，云退斜阳峰殊。

无题　1986.2.6

蓓芬在国内办理访问学者出国手续受阻。踏着厚厚的积雪赴 CSU International Office 补办手续有感：

雪压落基山，朔风刺骨寒。异国他乡客，踩雪踏斜阳。
僕僕徒奔波，无计分忧患。崎岖路难行，归心渡重洋。

落基吟　1989

自 1985 年 8 月至 1988 年 8 月访问科罗拉多州立大学从事云动力学和中尺度对流系统研究。在科罗拉多州住了三年，一家三口在异域他乡生活倒也安宁适意，除了家乡浙江东阳和南京，科林斯堡这个学院城是我连续居住时间最长的一个地方。临别时我将出国时弟弟叶家时送的一包家乡土撒在了大气科学系大楼门前那棵菩提树下留个纪念：

林深草青处宁静，风寒雪皑时归雁。
落基山麓柯林堡，莽莽荒原嵌绿荫。
道旁松鼠时时舞，湖中天鹅悠悠行。
庭院深深无墙围，春华秋实世人怜。
矫健野鹿奔斜阳，策马牧女驰日边。
蓝天白云红尘净，幽幽山坳闻啼莺。
连天西山雪峰寒，越岭焚风喷烈焰。
巧夺天工蓄水坝，涓涓清流滋生灵。
碧眼黄发非异类，偶逢行人笑相迎。
山乡人情重自然，浪迹天涯忆武陵。

感怀　1990.4.13

无星伴明月，有梦桃源觅。武陵匪盗绝，学府山寨立。
鳖化龙腾飞，鼠变虎逞威。闻者趋巷里，犹悸毒尾蠍。

偶感　1992.10.9(农历八月二十一日)

梦里乾坤大，茶中日月清。

笔耕度春秋，文章慰生平。

南京大学九十周年校庆：
五九届校友聚会游东郊 1992.6.1

岁月沧桑路漫漫，欣逢同窗惊秋霜。
东西南北难得聚，弃忧聊作少年欢。
似水流年去不还，片片白云过斜阳。
不论酸甜沉浮梦，且说天命祝安康。

龙王山 1992.6

1992年5月27日应邀参观南京气象学院多普勒雷达。午间暂息友人家，闻龙王山杜鹃竞啼有感：

大江东去水悠悠，龙王山北观云楼。
苍龙欲飞风不起，杜鹃竞啼春难留。
春难留，水悠悠，权财名位莫强求。
雀跃蛙鸣谋生计，东篱采菊自忘忧。

晨起 1992.6.10

蛰居避尘嚣，警梦群雀噪。
举头望井天，低头吟离骚。

水调歌头 1994.6

大气科学系建系五十周年庆典期间，五九届老同学毕业三十五周年聚会，6月5日游镇江归来。晚宴间闻长春陈永枢朗诵苏东坡《水调歌头》（明月几时有）词有感，遂步其调试填之：

年少何处寻，把酒问青天。
转眼三十五载，风尘染霜鬓。
我欲乘风归去，醉里倒转乾坤，
高歌康定情。驱车辞金陵，
访古觅幽境。
金山寺，江心焦，北固亭，不应有憾，
今日兴会值千金，明朝散泊如云，
盼五十年再聚，扶杖重登临，
但愿人人健，松竹年年青。

后记：2002 年我和蓓芬应吉林省人工降水办公室主任汪学林之邀赴长春讲学，老同学前来探望，汪学林设晚宴款待。席间陈永枢取出八年前我寄给他的上述词的手稿，当众朗诵起来，依然是情真意切，声情并茂。他从吉林省气象台台长岗位退休后，担任长春市老干部合唱团团长，颇为风光。

五九吟　1994. 6

大气科学系建系五十周年，五九届老同学聚会，6 月 6 日在西苑宾馆举行晚宴。席间闻合肥王相文唱“拉兹之歌”有感，据他说当年失恋时常唱这首流浪者之歌。如今面对诸多老同学和当年的初恋情人演唱这首歌，真的是别有一番滋味在心头。

五月石榴红似火，喜鹊喳喳意若何？
欢庆五十气象新，五九子弟聚金陵。
去时雏燕凌空舞，归来老骥伏枥悟。
三十五载几坎坷，南园北楼觅旧故。
兴来结伴游金山，斗酒十千酣高歌。
跑马山，流浪汉，声声醉人黯神伤。
无那风烈马迷路，桃花逐水无寻处。
丝丝情愫丝丝雨，世间何处是坦途。
明月自古有圆缺，人生难得心宽容。
举杯祝酒意未尽，频频相约九九唔。

渔家傲　1994. 6

姜效泉自叙羁留内蒙古一十五载，思归心切，婉谢秀丽胡女，终未成家。年四十只身返归江南，年近半百方得一儿，疼爱不已，惟窘于参加学校家长会，盖因恐被误认为“爷爷”！闻而感慨系之：

塞外风疾秋色异，碧云衰草连天际。奶茶一碗家万里，归心切，平沙落雁孤城闭。
乡魂悠悠谢胡姬，不惑游子返故里。西子湖畔创新业，思范蠡，淡泊书生老无泪。

注：范蠡施美人计，荐西施助勾践灭吴，遂辞官退隐江湖，自号鸱夷子皮，意谓随遇而安也。

送阮贤舜返长春　1994. 6

6 月 6 日东北师大阮贤舜从夫子庙花鸟市场购得画眉鸟一匹，桂花树苗四棵，自鸣得意，连夜返回长春。那画眉挺识趣，终日在阳台鸣叫不已，仿佛满嘴吴韵越语，阮喜不自禁，遂为其换一圆形大鸟笼居住，且甚得小外孙喜，戏作小诗一首赠之：

勤勤恳恳兢兢业业，切切实实营营。
不慕苍鹰击长空，无意伴蝶花间舞。
点点滴滴勤采集，白山黑水未言苦。

江南月桂关外情，芳草天涯无归路。
惊西风流年暗渡，倚楼极目淡云过。
且得意儿孙绕膝，莳花遛鸟自欢娱。
冷不丁悄无人处，频与画眉乡音语。

后记：2002年范蓓芬和我应邀赴长春讲学，抽空打电话给东北师大环科系询问阮贤舜家的电话，对方沉默片刻问："你是他什么人？""是老同学出差来长春，想去看望他。"回答称："他已于五年前去世了！"我愕然无语，想不到竟会是这样！后据陈永枢讲，阮晚年嗜酒，患肝腹水去世。

丑奴儿　1994.10

游栖霞，观枫林，烟霭沉沉，弃物遍野，颇感扫兴，步辛弃疾，"丑奴儿"词调试填之：

少年识尽愁滋味，欲醉还休。
欲醉还休，为惧酒后说嘴漏。
而今愁去天方秋，爱看层林。
爱看层林，却恼丹枫染尘垢。

注：借酒浇愁，古来有之。然1970年春节在溧阳农场时值深挖"五一六"高潮期间，问一四年级同学为何春节不去一同喝酒？答曰：怕酒后失言！该班一朱姓同学就是因为在寝室骂了一句"都是江青这婊子搞的！"而被打成现行反革命，终于被逼疯而遣返回乡，断送了一生前程！

碧云亭　1994.10

巴山楚水秋万里，吴天碧云飘无羁。
丹枫古寺钟一声，水随云去无消息。

无题　1995.3.25

人云范进中举难，生计无着典家当。
苍天有情遣周翁，凉酒无心醉痴汉。

述怀　1995.10

风雨匆忙六十秋，去日苦多频回首。
游丝如花窥霜鬓，耕云驱雾未言愁。
漫道此身非我有，学剑无意觅封侯。
千跃百转山溪笑，迤逦归海自在流。

北大山居室装修自娱歌　1996.4.20

九华紫金远，鸡鸣北极近。余脉北大山，辟地营新村。
五台倚清凉，鼓楼扶钟亭。斗室初装修，劳神心安定。
宁静望流云，淡泊听雨声。井天映碧树，斜阳绘风屏。
梧桐鸣斑鸠，喜鹊若比邻。报晓白头翁，八哥仿童吟。
耳顺闻天籁，夜阑月色清。梦醒虫切切，倍觉小院静。
晨起练太极，双剑舞蹁跹。不觉风寒侵，地气暖无言。
无言育生机，百草自有情。报谢三春晖，蜂蝶舞芳庭。

君子兰　1996.5.1

手植君子兰六载，不以冷落际遇为怀，今春蓦然开花十二朵，艳若骄阳，感慨系之：

兰花真君子，开合自有时。冷眼望世情，淡泊炎凉事。
清寒守洁土，寂寞不移志。春风吹暖雨，真心慰相知。

琅琊山　1996.10

醉翁之意不在酒，峰回路转觅深秀。
跃上天门登云阁，回眸古刹林壑幽。
世间醉物岂惟酒，财色权位竞相诱。
散淡诗翁山林醉，清风朗月云鹤游。

春联　1997.1

辞旧岁，岁岁明月依旧，
心系北美远涉重洋漂泊儿。
迎新年，年年照样更新，
身栖江南遥望天涯彩云归。

冬练太极剑　1997.1.14

岁寒风寂暮云驻，锦翠华年悄然去。
心宁气沉练太极，仗剑凝目刺天破。

水仙　1997.2.26

1月初友人送来两个水仙花球茎。夫妇俩细心照料月余，现已亭亭玉立，繁花似锦。两棵水仙各有九个花柱，分别开出五十五朵和三十三朵洁白似玉的水仙花，清香满堂，甚是喜人。九九加八八，颇感吉祥：

冰肌无暇披露寒，清风摇曳沐春光。
玉立亭亭凌波仙，洁身净心自芬芳。

扫墓　1998. 6. 6

南峰林幽百叶旺，东江水碧歌山吟。
勤教一生桃李艳，清风两袖桑梓情。

后记：先父叶光球（梦耕）先生在世时有一首怀念先母楼文涛（彩娟）女士的诗：题为《歌山吟》。

行近歌山便怕行，溪云村树总关情。
路当南北几往来，事异悲欢一死生。
卅载光阴成梦蝶，四方儿女似浮萍。
衰年已觉无归着，是处遗踪入眼明。

当年先父在东阳中学任教，近乡情更怯的心情做儿女的自然最能体会。站在小歌山东南侧的观鱼亭旁的那块巨岩上，向西北五里远处便可望见先母的娘家东楼，向南十里远就是我们的家，象山干。而且还能望见先母生前筹建的位于鹅头颈的一座颇具规模的家族坟墓。1951 年先母突然病逝便葬于此，后被迫迁至大坟山，留下一段不堪回首的记忆。1978 年先父去世也安葬在大坟山，当我和弟弟叶家时去县里报丧时，县教育局局长说：叶（光球）先生一生从教，为东阳县（市）的教育事业做出了重要贡献。东阳中学派了一位教师驾车将灵柩在我等护送下运回故里安葬。离城时，邻里十数人持香烛沿东街至南街，一直送出城至南街口。当灵车抵达象山干时，本家族数十人聚集在桃源塘沿迎候。按照乡风习俗：将棺木抬至大坟山预先建成的墓地，当棺木送进墓穴临将封墓时，我跪在墓穴里，突然感到父亲将永远躺在这暗无天日的墓穴中，一种莫名的悲伤涌上心头。我双手拍打着棺木，撕心裂肺地呼唤着远去的阿爹，整个山谷被震撼了。当堂兄弟将我扶出墓穴，我方才感到全身酥软乏力，回到村里，在前厅摆了十数桌酒席，按照习俗，我代表家属向丧礼主持人、从邻村请来的长辈和八位抬棺木的伙伴敬酒，表示感谢。整个丧事办得隆重而平和，由弟弟叶家时和他的伙伴及本家族人协同操持，成为这个小山村多年未见的一桩盛事。大姐叶竹坚、二姐叶春晓一同赶回来送丧。当时还在东北师大任教的哥哥叶家明（后调至浙江大学任教）虽然未能及时赶回来，但却汇来 250 元钱，这接近于我当时五个月的工资，承担了大部分的丧葬费用。

苏幕遮　2002. 5. 21

南京大学百年校庆，5 月 21 日大气科学系在向阳渔港举行盛大晚宴。数十届校友欢聚一堂，盛况空前。师生系友相聚，惊喜激动，叙长问短，情谊深重：

百年庆，千人宴，
向阳渔港，师友喜相逢。
吴韵越语频祝酒，醉眼相对，依稀旧音容。

追往事，恍若梦，
桃红李白，依然笑春风。
昔日重现思无绪，聚散匆匆，明日各西东。

校训吟　2002.6

百年校庆，解读南京大学校训“诚朴雄伟，励学敦行”有感：

诚信为本树人生，朴实无华修身性。
雄健擎托凌云志，伟宏创业天地新。
励志攀登科学峰，学海无涯汲华精。
敦睦和衷齐奋博，行路万里济苍生。

注：此诗曾发表在《南京大学报》2004年12月20日第四版。

江城子　2002.6

5月21日，南京大学百年校庆，大气科学系五九届校友聚会游阅江楼有感：

狮子山头阅江楼，原有记，本无楼。
风雨苍黄，悠悠六百秋。
时逢盛世百业旺，树名楼，靓神州。
携来百侣望江流，宏桥伟，山川秀。
雾锁琼楼，恍若天上游。
惯看行人步匆匆，抚霜鬓，莫寻忧。

鹧鸪天　2003.3.30

（二）游千岛湖

绿荫葱茏青黛光，散落翠珠碧玉盘。
百丈龙山观云亭，千年湖底眠淳安。
孔雀靓、红鲤欢，倚天天锁锁湖江。
醉客忘返迷津渡，千岛湖水海蓝蓝。

鹧鸪天　2003.9.12

大气科学系退休协会组织秋游，赴溧水无想山天池游览，尽兴而归。

无想山清天池幽，无尘碧水映竹楼。
松影照水静无语，自在黄莺鸣枝头。
茶榭外，秋千旁，一个渔翁一钓钩。
冷暖深浅谁先知，白云悠悠风轻柔。

苏幕遮 2004

美国大华府樱花节

蔚蓝天，青草地，樱花似云，风卷雪纷飞。
平湖落雁鸭戏水，春风天涯，何处不芳菲。
追梦乡，思故园，年年岁岁，好景人欲醉。
月朗风清休恋依，彩云当归，太湖天地催。

大气科学系六十周年系庆 2004.10.30

六十华诞新气象，总是难忘旧时光。
四方学友齐欢聚，弦歌一堂叙衷肠。
红山炼钢炼赤胆，栖霞垦荒广积粮。
欲除四害震天吼，跃地卫星满地放。
毋论荒唐莫惆怅，热血少年心向党。
与时俱进风雨后，寄语新秀读书郎。
长江后浪推前浪，珍惜科学发展观。
心忧天下树正气，振兴中华国运昌。

老年运动会 2006.9

古稀今不稀，耄耋不称奇。八百白头翁，千米健步飞。
投壶掷金镖，宁神练太极。抖擞老精神，忘年竟儿戏。

高淳游子山 2006.10.23

游子山庙游子心，浪迹天涯故乡情。
装点江山好风光，金佛玉宇夕照明。

注：据称游子山宏伟壮观的庙宇和顶天立地的金佛系旅居比利时的华侨捐资兴建。

重阳七十寿 2006.10.30

大气科学系领导邀七十岁老教师聚会，祝寿庆重阳。新任系主任杨修群教授在会上概述拓系建院的规划方略，闻之颇为感奋：

天高云淡绘清秋，尊师敬老庆重九。
更新气象展宏图，伏枥老骥望新秀。

波多麦克山林居　2009

东西南北森林茂，但见群鹿日日来。
黄杨紫槐护青坪，玉兰天竺红墙绕。
左邻右里逾百米，时闻犬吠罕见人。
且喜玲珑小红鸟，频啄门扉笃笃敲。
启门欲问何所需，轻歌曼舞枝头俏。
参天古树隔风尘，心远地偏望云飘。
春湖化雨春光好，秋风落叶也逍遥。
最是孙儿放学归，雀跃虎跳添热闹。
古稀老人天伦乐，夕阳穿林分外娇。

童心本善　2009.10

蔚蓝天上白云飞，秋风落叶铺满地。
公公扫叶清道路，身旁跟着卫斯理。
忽然惊呼一蠖儿，躺在路上好可怜。
我去取水来救它，不然它会死掉的。
我说不用不用怕，泥土里面是它家。
我们就来帮助它，轻挑蚯蚓花丛下。
那里有水有吃的，不饥不渴还有家。
孙儿这才放下心，撒开小腿去玩耍。
幼稚园里育仁爱，从小懂得悯生灵。
公公欣喜乐开花，宝宝可爱心善佳。

黄士松先生九十华诞　2009.11.27

寿比南山不老松，冬夏长青度春秋。
福如东海万顷浪，桃李芬芳逾神州。

后记:前些年在一次郊游时，黄士松先生悄悄告诉我他的一则故事。他九十一岁时，按习惯在楼下阳光广场练步。走着走着，心血来潮，突发奇想，走到单杠下，想一展身手，纵身一跃双手抓住单杠。怎奈毕竟年事已高，加上当年又截去几个指头，终于力不从心，抓不住单杠滑了下来，一屁股摔在沙坑上。当时心里一阵紧张:坏了！一定骨折了。定了定神，慢慢地爬了起来，抖抖身子，觉得还好，不怎么痛。于是慢慢地走回家，拿出紫药水，在手臂上磨破皮处擦擦。夫人汤敏明老师看见问:“怎么？皮擦破了?”“嗯，擦破了。”不一会，女儿回来，年轻人眼快，一眼就看出破绽:“啊呀！怎么啦！爸！跌跤了？裤子都破了呦!”老两口这才发现膝盖上的裤子都跌破了。黄先生这才如实交代。我听了忍俊不住，笑着说，黄先生这种事以后还得谨慎为好，九十一岁不比一十九岁当年。但是，我从心里佩服老师的豁达乐观，颇有点老顽童的味儿。如今黄先生已九十七岁高龄，依然头脑清晰，仍坚持步行锻炼。前年大气科学学院建系七十周年纪念大会上，九十五岁高龄的黄士松先生上台发言，不用稿子，声若洪钟，一口气讲了大气系发展三个里程碑时段，思路清晰，条理分明。我常说，我们的老系主任黄先生健康长寿，是我们大气系的福分，也是我们后辈学习的榜样。黄先生是一个重情义的人，1962 年我和蓓芬结婚时，在南园十二舍借了一间房子做新房，在简朴而热闹的婚礼上，黄先生和汤老师特意送了一本精美的大相册，这使我们感动，在那个年代，这弥足珍贵，我们时时铭记在心。衷心祝愿我们的老师黄士松先生乐观健康，淡定从容，勇闯百岁大关！

海洋城休闲滨海居　2009

城东呼啸大西洋，城西门前临海湾。
咫尺天涯风景异，南北一线似海塘。
翻空白鸥时时舞，掠水快艇穿梭忙。
难得佝偻三劲松，凌风傲雪守海疆。
游艺乐园熙熙攘，喷泉泳池不稀罕。
高尔夫球富人玩，赌场不是好地方。
购物练步尝海鲜，休闲似比上班忙。
最诱人处海沙滩，逐浪垒沙浴日光。
人间仙苑何处有，观海听涛心无疆。
夜来渔火点点闪，自在潇洒各喜欢。
晨起宁神练太极，车库侧旁沙沙响。
一只海龟三五磅，欲觅产房迷方向。
捧至岸边致歉意，吾等占地尔彷徨。
左边那片保护带，产卵育儿最平安。
海龟小眼幽幽亮，迅疾入水潜海湾。
人类欲望千千万，环境友好莫遗忘。
留得蓝天青山在，芸芸众生俱兴旺。
长治久安科学观，和谐梦园有希望。

童心　2010.6

小外孙 Wesley · 王永正六岁，在幼稚班精心制作了一个珍珠手链，并郑重其事地送给妈妈叶黎，作为“母亲节”礼物，至情至性，甚是可爱。

光彩夺目珍珠链，小手灵巧倾真情。
童雅无暇赤子心，大爱无疆献娘亲。

南京大学大气科学学院
旅居美国大华府地区校友野餐聚会
2010.7.21

南京大学大气科学学院(系)旅居美国大华府地区的校友每年一度在马里兰黑山公园举行野餐聚会。我有幸参加过几回,与会的有阔别数十载的老同学老同事,也有事业有成、落地生根的中年才俊,更多的是素未谋面的年轻学子。大家欢聚一堂,共叙南大乡情。气氛祥和,其乐融融。一位年轻系友带着母亲和未满周岁的宝宝前来与会。她坐在对面望着我们悄声询问身旁的一位中年校友:他们俩是谁?那位中年校友故意拉长声调回答:是老师的老师!引来一阵友善的哄笑。那位年轻的妈妈脸都红了。

赤日炎炎暑气旺,难得今日清风凉。
五十载老友新朋,四万里异国他乡。
扶老携幼喜洋洋,欢聚一堂马里兰。
天蓝蓝兮云淡淡,路漫漫兮回首望。
忆昔抚今感慨多,说三道四论短长。
更有新秀读书郎,意气风发热心肠。
牛排苞米烧烤香,凉面披萨中西餐。
瓜果糕点花色多,啤酒饮料碰杯忙。
同是天涯拓荒人,相逢何必曾相识。
万里幸会皆是缘,南大老乡新气象。
不谈海外多浪漫,且说仙林好风光。
相约明年重聚首,繁花硕果庆安康。

晨起　2010.11.20

天高云淡秋意浓,霜风渐劲染丹枫。
朝阳斜照西窗外,万木凋零一树红。

晨练　2011.11

步履蹒跚未觉慢,总嫌时光太匆忙。
天旋地转韶华去,翁媪年迈当自强。

海归谣　2012.1.24

归去来兮归去来,仙林新居新风采。
落地生根育新秀,叶落归根情常在。
仰天大笑出门去,吾辈原本中国来。

跨洋越海仙人慕，朝发北美暮上海。
风驰电掣高铁快，轻轨勾联城内外。
山明水秀清净处，蓝天白云宽心怀。
校园安宁生机旺，社区温馨敬老迈。
南大和园仙林居，修身养性赛蓬莱。

雪灾　2012. 2. 7

2 月 5 日开始下大雪，积雪达 24～34 英寸，最厚 138 厘米，创 110 年来最大纪录。停电停水停空调，全家奋力铲雪，打通门前百米通道，驱车赴 DC 有电处“避难”。7 日下午恢复供电，返家安顿。

大雪纷纷飘，林间静悄悄，千里吹雪雁，方树樱花俏。
景俏别陶醉，雪里藏冰刀，群鹿无踪影，玉兰折断腰。
停电又断水，无端添烦恼，空调不运作，宅房似冰窖。
无水冲马桶，无电供电脑，电讯皆中断，雪海一孤岛。
穷极则思变，困久欲脱逃，举家齐出动，奋力铲雪妖。
孙儿卫斯理，年幼胆气豪，得亏吹雪机，临难不辞劳。
铲雪赛猛牛，吹雪冲天笑，奋战两小时，百米劈通道。
驱车赴 DC，有电旅社找，充饥方便面，矿泉有饮料。
夜阑人无寐，缘何灾频扰，怨天天不应，人贵自知明。
天大大无边，人定难胜天，逆天无善果，顺天有奇招。
节能低排碳，植树森林茂，生态持平衡，发展重环保。
天人合一统，和谐路迢迢。

张家界　2012. 5

（一）天门山

天门索道天下奇，鬼谷栈桥人间稀。
天门洞开迎客笑，九九九级登天梯。

（二）宝峰湖

峰伟林静育清幽，湖碧水蓝情意柔。
阿哥阿妹唱不尽，人间瑶池仙苑秀。

（三）武陵源

百龙天梯耸云霄，别有天地土家傲。
长问桃源何处觅，武陵源深藏天骄。

（四）金鞭溪

金鞭出炉冲天吼，林深溪清静幽幽。
艳阳高照游人旺，奇峰幽谷觅清秋。

苏幕遮　2013. 1. 26

墨西哥坎昆海滨宾馆

白沙滩，碧海湾，椰林婆娑，泳池戏水欢。
异域风情情无异，亚欧美餐，任凭君品尝。
人慵懒，浴日光，小舟舢板，奋搏击海浪。
人间瑶池百十千，脱尘超凡，坎昆醉忘返。

黎明 POTOMAC 新居　2013. 1. 30

紫气东来朝霞艳，苍松翠柏御风寒。
居高不熬睦邻善，天清气爽夕阳红。

克鲁兹　2013. 3

挪威鲸轮爱匹克，举世无双巨无霸。
十五万三千吨位，旅客乘员五千八。
首尾相距三百米，上下船高十九层。
中央大厅九层塔，金碧辉煌富贵相。
舞场画廊演艺厅，泳池球场健身房。
休闲摄影大卖场，酒吧白费西餐厅。
美味佳肴任君尝，切忌过量保健康。
吃喝玩乐各喜欢，赌场自有人向往。
八方来客乐逍遥，千余躺椅浴日光。
俨如一座小城镇，娱乐健身睡稳安。
诺亚方舟现代版，长风破浪驶大洋。
世事苍茫置脑后，了无牵挂游四方。

大西洋葡属海岛芬切尔　2013. 4

明珠溅落大西洋，风光旋旎疑方丈。
紫英满树香满路，行道拼画行人畅。
绿荫葱茏城中藏，奇花异树竹掩墙。

天鹅赋闲浮绿水，仰饮飞泉巧模样。
群鸽飞舞惹人欢，饲食嬉闹似家养。
醉翁流连不忍去，忘却仙苑是他乡。

巴塞罗那 2013.4

巴塞罗那久闻名，首都造访后清明。
西山北峰屏风张，东南海湾天水宽。
观光车上观市容，耳机导游说端详。
多种语言任尔选，中文解说数七档。
边引边望边听讲，省力省钱省时光。
最振精神北高峰，悬崖膝上建教堂。
俯视全城气势宏，云蒸霞蔚欲飞翔。
富丽堂皇艺术宫，目不暇接费思量。
世上胜境谁沾先，寺庙道观洋教堂。
脱尘超凡出家去，醉卧云松亲自然。
清净无为修身性，返璞归真寻梦乡。

挪威峡湾 2014.6

诺亚方舟载我行，劈波斩浪到天涯。
碧海蓝天心无羁，白云青山有人家。
挪威峡湾风光奇，雪冠飞瀑情依依。
山重林茂疑无路，峰回水转又一湾。
渐行渐远渐宽广，落日红霞漫天燃。

苏幕遮 2014.9

生日感怀

2014年9月14日是农历八月二十一日，我78周岁生日，次日9月15日是蓓芬77周岁生日。晚餐时在南大仙林校区六食堂用餐时，买了两碗长寿面两个人吃得津津有味，自得其乐。同桌一女生见了说："真想不到你们有这么大年纪，比我爷爷奶奶都老，真的看不出来，好精神哟！"我们听了喜笑颜开，"谢谢你啦！我们最爱听这种话。比吃一块大蛋糕还开心"。改日，在校门口西侧等候校园车赴大气科学学院，一位环境科学学院的来自沈阳的女研究生在闲聊中说，"大家都说你们老两口手拉手在校园里来来回回挺恩爱的"。我们听了笑道："啊呀！主要是怕跌跤。"现在年岁大了，眼神不济，腿脚又不大灵便，特别是在过街地道两端那三十五级台阶，尤其是傍晚，每次都小心翼翼，紧握老伴的手，亦步亦趋，生怕踩空滑跌。回想五十五年前当我们像你们这般年轻的时候，在校园里树荫灯下谈心论学，那个年代，哪好意思手拉手在校园里走啊！半个多世纪，风风雨雨，匆匆忙忙，来不及细想，不知不觉忽然就这么老了。往事如梦、如潮、如烟云，都已成往事，也不用多想。现在我们老人间有句口头禅，快乐过好每一天，少给子女添麻烦。何况，我们的女儿叶黎尚远在太平洋彼岸。不！大西洋西岸的马里兰！谈何容易。空巢老人当自强。享受生活别伤感。试作词《苏幕遮》一首以自勉。

静养寿，动健身。动静调谐，寿康得益彰。
人生跋涉多沟坎，风霜雨雪，寒暑一肩担。
敬天命，重平淡，老伴相伴，宁静望夕阳。
人道暮年欲何求，世事纷繁，养生育心安。

苏幕遮　2015. 2. 8

仙林漫步

智不惑，勇无惧，仁者无忧。古道步斜阳。
暮鼓晨钟世外音，倦鸟归巢，仙林栖霞光。
知足乐，知趣安。知进知退，知己知人善。
八全九美足珍贵，天地无全，何事徒伤感。

除夕偶感　2015. 2

空巢老人当自强，享受生活别惆怅。
平和安康每一天，少给女儿添麻烦。

苏幕遮　2015. 9

喜闻乐见

欣闻女儿叶黎在马里兰大学培训班学习三个月后，连闯三关，一举通过美国全国和马里兰以及佛吉尼亚和华盛顿 DC 三场 Brocker 资格考试，奠定了开拓莱蒙克斯集团总部委托承办的连锁子公司（Universal）的技术基础。不惑之年，敢于闯关，勇于担当，坚毅可嘉。

莫懈怠，别偷懒，勤奋耕耘，金秋硕果香。
人生征途无捷径，披荆斩棘，拓荒励志强。
心不贪，情无妄，诚信厚道，兴业勇担当。
风生浪涌等闲事，扬帆起航，劈波迎朝阳。

观电视节目《三国演义》有感　2015. 10

诸葛一生唯谨慎，险来难免演空城。
七擒孟获宁边陲，六出岐山气若虹。
世界精彩多无奈，谋事在人成在天。
明知不可而为之，弱国安邦谈何易。

以攻为守良苦心，几人体察个中味。

秋游大觉寺 2015.10

大气翁妇结伴游，七老八十踏清秋。
镜湖风光大觉寺，气势恢宏星云筹。
罗汉十八顶天地，风雨长廊行无忧。
大雄宝殿瑞气旺，恍若雷音下神州。
超凡入圣星云师，海峡两岸勤奔波。
兴建寺庙扬佛法，普渡众生壮志酬。
捐资南大三千万，助学重教星云楼。
看得清，想得透，大觉大悟放得下。
雁渡寒潭不留影，人历一世竟有痕。
积德行善暖人心，湖光山色总是情。

除夕偶感 2016.1.7

空巢老人当自宽，力衰智沌心莫烦。
清净安宁修身性，行云流水任自然。

南京大学重九祝寿会 2016.9

风雨苍黄八十秋，耳背目眩不寻忧。
用心尽力抗顽疾，宁神量力健步走。
自知自信图自在，平静平和谋平安。
且喜夕阳无限好，无暇品味黄昏近。

九三诗翁邹进上颂 2017.10

2017 年国庆中秋双节期间，93 岁高龄的邹进上教授打电话来祝贺双节节日快乐，神清气爽，笑声朗朗，还即兴赋诗，庆贺国庆中秋佳节，情深意长。感奋之余，试作小诗一首颂之：

大气邹公一诗翁，笑逐颜开老欢童。
自云我是九〇后，目清体健耳未聋。
欲问我从哪里来，洞庭湖南武陵东。
庭前膝下何所有，英雄楷范志气宏。
执教气象全天候，大气气候伴天动。
十年军旅坎坷路，荒漠边陲报国忠。
两弹一星扬国威，气象保障树军功。

踏遍青山人未老，讴歌时代气如虹。
大气学院多情种，满目彩霞满江红。
祝君安康神志旺，百岁闯关步轻松。

偶感　2018.1.1

耄耋老人费思量，气急愁闷烦没完。
安康源自心宁静，健身养心重恬淡。

人与自然，和谐相处，其乐融融，落基山野鸽，天真活泼，亲近可爱。

1937 年浙江丽水处州中学校园
先父叶光球先生先母楼文涛女士
作者时年 1 周岁

1959 年南京大学毕业前夕

1986 年美国落基山艾虚坪 SnowMass 国际会议上作学术报告

1987 年美国科罗拉多州立大学大气科学系主楼前合影

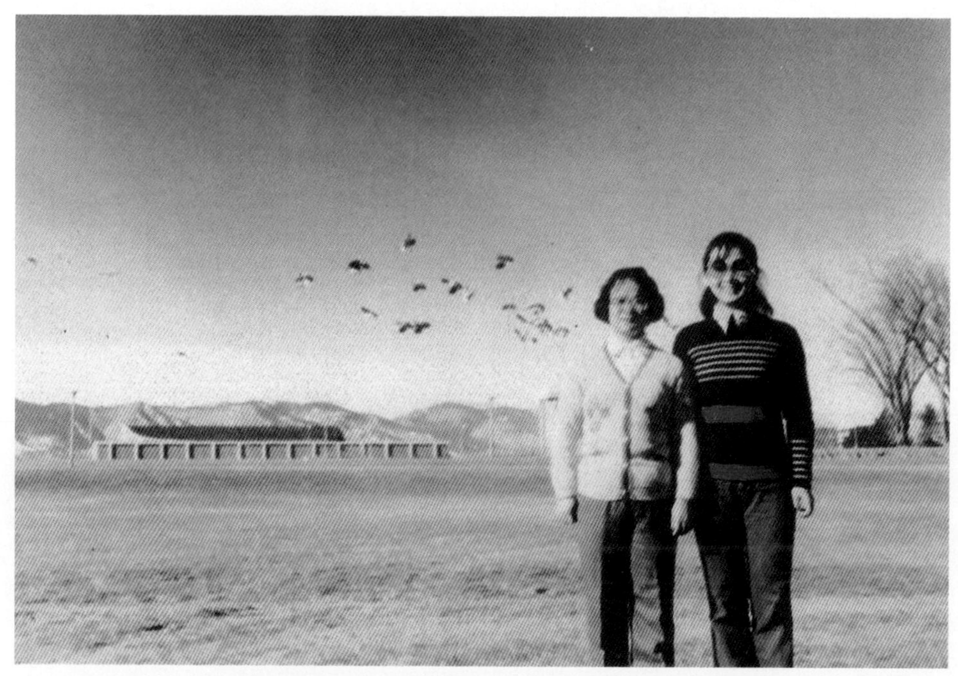

1986 年隆冬季节,成百上千大雁汇聚 **CSU** 校园广场,
漫天飞舞,满地呼唤,寒冬腊月,热闹非凡,蔚为壮观

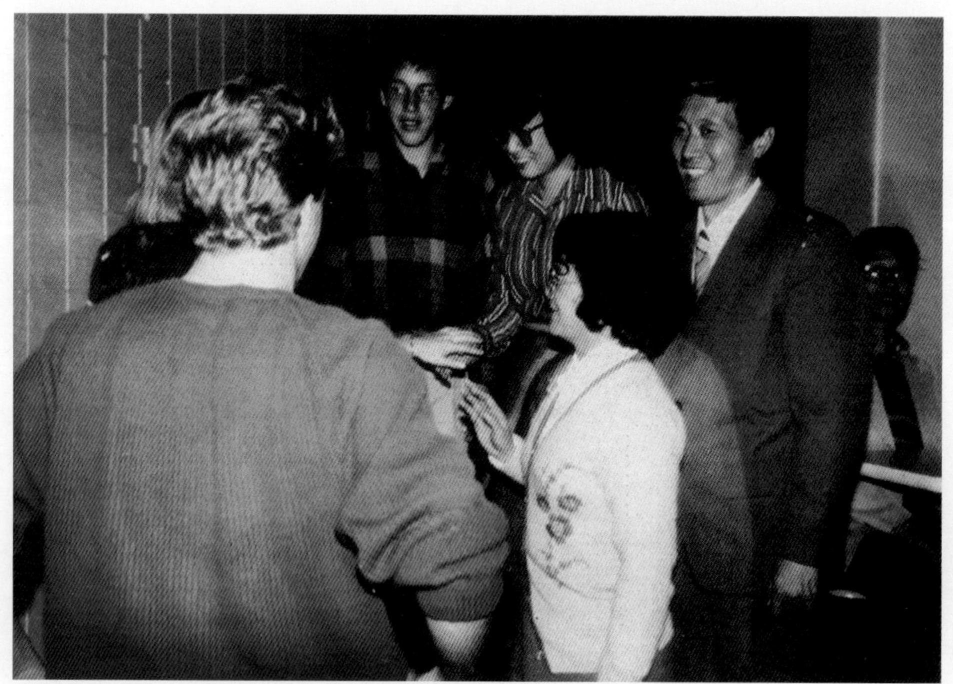

1987 年国庆节女儿叶黎和应邀参加国庆招待会的同学合影

1987 年加拿大温哥华不列颠哥伦比亚大学校园(参加国际会议)

1987 年美国科罗拉多州立大学住所楼下(学校体育馆侧旁),女儿叶黎

2014 年北欧旅游合家欢
左起:叶家东、范蓓芬、女婿王立明背上小外孙 Wesley · 王永正、女儿叶黎

2014 年北欧旅游天青海蓝心无羁